Lecture Notes in Computer Science   4822

Commenced Publication in 1973
Founding and Former Series Editors:
Gerhard Goos, Juris Hartmanis, and Jan van Leeuwen

Dion Hoe-Lian Goh   Tru Hoang Cao
Ingeborg Torvik Sølvberg
Edie Rasmussen (Eds.)

Asian Digital Libraries

Looking Back 10 Years and Forging New Frontiers

10th International Conference on
Asian Digital Libraries, ICADL 2007
Hanoi, Vietnam, December 10-13, 2007
Proceedings

 Springer

Volume Editors

Dion Hoe-Lian Goh
Nanyang Technological University
Wee Kim Wee School of Communication and Information
31 Nanyang Link, Singapore 637718
E-mail: ashlgoh@ntu.edu.sg

Tru Hoang Cao
Ho Chi Minh City University of Technology
Faculty of Computer Science & Engineering
268 Ly Thuong Kiet Street, District 10, Ho Chi Minh City, Vietnam
E-mail: tru@cse.hcmut.edu.vn

Ingeborg Torvik Sølvberg
Norwegian University of Science and Technology (NTNU)
Department of Computer and Information Science
Sem Sælands vei 7-9, 7491 Trondheim, Norway
E-mail: ingeborg@idi.ntnu.no

Edie Rasmussen
The University of British Columbia
School of Library, Archival and Information Studies
6190 Agronomy Road, Vancouver, British Columbia, V6T 1Z3, Canada,
E-mail: edie.rasmussen@ubc.ca

Library of Congress Control Number: 2007940530

CR Subject Classification (1998): H.3, H.2, H.4.3, H.5, J.7, D.2, J.1, I.7

LNCS Sublibrary: SL 3 – Information Systems and Application, incl. Internet/Web and HCI

ISSN   0302-9743
ISBN-10   3-540-77093-3 Springer Berlin Heidelberg New York
ISBN-13   978-3-540-77093-0 Springer Berlin Heidelberg New York

Springer is a part of Springer Science+Business Media

springer.com

© Springer-Verlag Berlin Heidelberg 2007
Printed in Germany

Typesetting: Camera-ready by author, data conversion by Scientific Publishing Services, Chennai, India
Printed on acid-free paper   SPIN: 12199600   06/3180   5 4 3 2 1 0

Preface

The International Conference on Asian Digital Libraries (ICADL) is one of the leading international conferences in digital libraries research. The conference has come a long way since its inception in 1998 as the First Asia Digital Library Workshop held in Hong Kong. Since then, the conference has traveled across the Asian continent and has been hosted by Taiwan (ICADL 1999), Seoul, Korea (ICADL 2000), Bangalore, India (ICADL 2001), Singapore (ICADL 2002), Kuala Lumpur, Malaysia (ICADL 2003), Shanghai, China (ICADL 2004), Bangkok, Thailand (ICADL 2005), and Kyoto, Japan (ICADL 2006).

The 2007 edition of the conference marks an important milestone in the ICADL series. Into its tenth year, the conference matured into a significant gathering of practitioners, researchers, educators and policy makers from diverse disciplines sharing a common interest in advancing digital libraries research in Asia. ICADL 2007 was held in Hanoi, Vietnam during December 10–13, 2007, with the theme, "Asian Digital Libraries: Looking Back 10 Years and Forging New Frontiers." The theme reflects upon the growth of the digital libraries community and explores new areas that the community could delve into in the coming years.

ICADL 2007 attracted 154 papers from 36 countries across Asia-Pacific and Europe. These papers were reviewed by an international Program Committee of 80 members who are established researchers in various areas in digital libraries. The size of the committee was determined by the need to ensure the acceptance of good quality papers as well as to reach out to a greater audience beyond the Asian region. Additional reviewers were also recruited to assist in the sharing of the review workload. Of the 154 submissions, 41 were accepted as full papers. A further 15 short papers and 10 posters were also accepted to provide a program that allowed for more participation and interaction within the digital libraries community. The full papers, short papers and extended abstracts of posters are included in this proceedings volume.

The conference consisted of 15 sessions across three days, divided into three parallel tracks. A diverse range of topics were presented, reflecting the conference's theme. These included information retrieval and mining, user interfaces and evaluation, information management, digital archives, multimedia digital libraries, and social media. In addition, ICADL 2007 featured three tutorials, three keynote addresses and three invited talks. The conference also featured a panel on I-Schools in the Asia-Pacific region as well as one on fostering collaboration among researchers using a speed-dating format. Also for the first time, the conference included two sessions on European digital libraries initiatives. The primary goal of these sessions was to provide a platform for digital libraries researchers in Europe to share their work with their Asian counterparts, and in the process, explore opportunities for conducting joint research.

We would like to thank everyone involved in ICADL 2007. Special thanks go to all members of the Organizing Committee for ensuring that every aspect of the conference was taken care of. We also thank the members of the Program Committee and the additional reviewers, who worked hard at completing all their reviews promptly. The Chair of the ICADL Steering Committee, Ee-Peng Lim, and the Steering Committee members must also be thanked for their leadership and support. We acknowledge the help of the many volunteers who worked behind the scenes to make this conference a success. Finally, thanks to all the authors who submitted their camera-ready papers on time, and to all ICADL 2007 delegates for participating in the conference. We look forward to the ongoing development of the ICADL series and hope that it continues to be a platform for the exchange of significant ideas in the field of digital libraries.

October 2007

Dion Goh
Tru Cao
Ingeborg Sølvberg
Edie Rasmussen

Conference Organization

The Tenth International Conference on Asian Digital Libraries (ICADL 2007) was hosted by the National Center for Scientific and Technological Information (NACESTI), the National Library of Vietnam (NLV), General Science Library of Ho Chi Minh City (GSL), the Library and Information Center of the Vietnam National University Hanoi (LIC/VNUH), Hanoi University of Technology (HUT), Vietnam Association on Scientific and Technological Information and Documentation (VASTID), Vietnam Library Association (VLA), Vietnam Association for Information Processing (VAIP), and other organizations.

ICADL 2007 was supported by the Ministry of Science and Technology (MOST), the Ministry of Culture and Information (MOCI), and the Ministry of Education and Training (MOET) of Vietnam.

Organizing Committee

General Chair	Ta Ba Hung, NACESTI, Vietnam
Vice Chair	Cao Minh Kiem, NACESTI, Vietnam
Program Co-chairs	Dion Goh, NTU, Singapore
	Tru Cao, HCMUT, Vietnam
	Ingeborg Sølvberg, Norway, Norwegian University of Science and Technology, Norway
	Edie Rasmussen, University of British Columbia, Canada
Tutorial Chair	Yin-Leng Theng, Nanyang Technological University, Singapore
Publication Co-Chairs	Chu Keong Lee, Nanyang Technological University, Singapore
	Nguyen Tien Duc, National Center for Scientific and Technological Information, Vietnam
Publicity Co-chairs	Cao Minh Kiem, National Center for Scientific and Technological Information, Vietnam
	Quan Thanh Tho, Ho Chi Minh City University of Technology, Vietnam
Financial Chair	Tran Thu Lan, NACESTI, Vietnam
Local Arrangements Chair	Phung Minh Lai, NACESTI, Vietnam
Steering Committee Chair	Ee Peng Lim, NTU, Singapore

Program Committee

Akira Maeda, Ritsumeikan University, Japan
Alton Chua, Nanyang Technological University, Singapore

Michael Nelson, Old Dominion University, USA
Ming Zhang, Peking University, China
Min-Yen Kan, National University of Singapore, Singapore
Mitsuharu Nagamori, University of Tsukuba, Japan
Ngoc-Thanh Nguyen, Wroclaw University of Technology, Poland
Nguyen Ngoc Binh, Vietnam National University, Vietnam.
Nguyen Thanh Thuy, Hanoi University of Technology, Vietnam
Noriko Kando, NII, Japan
Ohm Sornil, National Institute of Development Administration, Thailand
Patrick Weiguo Fan, Virginia Polytechnic Institute and State University, USA
Paul Nieuwenhuysen, Vrije Universiteit Brussel, Belgium
Paul Janecek, Asian Institute of Technology, Thailand
Pavel Braslavski, Russian Academy of Sciences, Russia
Peter Jacso, University of Hawaii, USA
Pimrumpai Premsmit, Chulalongkorn University, Thailand
Praditta Siripan, National Science and Technology Development Agency, Thailand
Pu-Jen Cheng, National Taiwan University, Taiwan
Ray Larson, University of California, Berkeley, USA
Richard Butterworth, Middlesex University, UK
Richard K. Furuta, Texas A&M University, USA
Rick Kopak, University of British Columbia, Canada
Robert Allen, Drexel University, USA
Rudi Schmiede, Technische Universität Darmstadt, Germany
Sachio Hirokawa, Kyushu University, Japan
Sally Jo Cunningham, Waikato University, New Zealand
Schubert Foo, Nanyang Technological University, Singapore
Shigeo Sugimoto, University of Tsukuba, Japan
Shiyan Ou, University of Wolverhampton, UK
Shuigeng Zhou, Fudan University, China
Stuart Weibel, OCLC, USA
Sung Hyon Myaeng, Information and Communications University, Korea
Sunghyuk Kim, Sookmyung Women's University, Korea
Takashi Nagatsuka, Tsurumi University, Japan
Thomas Baker, Goettingen State and University Library, Germany
Trond Aalberg, Norwegian University of Science and Technology, Norway
Uta Priss, Napier University, UK
Wai Yeap, Auckland University of Technology, New Zealand
Yan Quan Liu, Sourthern Connecticut State University, USA
Youngok Choi, Catholic University, USA

Steering Committee

Chair: Ee Peng Lim (NTU, Singapore)
Shalini Urs (University of Mysore, India), (Vice-Chair)
Hsinchin Chen (University of Arizona, USA), (Advisor)
Ching-Chih Chen (Simmons College, USA), (Advisor)

Christopher Yang (Chinese University of Hong Kong, China)
Hsueh-Hua Chen (National Taiwan University, ROC)
Key-Sun Choi (Korean Advanced Institute of Science and Technology, Korea)
Sung Hyon Myaeng (Information and Communications University, Korea)
Schubert Foo (Nanyang Technological University, Singapore)
Zawiyah Baba (National Library of Malaysia, Malaysia)
Tengku M.T. Sembok (Universiti Kebangsaan Malaysia, Malaysia)
Zhaoneng Chen (Shanghai Jiaotong University, China)
Qihao Miao (Shanghai Library, China)
Edward Fox (Virginia Tech, USA)
Pimrumpai Premsmit (Chulalongkorn University, Thailand)
Shigeo Sugimoto (University of Tsukuba, Japan)
Katsumi Tanaka (University of Kyoto, Japan)

Sponsors

Ministry of Science & Technology of Vietnam
IEEE Technical Committee on Digital Libraries

Table of Contents

Information Retrieval Techniques I

Multilingual Techniques

Information Seeking and Use

European DLs I

Multimedia Digital Libraries

Information Retrieval Techniques II

European DLs II

Digital Library 2.0

Information Mining I

User Interfaces

Information Mining II

Digital Libraries and Education

Information Organization

Posters

Multicultural and Globalized Digital Libraries: Digitizing and Empowering the "Other"

(Extended Abstract)

Clara M. Chu

UCLA Department of Information Studies
210 GSE&IS Building, Box 951520
Los Angeles, CA 90095-1520
cchu@ucla.edu

Abstract. Our multicultural societies in a globalized context offer the opportunity to forge new frontiers in digital libraries. Focusing on the cultural diversity of the Asian region, this paper examines what should be digitized, who should be involved in the digitization efforts, and what access issues need to be considered. More specifically, documenting the experiences of the Asian diaspora and ethnic minorities will be discussed, engaging critical theories and multicultural scholarship. By problematizing the cultural production of digital libraries as an act of nostalgia, of inclusion and exclusion, and of racial, social and sexual differentiation, we can unpack the role that digital libraries play in the creation of communities in our imaginary and in the perception of space and place from those objects we select to digitize. The paper concludes with a call for decentering digital libraries and digitizing the "Other" as an act of empowerment and representation.

Keywords: Multicultural digital library, Asian diaspora, minority communities, empowerment.

1 Extended Abstract

Information professionals and policy makers who recognize the multicultural make-up of our societies in a globalized context understand the opportunity to open up and advance new frontiers/spaces in digital libraries. Here, the term digital library is used broadly to not only refer to local, national, regional, or world digital collections of its diverse cultural and scientific heritage (books, films, maps, photographs, music, etc.), but to also include original materials that traditionally would make up digital archives (i.e., both works and documents). The reason for this breadth of scope is that a critical multicultural audit of digital libraries, or most information institutions/services for that matter, would likely reveal that underrepresented communities, such as cultural minorities and diaspora communities, have been overlooked and still require to engage in basic activities, such as documenting their cultural practices, recording oral accounts or digitizing original documents.

Asia is a region with extraordinary ethnic and cultural diversity. It is the most linguistically diverse area of the world, and its religious diversity is largely a result of

D.H.-L. Goh et al. (Eds.): ICADL 2007, LNCS 4822, pp. 1–4, 2007.

contact with traders, missionaries or colonial settlers [1]. According to the Embassy of Socialist Republic of Vietnam in Canada, in Vietnam alone, there are "fifty-four ethnic groups, each with their own traditions, festivals, clothing, songs and dances" [2]. Adding to the complexity of an evolving multicultural or cultural diverse society is that cultures are dynamic and they interact with each other. Thus, cultures have been erased, fused, re-emerged or stayed static as a result of movements of peoples through migration, colonialism, assimilation, war and intermarriage, for example. These events have created creolized, transnational, diaspora, minority (e.g., linguistic, ethnic, racial, religious, etc.), and ethnically- or racially-mixed peoples and communities. Depending on the complexity of the situation we may encounter individuals who are twice minority, that is, they belong to two minority sub-cultures (e.g., Muslim Indian in Singapore, Japanese Brazilian woman in Japan). In their work on Multiculturalism in Asia, He and Kymlicka [3] note that "managing diversity is therefore key to political stability in the region. The centralized, unitary 'nation-state' model adopted by postcolonial states appears increasingly unable to meet this challenge....Whatever the explanation, Asia is witnessing the rise of 'identity politics'. People are mobilizing along ethnic, religious, racial, and cultural lines, and demanding recognition of their identity, acknowledgement of their legal rights and historic claims, and a commitment to the sharing of power."

In defining "cultural diversity" UNESCO [4] declares that "culture should be regarded as the set of distinctive spiritual, material, intellectual and emotional features of society or a social group, and that it encompasses, in addition to art and literature; lifestyles, ways of living together, value systems, traditions and beliefs." Such a view of cultural diversity emphasizes an aesthetics/arts/humanities perspective, which does not reveal the adversity faced by minority and diaspora communities. The discourse on cultural diversity needs to shift to recognize racism, geopolitical and economic interests, and that oppressed cultures are likely to be marginalized, tokenized, erased, overlooked or made invisible. By going beyond celebratory multiculturalism, an Information Society by creating multicultural and globalized digital libraries can enter into a new frontier in defending human identity, dignity and expression.

Documenting the heritage and experiences of Asian diasporas and ethnic minorities will be discussed using multiple critical lenses. These include the work of the following scholars or scholarship:

- Paulo Freire - through literacy that individuals find their own voice and can eradicate cultures of silence that keep them oppressed or aim to assimilate them
- Juan Pérez de la Riva - *"historia de la gente sin historia"* > giving voice, memory and existence
- Peggy McIntosh - White Privilege
- Critical Race Theory - counter-narratives
- James A. Banks - cross-cultural perspectives (insider/outsider), multicultural citizenship

The role of the digital library, like many information institutions/services, is an essentializing one that celebrates and reproduces the 'grand narratives' of the nation-state. By problematizing the cultural production of digital libraries as an act of nostalgia, of inclusion and exclusion, and of racial, social and sexual differentiation, we can unpack the role that digital libraries play in the creation of communities in our imaginary and in the perception of space and place from those objects we select to digitize. In recognizing that digital libraries are socially constructed, as information professionals and policy makers, we have the responsibility and power to re-envision a more diverse and representative digital library. Thus, multicultural and globalized digital libraries would guarantee the right for all cultural voices to be included, would acquire the necessary funding to pursue multicultural projects, and would enable underrepresented voices to speak for themselves, to determine what cultural heritage and community (including individual) experiences are to be digitized, and how the information will be accessed (limited or full access, language, interface, etc.). Taiwanese writer and cultural critic Ying-tai Lung [5] also recognizes that culture has to be done bottom up and notes the need and value of not just the grand narratives but also counter narratives in learning about one's national culture.

> The more you know about your own culture, the more you realize that any great culture is made up of multiplicities. Chinese culture itself is made up of multiculturalism. For instance, Confucianism is only one of many elite traditions against a huge domain of folk traditions. So, to understand Chinese culture fully, you need to know its complexity.

This complexity is not contained within a nation-state's geographic borders, but is further complicated by the existence of diaspora communities. "What diaspora implies is not only a movement across the borders of a country, but also the experience of traversing boundaries and barriers of space, time, race, culture, language and history" [6]. Diaspora communities are challenged as being neither as well as both an insider and outsider at the same time.

The paper concludes with a call for decentering digital libraries and digitizing the "Other" as an act of empowerment and representation. The decentering of digital libraries enables underrepresented groups to be at the table, to be heard and to take center-stage in digitizing and accessing their cultural heritage and experiences [7]. This democratic process ensures that there is the undoing of the Other which is considered essential by historian, social critic and activist Howard Zinn who found that the story of ethnic minorities in the United States was invisible in history books, thus "the consequence of these omissions has been not simply to give a distorted view of the past but, more importantly, to mislead us all about the present" [8]. In conclusion, digital libraries are not neutral but privilege selected knowledge and information systems because their development is influenced by human and economic interests. Because recorded knowledge and information systems reflect the power and social relationships within society, we need to further the purpose of cultural diversity, which is to preserve diverse forms of cultural expression, enrich society and have people engage with each other. Actions plans to develop multicultural and globalized digital libraries should require all information professionals to design cultural delivery systems that are oppositional to racist, sexist, homophobic, and neoliberal practices which are institutionalized in the cultural practices and systems in our Information Society.

> One has to begin to lose memory, even small fragments of it, to realize that memory is what our entire life is made of. A life without memory wouldn't be life, just as an intelligence without means of expression wouldn't be

intelligence. Our memory [cultural record, digital library] is our coherence, our reason, our action, our feeling. Without it we aren't anything.

Luis Buñuel, *My Last Sight*

Source: http://www.webdelsol.com/istavans/is-1.htm

Acknowledgments. I wish to thank the Generalitat Valenciana (Spain) for its financial support during the research and writing of this work, as well as the Departamento de Historia de la Ciencia y Documentación, Universitat de Valencia, which has hosted me as a Visiting Researcher.

References

1. He, B., Kymlicka, W.: Introduction. In: Kymlicka, W., He, B. (eds.) Multiculturalism in Asia, pp. 1–21. Oxford University Press, New York (2005)
2. Embassy of Socialist Republic of Vietnam in Canada. Visit Vietnam, http://www.vietnamembassy-canada.ca/html/visit.html
3. He, B., Kymlicka, W.: Introduction. In: Kymlicka, W., He, B. (eds.) Multiculturalism in Asia, p. 3. Oxford University Press, New York (2005)
4. UNESCO Universal Declaration on Cultural Diversity (November 2, 2001), http://www.ohchr.org/english/law/diversity.htm
5. Lung, Y.T.: Cultivating Culture from the Bottom Up. Excerpts from Interview by Cheong Suk-Wai (August 29, 2004), http://www.tamilnation.org/culture/bottomup.htm
6. Zhang, B.: Identity in Diaspora and Diaspora in Writing: The Poetics of Cultural Transrelation. J. Intercultural Studies 21, 125–142 (2000), http://www.informaworld.com/smpp/contentcontent=a713678940db=all
7. Malik, K.: The Meaning of Race: Race, History and Culture in Western Society. MacMillan Press, London (1996)
8. Zinn, H.: The Missing Voices of Our World. TomDispatch.com (November 15, 2004), http://www.tomdispatch.com/post/2003/howard_zinn_the_missing_voices_of_our_world

From Content Organization to User Empowerment

Bjørn Olstad

Prof. NTNU Trondheim, Norway,
CTO FAST
bjorn.olstad@fast.no
http://www.ntnu.no
http://www.fastsearch.com

Abstract. If the network has become the computer, search is in the process of becoming its interface. This transformation impacts the design of future digital libraries. On the content side, innovations in contextual search are driving a new precision level compared to existing search paradigms inherited from the web. On the user side, search will have an equally profound impact. Closed loop designs connecting social computing and search is transforming libraries from a static repository to a dynamic learning and collaboration space.

Keywords: Search, contextual search, contextual navigation, analytics, conversational relevance, social computing, digital libraries.

1   Search Driven Service Architectures

Search is emerging as the orchestration framework for user-centric architectures, offering innovative functionality for re-engineering both the user experience and the data access layers that support it. In the world of the Web, the evolution from Web 1.0 toward Web 2.0 has seen a global transition from monolithic services, centralized content models and managed communities, to a new democracy of empowered users, personalized information access, and user-driven communities.

In this new ecosystem of information, evolution favors services that are "mashups" of component functionalities, resulting in complete interaction environments that are focused by user intent and customized to specific tasks or discovery goals. This Lego-model of information sources and services opens up many new opportunities for service orchestration within digital libraries. Capabilities in search driven service architectures include: the ability to connect to open components/services; the ability to track and utilize behavioral information; the ability to integrate components loosely to support "agile" development cycles; the ability to bridge information between independent sources; the ability to connect user experiences to social networks; and the infrastructure to provision these services through multiple channels as appropriate to the user.

2   Content Organization Based on Contextual Metadata

Contextual search developments enable searches to be restricted to structural parts of documents, such as paragraphs, sentences, patent claims, formulas or any entity of

D.H.-L. Goh et al. (Eds.): ICADL 2007, LNCS 4822, pp. 5–6, 2007.

interest. Documents are no longer a blob of information serving as the atomic retrieval unit. Hence, content organization can focus on both *global* and *contextual* metadata. Global metadata includes author, publication year, journal etc. Contextual metadata will on the other hand typically be derived by automated means producing concepts, entities and numerical quantitation. These metadata can be encoded as XML markup within the semantic context they occur in the document. Before, names of people, locations, formulas, dates, scientific concepts were blended together and effectively lost in the search index; now, they are contextually related facts waiting to be exploited. The richness of the surrounding context drives significant precision improvements for search, discovery and analytics across contextual metadata.

3   The User Revolution: Search Is the Portal

If the network has become the computer, search is in the process of becoming its interface. In doing so, search introduces a paradigm shift within digital libraries from content organization to user empowerment. Traditionally, both digital libraries and supporting technologies have focused on the aggregation, organization and management of information. Going forwards, innovation will increasingly focus on user centric approaches to information management. Search is fundamentally a user centric technology. The core purpose of search is to decode the information provided by content authors and *reorganize* the information in a factual and relevant way according to the users's intent and context. Hence, search is emerging as the key technology for transitioning content centric services to a user centric paradigm.

Search is no longer a small box inserted in portal frameworks. Future information portals need to capture intent and context from users in order to improve precision and user efficiency in real-time. The future information portal is driven by matching algorithms. The future information portal *is* search. Search driven portals provide an algorithmic opportunity to engage in smarter communication with users. The user experience can be dynamically shaped by joining, integrating and re-configuring data to present customized views and analytics of information from many different formats and sources.

Furthermore, closed loop designs connecting social computing and search is transforming libraries from a static repository to a dynamic learning and collaboration space. Social searching will change methods for content organization, metadata management and workflow around digital libraries.

Archival Tools to Match the Web:
Open, International, Comprehensive

Gordon Mohr

Internet Archive, 4 Funston Ave, San Francisco, CA, 94129, USA
gojomo@archive.org

Abstract. Together with a number of national libraries, the Internet Archive committed itself in 2003 to international collaboration to create open source tools and standardized formats for web archiving. This project was motivated by our experience as home to over 100 billion archived web resources dating back to 1996, and as a partner to memory institutions building thematic web archives. Resulting tools include the *Heritrix* archival web crawler/harvester, the *Wayback* archive browsing service, and the *NutchWAX* archive full-text index and query utilities. A standard ingest/archival format for web resources called *WARC* has also been developed. Software with full source code is free to download and reuse, and organizations worldwide have adopted and contributed to these tools. Working with large collections remains a challenge, and the web itself is constantly growing and changing, so we continue to seek international cooperation to expand and improve this web archive tool set.

Keywords: World wide web, internet, harvesting, crawling, archives, indexing, search, HTTP, open source, collaboration.

1 Introduction and Background

Starting in 2003, the Internet Archive began developing open source tools for web archiving, with the support and assistance of many national libraries.

The Internet Archive is a non-profit Internet library based in San Francisco, California, USA. We are known for our 'Wayback Machine', offering public access to a web site archive of over 100 billion captured URLs dating back to 1996. We also have a leading role in the Open Content Alliance mass book digitization effort, and host popular free audio and video content collections, including thousands of live music shows and educational presentations.

The bulk of our web collection has come from raw content donations by a commercial partner, Alexa Internet. However, in 2003, the Internet Archive, together with the national libraries which would go on to form the International Internet Preservation Consortium (IIPC), determined there was a need for new tools and standards, built in an open collaborative model, for web archiving. Prior tools lacked the flexibility, archival focus, and unencumbered licensing possible with an open source approach.

D.H.-L. Goh et al. (Eds.): ICADL 2007, LNCS 4822, pp. 7–8, 2007.
© Springer-Verlag Berlin Heidelberg 2007

2 Tools

Development of these tools began with a crawler, *Heritrix*, for harvesting web content. They have grown to also include a standard format, *WARC*, for storage and interchange of web content; a browsing service, *Wayback*, for viewing archived content; and search utilities, *NutchWAX*, for full-text indexing and querying of archived content using only free software. All are now available for free download and use, with full source code. Software is primarily implemented in cross-platform Java, and available for embedding and customization for other projects.

Heritrix **archival crawler.** Heritrix was designed for faithful and complete content archiving, with a high level of configurability and customization. At the Internet Archive and elsewhere, Heritrix has been used for crawls of various frequencies – daily, weekly, monthly, quarterly, yearly, or one-time – and sizes – from a few thousand captured URLs to billions. Heritrix may be remote-controlled by other software or incrementally extended with new code modules.

WARC **archival file format.** For over a decade, the Internet Archive stored captured web content in its own simple concatenated-responses format, called ARC. To better handle collection-time metadata, duplicate-reduction, format evolution, and other related storage needs, the Archive and other libraries designed a successor format, called WARC, now under consideration as an international ISO standard.

Wayback **archive browsing service.** Wayback software allows URL-based lookup and browsing of archived web content, in a browser, as if viewing the original website near a desired time. Wayback has enabled access to multi-billion-URL collections.

NutchWAX **full-text index and search utilities.** NutchWAX adds Google-style search to web archives, based on other open source projects. These include Lucene, a raw full-text search engine; Nutch, an adaptation of Lucene for web content; and Hadoop, a system for parallelizing large processing jobs. NutchWAX is being used in distributed configurations at the Internet Archive to provide search in collections containing over a billion URL captures.

3 Future Directions

The Internet Archive actively maintains each of these software projects, and improvements are often sponsored or contributed by our library partners and other software users. However, web archiving software continues to face serious challenges as the web grows in size, in diversity of content types and technologies, and in the level of adversarial content manipulation (web spam).

The collaborative, open source, international approach has worked to efficiently built a shared base of web archive capabilities, and we seek new collaborations to continue this progress. We hope the Asian library community will find these tools useful, report feedback from their experiences, and join us to build new functionality.

Digital Archiving: Making it Happen

Defining a Web Archiving Strategy

Kristine Hanna

Internet Archive
kristine@archive.org

Kristine Hanna, Director of Web Archiving Services at Internet Archive, will discuss the significance of web archiving, the challenges libraries, archives and memory institutions face in the digital age, as well as some of tools and best practices currently in use to create a successful web archiving strategy.

The Internet Archive, located at http://www.archive.org/index.php on the web, has been involved in web archiving since 1996 when the organization was founded as an internet library to provide permanent access for researchers, historians and the general public to the world's cultural artifacts. Additionally the web group at the Archive http://wa.archive.org/ works with institutions to created focused collections through crawling services and Archive-It, a web based application.

The Internet Archive is a founding members of the IIPC (International Internet Preservation Consortium http://netpreserve.org/about/index.php, and we work closely with national libraries and archives from around the globe to develop open source tools and document best practices for web archiving.

Libraries and archives have long collected information that serves scholars, historians and the general public in understanding history, culture, and society. So much of today's information is easily found on the world wide web – web pages have replaced hard copy newsletters, blogs are today's diaries, many government forms and documents are more readily accessible on the web than they are in paper form. And there is one sure thing about the World Wide Web: like the weather, it will change. An estimated 44 percent of Web sites that existed in 1998 vanished without a trace within just one year. The average life span of a Web site is only 44 to 75 days.

With the rapid growth of the Internet and the Web, hundreds of millions of people around the world have grown accustomed to using the web as their primary resource to acquire information. The availability of this electronic information is taken for granted. It is a fallacy that if something is on the web it will be there forever. There's an urgent need for people to understand that that web is who we are. It's our culture and our social fabric, and we don't want to lose any of it. What is here today might be gone tomorrow.

As part of an effort to appropriately document and capture today's information for tomorrow's use, institutions must adopt a web archiving strategy. However, for many institutions, the prospect of capturing and storing web sites or entire web domains is a daunting prospect. This presentation will include an overview of web archiving at Internet Archive with our 60+ partners in national, state, university and public libraries. We'll take a look at why our partners are invested in web archiving, how they are applying web archiving in their particular institutions, some of the challenges

D.H.-L. Goh et al. (Eds.): ICADL 2007, LNCS 4822, pp. 9–10, 2007.

they face, what benefits they have seen, the tools they use, their lessons learned, as well as what they are looking forward to in the future.

We'll also take a look at "Best Practices" around web archiving, what our partners have uncovered, what they are recommending, what we have learned. We'll review some of the questions partners ask when planning a digital archiving strategy. What are the biggest challenges your institution faces in archiving and preserving digital materials? What is prompting your institution to look for a web archiving service? Is there a specific reason or event? How is your institution using the collections you have gathered?

Lastly, we'll provide an overview of the digital stewardship role that the Internet Archive is playing re: archiving and preserving data from around the world, with our hopes and objectives for the future, including the roles that other international institutions could play in furthering the mission.

Internet Archive is a member of the American Library Association, Society of American Archivists and is officially recognized by the State of California as a library. All collections are publicly accessible with free access at www.archive.org.

The Internet Archive is a non-profit organization founded in 1996 as an internet library to provide permanent access for researchers, historians and the general public to the world's cultural artifacts. The Archive collaborates with federal and internationl institutions including the U.S. Library of Congress, National Library of Australia, the Bibliothèque Nationale de Franc and the Smithsonian, plus over 45 national, state, university and public libraries and archives. The Internet Archive has the largest public web archive in existence with over 2 petabytes of data containing 85 billion pages from over 65 million websites in 37 languages. The Internet Archive recently shipped 1.5 petabytes of data to our partner, Bibliotheca Alexandrina, in Egypt; a leading institution of the digital age and a center for learning, tolerance, dialogue and understanding.

Information Access
Through Digital Library Systems

Maristella Agosti

Department of Information Engineering, University of Padua, Italy
agosti@dei.unipd.it

Abstract. The talk presents an interpretation of the evolution of the events and trends in the information access area. Focusing mainly on the last twenty years, particular attention is payed to the digital library system which needs to be envisaged and designed to support the end user in accessing relevant and interesting documents.

1 Information Access

The term *information access* identifies the activities that a person – the *user* – has to conduct to choose, from a collection of documents, those that can be of interest to him to satisfy a specific and contingent information need. The three main actors and aspects that information access needs to address are: 1) user, 2) collection of documents, and 3) access, that is a function or model used in retrieving and accessing documents.

Where the user is the central actor of the situation, the collection of documents is the source from which documents can be extracted in the hope they are of interest, and access is the function that transforms a user information need into a set of documents that are supposed to satisfy the user information need.

When the collection of documents grows and reaches a size that makes a manual inspection of the documents prohibitive, the construction and management of the collection together with the application of the access function are managed in an automatic way through a digital library system. In parallel to the growth of the size of the collection, many other changes happened both on the size of the individual documents and on the type of the documents which form the collection. Now the size of a document can span from that of a textual abstract of just few hundred words to that of a complete book or a video where multimedia representations need to be managed and accessed by the final user. So the final user is confronted with a new and more complex context of work, where the diversity of the types of documents can increase together with the rise of the size of individual documents.

This means that the context of reference in building and using a digital library system changes over time requiring the creation of new architectures and approaches. The talk critically analyzes the evolution of the modeling of the information access systems, mainly in the last twenty years, relating the general analysis to conducted experiences[1].

---

[1] M. Agosti (Ed.). *Information Access through Search Engines and Digital Libraries.* The Information Retrieval Series, Vol. 22, Springer, Heidelberg, Germany, 2008.

D.H.-L. Goh et al. (Eds.): ICADL 2007, LNCS 4822, pp. 11–12, 2007.

2 Evolution of Information Access Systems

Early Days. In the early days of computer science, the common approach was to consider the specific type of documents constituting the collection and to manage and design the system and applications around it. The attention of the system designer was concentrated on the specific type of documents, mostly because the available technology limited the possibilities of representation and management only to textual documents.

1977–1986: The Last Decade of Centralized Systems. The attention of the system designer was prevalently focused on the textual collection of documents more than on the user and the specific function or model to implement in the system. Still in this decade systems were able only to manage a single type of documents, and the systems were named in accordance with the specific type of document collections they were designed to manage in a specialized way. It was in those years that a new generation of library automation systems started to be designed with the purpose of enabling the managing and accessing of a combination of different types of data: structured catalogue data together with unstructured data representing the contents of the documents. This new type of library automation systems, that can be considered the "ancestors" of present days *digital library systems*, also supported the interactive retrieval of information.

1987–1996: Towards a User-Oriented Decentralized Environment. At the beginning of the decade, researchers took new directions, trying to support the users with new types of access models or with a combination of different models. New approaches to information access were proposed. Some of those proposals were based on the hypertext paradigm, and those ones made use of the links that exist among documents and descriptive objects. Some successful approaches used a two-level architecture to represent on one hand the collection of content objects, and on the other the content representation structure.

1997–2006: The Overwhelming Amount of Digital Documents. During this decade it became clear that it was necessary to face the continuous growth of diverse collections of documents in digital form, so major efforts faced different aspects related to the growing of diversified digital collections. Multimedia access is the other relevant area that researchers started to face in a systematic way and for different media during the decade. The complexity of the management of collections of multimedia digital documents can be faced in particular for information access purposes, but also from a general architectural point of view, that is, the area of *service oriented* digital libraries and digital library systems.

From 2007 On. The new decade that starts this year is opening up new exciting challenges, both on the side of architectures and on the contents for the construction of a new generation of digital library systems.

How to Prepare a European Digital Library

Olaf D. Janssen

The European Library Office, Royal Library of The Netherlands,
Prins Willem-Alexanderhof 5, The Hague, The Netherlands
olaf.janssen@theeuropeanlibrary.org

Abstract. This presentation shows how the joint efforts of the national libraries of Europe over the past 20 years have paved the way for the creation of a European Digital Library; currently a collaborative platform for European museums, archives and libraries, but in future also a webservice for end-users to discover Europe's heritage on an unprecedented scale. This presentation sets out the recipe for the first construction phase (2007-2008) and discusses the ingredients that are needed to build an operational European Digital Library from 2009 onwards.

Keywords: Digital library development, Europe, computer-supported collaborative work, project management, strategic information systems planning, systems development.

1 Introduction

The European Library[1] is a multilingual portal currently offering access to over 200 million resources in the 47 national libraries of Europe[2]. It offers free federated searching and delivers both bibliographical and digital objects - some free, some priced. The European Library is a service of CENL[3]. Established in 1987 this organisation has increased and reinforced the role of national libraries in Europe, in particular in respect of their responsibility for maintaining the national cultural heritage.

The European Library, launched in 2005, is not only a website for end-users, but - by offering a strong framework for collaboration - has also tremendously accelerated technical & interoperability standardisation, trust and understanding between the national libraries of Europe. As a direct result of this The European Library has established a track record of successfully implementing and using some of the vital ingredients for a European Digital Library. These include A) a cooperative organisational network, B) a technological platform based on creating, maintaining and conforming to common standards in i) data harvesting & access protocols, ii) metadata and C) multilingual access.

In 2004 Google announced its plans to digitise millions of books from 5 major Anglo-American academic libraries. Initiated by fears that Google's initiative could

---

[1] The European Library – http://www.theeuropeanlibrary.org
[2] The 47 national libraries of Europe – http://www.theeuropeanlibrary.org/libraries
[3] CENL – The Conference of European National Librarians – http://www.cenl.org

D.H.-L. Goh et al. (Eds.): ICADL 2007, LNCS 4822, pp. 13–14, 2007.

create a bias towards Anglo-American language and culture, Europe quickly united to mobilise funds for the digitisation, preservation and accessibility of European cultural heritage and the creation of a European Digital Library, planned to include millions of digital works from libraries, museums and archives by 2010.

It was decided that this European Digital Library should not be constructed from scratch, but should center around pivotal initiatives such as CENL and The European Library, with their proven track records.

Within their own domains, museums, archives and libraries each have well established routines regarding object descriptions and accessibility, as well as dealing with technical, human, usability and business issues. Respecting these practises means that in creating a unified European Digital Library no single solution can be imposed from above. The only way to tackle the fragmented cultural heritage map of Europe is – similar to Europe's national libraries – by *creating consensus, trust and interoperability* between the 3 domains.

For this purpose in summer 2007 The European Library and CENL set up a 84 partner cross-domain strategic network[4] to establish large scale dialog between the archival, library and museum sectors. Such new collaboration on this scale is a significant step forward.

Running for two years, this network will investigate the political, human, sustainability, organisational, interoperability, technical and semantic ingredients that are needed to create a multilingual service for accessing digital content from Europe's distributed and varied cultural institutions.

The project will also develop a fully working portal prototype, with interoperable multilingual access to at least 10 collections from each of the three domains. This proof of concept will attempt to identify 2 or 3 European themes where there is digitised material that can demonstrate the power of this cross domain initiative.

By 2009 the network will result in a larger and more visible community of archivists, librarians and museum people committed to making content available and interoperable. It will also deliver a roadmap showing what needs to be achieved by when and a proposal for one or more separately financed practical implementations of a European Digital Library from 2009 onwards.

---

[4] EDLnet Thematic Network - http://www.digitallibrary.eu/edlnet/

Evaluation of Hospital Portals Using Knowledge Management Mechanisms

Chei Sian Lee, Dion Hoe-Lian Goh, and Alton Yeow-Kuan Chua

Division of Information Studies, Wee Kim Wee School of Communication and Information
Nanyang Technological University, Singapore 637718
{leecs,ashlgoh,altonchua}@ntu.edu.sg

Abstract. Hospital portals are becoming increasingly popular since they play an important role to provide, acquire and exchange information. Knowledge management (KM) mechanisms will be useful to hospitals that need to manage health related information, and to exchange and share information with their patients and visitors. This paper presents a comprehensive analysis of knowledge management mechanisms used by 20 hospital portals from North America and Asia to access, create and transfer knowledge. We developed a systematic and structured approach to evaluate how well the portals captured and delivered information to patients and visitors about the hospitals' business processes, products, services, and customers from the perspective of three KM mechanisms (i.e. knowledge access, knowledge creation and knowledge transfer). Our results show that our selected hospital portals provided varying degrees of support for these KM mechanisms.

Keywords: Portals, hospitals, checklist, knowledge management framework, knowledge access, knowledge creation, knowledge transfer.

1 Introduction

The number of people looking for health information online is increasing rapidly and so is the demand for hospital portals [5]. Clearly, providing up-to-date and relevant information is an important mission for hospitals since misplaced truths may influence serious health decisions. As such, effective information storage and management to ensure that the contents on the hospital portals are timely, credible and accurate is a responsibility of these hospitals. Past studies have shown that knowledge management (KM) is an effective strategy to store and manage information in an organization [11]. Specifically, KM helps in identifying strengths and weaknesses, problem solving, dynamic learning, creating opportunities, and strategic planning. In addition, KM provides the process to help both organizations and users to capture, store, organize and share the knowledge within and across communities effectively [14]. Hence, it seems intuitive to relate KM mechanisms to studies on hospital portals since portals are considered to be tools that revolutionize the access to information and knowledge [1].

Undoubtedly, the Internet enables individuals to access a huge amount of information and knowledge. Thus the ability to seek, share and evaluate information

D.H.-L. Goh et al. (Eds.): ICADL 2007, LNCS 4822, pp. 15–23, 2007.

through various mechanisms has become a key requirement for the success of any website [15]. However, despite the proliferation of Internet portals, to the best of our knowledge, there are no standardized evaluation guidelines found in the literature to guide hospitals in developing their portals, especially those designed with a KM agenda.

The objective of this study is to evaluate various interactive features used by hospital portals to facilitate KM and collaboration between hospitals and users. We examined hospital portals from two geographical regions (i.e. North America and Asia). We developed a systematic and structured approach to evaluate how well the portals capture and deliver information to the users about the hospitals' business processes, products, services, and customers from a KM perspective. The focus is on three KM mechanisms, namely, knowledge access, knowledge creation and knowledge transfer.

2 Literature Review

Various definitions of Internet portals exist. Some studies defined Internet portals as single-point-access software systems to provide easy and timely access to information and to support communities of knowledge workers [11]. Other studies defined Internet portals as a one-stop solution to the information problem created by the World Wide Web that increases the access to information in a specific domain [9]. Finally, there are other studies that viewed Internet portals as tools to revolutionize access to information and knowledge [1]. Here, we view portals as KM tools and propose that KM mechanisms will be useful to hospitals in accessing, creating and transferring knowledge via their portals.

Various methods of assessing and evaluating websites and portals have been conducted in past studies. Fritch [6] also proposed a set of heuristics, tools and systems to help evaluate health information on the Internet. For portal quality evaluations, Dragulanescu [3] proposed the use of total quality management specific tools and techniques. Factors such as accuracy, authority, coverage, currency, density, interactivity, objectivity, and promptness were considered. Another study done by [8] on healthcare portal evaluation focused on credibility evaluation. Nah et al. [13] evaluated how e-commerce and financial service websites disseminate, acquire and share knowledge using KM mechanisms. The review of past portal evaluation studies showed that there is little work done in establishing guidelines for hospitals in implementing KM mechanisms in portals, and our present work is therefore timely.

3 Research Model

We extended [13] model to examine the KM mechanisms that will be in effective in hospital portals. From our review of the literature, we modified the KM mechanisms in [13] to reflect the mechanisms that are appropriate and significant for hospital portals. Hence, our modified model as shown in Figure 1 consists of three main elements: Knowledge Access, Knowledge Creation, and Knowledge Transfer. The *Knowledge Access mechanism* refers to the mechanisms through which the users get access to the portal and information on the portal [2]. The *Knowledge Creation mechanism* refers to the process of capturing users' information such as demographics, preferences and behavioral behaviors [13] and creates new knowledge that will benefit the portal providers and the users [14]. The *Knowledge Transfer mechanism* refers to the mechanisms that allow the portal providers to foster user-to-user and provider-to-user sharing of knowledge [13].

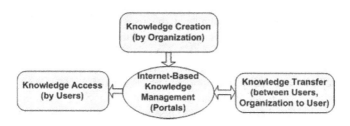

Fig. 1. Research model

4   Methodology

4.1   Data Collection

We sampled 20 hospital portals from North America and Asia (see Table 1). We focused on these two regions because of the availability of a large number of hospital portals for selection. Portals that are not available in English were excluded. We selected hospitals portals based on a combination of the following: 1) they appeared in the Best Hospitals 2005 survey conducted by US News & World Report, Best Health [21], 2) they were recipients of prestigious awards such as the eHealthcare Leadership Awards [12], 3) they received high ratings from third-party agencies such as alexa.com, 4) they provided self-regulating policies and third party seals such as HON Code and TRUSTe, 5) they were easily accessible via popular search engines such as Google, Yahoo and MSN.

4.2   Evaluation Checklist Formulation

Our evaluation checklist consists of 52 items that were derived by reviewing the features available on the selected portals and with reference to past studies. These checklist items were grouped under sub-dimensions based on similar functionalities. These collections of sub-dimensions were further grouped into dimensions. Finally, the dimensions were grouped under the three KM mechanisms. Based on the three modified KM mechanisms in our research model, 13 dimensions were formed.[1]

4.3   Portal Evaluation Approach

Each portal was evaluated by two members of the research team. For each portal, the evaluation results obtained were compared and any discrepancies were eliminated by

---

[1] Due to space constraints, we do not provide the evaluation questionnaire containing the 13 dimensions. The authors can be contacted for this if necessary.

combined re-assessment. Cohen's Kappa test was also conducted to measure the agreement of each question or checklist item by two evaluators. The results of test ranged from 0.77 to 1.0 and suggest a high degree of agreement among evaluators. Our portal evaluation approach consisted of the following three steps.

Step 1: Rating Scheme. For each checklist item, a rating of 1='Yes' or 0='No' based on whether those features are supported by the portals was assigned.

Step 2: Weighting Scheme. This study adopts the technique of assigning the weighting criteria as applied by [4]. They suggested the usage of the "Delphi" technique in assigning weights. The Delphi technique is a qualitative technique that requires the convergence of expert opinions which is anonymously and possibly subjective to the underlying criterion in an attempt to produce more precise results. In this work, we modified the technique to adopt the approach of [10] and [7]. Here, each evaluator in the group assigned weights between a scale of 1 to 5 to each sub-dimension independently, with five being the most important and one being the least important. When wide discrepancies occurred, the group discussed until a general agreement was reached. Finally, an average was taken to obtain single numerical weighting value for each sub-dimension. Following this, the final weighting is divided with the number of questions or checklist items.

Step 3: Scoring Finally, after ensuring that the evaluation results obtained are unbiased and accurate as possible, the scores for each sub-dimension were computed by multiplying the ratings of 1 or 0 with the weightings and summing them up.

5   Analysis and Findings

In this section, we present our findings on how well our selected hospital portals scored for each of the KM mechanisms.

5.1 Knowledge Access (KA)

KA refers to the mechanism through which users access the portal and its information. The results for each of the sub-dimensions of KA mechanism are shown in Table 1.

Access to Portal (AP). All hospital portals are accessible from the top search engines which rank these portals as most popular or the best-known pages when search terms such as "best hospitals" or "best hospital websites" followed by the country name or city name were used. Alternately, some of the portals were selected from the list of hospital websites maintained by Massachusetts General Hospital and Harvard. This was retrieved from search engines by querying for "listing of hospitals websites worldwide".

Searching. We further examined 2 sub-dimensions: Query (Q) and Results (R). With the exception of MohanRao Memorial Hospital, India, the query feature is existent on all other portals. However, advanced query features such as to expand or modify search results are usually not available.

Table 1. Evaluation results of the knowledge access mechanism

	Hospitals	AP	Q	R	B	ICU	ICO	A	IP
N. America	Alegent Hospital, US	5.0	1.2	0.0	3.0	3.3	4.3	0.0	3.0
	Cleveland Clinic, US	5.0	2.4	3.0	3.0	2.2	4.3	0.8	4.0
	Comer Children's Hospital, US	5.0	2.4	1.5	1.0	0.0	0.0	0.8	1.0
	Henry Ford Hospital, US	5.0	1.2	0.0	3.0	3.3	4.3	0.0	3.0
	John Hopkins Hospital, US	5.0	2.4	3.0	2.0	0.0	4.3	0.0	4.0
	Mayo Clinic, US	5.0	1.2	0.0	3.0	2.2	4.3	0.0	3.0
	Mt.Sinai Hospital, CA	5.0	1.2	1.5	1.0	0.0	4.3	0.0	1.0
	St Michael Hospital, CA	5.0	1.2	0.0	1.0	0.0	0.0	0.0	1.0
	Stanford Hospital, US	5.0	2.4	1.5	2.0	0.0	4.3	0.0	2.0
	The Ottawa Hospital, CA	5.0	1.2	0.0	1.0	0.0	0.0	0.8	1.0
Asia	Alfred Hospital, AU	5.0	1.2	1.5	0.0	0.0	0.0	0.0	0.0
	King Faisal Specialist Hospital, SA	5.0	1.2	0.0	1.0	0.0	0.0	0.8	1.0
	MohanRao Memorial Hospital, IN	5.0	0.0	0.0	1.0	0.0	0.0	0.0	1.0
	National University Hospital, SG	5.0	1.2	0.0	2.0	0.0	4.3	0.0	1.0
	Pantai Group of Hospitals, MY	5.0	1.2	0.0	0.0	0.0	0.0	0.0	1.0
	Royal Adelaide Hospital, AU	5.0	1.2	0.0	1.0	0.0	0.0	0.8	1.0
	Starship Children's Hospital, NZ	5.0	1.2	0.0	1.0	0.0	0.0	0.0	1.0
	Tan Tok Seng Hospital, SG	5.0	1.2	0.0	1.0	0.0	4.3	0.0	0.0
	United Family Hospitals, CN	5.0	1.2	0.0	0.0	0.0	0.0	0.8	0.0
	Wockhardt Hospitals, IN	5.0	1.2	0.0	1.0	0.0	0.0	0.0	4.0

Note: AU: Australia, CA: Canada, CN: China, IN: India, MY: Malaysia, NZ: New Zealand, SA: Saudi Arabia, SG: Singapore, US: United States of America

Browsing (B). We found that three hospital portals, Pantai Group of Hospitals Malaysia, Alfred Hospital Australia, and United Family Hospitals China did not provide any browsing features. In general, most of the US portals provided more variety of browsing features than hospitals portals in Asia.

Personalization and Customization. We further examined 2 sub-dimensions: Information Customized by User (ICU) and Information Customized by Organization (ICO). Only 9 of the hospital portals provided such features. Henry Ford Hospital US, and Alegant Hospital US are examples of hospitals that provide personalization and customization features.

Accessibility (A). Only six hospital portals from both geographical regions provide accessibility feature, through either multilingual support or multiple interfaces (i.e. intensive and low graphics, text interface).

Information Presentation (IP). Majority of the hospital portals and all portals from North American region provided at least one medium to aid in information presentation. Wockhardt Hospital India, John Hopkins Hospital US, and Cleveland Clinic US fulfill all the requirements in this dimension.

5.2 Knowledge Creation (KC)

KC refers to the mechanism through which an organization acquires information from users and creates new knowledge in the enterprise repository for the benefit of the organization and users. The results obtained from each of the sub-dimensions of the KC mechanism are shown in Table 2.

Acquisition of User Information (ACU). More than half of the hospital portals provided features to capture information from users.

Feedback (FB). This is an important feature and is listed as one of the mandatory features that a portal must have. All the hospital portals surveyed managed to obtain a full score in providing feedback.

Domain Data Acquisition (DDA). Only Chicago Comer Children's Hospital US, John Hopkins Hospital US, Mt. Sinai Hospital Canada, United Family Hospital China, and Pantai Group of Hospital Malaysia portals offered one of the domain data acquisition features.

Table 2. Evaluation results of the knowledge creation mechanism

	Hospitals	ACUI	FB	DDA
N. America	Alegent Hospital, US	4.0	4.7	0.0
	Cleveland Clinic, US	4.0	4.7	0.0
	Comer Children's Hospital, US	0.0	4.7	1.0
	Henry Ford Hospital, US	4.0	4.7	0.0
	John Hopkins Hospital, US	2.7	4.7	1.0
	Mayo Clinic, US	4.0	4.7	0.0
	Mt.Sinai Hospital, CA	2.7	4.7	1.0
	St Michael Hospital, CA	1.3	4.7	0.0
	Stanford Hospital, US	0.0	4.7	0.0
	The Ottawa Hospital, CA	0.0	4.7	0.0
Asia	Alfred Hospital, AU	0.0	4.7	0.0
	King Faisal Specialist Hospital, SA	2.7	4.7	0.0
	MohanRao Memorial Hospital, IN	1.3	4.7	0.0
	National University Hospital, SG	0.0	4.7	0.0
	Pantai Group of Hospitals, MY	0.0	4.7	1.0
	Royal Adelaide Hospital, AU	1.3	4.7	0.0
	Starship Children's Hospital, NZ	0.0	4.7	0.0
	Tan Tok Seng Hospital, SG	1.3	4.7	0.0
	United Family Hospitals, CN	2.7	4.7	1.0
	Wockhardt Hospitals, IN	2.7	4.7	0.0

5.3 Knowledge Transfer (KT)

KT refers to the mechanism through which the knowledge is transferred or shared between the organization and users and among the users. The results obtained from each of the sub-dimensions of the KT mechanism are shown in Table 3.

Online Collaboration. We further examined 3 sub-dimensions: Collaboration from Organization to User (COU), Collaboration Between Users (CBU), and Synchronous Support (SS). Overall, few of the hospital portals offered online collaboration features. Those that did are Alegent Hospital US, Henry Ford Hospital US, and MohanRao Memorial Hospital India that employed 'Ask an Expert' feature which allowed users to ask questions about specific topics. Only Johns Hopkins Hospital US provided features such as discussion forums and online groups to promote collaboration between users.

Information Alerts (IA). Our findings show that with the exception of Royal Adelaide Hospital, Australia, all hospital portals managed to have at least one feature to alert users to any new information.

Users Support (US). Three hospital portals, King Faisal Specialist Saudi Arabia, Alfred Hospital Australia, and Chicago Comer Children's Hospital US failed to provide any user support features. Mayo Clinic, Cleveland Clinic, and John Hopkins from the US, scored the highest among all portals in this category by providing four of five features. None of the portals provided usage demos or tutorials to guide users on their services provided.

Resource Sharing (RS). Except for Alfred Hospital Australia, all sampled hospital portals managed to have at least one feature for resource sharing. Wockhardt Hospital India was the only portal that fulfilled all the requirements in this dimension.

Table 3. Evaluation results of the knowledge transfer mechanism

	Hospitals	COU	CBU	SS	RS	US	IA
N. America	Alegent Hospital, US	2.3	0.0	0.0	2.1	0.7	2.4
	Cleveland Clinic, US	0.0	0.0	1.7	2.1	2.7	2.4
	Comer Children's Hospital, US	0.0	0.0	0.0	0.5	0.0	1.2
	Henry Ford Hospital, US	0.0	0.0	1.7	1.6	0.7	2.4
	John Hopkins Hospital, US	0.0	0.9	1.7	2.1	2.7	2.4
	Mayo Clinic, US	2.3	0.0	1.7	1.0	2.7	2.4
	Mt.Sinai Hospital, CA	0.0	0.0	0.0	1.0	0.7	2.4
	St Michael Hospital, CA	0.0	0.0	0.0	1.6	1.3	2.4
	Stanford Hospital, US	0.0	0.0	0.0	0.5	1.3	2.4
	The Ottawa Hospital, CA	0.0	0.0	0.0	1.6	0.7	1.2
Asia	Alfred Hospital, AU	0.0	0.0	0.0	1.0	0.0	0.0
	King Faisal Specialist Hospital, SA	0.0	0.0	0.0	2.1	0.0	1.2
	MohanRao Memorial Hospital, IN	2.3	0.0	1.7	1.0	0.7	2.4
	National University Hospital, SG	0.0	0.0	0.0	1.0	0.7	1.2
	Pantai Group of Hospitals, MY	0.0	0.0	0.0	1.6	0.7	1.2
	Royal Adelaide Hospital, AU	0.0	0.0	0.0	0.0	0.7	1.2
	Starship Children's Hospital, NZ	0.0	0.0	0.0	0.5	0.7	1.2
	Tan Tok Seng Hospital, SG	0.0	0.0	0.0	1.6	0.7	1.2
	United Family Hospitals, CN	0.0	0.0	0.0	1.6	0.7	1.2
	Wockhardt Hospitals, IN	0.0	0.0	0.0	0.5	0.7	3.7

6 Discussion and Conclusion

Results indicate that all selected 20 hospital portals utilized a combination of KA, KC and KT mechanisms. Overall, our results also suggest that North America hospital portals appeared to perform better than Asia hospital portals in terms of meeting the KM evaluation criteria. Specifically, the top seven hospital portals that were able to utilize KM mechanisms to access, create and transfer knowledge effectively and efficiently were all from North America.

Unsurprisingly, our results show that the KA mechanism was more prevalent than KC and KT mechanisms in most hospital portals. The most commonly available feature to support the KA mechanism is providing portal access to users via search engines. The most commonly available feature to support the KC mechanism was getting feedback from users. This finding is also consistent with the findings of Nah et al.[13] in their study on financial websites. The most uncommonly available feature for creating knowledge is acquiring domain/subject specific data from users. The most commonly available features to support KT mechanism is via resource sharing in terms of providing catalog information, external links to other websites and viewing information contributed by other users. Surprisingly, we find that supporting the KT mechanism via online collaboration among and between users is still lacking in many portals.

Our research has provided several important contributions. One major contribution is a comprehensive and systematic evaluation checklist that includes 13 dimensions and 52 items for hospital portals using KM mechanisms. We have thus taken an important first step to expand our knowledge on the relevance of KM mechanisms for hospital portals. This checklist may even be useful to portals outside the hospital domain that have been designed with a KM agenda. Second, we believe that the checklist and results obtained can be utilized by hospitals to develop highly interactive portals to meet users' expectations. Stakeholders should however decide whether these gaps should be addressed given the objectives and scope of their respective portals. In addition, the evaluation checklist does not include design and usability issues as these are sufficiently addressed in the literature. Developers should therefore use our checklist in conjunction with established usability instruments and guidelines during portal implementation.

There are two main limitations in our study. First, the selection of portals was limited to hospitals from North America and Asia that are available in English; this may prevent generalization. Future work can look into evaluating portals in other languages, and from other regions. Second, there are currently many portals that also provide healthcare information but are not operated by hospitals. Some of these are operated by the government (e.g. US Food and Drug Administration) and some by other business organizations (e.g. Healthcentral.com). Future work can look into evaluating these non-hospital portals and examine the differences between them and hospital portals in terms of accessing, creating and transferring health related information and knowledge.

Acknowledgments. The authors would like to thank Aryu Novieta Sunyoto, Thimme Gowda Rashmi and Omair Ahmed for assisting in data collection.

References

[1] Cloete, M., Snyman, R.: The enterprise portal - is it knowledge management? Aslib Proceedings 55(4), 234–242 (2003)

[2] Davies, J., et al.: Next Generation of Knowledge Access. Journal of Knowledge Management 9(5), 64–84 (2005)

[3] Dragulanescu, N.: Website Quality Evaluations: Criteria and Tools. International Information & Library Review 34(3), 247–254 (2002)

[4] Edmonds, L.S., Urban, J.S.: A method for evaluating front-end life cycle tools. In: Proceedings of the IEEE International Conference on Computers and Applications, pp. 324–331. Computer Society Press, Los Alamitos, CA (1984)

[5] Fox, S., Fallows, D.: Internet Health Resources. Retrieved (January 16 2003) (2006), from http://www.pewinternet.org

[6] Fritch, J.W.: Heuristics, tools and systems for evaluating Internet information. Online Information Review 27(5), 321–327 (2003)

[7] Goh, D.H., Chua, A., Khoo, D.A., Khoo, E.B., Mak, E.B., Ng, M.W.: A checklist for evaluating open source digital library software. Online Information Review 30, 360–379 (2006)

[8] Kim, P., Eng, T.R., Deering, M.J., Maxfield, A.: Published criteria for evaluating health related websites: review. British Medical Journal 318(7184), 647–649 (1999)

[9] Kotorov, R., Hsu, E.: A model for enterprise portal management. Journal of Knowledge Management 5(1), 86–93 (2001)

[10] Loo, R.: The Delphi method: A powerful tool for strategic management. Policing: An International Journal of Police Strategies & Management 25, 762–769 (2002)

[11] Mack, R., Ravin, Y., Byrd, R.J.: Knowledge Portals and the emerging digital knowledge worklplace. IBM Systems Journal 40(4), 925–955 (2001)

[12] Medseek. Healthcare Organizations Receive Prestigious Leadership Awards in eHealth, Retrieved from (February 8, 2006) (2005), http://www.medseek.com/1281.cfm

[13] Nah, F.F., Siau, K., Tian, Y.: Knowledge Management Mechanisms Of Financial Service Sites. Communications of the ACM 48(6), 117–123 (2005)

[14] Smith, L.C.: Knowledge discovery, capture, and creation. American Society for Information Science 26(2), 11 (2000)

[15] Tabatabai, D., Shore, B.M.: How experts and novices search the Web. Library & Information Science Research 25(2), 222–248 (2005)

Supporting Student Collaboration for Image Indexing

Palakorn Achananuparp, Katherine W. McCain, and Robert B. Allen

College of Information Science and Technology
Drexel University
pkorn@drexel.edu, kate.mccain@ischool.drexel.edu,
rba@drexel.edu

Abstract. We describe the Image Tagger system – a web-based tool for supporting collaborative image indexing by students. The tool has been used in three successive graduate-level classes on content representation. To fully satisfy the class' requirements and provide support for student indexing activities, it was designed and developed iteratively in accordance with the feedback and suggestions from the students as well as the instructor. The tool was well received by most students. They expressed a positive opinion toward collaboration support and thought it enhanced the overall learning experience in the class' image indexing project.

Keywords: Collaboration, Digital Library Education, Image Indexing, Metadata, Repositories, User Interfaces.

1 Introduction

Quality metadata remains a cornerstone of effective digital libraries. While social tagging systems, such as Flickr, are very useful for providing coarse labels (e.g., [9]) high-quality, consistent, labels still require trained professionals. We believe that metadata development tools are best be implemented as a digital repository. In addition, for teaching purpose, those tools should support collaboration among groups of students. However, existing repository tools such as Fedora DSpace, and Greenstone are not well designed for that.

We report on the design, implementation, and evaluation of a educational Image Tagger. This has been used in image indexing projects in three successive terms of a graduate-level course on Content Representation. The class' image indexing project is intended to make the students aware of the challenge and issues in representing non-textual materials such as images, give them experience in working with established controlled vocabularies currently used for image indexing, and allow students to work both individually and in a collaborative group mode. This indexing tool is an integral part of a larger repository system for both text and non-textual resources.

To customize the digital libraries to task-specific requirements, several researchers have adopted a user-centered design approach. In this approach, the objectives are to understand how the users perform their information seeking activities within the digital libraries scope and to expand the scope of the digital libraries to better support the tasks in context. Our project shared the same design philosophy. We believe that the best way

D.H.-L. Goh et al. (Eds.): ICADL 2007, LNCS 4822, pp. 24–34, 2007.

to provide support for student collaboration can be achieved by involving the stakeholders early in elicitation and design phases. A few collaborative indexing systems have been developed e.g., [8, 11] but these systems are focused on free-text annotation we emphasize the use of controlled vocabulary in image indexing.

2 Goals and Requirements

We are investigating effective ways to teach library and information science (LIS) students about repositories and metadata. Simultaneously, we aim to leverage such technology to improve the integration of digital library components into existing classroom. This leads us to the development of Image Tagger, a tool for supporting students' image indexing tasks. To achieve this, we have worked closely with the potential stakeholders -- the instructor and the students -- to develop the requirements and specify the design of the tool. One of the major requirements is to provide a facility that supports student collaboration in image indexing project. As the class usually conducted the indexing projects with the groups of online students, it was crucial that the working environment had to be specifically designed to support online collaboration. Lastly, as most LIS students had moderate computer proficiency, the tool had to be simple enough to use and required a minimum amount of installation.

We began an early version of Image Tagger prototype in fall 2005. The main users were LIS students enrolled in the class; the instructor (who is the second author on this paper) was also a key user, since she needed to monitor student activity and trouble-shoot indexing performance. The prototype was primarily intended to allow the students to view a set of images and index them with appropriate metadata through a web interface. In addition, the students should be able to search through images indexed by other group members through a simple interface. The feedback we got from the students during the first term was very positive; most of them liked the simplicity of the tool for doing their image indexing project. We have iteratively refined the prototype in the subsequent terms. The instructor provided us with requirements to make the tool a better fit for the class projects and to support instructional activities. Other requirements were gathered directly from comments and suggestions of the students after they had used the tool throughout the term.

At the end of each term, we anonymized and analyzed the recorded discussions of students when they were working on their group indexing assignment. This has helped us to understand student collaboration patterns and provided us with requirements that we later developed into the subsequent version of the prototype.

3 Image Tagger

Here, we describe the specific features of the current version of Image Tagger tool, focusing in particular on those aspects which have been essential for the class project. Particularly, we try to come up with collaborative indexing features, which are still lacking in existing digital library tools. Image Tagger was developed in a Java Server Pages (JSPs) platform using MySQL as a backend database. It consists of the following modules: login and authentication, image handler and storage, metadata generation, image metadata search, and group tools. The developer (first author) was responsible for the prototypes and providing technical support to the class.

We used a custom metadata set tailored for these photographic materials. The metadata fields include ID number, title (text included in photograph), textual description on the front and the back of the photograph's cardboard mount, city and country (derived from descriptions with preferred names derived from Getty's *Thesaurus of Geographic Names* (TGN) [3]), subject terms based on the *Thesaurus for Graphic Materials* (TGM) [4], subject terms based on the *Art and Architecture Thesaurus* (AAT) [2], and notes.

3.1 Individual Indexing

After students log in, they are directed to an assignment page which displays a set of 15 images the individual student is assigned to index. The purpose of the individual indexing assignment was to allow the students to become familiar with images' subject matter and the controlled vocabularies used for subject indexing and control of city and country names. In this way, the students would be prepared to contribute to a group indexing exercise. The system did not provide a direct link to access terms in the controlled vocabularies. Instead, the students are told where to access the thesauri via their web browser. The student can return to a particular image and modify the metadata as often as desired within the individual indexing portion of the assignment.

Fig. 1. An example of a page for a single image with candidate terms suggested by group members and discussion about which of those terms should be selected

3.2 Photo Pages with Overviews of the Group Members Individual Reponses

After completing individual indexing activity, the students are given access to their group's page. The group page displays thumbnails of the same set of images the students had previously indexed in their individual assignment. In this second stage, the members of each group work together to create the consensus metadata for their 15 images. To help the group manage their workflow, the thumbnail display includes visual cues showing the indexing status of each photo. A borderless image shows that the image had not been indexed yet. A dotted border indicates that the image is currently being indexed while a solid border indicates a completely indexed image with all metadata fields completed. To prevent group members from accidentally overwriting one another's works, an image could be indexed by only one student at a time. In addition to the group's image pool, this page also displays the recent comments made by group members on the particular images.

3.3 Displaying Common Indexing Terms

Clicking on a thumbnail in the group page directs the students to the image indexing page (Figure 1). This page displays a high-resolution image along with its associated metadata (if any). Group members are also able to discuss the indexing task by posting comments (lower left of figure). A major design consideration for image indexing page is to support students working toward consensus metadata. Image Tagger provides summaries of controlled vocabulary terms used by each group member to index a particular image (right side of figure). Each summary displays a list of terms used by the group members in their individual indexing assignments along with the frequency of use. In this particular example, rice paddies/Rice Paddies assigned from the Thesaurus for Graphic Materials (TGM) four times (twice with and twice without capitalization).

3.4 Exploring the Group's Indexing Terms

Additional information on subject term choices is provided to the group as two aggregated listings of all terms assigned to all 15 images — the subject term pool and subject term cloud. These two views are intended to help the students efficiently explore their group's indexing space. The subject term pool shows the subject terms the group members used in their individual indexing assignments. It inverts the image/term relationship, displaying a list of subject terms used by group members in alphabetical order. Each row contains a subject term, the images indexed with this term, and the names of the indexers. The list is organized according to the controlled vocabularies. The students can choose to view subject terms in either of the two controlled vocabularies – the Thesaurus for Graphic Materials (TGM), or the Art and Architecture Thesaurus (AAT). Figure 2 shows a screenshot of the subject term pool page.

3.5 Searching Image Metadata

The students have access to a simple search interface with two search options available. One option is to search for one or more subject terms in the two subject fields (TGM and AAT). The second is to search for text in the remaining metadata fields, such as title, city, country, notes, etc. Once the query is submitted, the system will display the retrieved images along with the associated metadata.

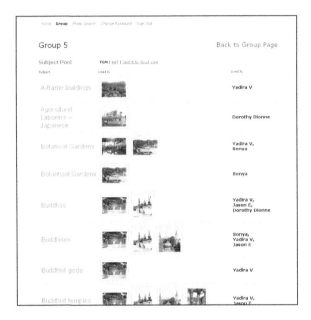

Fig. 2. Subject term pool

3.6 Tools for the Instructor

Image Tagger also provides basic tools for the instructor. The instructor can keep track of the groups' progress and discussions using the same tools as the students, but with full access to all students' work. For example, the instructor can view the index metadata using a search interface or access the specific group page to inspect the students' discussion and indexing term assignments. The instructor can also view the activity log to see the participation rates of individual students. The activities tracked by the current system include sign in, sign out, annotate, and comment.

4 Use in Class

4.1 Class Procedure

The Image Tagger tool has now been used in a class on Content Representation for three consecutive terms. The initial version was introduced in the Fall 2005 Content Representation on-campus class as a relatively simple tool that supported image metadata generation and subsequently (Spring/Fall 2006) in two online classes. In all three classes, collaboration was accomplished primarily via threaded discussion on Blackboard, although the on-campus class also was given time for in-class group discussions as well.

The project organization varied slightly from term to term. To simplify the scope of our discussion here, we will focus on the procedure used in the last two terms in which collaboration support was totally online. The image indexing project was designed as a combination of individual and group activity to give each student some experience with the challenge of image indexing, working with different established controlled vocabularies, and also with seeing how others dealt with the same issues. After an initial orientation, assigned reading, and writing assignment, individual students worked on the images with general troubleshooting discussion (discussion thread) and feedback from the instructor. By Week 7, all students had completed their initial set of 15 images and had begun the group indexing assignment. Students were organized into groups of 4-6 students, all of whom had worked on the same set of images. Each group had to work together to develop consensus metadata for each of their assigned 15 images.

There were about 30 students enrolled in each of the classes (Spring/Fall 2006). All were online students who resided in different geographic locations and time zones. Students had two or three different modes of collaboration available depending on the academic term. They could use assigned discussion threads in the Blackboard course shell, real-time/recorded group chat, or the comment field in the Image Tagger (in Fall 2006). By the time students began working on their individual projects, the course had covered the basics of general resource surrogation and image representation. They were given minimal guidelines – decide what was "index-worthy" in the image and then find a reasonable number of appropriate terms in the source vocabularies. The goal as stated in the assignment was to give the student the full experience of facing an image without prior discussion of its content, difficulty, historical background, etc. and without extensive supervised indexing training. At the group indexing stage, students were expected to establish a consensus representation based on group discussion. Had there been time, this would have been followed by cross-group retrieval experiments.

At the end of the course, each student submitted a reflection paper commenting on the project and his or her experiences and insights. The students were not graded on the "correctness" of their individual or group indexing but on their understanding of the process and the issues involved in image representation they acquired.

4.2 Materials

The images used were taken from a collection of 19th century photographs of Asian countries, including India, China, Japan, and Thailand, held by The Free Library of Philadelphia. Most of them are black-and-white photographs. Image content included landscapes, architectural details, portraits, and local events. This material was used because the photos are of inherent interest, were out of copyright, and illustrated some of the challenges in using Western-centric controlled vocabularies.

4.3 Observations on Indexing Activity

We observed two different trends in the collaborative indexing activity. If there was any commonality of term use among group members (frequently assigned, even if some slight modification), the group consensus terms were taken from the most frequently used terms. Otherwise, terms were negotiated or new indexing terms were selected by the groups after discussion and exchange of views on the "best" level of indexing, "index-worthiness" of the image feature or concept, etc.

Most students found the assignment of terms from controlled vocabularies the most challenging part of the project. The subject matter of the images (19th century Asia) was unfamiliar to almost all students, and not well dealt with by the AAT and TGM, which have a known Western bias ([6] pp 84-91). Consistent with Jörgensen's "naïve" users [5], the students reported greater consistency and comfort dealing with generic objects and scenes (e.g., elephants, temples) and specific named entities (e.g., Kirifuri Falls at Nikko National Park) and more problems in agreeing upon abstract objects and concepts. The inclusion of a notes field in which they could describe the image in natural language terms, frequently informed by additional research, was an important adjunct to the controlled descriptor fields. Several students commented on the benefits of the group indexing module – if only to know what thought processes others followed when deciding what to index in a given image.

Tables 1 and 2 show the numbers of indexing terms used by each group. The students in both classes (Spring/Fall 2006) were assigned the same set of 15 images in their individual as well as group assignment. Quantitatively, students in both classes had similar indexing patterns overall. This might be explained by the fact that most students in both classes were unfamiliar with Asian culture. Thus, the choices of indexing terms were limited by their background knowledge of the subject matter.

We initially expected groups to consistently use a similar set of indexing terms. However, our preliminary analysis showed that the choices of consensus terms made by the two groups are quite different. Although the groups used similar numbers of indexing terms, the assignment of indexing terms between groups was not very consistent. We used Rolling's measure [10] to evaluate indexing consistency between groups. For example, an average Rolling's measure between Group 1 of Spring 2006 and Group 1 of Fall 2006 is 47.3% for TGM terms and 21.1% for AAT terms, while an average Rolling's measure between Group 2 of Spring 2006 and Group 2 of Fall 2006 is 52.1% for TGM terms and 51.1% for AAT terms. The result is similar to many studies of inter-indexer consistency that a high consistency level is hard to attain, even among trained indexers (e.g., [7]). Although the students tended to agree about the index-worthy image content, the exact indexing terms selected often differed.

Table 1. Comparison of the number of TGM terms selected across groups

Group	Spring 2006				Fall 2006			
	Max	Min	Mean	SD	Max	Min	Mean	SD
1	11	3	6.1	1.8	10	2	4.8	2.2
2	3	1	2.3	0.7	4	1	1.9	0.8
3	11	4	6.7	1.9	9	3	7.1	1.4
4	12	4	7.3	2.4	12	4	7.5	2.5
5	10	3	5.6	2.1	18	5	11.3	3.4
6	18	7	12.6	3.4	10	2	4.1	1.8
Overall	18	1	6.8	3.8	18	1	6.1	3.7

4.4 Observations on Collaboration

Since the collaborative indexing feature was recently implemented, the collaboration data was derived from the Spring 2006 class. The collaboration process focused on

Table 2. Comparison of the number of AAT terms selected across groups

Group	Spring 2006				Fall 2006			
	Max	Min	Mean	SD	Max	Min	Mean	SD
1	7	2	4.6	1.5	11	3	5.5	1.9
2	4	1	1.9	1.0	4	0	1.5	0.9
3	12	3	7.3	2.8	16	5	9.0	3.0
4	15	5	8.9	1.7	15	6	8.7	2.4
5	12	4	6.6	2.4	22	5	11.0	4.8
6	25	11	17.9	4.7	22	5	11.0	4.8
Overall	25	1	7.9	5.7	22	0	7.8	4.7

voting and validation of thesaurus subject terms. The groups used two different strategies to collaborate on indexing images. Some groups used a "divide-and-conquer" strategy by assigning subsets of images to individual group members and the assignees to decide on consensus metadata. Afterward, the group as a whole reviewed the consensus metadata, discussed any problems and made appropriate final changes. The second strategy was to work together on each of the 15 images. Group members were asked to specify indexing terms they would choose to include in the consensus metadata for each image and their rationale for choosing so. After the further discussion within group, the group voted and approved terms were selected. To validate specific subject terms, the common practice among the groups was to cite the external sources that they found relevant from their own research. This ranged from print and digital reference sources and tourist information in the Web to personal resources and experience (in each class at least one or two students had visited one or more of the sites photographed or was generally familiar with the geographic area). On a photograph of boats labeled "Houseboats—Thailand" discussed by three students:

#1 *If we use "shantyboats," we shouldn't also use "houseboats" because shantyboats is NT to houseboats. If we use "waterfronts" in AAT, we should also use it in TGM. We should choose one of "harbors" and "bodies of water"--"harbors" is narrower than "bodies of water." Whichever we choose, we should use it in TGM as well ("harbors" and "bodies of water" are both TGM terms).*

#2 *Are we certain that they are "Shantyboats"? I would be more comfortable w/ just using "Houseboats". Also, I agree with "Harbors", "Bodies of water" seems too broad to me.*

#3 *I'm not sure I agree here… at least on the "harbors" matter. I don't think we can tell from this whether we are looking at a river, a harbor, a bayou, etc. I'd prefer sticking with "Bodies of water." I feel confident that these are shantyboats (at least the vessels in the foreground are). However, because the AAT has "houseboats" as a preferred term from "shantyboats," and because it appears "shantyboats" is a primarily English word, I'm fine with using "Houseboats" for both.*

These examples provide clear evidence for the value of collaboration as a pedagogic tool reinforcing for principles of metadata development that had been discussed in class.

As there were multiple options for collaboration, each group was able to decide how to communicate. Asynchronous collaboration was used by most groups. This consisted of using Image Tagger's commenting feature to some extent and Blackboard Discussion to discuss and vote on indexing terms. Many groups began the discussion by copying all indexing terms available in the indexing page for different images and pasting them into a Blackboard discussion thread. After that, each group member would choose indexing terms and begin the discussion of the acceptability of the term for subject indexing of that image. In addition, the Blackboard Discussion was also used to communicate on group administration, e.g., breaking down the tasks, setting up a schedule, etc., as well as supporting extensive discussion for some groups. Students also shared external hyperlinks with one another. These links often contained useful information that could help them identify the images. One group put all terms into a set of Excel spreadsheets and circulated. A few groups used synchronous real-time chat as the primary channel of communication.

Table 3. A quantitative summary of student collaboration by group

Group	Comments on Image Tagger		Posts on Blackboard	Total collaboration
	Total	Mean		
1	107	7.13	122	229
2	1	0.07	319	320
3	15	1.00	48	63
4	86	5.73	133	219
5	58	3.87	116	174
6	231	15.4	135	366

Table 3 summarizes collaboration based on the number of comments/discussions posted on Image Tagger and the Blackboard Discussion Boards. Most comments posted on Image Tagger's indexing page were related to subject-term voting and validation while discussions in Blackboard forum included subject-term discussion, group administration and other topics. Notice that Group 2 relied heavily on Blackboard forum for collaboration and Group 3, while having the smallest number of asynchronous collaboration, used real-time chat (data not shown).

Inter-group collaboration occurred occasionally and we did not restrict the students from access other groups' threaded discussion, thus they were free to read and even participate in the discussions (but not the group chat) outside of their own group. One example of inter-group collaboration started from the use of the subject cloud. One student looked through the cloud to explore what other people did in their indexing assignment. She discovered an image of the Fujiya Hotel in Miyanoshita, Japan, which was indexed by another group. Since she had been there, she commented on that image. Students also monitored discussions to catch any instructor feedback to specific questions from other groups in a type of vicarious learning.

4.5 Feedback and Evaluation

At the end of the project, the students participated in a discussion thread to talk about their experience with the tool and the project in general. Out of 14 students during

Fall 2006 who participated in the discussion, 5 of them stated that they really liked working with the tool and did not see any problems with it. 2 of them reported the specific bugs that need to be fixed and the remaining 7 proposed possible features and improvements they would like to see in the future version. Overall, the students found the tool easy to use and support their collaboration process well. One student commented on the value of real-time discussion available to them in Blackboard as it worked best for their group's collaboration process.

From the instructor's point of view, the interface worked reasonably well in terms of monitoring student input and catching potential problems (e.g., appropriate content in metadata fields, appropriate use of TGN terms for city and country, addition of subheadings to TGM terms) early. The initial version of the activity log was somewhat useful in gauging the level of student activity but requires future work to make it easier to use and to provide a greater level of detail.

Although we initially provided a simple space to allow the students to post discussion on the page associated with a particular image, the functionality of a full-featured discussion board was still preferred by the students in the overall collaborative activity. Different modes of collaboration also affect the use of the tool and the group strategy. As we mentioned earlier, some of the group thought a synchronous collaboration, real-time chat, worked well for them and successfully made use of it for their collaboration, while the other groups preferred asynchronous collaboration over synchronous one for various reason.

5 Conclusions and Future Work

We describe Image Tagger, a digital library tool built to support image indexing instruction in a collaborative context. Taking a user-centered design approach, the tool was developed iteratively in accordance with feedback and suggestions from the students and the instructor. The tool has supported image indexing class projects in three academic terms. Most students expressed positive opinions about the tool. They commented that Image Tagger and the indexing assignments offered them a unique learning experience. They learned by direct interaction with other students in their indexing group and from vicarious observation of the discussions of other groups. Moreover, many of them thought the tool worked well for virtual group work and helped them in their collaborative deliberations. The students made some suggestions, outlined below, for how to improve the tool based on the features they would like to see in the future.

The deliberation and decision process can be made more robust. One solution is to provide a voting interface directly for each candidate terms. For example, each candidate term might have a checkbox by its side. A checked status on a checkbox indicates that the students have voted to accept that term. In addition, the students might specify the rationale behind their votes. For example, the terms might be approved because they represent main objects or peripheral objects or describe certain events or concept in the image. Next, the terms might be chosen because of their ease of searching. If the terms are rejected, the students might supply specific reason for voting so. For instance, they might be too specific or too broad.

To fully support group collaboration, the other collaborative tools, such as real-time chat room, forum, and file sharing, were available to students via the Blackboard system. Because these tools were outside the Image Tagger environment, the students had to sort through multiple places to collaborate with their peers. In critiquing the current version of the tool, several students expressed their preference to be able to work in a unified collaborative environment. We hope to find ways to add these other collaborative services to Image Tagger. Finally, we are extending the Image Tagger to incorporate more administrative functions [1]. By continuing to emphasize student-friendly design we hope to develop a tool that will be effective for teaching students about the administration of repositories.

Acknowledgements. This project was supported by the Institute of Museum and Library Services (IMLS) Grant RE-05-05-0085-05 on developing a Model Curriculum for the Management of Digital Information. We thank the Free Library of Philadelphia the photographic materials used in the students' project and the students in Content Representation class at Drexel University for their participation.

References

1. Achananuparp, P., Allen, R.B.: Developing a Student-Friendly Repository for Teaching Principles of Repository Management, DigCCurr Symposium. Chapel Hill, NC (2007)
2. Getty, Art & Architecture Thesaurus, http://www.getty.edu/research/conducting_research/vocabularies/aat/
3. Getty Thesaurus of Geographic Names, http://www.getty.edu/research/conducting_research/vocabularies/tgn/
4. Library of Congress, Prints and Photographs Division, Thesaurus for Graphic Materials, Cataloging Distribution Service, Library of Congress, Washington, D.C (1995)
5. Jörgensen, C.: A conceptual framework and empirical research for classifying visual descriptors. JASIS 52(11), 938–947 (2001)
6. Jörgensen, C.: Image Retrieval: Theory and Research. Scarecrow Press, Lanham, MD (2003)
7. Leininger, K.: Inter-indexer consistency in PsychINFO. Journal of Librarianship and Information Science 32(1), 4–8 (2000)
8. Marias, H., Bharat, K.: Supporting cooperative and personal surfing with a desktop assistant. In: UIST 1997. Proceedings of the ACM Symposium on User Interface Software and Technology, pp. 129–138. ACM Press, New York (1997)
9. Peterson, E.: Beneath the metadata: Some philosophical problems with Folksonomies. DLIB Magazine (November 2006)
10. Rolling, L.: Indexing consistency, quality and efficiency. Information Processing and Management 17, 69–76 (1981)
11. Shroeter, R., Hunter, J., Kosovic, D.: Vannotea - A collaborative video indexing, annotation, and discussion system for broadband networks. In: K-CAP, Workshop on Knowledge Markup and Semantic Annotation (2004)

Analysing HTTP Logs of a European DL Initiative to Maximize Usage and Usability

M. Agosti[1], G. Angelaki[2], T. Coppotelli[1], and G.M. Di Nunzio[1]

[1] University of Padua, Italy
{agosti,coppotel,dinunzio}@unipd.it
[2] The European Library, The Netherlands
Georgia.Angelaki@KB.nl

Abstract. In the context of an ongoing collaboration conducted between DELOS, the European Network of Excellence on Digital Libraries, and The European Library, we discuss how both the analysis of the Web log data of The European Library service and a user study can contribute to the personalization of services for such a system.

1   Introduction

The European Library is a non-commercial organisation[1]; it provides the services of a physical library and allows searching through the resources of many of the European national libraries, where the available resources can be both digital or bibliographical, e.g. books, posters, maps, sound recordings and videos. This paper presents results that have been achieved up to now on the analysis of who the users of The European Library are as well as how its functionalities are perceived by its users. The analysis involves two different and complementary strategies: 1) Web log analysis; and 2) user study.

HTTP logs of The European Library server are used to reconstruct the users sessions and to study the users behaviour. On the one hand, the effective number of users can be estimated and data can be obtained about users mean sessions length by using HTTP logs and by eliminating the access of crawlers. It is also possible to estimate if the users use advanced search functionalities or if they prefer to use the portal to search documents with the minimum effort, and obtain data about their geographical distribution. On the other hand, it is possible to understand how users exploit The European Library portal, what they expect from it, and what they would like to get by using additional information like user questionnaires. With the use of questionnaires, knowledge about their level of satisfaction can be obtained along with recommendations and hints about possible improvements.

2   The Initiative

The European Library initiative aims at providing a *"low barrier of entry"* for the national libraries that should be able to join the federation with only minimal

---

[1] http://www.theeuropeanlibrary.org/

D.H.-L. Goh et al. (Eds.): ICADL 2007, LNCS 4822, pp. 35–44, 2007.

changes to their systems [1]. With this objective in mind, The European Library service is constituted by three components:

- a Web server: which provides users with access to the service;
- a central index: which harvests catalogue records from national libraries, supports the *Open Archives Initiative Protocol for Metadata Harvesting (OAI-PMH)*[2], and provides integrated access to them via *Search/Retrieve via URL (SRU)*;
- a gateway between SRU and Z39.50: this also makes national libraries accessible through SRU which would otherwise only be accessible through Z39.50[3].

In addition, the interaction between the portal, the federated libraries and the user mainly happens on the client side by means of an extensive use of Javascript and *Asynchronous JavaScript Technology and XML (AJAX)*[4] technologies. Once the client, which is a standard Web browser, accesses the service and downloads all the necessary information from the Web server, all the subsequent requests are managed locally by the client. The client interacts directly with each federated library and the central index, according to the SRU protocol, makes separate AJAX calls towards each federated library or the central index, and manages the responses to such calls in order to present the results to the user and to organize user interaction.

3   Web Log Analysis

The analysis was performed on seven months of The European Library Web log files, starting from October 1st 2006 to April 30th 2007. The structure of the Web logs conforms to the W3C Extended Log File Format [2].

This kind of log contains, among other things, the following useful information: 1) the *Internet Protocol (IP)* address and the user-agent which allow the identification of single users [3]; and 2) the referrer field, a *Uniform Resource Locator (URL)* address which communicates the last page viewed by the user, which can be used to know how visitors arrive at The European Library service.

The European Library Web logs also contain the cookie[5] saved on the client which reports extra information: 1) the language selected by the user during the navigation of the service; 2) the collections of documents selected during the query or query refinement; 3) the identifier of the session assigned by the server to a specific user.

A methodology for analysing these log files has been developed. It requires the use of a parser and a database for storing the data. Initial specifications of this database application were presented in [4]. The database enables separation

---

[2] http://www.openarchives.org/OAI/openarchivesprotocol.html

[3] http://www.loc.gov/z3950/agency/

[4] http://www.w3.org/TR/XMLHttpRequest/

[5] Cookies are plain text information stored locally by the client. The stored data are initially sent by a Web server to a Web client and then are sent back to the server on subsequent requests.

of the different entities recorded and facilitates data-mining and on-demand querying of the data.

3.1 General Information from the Analysis

The following data include all the requests and sessions that reached the portal, even those which can belong to automatic crawlers and spiders. These data can give a a first estimation of the trend of the traffic volume: a total of 22,458,350 HTTP requests were recorded in the log files, with a monthly average of 3,208,336 HTTP requests, a daily average of 105,936 requests, and an hourly average of 4,414 requests. Since The European Library service is a 7 days and 24 hour service with users from all over the world, it can be considered a busy service that answers to an average of 74 requests per seconds, 24 hours a day/7 days a week.

This traffic is generated both by human users and by automatic software agents. Before starting any personalization analysis, it is then mandatory to identify human requests from others. A simple analysis performed on the user-agents included in each request allows a first distinction between human requests and software-agent requests. The experimental data are reported in Table 1. The fields contained in the user-agent are decided by each browser and different versions of the same browser can have different fields. As a consequence, an automatic elaboration of this data is a hard task. However, the most used browsers use a standard-like string that allows the identification of such browsers. For example, the following user-agent, that is the most recurrent one in the analyzed logs, corresponds to Internet Explorer 6.0 (MSIE 6.0):

```
Mozilla/4.0+(compatible;+MSIE+6.0;+Windows+NT+5.1;+SV1)
```

There is a heterogenous quantity of data that browsers can store in this field. For example, it is possible to find complex user-agents such as:

```
Mozilla/4.0+(compatible;+MSIE+6.0;+Windows+NT+5.1;+SV1;+.NET+CLR
+1.0.3705;+.NET+CLR+1.1.4322;+Media+Center+PC+4.0;+.NET+CLR+2.0.50727)
```

Nevertheless, the aim at this point is not to correctly identify the meaning of each field, but to use this information to distinguish a human from a crawler. To this aim, the first step is that of tagging as crawlers all the user-agents that contain key terms like crawler, robot, spider, etc. Other crawlers are identifiable because they disclose themselves in the user-agent, for example:

```
Mozilla/5.0+(compatible;+Yahoo!+Slurp;
+http://help.yahoo.com/help/us/ysearch/slurp)
```

is the well known Yahoo! crawler. There are some repositories that contain up-to-date lists of known crawlerse[6].

After crawler identification, all the user-agents generated by internal usage of the server can be tagged as software agents. In this category it is possible to find

---

[6] One of those repositories is available at the URL: http://www.botsvsbrowsers.com/

Table 1. Distribution of requests on the basis of user agents

origin	requests	percentage
human	17,006,566	75.72%
crawler	2,904,134	12.93%
software	2,460,867	10.96%
non standard	86,783	0.39%
total	22,458,350	100.00%

user-agents that correspond to requests made by: Verity *Information Retrieval System (IRS)* (the IRS that handles query execution in The European Library), Java or PERL applications (`Java/1.4.2_04`, `libwww-perl/5.801`), and other software tools like the MS FrontPage HTML editor. Some user-agents are visibly faked to resemble that of known browsers and then are tagged as non-standard; for example

`Mozilla/4.0+(compatible)` and `Mozilla/4.0+(compatible;+MSIE+5.00;)`

lack required information like the browser or the operating system. The remaining user-agents are manually checked and are tagged as human when they seems to have no anomalies.

The high percentage of human users obtained with this technique is partially biased by the behaviour of some crawlers that successfully fake a user-agent and are recognized as a human user. Despite this, the produced estimation of human requests can be considered good.

Other statistical information that is computable using these logs regards operative systems and browsers. The products of Microsoft are by far the most used: Windows alone is used by about 74% of the users; this tendency also affects the situation found in the browser analysis, with Internet Explorer as the most used browser, since it is used by 60% of the users. However, there has been an increase in the use of Mozilla Firefox, compared to a preliminary analysis performed in previous months of the logs (from November 2005 to January 2006) as reported in [4]; currently Mozilla Firefox is used by 13% of the users.

3.2   Session Reconstruction

With the term session a set of requests is intended that are performed by a single user during the browsing activity. Because a user is supposed to access the portal more than once during the analysed period, a time-out is applied to distinguish different sessions of the same user. When the user remains inactive for more then 15 minutes, his session is terminated and a new one is created when the next request is performed.

The reconstruction of sessions is an important process that allows the identification of single users (either humans or software-agents), their tracking during

the portal navigation and eventually their grouping on the basis of similar behaviors. Moreover, the session identification is an intermediate step that has to
be performed in order to distinguish recurrent users from bouncers. We have
proposed two different methodologies for reconstructing sessions: 1) a heuristic
technique [3] that allows the identification of a single user using the IP address
and the user-agent; and 2) an exact technique that takes advantage of the information contained in the cookies to reconstruct the sessions.

The first technique assigns each HTTP request to a session, including sessions
of users that do not accept cookies. In this way, a great number of sessions
(690,879) is reconstructed, but the drawback is that a large portion of them
is made up of an extremely reduced number of requests, hence they are not
exemplificative of human navigation behaviour. However, it is possible to analyze
the different kinds of software agents that interact with the Web server. For
example, there is a significant number of sessions that do not contain any HTTP
request of a Web page, while the first request of a human user would clearly be
a Web page.

The second approach to session reconstruction is therefore preferred when an
analysis of human sessions is required, and it takes advantage of cookies to identify sessions. The European Library cookies contain a unique identifier, named
TELSESSID, assigned runtime by the PHP[7] interpreter of the Web server each
time a session is started by a user. This identifier is important for two reasons: 1)
it more precisely distinguishes users that are hidden behind a proxy (therefore
with the same IP); and 2) it enables human users to be separated from the other
users of the portal because many of the automatic crawlers and softwares do not
memorize cookies. The requests of the same session have the same TELSESSID
value in the cookie field; in order to be coherent with the previous analysis, we
decided to consider two sessions distinct when the delay between two requests
of the same session is greater than 15 minutes. This approach cannot take into
consideration sessions of users which do not enable cookies in their browser. Despite this drawback, the results are more accurate than those obtained with the
first approach of sessions reconstruction. A deeper analysis and comparison of
the results obtained with these two different approaches is under way and has
already allowed the identification of odd behaviour in the sessions of users from
some countries.

In the analysed period of time, we were able to reconstruct 209,900 different
sessions on the basis of the cookies content. There is a significant number of
sessions, almost 45% of the total number of sessions, which last more than 60
seconds regardless of the number of requests per session. Therefore, an analysis
of the sessions which last more than 60 seconds and have more than 100 requests
has been computed separately, since these sessions are valuable for the analysis
of users for personalization purposes and to give an answer to the points of
interest. Results shown in Figure 1 are important since they confirm that users
do not only have a look at the home-page of The European Library portal,
but they spend some time on the portal, interacting with it (more than 100

---

[7] http://www.php.net/

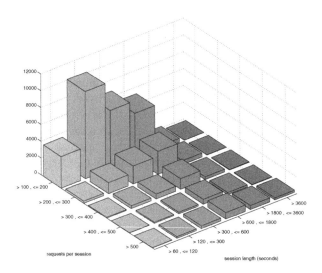

Fig. 1. Sessions (cookies) which last more than 60 seconds with a number of requests per session > 100

HTTP requests) and analysing the results (the majority of sessions last between 2 minutes to 10 minutes). Therefore, it seems reasonable to think that there is a sizeable number of users that do not only come across the portal but that actively use the functionalities that it offers, users that are really interested in the search of documents which are written in their own mother language and also in other languages.

3.3   Session Provenance

In this section, the analysis of sessions classified according to the different geographical areas are shown. The abbreviations that are used in the graphs are those adopted by the ISO 3166 standard[8], which is a three-part geographic coding standard for coding the names of countries and dependent areas, and the main subdivisions thereof.

The nations with the highest number of sessions reconstructed using the cookies are shown in Figure 2. Most of the accesses come from European nations that are active members of the The European Library service and, in particular, there is a noticeable increase in the number of accesses from the countries that recently joined the initiative. The United States of America are second in this list; however, we recorded a significant decrease in the number of sessions if compared with sessions reconstructed with HTTP requests. This fact indicates that the majority of crawlers have an IP address belonging to the geographic area of the United States.

---

[8] http://www.iso.org/

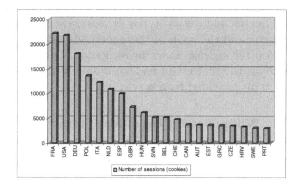

Fig. 2. The nations with the highest numbers of sessions (cookies) are shown

3.4 Advanced Search Usage

The users of The European Library have two different choices for performing advanced searches: use the advanced search functionality or personalize the set of searched collections. While we do not have any information on the first choice, the information on the collections selected by the user is saved in a specific field of the cookies and the analysis of this variable makes the analysis of the personalized searches possible.

In general, we can observe that users usually use the default collection selection instead of explicitly selecting which collections have to be searched; to give an idea, the ratio between explicit selection and the default selection is less than 1 to 10. However, the following analysis focuses only on those collections that are explicitly selected by the user and does not consider the collections assigned by default to the user. The study of these collections allows us to understand the behavior of users that are actually refining the query. The selection is equally distributed over these collections, with a mean of 1,708 selections, and a maximum of 11,000. Thus, it appears that only a reduced number of users actively selects a different set of collections; therefore, it is important to accurately select the initial set of collections in order to have a better exploitation of The European Library. Moreover, if this number of users corresponds to those users who perform an advanced search, then the behavior is comparable with that of other online services where only a limited number of users effectively uses advanced search tools.

4   User Study

Users surveys are a valuable method for understanding user behaviors in different situations. However, surveys usually require a significant amount of time and effort; for this reason, an accurate design of the process of studying users has to be carried out. Extensive methods can be used to broadly represent a population of users and investigate some characteristics of interest such as: the background, the information need of the users, and the level of their satisfaction given a service. Surveys or questionnaires are types of extensive methods that can be used to interview users, and the goals of these methods can be those of learning how to better develop the service under investigation. Questionnaires can also provide simple feedback to build up an understanding of the different way users perceive the search tools provided by a service.

4.1 Study of the Willingness of Users for Interactive Search

Here, we want to discuss the kind of activities that users perform during the usage of The European Library service. We are particularly interested in studying the willingness of users for iterative search and the satisfaction of search tools offered by the service. Searching means both a simple action like specifying terms of interest, and a complex action like browsing results and iterating the search using more focused terms. In The European Library, we also have information about user preferences on collections during the search of documents: the country of the collection of the national library, the subject of the collection, and so on. The final aim of the analysis will be that of combining the observations carried out with the questionnaires and the results carried out with the log analysis.

In the following, the results of a controlled study are presented. It was decided to conduct a controlled study, because previous studies on logs and observation in naturalistic settings, combined with interviews, seem more scientifically informative with respect to each of the two types of studies when conducted alone[5]. The final aim of the study is to gain insights on a specific group of data, and to use them in a more general way.

4.2 Controlled Study Set Up

A controlled study was conducted for a group of users who were asked to freely crawl and navigate The European Library Web site and, after that, to fill in the questionnaire provided by The European Library to report and describe their impressions[9]. The goal of this controlled study was to combine the data in the Web log files of the sessions of the people who compiled the questionnaires with those that were reported in the questionnaires with the aim of gaining insights from data on user sessions and judgements in the questionnaires to be used for personalization purposes.

The first important question of the questionnaire that gives a first impression about how users use The European Library portal is question Q3 which requires choosing between the following answers: 1) I prefer to use the Simple Search facility; 2) I prefer to use the Advanced Search facility. Results shows that 81% of users prefer the advanced search facilities. In Figure 3a and 3b, the percentage of users that agree with the following sentences are shown: (Q5a) An advanced search automatically searches all the collections held by The European Library; (Q5b) I usually use the *"choose your own collections"* option before undertaking an advanced search.

---

[9] The European Library. Questionnaire on The European Library's Portal, 2006.

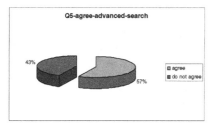

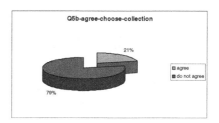

(a) User agreement on advanced search tools. Question Q5a.

(b) User agreement on using their own collections. Question Q5b.

Fig. 3. Users agreement on search tools and collections

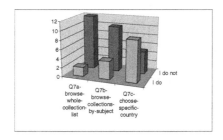

Fig. 4. How users use the *"choose your own collections"* option. Question Q7.

The numbers indicate that there is almost a complete disagreement among users about how many collections are used during an advanced search. Moreover, most of the people using advanced search tools do not make use of the option that allows users to select their preferred collections. However, there is still a significant number of users (14%) who do not know after this experiment how to use the *"choose your own collections"* option (question Q6 of the questionnaire). In Figure 4 the number of users who use and how they use the *"choose your own collections"* option is shown. The possible answers were (question Q7 of the questionnaire): (Q7a) I browse the whole collections list to find relevant collections; (Q7b) I use the *"browse the collections by subject"* option; (Q7c) I choose a specific country and look at collections; (Q7d) I use the *"search collections by description"* option.

For those users who do use this option, few of them like to browse the whole list of collections (Q7a) or browse the subject of the collections (Q7b). Instead, the majority of users prefer to browse the collection choosing the countries of interest (Q7c). None of them, not shown in the figure, have ever used the search collection by description option.

In Figure 5a and 5b, two different types of evaluation of the portal are shown from the point of view of the speed of search and the overall rating. Half of the users are not satisfied with the portal speed of searching; however, and most importantly, more than 80% of the users at least rate the portal as good.

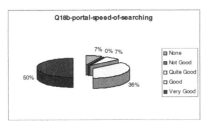
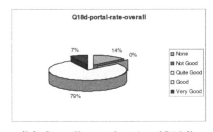

(a) Speed of portal rating (Q18b). (b) Overall portal rating (Q18d).

Fig. 5. Web portal rating. Question Q18.

5 Conclusions

Preliminary analysis using both HTTP requests and user surveys has shown that users of The European Library come from different geographical areas, especially from countries that recently joined the initiative, some of them are willing to spend some time to perform advanced searches and to select collections of documents different from the default ones. Further analysis is underway to better combine the results of the HTTP logs analysis with that of the user study to give a better understanding on the usage of The European Library service and give directions towards an innovative personalization of the service.

Acknowledgements

The work reported in this paper is conducted in the context of a joint effort of the DELOS Network of Excellence on Digital Libraries and the The European Library project. The work has been partially supported by the DELOS Network of Excellence on Digital Libraries, as part of the Information Society Technologies (IST) Program of the European Commission (Contract G038-507618).

References

1. van Veen, T., Oldroyd, B.: Search and Retrieval in The European Library. A New Approach. D-Lib Magazine 10 (2004)
2. Hallam-Baker, P., Behlendorf, B.: Extended Log File Format, W3C Working Draft WD-logfile-960323 (1996), http://www.w3.org/TR/ WD-logfile.html
3. Agosti, M., Di Nunzio, G.: Web Log Mining: A study of user sessions. In: PersDL 2007. Proc. 10th DELOS Thematic Workshop on Personalized Access, Profile Management, and Context Awareness in Digital Libraries, Corfu, Greece, pp. 70–74 (2007)
4. Agosti, M., Di Nunzio, G., Niero, A.: From Web Log Analysis to Web User Profiling. In: Thanos, C., Borri, F., Candela, L. (eds.) DELOS 2007. LNCS, vol. 4877, pp. 121–132. Springer, Heidelberg (2007)
5. Ingwersen, P., Järvelin, K.: The Turn. Springer, The Netherlands (2005)

Digital Library Evaluation Make Simple with Nielsen's Design Heuristics: Design Compliance and Importance

Yin-Leng Theng, Maggie Yin, Norasyikin Binte Ahmad Ismail,
and Nureza Binte Ahmad

Wee Kim Wee School of Communication and Information
Nanyang Technological University, Singapore
{tyltheng,yinm0002,nora0004,nu0002ad}@ntu.edu.sg

Abstract. Using Nielsen's well-established Heuristic Evaluation normally used for gathering *qualitative* feedback, this paper describes a user study conducted on the National Library Board's Digital Library (NLBDL) in Singapore to gather *quantitative* feedback on users' perceptions regarding compliance/violation of design heuristics implemented, and draw recommendations for design refinement.

1 The Study

For digital libraries (DLs) to realise their full potential, the design of a DL system needs to take into account the needs and preferences of users in the community. This is important as end-users are typically individuals who may not have particular skills in information retrieval, and are accessing library resources from their own desks, without support from a librarian. Hence, the design of DLs should be intuitive, flexible and easy to use, and usability evaluation plays an important role in DL design to ensure minimal effort by users using the system [5].

Previous studies [e.g. 5, etc.] on the National Library Board's Digital Library (NLBDL; http://www.nlb.gov.sg/) focused on conducting *detailed* and *time-consuming* usability inspection techniques such as Claims Analysis [2] to detect usability problems. In this paper, we describe a user study conducted on the NLBDL using Nielsen's well-established Heuristic Evaluation [4] to gather *quick, first-cut* feedback on users' perceptions regarding compliance/violation of design heuristics, and draw recommendations for design refinement. The objectives of this study were:

- Objective 1. To investigate users' perceptions of design heuristic compliance and/or violation when applied to NLBDL, and rank these heuristics in order of importance; and
- Objective 2. To identify common problems faced by users while using the NLBDL, and propose refinements to the design features and improve the usability of the NLBDL.

Target Respondents
Three main types of libraries come under the National Library Board (NLB): regional, community and children's libraries. For this study, we selected users of the NLBDL at

D.H.-L. Goh et al. (Eds.): ICADL 2007, LNCS 4822, pp. 45–48, 2007.

Jurong Regional Library (JRL), being the largest public library in Singapore, with a total floor space of 12,020 square metres. It has half a million collection of books, magazines, audio-visual materials, microfilms and newspapers, with an average visitorship of 200,000 per month.

Data Collection and Protocol
The survey was collected on 22 February 2006. Our target participants were those aged 15 years old and above, as they represented a generation of tech-savvy, internet-connected population according to the latest Infocomm Development Authority Annual Survey on Infocomm Usage in Households and Individuals for 2004 (http://www.ida.gov.sg/idaweb/factfigure; accessed 29 Jun 2007), and hence represented a pool of current and potential users of the NLBDL. To identify potential participants, we approached those who were using the multimedia terminals at the Jurong Regional Library or had used the NLBDL before.

Hourly announcement was made by the library staff inviting participation to the survey. A small token of appreciation was given to the participants. Participants took an average of fifteen minutes to complete the survey. A total of 100 took part in the survey.

2 Findings and Analyses

2.1 Profiles of Respondents

The sample population was divided almost equally with 44% males and 56% females. 64% of the participants were between 18-24 years old, and 36% were between 25-34 years old. Of the 100 participants, 63% were self-reported novice users, while 37% were intermediate users.

2.2 Degree of Compliance of Design Heuristics

Table 1 shows participants' comments in response to Question 1 on the degree in which H1 (visibility of system status) was well-implemented in NLBDL. We made the following assumptions: (i) responses marked "strongly disagree" (SD) and "disagree" (D) suggest negative comments/violation of the design heuristic; (ii) responses marked "neutral" (N, Column 5) were discarded; and (iii) responses marked "Agree" (A) and "strongly agree" (SA) suggest positive comments/compliance. Column 8 computes the Compliance Index (CI) by multiplying frequencies in Columns 3 and 4 with "-2" and "-1"; Column 5 with "0"; Columns 6 and 7 with "1" and "2". As an illustration in Table 1, total CI for HI is 62.5, with CI = 82 for sub-heuristic by statement H1i that NLBDL provides indication that an application was processing, CI=70 of indicator being appropriate, and CI=35 that the processing speed of the indicator was fast.

Similarly, the compliance indices of the rest of 9 heuristics were computed in this manner. Table 2 shows the total compliance indices of the 10 heuristics in decreasing order. Overall, the NLBDL interface design was found by the participants to adhere to most of Nielsen's ten heuristics. Of the ten heuristics, 6 heuristics (H1, H2, H3, H4, H6, H7) were found to be highly rated by the respondents (CI>60).

Weak compliance/violation of the heuristics was perceived in H5, H8, H9 and H10. H5 and H9 were rated with a compliance index of 47, implying more could be done in NLBDL to make error prevention and correction, and help to recognize, diagnose and recover from errors more explicit.

It seems that NLBDL might not be providing sufficient help and documentation. Congruent to findings from [3] that novice users were confused and overwhelmed as they were unfamiliar with the library web pages, novice participants in our study could also be frustrated and confused by the lack of help and documentation. They felt overwhelmed when faced with a new system. Therefore, NLB should look into improving their help and documentation feature. This could be due to the fact that the NLBDL users prefer online help as the customer service counter in JRL is located far from the multimedia terminals. Ambiguous representations of interface elements such as icons, toolbars, dialogues, and cursors can become a barrier to users' experience of effective navigation of the system.

To reduce such mismatch between users' mental models and design implementation, an effective interface design should provide clear and visible help and documentation, as suggested by Nielsen's heuristics H10.

Table 1. Participants' Feedback on H1 (Visibility of system status)

HI	Visibility of system status	SD	D	N	A	SA	CI
i.	NLB's Digital Library provide an indicator (e.g. an hourglass icon or a status bar indicator which shows system is running, etc) that an application is processing	0	1	24	67	8	82
ii.	Indicator given is appropriate	0	2	32	60	6	70
iii.	Processing speed of indicator is fast	2	10	43	41	4	35
	Total Compliance Index						**62.5**

Table 2. Total Compliance Indices of the 10 Heuristics (Decreasing Order)

No	Nielsen's Heuristics	Compliance Index (CI)
H2	Match between system and the real world	88.5
H6	Recognition rather than recall	80.7
H4	Consistency and standards	77.0
H3	User control and freedom	75.8
H7	Flexibility and efficiency of use	64.5
H1	Visibility of system status	62.5
H8	Aesthetic and minimalist design	57.0
H10	Help and documentation provided by the NLBDL	54.0
H5	Error prevention and correction	47.0
H9	Help the users in recognizing, diagnosing and recovering from errors	47.0

2.3 Importance of Heuristics

Table 3 tabulates the "importance index (II)" computed from participants' ratings for each of the ten heuristics. Similar to CI, II is computed by multiplying frequencies of "strongly disagree" and "disagree" with "-2" and "-1" respectively; "neutral" with "0"; "agree" and "strongly agree" with "1" and "2" respectively. As in CI, the range of the II lies between -200 and 200.

The top three heuristics with the highest II were H7 (II=112), H3 (II=110), and H5 (II=102) suggesting that flexibility and user control are important in ensuring positive user experience. This seems to concur with Borgman (2003) who advocates that minimum criteria for usability are that systems should be easy to learn, flexible, adaptable and efficient for the task [1]. On the other hand, respondents rated H6 (II=62) and H8 (II=54) the least important.

3 Discussion and Conclusion

This paper describes a *quantitative* usability technique complementing Nielsen's well-established Heuristic Evaluation normally used for gathering *qualitative* feedback. With 100 participants, we were able to gather *useful, quantitative* comments regarding compliance/violation of design heuristics implemented on NLBDL, and draw recommendations for design refinement.

This is pilot study using a heuristic-inspired survey instrument. Future research could significantly expand the sample size, lengthen the survey period, and obtain responses from participants from more diverse backgrounds.

Acknowledgements

We would like to thank the National Library Board for allowing us to conduct a survey, and special thanks to the staff at Jurong Regional Library for their support and hospitality. This project is supported by NTU's AcRF grant (RG8/03).

References

1. Borgman, C.L.: Designing digital libraries for usability. In: Bishop, A.P., Van House, N.A.V., Buttenfield, B.P. (eds.) Digital Library Use: Social Practice in Design and Evaluation, pp. 85–118. MIT Press, Cambridge, Mass (2003)
2. Carroll, J.M.: Making use: scenario-based design of human-computer interaction. MIT Press, Cambridge, Mass (2000)
3. Crowley, G.H., Leffel, R., Ramirez, D., Hart, J.L.: User perceptions of the library's Web pages: a focus group study at Texas A&M University. Journal of Academic Librarianship 28(4), 205–210 (2002)
4. Nielsen, J.: Usability problems through heuristic evaluation usability walkthroughs. In: Proceedings of ACM CHI 1992 Conference on Human Factors in Computing Systems, pp. 373–380. ACM Press, New York (1992)
5. Theng, Y.L., Chan, M.Y., Khoo, A.L., Buddharaju, R.: Quantitative and Qualitative Evaluations of the Singapore National Library's Digital Library. In: Theng, Y.L., Foo, S. (eds.) Design and Usability of Digital Libraries: Case Studies in the Asia Pacific, pp. 334–349. Idea Group Publishing, USA (2005)

On Building a Full-Text Digital Library of Historical Documents

Szu-Pei Chen[1], Jieh Hsiang[1,*], Hsieh-Chang Tu[1], and Micha Wu[2]

[1] Department of Computer Science and Information Engineering
[2] Department of History
National Taiwan University,
Taipei, Taiwan
{gail,tu}@turing.csie.ntu.edu.tw, hsiang@csie.ntu.edu.tw,
wumc@ntu.edu.tw

Abstract. The National Taiwan University Library has built a digital library of historical documents about Taiwan. The content is unique in that it covers about 80% of all primary Chinese historical materials about Taiwan before 1895, and that they are all available in searchable full text, in addition to metadata. To make these materials more accessible to the research community, we have developed, in addition to full-text search and retrieval, a concept of regarding the set of documents retrieved by a query as a sub-collection, and have designed post-query classification methods to help users find the inter-relationships among documents and the collective meaning of a sub-collection. We have also developed techniques for term extraction for old Chinese and a data format for representing governmental structures. We hope that our system will help advance research in Taiwanese history, and will set a model for other similar endeavor.

Keywords: Historical documents, digital library, Taiwan, classification of query results.

1 Introduction

Starting from 2003, the National Taiwan University Library (NTUL) embarked on a major effort to systematically collect and digitize Taiwan related historical documents. The documents, numbered over 80,000, came from a wide range of sources, including imperial court archives, local and central judicial and administrative records, personal records of high-ranking officials, travel journals, diaries of influential people of the time, and land deeds. They were selected by historians, then typed, punctuated, proof-read, and supplemented with metadata records. Currently we have accumulated about 150 million Chinese words (characters), all in full text and searchable. They cover about 80% of all primary Chinese historical materials about Taiwan before 1895.

To our knowledge, there has never been such a collection, in both variety and magnitude, about Taiwan history available in searchable full text before. It should

---

* Corresponding author.

D.H.-L. Goh et al. (Eds.): ICADL 2007, LNCS 4822, pp. 49–60, 2007.

provide an exciting playground for anyone interested in pursuing research of Taiwanese history, scholars and laymen alike. To make these materials more widely available and easy to use, we have built a *Taiwan History Digital Library* (THDL). In addition to full-text and metadata search, we have also built referential tools to further facilitate their use. The tools that we have built so far include a Chinese-Gregorian calendar converter, corpus of names of people and locations, and charts of the evolution of local administrative structure with names of the officials and the duration of their terms.

While constructing THDL, we noticed that providing search/retrieval facilities alone is not sufficient for taking full advantage of its rich content. Most retrieval systems, when issued a query, ends at returning a list of relevant items. The question of how to make sense of the query results is usually left for the user to ponder. This way of representing results might be the best one can do if the returned items are independent objects and the meaning of each item is somewhat complete by itself. However, historical documents are often inter-related. For instance, a query may result in a list of reports from officials to the emperor and his responses, about a specific political incident. The documents may span over several years and represent several turns of events during its development. Thus, the retrieval results, if treated as a *collection*, may reveal meaning much more significant than the sum of its parts. Due to this observation, we have developed methods to *post-process* query results to show collection-level information, so that the intricate relationship among documents from the same query and their collective meaning can be investigated.

This paper is structured as follows. We give a (very) brief introduction of Taiwan's history, with emphasis on the political changes, in Section 2. Section 3 describes the content of THDL. Section 4 outlines the technical features and a new concept of treating the documents returned from a query as a sub-collection. A short discussion, including the differences between our work and other related work such as the Million Book Project, is given in Section 5.

Most translations from Chinese to English in this paper are done using *pinying*. The few exceptions are when the Wade-Giles translation is more commonly used. For instance, we use Taichung (Wade-Giles) instead of Taizhong (pinying) for 台中. All person names are presented as family name first, followed by given name. For instance, the family name of Zheng Chenggong is Zheng.

2 A Brief History of Taiwan

The first trace of human activities in Taiwan, according to archeological findings, dates back to at least 30,000 years. It is not clear if these pre-historic dwellers are the direct ancestors of the indigenous people of Taiwan, part of the Austronesian group, which form about 2% of the current population (numbered about 450,000, the rest are mainly descendants of Han Chinese).

The first mentioning of Taiwan in Chinese historical records was during the Three Kingdoms period (230 A.D.), although Han Chinese did not migrate to Taiwan in significant numbers until about 1,000 years ago. In 1624 the Dutch East India Company established a base at southern Taiwan and built the fort of Zeelandia at the location of present day Tainan. At about the same time (1626), the Spanish also came to northern Taiwan and built three forts. They were driven out by the Dutch in 1642. The Dutch were then driven out by the Ming general Zheng Chenggong (鄭成功, known in the west as Koxinga) in 1662. Koxinga and his descendants established the first Han Chinese government in Taiwan and used it as a base for their attempt to recover China mainland (then occupied by the Manchurian-lead Qing Dynasty, established in 1644 by overthrowing the Han-Chinese Ming Dynasty) for the Ming Dynasty. In 1683 Zhang Keshuang (鄭克塽), a grandson of Koxinga, surrendered to Qing, which established prefectural level governments in Taiwan and upgraded it to the Province of Taiwan in 1885. After loosing the Sino-Japanese, Qing ceded Taiwan to Japan in 1895. After being defeated in the Second World War, Japan returned Taiwan to China in 1945. The Chinese Nationalist government, after loosing the mainland to the communists, moved their seat to Taiwan in 1949. The island moved toward full democracy when the Marshall Law was lifted in 1987. (The State of War between the two governments did not officially end until 1991.) In the year 2000, Mr. Chen Shui-Bien of the Democratic Progressive Party was elected President, thus ending the monopoly of the Nationalist Party to the Presidency. Those who wish to learn more may consult the "History of Taiwan" entry of Wikipedia [1].

3   The Content of THDL

The content in THDL can be roughly divided into three categories: imperial court documents of Ming and Qing Dynasties, documents of local governments - in particular the Danxin Archives, and local land deeds. Together they yield a rather extensive picture of the political, sociological, and economic landscape of pre-1895 Taiwan, from the perspective of the central government to everyday people.

Our content has at least two other distinctive features. One is that most of our content are primary documents, and they cover at least 80% of all such Chinese materials. Other than diaries of important officials and travel journals, we did not include any "secondary" material such as memoir, biography, or scholarly research work. The reason is that we wish to present historical documents in their original form, with as little later interpretation as possible. (We remark that we are building a database of research work on Taiwanese history. This database, however, will serve a purpose different from that of THDL.)

The second distinctive feature is that all of the documents in the digital library are keyed-in as full text, with punctuation added, in addition to metadata. The availability of full text and full-text search sets THDL apart from any other database of Chinese historical documents that we know of. It also makes it an exciting research environment for finding associations among documents and among collections of documents that cannot be done with metadata alone. From the technological side, it serves as a good source for experimenting in text mining and other information techniques. Indeed, we have already developed tools for extracting terms, names, and dates [2].

In the following we describe the three categories of contents in THDL.

3.1   Selected Collections of Taiwan-Related Documents from the Imperial Governments of the Ming and Qing Dynasties

During the Ming and Qing dynasties (1368-1644 and 1644-1911), China extended its border to Taiwan and neighboring areas. Hence, the imperial governments of China produced a significant number of court documents involving Taiwan during the two dynasties, especially Qing. These documents are hidden in various imperial government archives that are kept in different institutions, some in China and some in Taiwan. Furthermore, most of them are never published and, if available for viewing, are only as microfilms. It is obvious that they are of fundamental importance for anyone who wishes to study Taiwan history during the Ming and Qing dynasties, but it is almost impossible to access them in their entirety. Thus, the Council for Cultural Affairs (CCA) of Taiwan commissioned the National Taiwan University Library (NTUL) in 2003 to make a comprehensive digitization of imperial court documents related to Taiwan. In this ambitious project, NTUL first collected copies of the imperial government archives of Ming and Qing dynasties from different libraries and archives, then collaborated with a team of historians (first lead by Professor Li Wen-Liang, then by Professor Wu Micha, both of National Taiwan University) to carefully select the documents within those archives that are related to Taiwan. In addition to creating metadata for those selected documents, the full text contents were also keyed-in. The full texts were then proof-read with punctuation added. In the two years that followed, more than 40,000 such documents were selected and over 35,000,000 characters of punctuated full text were produced [3].

Although the funding from CCA stopped after two years, NTUL continued looking for new sources of historical documents and has added at least another 4,000,000 characters since then.

The contents of this collection are selected from the Ming Reign Chronicles and the Qing Reign Chronicles (明實錄,清實錄), Palace Memorials (奏摺), Archives of the Grand Council (the Administration of Military Affairs) (軍機處檔案), the Imperial Decrees Archives (上諭檔), the Grand Secretariat Archives (內閣大庫), Archives of the Diary-Keepers (月摺檔, 起居注), Archives of the Imperial Palace (宮中檔), Diplomatic Documents (照會), officially edited local gazetteers (地方志), Court Edicts concerning Revolts (剿捕廷寄檔), and others. Altogether there are at least 235 different archives and collections that we have examined. The earliest document in this collection was written in 1388 and selected from the Ming Reign Chronicles, and the latest was written in 1911, just before the Qing dynasty ended.

This collection represents the history of Taiwan from the perspective of the Chinese imperial government. As an outlying island of a vast empire, Taiwan did not get mentioned often in the imperial court unless something bad, such as a rebellion or famine, had occurred. Fig. 1 gives a breakdown of documents according to the years that they were written. Indeed, each peak in the chart corresponds to an event that was crucial to Imperial China. For instance, the peaks of 1786 to 1788 reflects the most serious revolt ever occurred in Taiwan (the rebellion of Lin Shuangwen 林爽文). The peak of 1884 corresponds to the Sino-Franco War (1883-1885), during which the French invaded, unsuccessfully, northern Taiwan. The peak of 1895 indicates the Sino-Japanese War (1894-1895), which ended with Taiwan's secession to Japan.

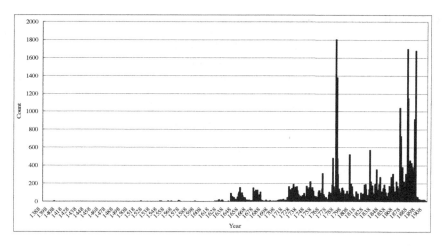

Fig. 1. Yearly count of documents in the Selected Collections of Taiwan-related Documents from the Imperial Governments of the Ming and Qing Dynasties

3.2   The Danxin Archives: Official Documents of the Danshui Sub-prefecture and the Xinzhu County (1789-1895)

The Danxin Archives is a collection of administrative and judicial records of the Danshui sub-prefecture and the Xinzhu County. The area involved covers the entire northern Taiwan, with the present day Miaoli at the southern end. The dates spanned from 1789 (the 54th year of the Qianlong's乾隆reign) to 1895 (the 21st year of the Guangxu's光緒reign). There are 19,281 documents, grouped into 1,143 cases. More than half of the cases were produced during Guangxu's reign, from 1875 to 1895 [4].

The Danxin Archives was organized and classified by the legal scholar Dai Yanhui (戴炎輝) into three categories: administrative, civil, and criminal [5]. Documents in each case are further arranged into a series in chronological order. The Danxin Archives is one of the only three pre-1911 Chinese local government archives known to exist. Different from official local gazetteers, the Danxin Archives provides a first-hand detailed account of the social life of citizens in the Qing dynasty. It is invaluable for anyone who wishes to study the political, economical, judicial, or administrative development of late 19th century Taiwan and China.

The original Danxin Archives is in the care of the National Taiwan University Library. There have been quite a few research articles and books based on Danxin Archives (see, for instance, [6]), mainly via studying a microfilm version produced by the University of Washington. In order to make this important material more available to the research community, NTUL embarked on a project to publish the full text, on the average of releasing 4 volumes a year. So far we have published 20 volumes, and the total is estimated to be about 36. We have also scanned all images (27,017), which will be incorporated into THDL. Currently we have proof-read and punctuated 11,611 full-text documents, all of which are available in THDL. There are also 11,242 associated metadata records.

3.3 Collections of Land Deeds of Taiwan

Until the beginning of the 20th century, land deeds were the only proof of ownership and transition of land in Taiwan. They were hand written and were usually prepared by a scrivener. In the early phase of the Japanese occupation of Taiwan, the Japanese government brought in western style land measurement mechanism and established a modern system to manage land and ownership, thus replaced the ad hoc land deed system. During the transformation of the land management system, the Japanese Governor-General made a concerned effort to record land deeds so that they can be ported to the new system. Thus, about 15,000 land deeds are included in the Archives of the Japanese Taiwan Governor-Generals (臺灣總督府檔案). It was estimated [7] that there are an additional 20,000 land deeds in the hands of libraries, museums, private collectors, and individual families which were not accounted for in the aforementioned archives. In 2003 and 2004, CCA commissioned the National Taichung Library (NTL) to collect and digitize (in full text) the hand-written copies of land deeds from the Archives of the Japanese Taiwan Governor-Generals. In this project, NTL keyed-in the full text of 15,901 land deeds from the Archives of the Japanese Taiwan Governor-Generals, 1,674 from published literature, and 157 from private collections. In addition, NTUL has digitized its own collections of land deeds which totaled at 3,667. Together, NTL and NTUL have collected more than 20,000 land deeds in Taiwan, all incorporated into THDL. We are adding another 3,000, which should be done in a few months.

While each land deed may have significance only to its owner, the collection as a whole provides a fascinating glimpse into the pre-1895 Taiwanese grassroots society. For example, the fluctuation of value of land reveals a great deal about the economic development of each region. Since many of the deeds were between indigenous people and Han immigrants, they also provide clues to the intricate relationship among the various peoples of Taiwan [8], the evolution of rights to land, and the gradual assimilation of the indigenous people (in particular the Pinpu平埔族群) into the Han society.

Land deeds usually have a fixed format, many of which differ only in the names of the parties involved, the names of the witnesses and scrivener, location and boundary of the land, and the date. They are thus ideal for experiments in term extraction and text mining. Indeed, we have already mined over 90,000 names of people and locations. Just people's names alone should be a valuable source for research in history.

4 Post-query Analysis and Referential Tools

The availability of full text opens an exciting new door for using primary important historical documents in research and teaching. While full-text search is a must-have, we try to explore other ways with the aim of building a research environment around these resources for historians and researchers of other disciplines.

4.1 Query Returns as a Sub-collection

Most document retrieval systems regard query results as a set of more or less independent documents. It is the user's job to go through the returns to see if any is relevant to her request. The facility of full-text retrieval sometimes makes the resulting set too large to manage. Thus some systems also provide query refinement mechanisms to further restrict the query results.

Historical documents have the feature that they are often *inter-related*. For example, while a single land deed may yield little meaning by itself, a series of land deeds of the same piece of property may reveal significance far greater than the sum of its parts. Indeed, historians rarely look at a single document alone (unless they are looking for something to refute a conjecture or to add support to an observation). They gather documents from different sources, try to figure out their *collective meaning* (relationships among the documents, and the meaning of the collection as a whole), and draw conclusions accordingly. Since THDL has already accumulated an unprecedented amount of full-text documents on Taiwan history under one roof, our next step is to represent the results from a query as a collection by itself and try to show the various possible relationships among the query result documents. This is done mainly through *post-processing* a query's returns as a sub-collection.

THDL features two facilities to help user find collective meanings of query returns, *multi-dimensional post-query classification* and *term frequency analysis*.

Post-query Classification. The post-query classification mechanism classifies documents of the resulting set of a query according to several predefined dimensions, which are metadata fields describing important background knowledge of documents such as dates, authors, and sources. After the resulting set of a query is returned (we call it a *sub-collection*), THDL classifies its contents according to year, author, and source on the left of the web page, and presents summaries of the documents themselves on the right. Fig. 2 shows the returns of the query *Lin Wencha* (林文察), a well-known Taiwanese native general during the Qing dynasty, with the timeline on the left. Among the 375 returned documents (all *zouje* 奏摺 - reports from officials to the emperor), 331 appeared between 1861 and 1864, the year when Lin was killed in battle. What is interesting is that 31 additional *zouje* mentioning Lin appeared after his death, until as late as 1906. An examination of these documents reveals a series of family tragedies involving his brother's (Lin Wenmin, also a Qing general) being wrongly accused of crimes and executed, his son's (Lin Chaodong) death in battle, and so forth. Thus a family's story unfolds in the sub-collection of a single query.

Note that the predefined dimensions are dependent on the characteristics and metadata of each collection, so different collections may need different dimensions. Our system also provides reordering facility, so that the user can examine the returned list of documents in any of the predefined dimensions. By presenting classifications of multiple dimensions, the user can switch from dimension to dimension and observe the distribution and behavior of each dimension. We also allow users to bookmark documents and create their own sub-collections.

Term Frequency Analysis. We have also built tools in THDL for term extraction, and have used the tools to build a corpus of over 90,000 names of people and locations. The names are used to provide term frequency analysis which further helps the user to explore and expand the sub-collections. Term frequency analysis tabulates the numbers of times terms appear in the sub-collection and presents them to the user (see Fig. 3). The user can use them to decide how relevant a person or a location is to the present query (and the associated sub-collection) and explore further.

Fig. 2. Query results of "*Lin Wencha*"

Through post-query classification and term analysis, we hope to provide users better ways to analyze the sub-collection retrieved from a query as a whole, rather than as individual documents.

4.2 Referential Tools

A historian always has referential resources within reach when she conducts research. To make THDL more useful, we have also built several referential tools and are in the process of building more. We describe some of them here.

Chinese-Gregorian Calendar Converter. Date is obviously among the most important information for history. The traditional Chinese calendar is lunar and does not correspond directly to the Gregorian calendar. During the dynastic era, the years were indicated by the reign title of an emperor, who may change the title from time to time. To complicate things further, Koxinga continued to use Ming reign even if it no longer existed (after being terminated by Qing). To maintain consistency, we use Gregorian calendar as the standard metadata format for dates and wrote a converter that made a daily correspondence between Gregorian and the Chinese calendars starting from the first day of Ming (1368/01/25) to the last day of Qing (1912/02/17). We also included part of the Japanese calendar for documents produced during the period of Japanese rule (1895-1945). The total number of dates in our database is about 250,000.

We also wrote a program to automatically recognize dates appeared in documents and convert them to Gregorian. A query interface for users to directly access the conversion table is available.

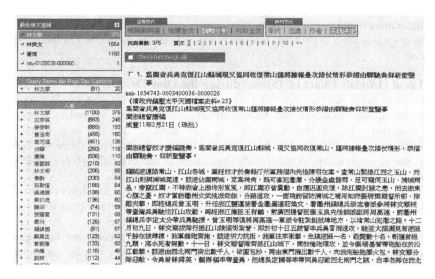

Fig. 3. Term frequency analysis of the sub-collection from the query "*Lin Wencha*"

Corpus of Names of People and Locations. *Who* and *where* are no less important to historical events than *when*. Thus we need ways to extract names of people and locations from our documents. While the names mentioned in the court documents are relatively easy to collect, since they usually only involve important officials and locations, those that appeared in local documents and land deeds pose a problem.

The challenge is further compounded by the fact that the Chinese language has no spacing between characters and the old scripts have no punctuation. Although there have been research of term segmentation for Chinese [11, 12], they are not directly applicable because the grammatical structure of old Chinese is quite different from that of the modern Chinese. Term frequency count won't do the trick either since many individuals only appear once (such as those appearing in only one land deed). The irregularity of names of the indigenous people, as opposed to the simple family-given name format of Han Chinese, adds more challenge.

To solve this problem we developed a *word-clip* algorithm to recognize proper nouns in Chinese documents [2]. The idea is to recognize existing relations between proper nouns and their context in a collection in which the documents have similar styles of writing. In such a collection, a specific type of proper nouns (such as locations) is often surrounded by similar leading phrases and ending phrases. We call such a pair of leading and ending phrases a *clip*. The characters within a clip usually form a proper noun.

Our algorithm starts with a set of known names to find useful clips. The clips are then used to catch more names. This process is iterated until it reaches saturation (very few new clips are generated). We have applied the word-clip method to two of the collections (except the Danxin Archives). Experimental results show that the average precision rate of identifying person names is about 50% and the estimated recall rate is about 75%. For location names the precision rate is 82% and the estimated recall rate is 84%. So far we have collected 90,948 person names and 3,496 location names in the two collections.

The corpus is used in the term frequency analysis utility mentioned in the previous section.

The Local Officials Chart of Taiwan during the Qing Dynasty. Knowing who was in charge of what at what time is very useful when studying history. We took an important source book on the local officials of the Qing Dynasty, Listing of Administrative Offices and Officials [9], and built a tool to help users find information about officials. The book lists all the administrative offices of Taiwan during the Qing dynasty, together with the names of the officials who occupied the offices, with their starting and ending dates.

Instead of just making a table, we designed a data format that allows one to fully utilize the information provided by the book. The design principle of the data format is to make each tuple as small as possible so that local changes can be made easily. This is necessary because not all records in the book are complete or accurate. However, the tuples have to contain enough fields so that they can be connected to provide answers to new types of queries. We have proved that the data format we designed is both minimal and sufficient [10].

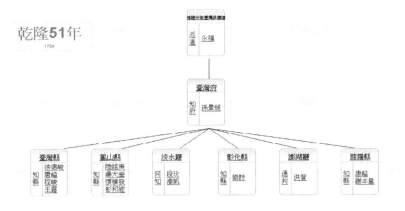

Fig. 4. Taiwan government structure during the year 1786

Our system provides three types of queries at the moment: *query by office*, *query by person*, and *query by year*. A *query by office* gives the list of names of people who ever held that office, in chronological order. This is also the way data were organized in the book. The other two types of queries, however, show how digitized data can provide a lot more than the source book was originally intended for. A *query by name* returns a chronology of all the offices of Taiwan that the named person ever held. A *query by year* returns a chart that shows the entire governmental structure of Taiwan of that year with the names of the officials for each office during that year (see Fig. 4). A click on the name of an office shows another chart that represents the internal structure of that office, also with names of officials holding positions within that office during that year. A click on any person appeared in the tree sends a *query by person* command to the system and generates the respective chronological chart accordingly.

5 Discussion

This paper describes THDL, the Taiwan History Digital Library of historical documents. In addition to incorporating the full texts of over 80% of primary Chinese documents about Taiwan before 1895, we have also developed different ways to facilitate THDL for research in history. They include full-text search, techniques and interfaces for classifying and exploring a query result as a sub-collection, term frequency analysis, and referential tools.

There are many ambitious digitization projects currently in progress, such as Google Books (see http://books.google.com/googlebooks/about.html) and the Million Book Project [13]. THDL is different from these projects in many ways. First, THDL is about a single domain – Taiwan history, and thus can utilize domain-specific knowledge and metadata to provide important features such as post-query classification. Second, we only select the contents in the archives that are related to Taiwan. In other words, we do not cover entire archives or books, only the parts that are relevant to our purpose. Third, we did not (could not) use OCR to obtain full texts because all the documents are hand-written, which renders OCR impossible. Thus all our texts were keyed-in. Lastly, there is no copyright concern since we only deal with ancient materials. Like the other two projects, we emphasize on the ease of use. In addition to providing full-text search, we want our interface to be at least as friendly as Google's, with additional emphasis on providing help *after* the query results are returned.

Another project that we should mention is Taiwan's National Digital Archives Program (NDAP) [14]. Although NDAP also deals with historical subjects about Taiwan, it does not emphasize on full texts. Thus we did not request/receive funding from NDAP for constructing THDL, except for building the full text of the Danxin Archives.

We are still building more contents, both primary materials and research work, about Taiwan history. We are also developing text mining techniques to provide better analytical tools of historical documents. We hope THDL can help push research on Taiwan history to a new horizon.

Acknowledgments. Fundings for THDL were provided by the National Taiwan University and the Council for Cultural Affairs for building the full text of the Imperial Court documents, National Science Council under grant NSC94-2422-H-002-001 for building the full text of the Danxin Archives, and NSC95-2221-E-002-277 for building the THDL system. Their generous supports are greatly appreciated.

Many people helped in creating the content of THDL. We thank Chiu Wan-Jung and her staff of the Special Collections Department of the NTUL, the National Taichung Library, Professor Lee Wen-Liang and his assistants of the Department of History of NTU. We also thank Lin Hsin-Yi who provided many comments, and the graduate students of the Laboratory of Digital Archives of the Department of CSIE of the National Taiwan University for building some of the tools.

References

1. http://en.wikipedia.org/wiki/Taiwan/History
2. Chang, S.P.: A Word-Clip Algorithm for Named Entity Recognition: by Example of Historical Documents. Master Thesis, National Taiwan University, Taiwan (in Chinese) (2006)
3. Chiu, W.J.: The Digital Project of Taiwan-Related Archives in Ming and Qing Dynasty. The Library Yearbook of ROC 2006. National Central Library, Taiwan, 128–129 (in Chinese) (2006)
4. NTU Library (in Chinese), http://www. 140.112.113.4/project/database1/database1_1.htm
5. Dai, Y.H.: Preliminary Remarks on Putting in Order the Qing Danxin Archives. Taipei Cultural Relics (in Chinese) (1953)
6. Allee, M.: Law and Local Society in Late Imperial China: Northern Taiwan in the Nineteenth Century. Stanford University Press (1994)
7. Wu, M.C., Ang, K.I., Lee, W.L., Lin, H.Y.: A Brief Introduction to the Integrated Collections of Taiwan-related Historical Records. CCA and Yuan-Liou Publishing, Taiwan (in Chinese) (2005)
8. Hong, L.W.: A Study of Aboriginal Contractual Behavior and the Relationship between Aborigines and Han Immigrants in West-Central Taiwan, vol. 1. Taichung County Cultural Center, Taiwan 5 (in Chinese) (2002)
9. Pan, C.W. (ed.): Taiwan Geography and History, Taiwan Provincial Literature Committee, Taiwan 9(1) (in Chinese) (1980)
10. Chang, J.T.: Model and Implementation for Representing Governmental Structures and Officials. Master Thesis, National Taiwan University, Taiwan (in Chinese) (2007)
11. Chien, L.F.: PAT-Tree-Based Keyword Extraction for Chinese Information Retrieval. In: Proceedings of 1997 ACM SIGIR Conference (SIGIR 1997), Philadelphia, USA, pp. 50–58 (1997)
12. Chen, H.H., Lee, J.C.: Identification and Classification of Proper Nouns in Chinese Texts. In: Proceedings of 16th International Conference on Computational Linguistics, Copenhagen, Denmark, pp. 222–229 (1996)
13. Reddy, R., StClair, G.: The Million Book Digital Library Project (2001), http://www.rr.cs.cmu.edu/mbdl.htm
14. National Digital Archives Program (2007), http://www.ndap.org.tw/index_en.php

Towards a Digital Archive for Handwritten Paper Slips with Ethnological Contents

A.C. Schering[1], I. Bruder[2], C. Schmitt[3], H. Meyer[1], and A. Heuer[1]

[1] Database Research Group, Dept. of CS, University of Rostock, Germany
[2] IT Science Center Ruegen, Putbus, Germany
[3] Wossidlo Archive, University of Rostock, Germany

Abstract. Contemporary digital libraries and archives of ethnological information focus mainly on document based storage and access methods for their data. However, our archive is designed to manage smallest pieces of information and can enable ethnologists not only to easily store and access their material, but also to derive new knowledge by combining existing data. In this paper, we present the first steps in building a digital archive for paper slips with ethnological contents from the 19th and the beginning of the 20th century. Along with the architectural and accessibility aspects of the *WossiDiA* system, we describe enhancements for efficient retrieval and for supporting modifications to access structures.

1 Introduction

The Wossidlo Archive embraces one of the most diverse collections of highly interconnected paper slips documenting regional folk culture, worldwide. It was founded by Richard Wossidlo, an ethnologist and ethnographer who studied and collected terms and definitions in the Low German language related to ancient customs with location-specific usage in the region of Mecklenburg, Germany, between 1883 and 1939. With the help of several hundreds of contributors, he labeled millions of slips of paper with cognitions from his field research. These slips are organized in small bundles which are physically represented by envelopes, stored in wooden boxes which are arranged in huge shelves.

This paper represents the latest results from the *WossiDiA (Wossidlo Digital Archive)* project, developing a digital archive for a collection of about two million slips of paper, about 30,000 envelopes, more than 1,200 boxes, and additional material such as correspondences and literature excerpts. Among the scanned images of the paper slips and documents we store a substantial number of corresponding metadata such as bibliographic and classification data, reflecting both the diverse aspects on the contents of slips and documents as well as their correlation and interdependence. The information encoded in the paper slips, such as the cultural fact, description, date and place of origin as well as contributor is of great interest to cultural scientists. To retrieve this information we employ full-text and hierarchical retrieval techniques, as used in most existing digital document archives. Wossidlo defined complex access mechanisms using detailed finding aids for different topics and a number of indexes and thesauri. The interesting feature of his slip collection is that he had not interlinked the papers

D.H.-L. Goh et al. (Eds.): ICADL 2007, LNCS 4822, pp. 61–64, 2007.

with just the access mechanisms, but even the papers among each other as well. Comprehensive information cannot be found in a single paper slip. It has to be compiled from slips and documents all across the archive. Therefore the true value of *WossiDiA* lies in the structures and relationships between facts and contents of the papers in conjunction with the comprehensive finding aids (e.g., shepherd is specified in the categories labour, rites, and magic (rites→shepherd rites→shepherd (labour)←shepherd magic←magic). The challenges we discuss particularly in this paper are modeling and implementation of the digital archive as well as providing means for efficient storage, complex retrieval, and structural modifications supported by data mining techniques within this framework.

2   Model, User Scenarios and Architecture

The first step in designing the digital archive as proposed by [2] was to identify the material to be included, such as scanned image data (slips of paper and aux- iliary documents), corresponding metadata, and detailed finding aids including indexes and thesauri. We have devised a conceptual data model to represent materials, digital data and metadata, as well as access structures on three levels. It defines (1) concrete entities, such as slips of paper, envelopes and boxes on the physical level, (2) digital images and their metadata, such as bibliographic data, descriptions, place and time of origin on content level, and (3) finding aids, taxonomies, indexes and thesauri on the access level. The metadata in the digital archive are stored in METS format [3]. This is especially useful for bibliographic metadata and can be further utilized to hold descriptive metadata recorded in formats such as DC, MODS, EAD, etc. Multimedia objects are modeled by using the multimedia metadata standard MPEG-7. Such multimedia objects are im- age data (slips) and image parts (special regions providing signature, keywords, names, dates, etc.) as well as audio data (voice recordings for particular slips). MPEG-7 fragments can be easily integrated into METS. Both data model and formatting issues are thoroughly described in [5]. The system design (Fig. 1), is based upon the data model described above. *WossiDiA* is implemented upon the Oracle 10g database system, which provides support to efficiently store and manage data and XML metadata. Furthermore, it features a data mining option as well as spatial extensions to store the geospatial data and to write mapping modules. The digital archive provides interfaces for three basic usage scenarios:

Manipulation comprises two tasks: (1) data and metadata input performed by the data entry staff using entry forms provided by the Protégé-Frames editor, specially enhanced to enter *WossiDiA*-specific data, and (2) structural modifica- tion of the metadata model collaboratively performed by professional researchers (see Sect. 3). Both user groups are supervised by the domain director who is in charge of the digital archive and the metadata model in particular.

Retrieval is the main usage scenario performed by public users and profes- sional researchers. The public user is only interested in searching and browsing the archive whereas the professional researcher also intends to alter structures

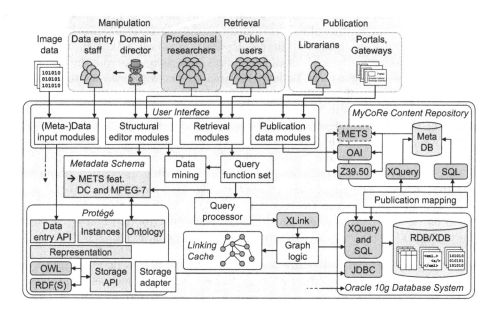

Fig. 1. *WossiDiA* Digital Archive architecture

(see above). The retrieval scenario requires the most efforts in *WossiDiA*, since it is responsible for putting together the various tiny pieces of information mentioned in the introduction. Full-text and hierarchical search mechanisms are used to provide the user with information about a particular topic. Using the various access structures he can obtain more information, e.g., about adjacent topics, from the same contributor, in the same time frame, or within a particular place.

Publication is the scenario used by librarians and archivists to retrieve the information of the archive in a combined manner, such as articles about particular topics. In contrast to the retrieval scenario, publication focuses on content generation according to specific standards and formats, including, but not limited to METS, DC, OAI, and Z39.50. Thus, the *WossiDiA* material can easily be integrated into libraries as well as portals and gateways in the field of cultural sciences. We use a MyCoRe content repository as publication subsystem to provide the formats mentioned above and to enable *WossiDiA* to be integrated into the web portal of the Library of the University of Rostock.

3   Enhancing the Architecture

Structure editing enhancements: The main problem in the course of creating an adequate system design is to give the professional researcher the opportunity to efficiently alter finding aids and taxonomies. For this purpose we have introduced the structural editor module to the system architecture. It enables the user to conduct modifications of the structures manually. Since the number of data and metadata records is extraordinarily high and the amount of relationships between the paper slips and their access mechanisms is quite substantial the professional researcher needs an additional help provided by data mining techniques [1]. Data mining algorithms are utilized to derive implicit, previously unknown, cultural information hidden in the interior of Wossidlo's collection all across the archive on every single slip of paper. To help the user to develop new and to adjust existing Wossidlo-specific finding aids we provide solutions to employ automatic data mining algorithms in the context of the structural editor module. In the course of that we use clustering analysis to identify homogeneous groups of information units upon which terms used in the respective finding aids are to be built. Another important data mining scenario is the identification of groupings in the data for retrieval purposes, e.g., to find contributors writing about a certain fact limited regionally and chronologically. This requires the integration of data mining aspects into the retrieval module as well. For that scenario, we use in particular the clustering algorithms k-means (partitional) and O-Cluster (hierarchical) [4], both featured by the Oracle database system.

Retrieval enhancements: WossiDiA includes a query processor for a domain specific query language based on XQuery. Some extensions, such as XLink and graph logic, are necessary because XQuery is primarily based on a hierarchical data model and thus cannot handle diverse graph structures efficiently, especially inter-slip relationships. A function set provides methods to query and to navigate our XML-based model. A linking cache is introduced to efficiently manage relationship information condensed to a very small size in order to keep as much links in main memory as possible.

4   Conclusions

The digital archive concept presented in this paper allows data storage, access, analysis, retrieval and presentation of this unique cultural material. Dealing with a huge number of smallest pieces of information which are highly interconnected, the implementation of *WossiDiA* requires unusual techniques in terms of conventional digital archives. Furthermore, the system will provide intelligent means enabling knowledge deduction and reclassification of finding aids.

References

1. Han, J., Kamber, M.: Data Mining: Concepts&Techniques. Morgan Kaufmann, San Francisco (2006)
2. Witten, I., Bainbridge, D.: How to Build a Digital Library. Morgan Kaufmann, San Francisco (2003)
3. The Library of Congress. Metadata Encoding and Transmission Standard: Primer and Reference Manual (2007), http://www.loc.gov/standards/mets/
4. Milenova, B.L., Campos, M.M.: O-Cluster: Scalable Clustering of Large High Dimensional Data Sets. In: ICDM, pp. 290–297 (2002)
5. Schering, A.-C., Bruder, I., Meyer, H.: Management of Highly Interconnected Information Units in the Digital Wossidlo Archive. In: Proceedings of the 19th GI-Workshop on Foundations of Databases, Germany (2007)

Automatic Classification of Web Search Results: Product Review vs. Non-review Documents

Tun Thura Thet, Jin-Cheon Na, and Christopher S.G. Khoo

Wee Kim Wee School of Communication and Information
Nanyang Technological University
31 Nanyang Link, Singapore 637718
{ut0001et,tjcna,assgkhoo}@ntu.edu.sg

Abstract. This study seeks to develop an automatic method to identify product review documents on the Web using the snippets (summary information that includes the URL, title, and summary text) returned by the Web search engine. The aim is to allow the user to extend topical search with genre-based filtering or categorization. Firstly we applied a common machine learning technique, SVM (Support Vector Machine), to investigate which features of the snippets are useful for classification. The best results were obtained using just the title and URL (domain and folder names) of the snippets as phrase terms (n-grams). Then we developed a heuristic approach that utilizes domain knowledge constructed semi-automatically, and found that it performs comparatively well, with only a small drop in accuracy rates. A hybrid approach which combines both the machine learning and heuristic approaches performs slightly better than the machine learning approach alone.

Keywords: Product Review Documents, Genre Classification, Snippets, Web Search Results.

1 Introduction

In recent years, we have witnessed tremendous growth of online discussion groups and review sites, where an important characteristic of the posted articles is their sentiment or overall opinion towards the subject matter. Researchers are turning their attention to a kind of non-topical classification called *sentiment classification* [9]. Research in automatic sentiment classification seeks to develop models (i.e. sentiment classifiers) for assigning category labels (positive or negative) to new documents or document segments based on a set of training documents that have been classified by domain experts.

In our previous work [8], a prototype meta search engine providing automatic sentiment classification was developed. It allows the user to specify a product name and subsequently categorizes the search results by the polarity of the desired reviews, such as *recommended* or *not recommended*. It can help the user to focus on Web articles containing either positive or negative comments. For instance, a user who is interested mainly in the negative aspects of a product (e.g. a digital camera) can look at Web articles under the negative review category.

D.H.-L. Goh et al. (Eds.): ICADL 2007, LNCS 4822, pp. 65–74, 2007.

For effective sentiment classification, non-review documents should first be filtered out so that further classification (i.e. sentiment classification) can focus on product review documents. We define a review document as a page that contains only a single review, since the sentiment classifier is designed to classify one review at a time. The Web search results from the meta search engine mainly consist of e-commerce Web pages selling the product, product specifications from manufacturing companies, on-line product review documents, etc.

This paper focuses on the filtering of product review documents from various documents in the Web search results. In this study, only snippets and not full text documents are used in the filtering process since full text documents would need more processing time. Determining whether a snippet is a review or non-review document is a challenging task, since the snippet usually does not contain many useful features for identifying review documents.

In the following sections, section 2 discusses related works of automatic text classification, section 3 presents our approaches for review classification and, finally, section 4 discusses future work and conclusion.

2   Related Works

Research in *automatic text classification* seeks to develop models for assigning category labels to new documents based on a set of training documents. For classification, documents are represented as sets of features from their content and style, called *document vectors*. Most studies of automatic text classification have focused on either "topical classification" classifying documents by subject or topic (e.g. *education* vs. *entertainment*), or "genre classification" classifying documents by document styles (e.g. *fiction vs. non-fiction*). A detailed introduction to automated text classification has been provided by Sebastiani [11].

Determining whether a snippet is a review or non-review document is considered as a genre classification problem. Documents (i.e. snippets) discussing the same topic (e.g. a digital camera) can be classified into different genres, such as *product specification* or *product review*. Compared to topical classification which mainly utilizes text features of documents, genre classification uses various document style features, such as *part-of-speech* and linguistic features (e.g., average sentence length), in addition to text features to analyze how documents are described. However, our study does not use document style features because snippets are too short to analyze them. Thus our approach mainly uses text features from summary text, in addition to the URL and link title. We have performed a preliminary study [12] on this problem and this paper discusses further extensions of the study by incorporating more snippets from various products and exploring effects of various feature selection approaches, such as phrase terms (n-grams) and feature reduction, on the classification.

For genre classification, most researchers use full-text documents rather than summary documents, such as snippets. For instance, Finn, Kushmerick, and Smyth [4] investigated a genre classification, which decides whether a document presents the opinion of its author or reports facts (i.e. genre of subjectivity). C4.5, a decision tree induction program [10], was used with various text features: *bag of words* (unigrams), *part-of-speech*, and hand-crafted shallow linguistic features. For the *part-of-speech* approach, a document is represented as a vector of 36 *part-of-speech* features, expressed as percentages of the total number of words for the document. They argued that the *part-of-speech* approach provided the best accuracy when the learned classifiers were generalized from the training corpus to a new domain corpus. As another work, Kessler, Nunberg, and Schutze [7] studied automatic detection of text genre using logistic regression and neural networks techniques. The genres they investigated were *reportage, editorial, scientific/technical, legal, non-fiction*, and *fiction*.

Boese and Howe [1] investigated the effects of Web document evolution on genre classification. They reported that documents in some genres change rarely, and the genre classifier trained with an old corpus performed well on recent Web pages, with only a small drop in accuracy rates. From their study, we may argue that genre classification using document style features is not significantly affected by document evolution, compared to topical classification using text features that change over time. Since our study does not use document style features, the genre classifier learned from our study may be affected by document (snippets) evolution. Other genre classification works are well summarized in [3].

Some researchers have developed classification/clustering tools to categorize Web search results to help users locate relevant and useful information on the World Wide Web. For the classification/clustering they generally use the snippets from the search engine to provide reasonable response time to the user. Chen and Dumais [2] designed a user interface that automatically groups Web search results into predefined topical categories such as *automotive, local interest*, using a machine learning algorithm, SVM. The tool devised by Zeng, He, Chen, Ma and Ma [14] provides clustering of Web search results, and uses salient phrases extracted from the ranked list of documents as cluster names. For instance, with a query input, *Jaguar*, the generated cluster names are *Jaguar Cars, Panthera onca, Mac OS, Big Cats, Clubs*, and *Others*. Vivisimo (http://vivisimo.com) is an example of an operational clustering tool for Web search results. These tools, however, focused mainly on topical categorization—categorizing documents by subject or topical area.

3   Review Classification

This study is conducted with a dataset of 1200 documents (i.e. snippets). The first 800 documents are used for training and testing of the machine learning model with 10-fold cross validation. The remaining 400 documents are kept as unseen documents for final evaluation of the approaches.

A search engine, Google, is used in this study to gather the snippets of 1200 documents by submitting around 120 queries. The queries are submitted in the format of a product name followed by the key word "Review". For example, the query "Dell XPS M1710 Review" is used for the product "Dell XPS M1710". When the results are returned by the search engine, they are manually classified as either review or non-review documents. The manual analysis of the content is done by following the URL of the snippets and reviewing the full text. If the content is found to be a user or an expert review with ratings, it is classified as a review document. In addition, a comprehensive full-review without rating is also classified as a review document. The documents with product specifications, multiple brief reviews, list of review links or non-review-related contents are classified as non-review documents. In this study, the domain of electronic products is selected and products such as digital camera, mobile phone, MP3 player, PC, PDA, notebook, printer and monitor are included.

3.1 Machine Learning Approach

In the study, SVM [5] is used as a machine learning approach and the various components of the snippets are experimented as document features for effective classification. Unigrams (individual words), n-grams (phrases), and feature reduction are also explored to improve the accuracy. As the input to SVM, the text is converted into bags of terms (called document vectors), which are stemmed using Porter's stemming algorithm [6] after removing the stop words. Term Frequency (TF) is used as a weighting factor for the terms.

The terms are extracted from the title, the summary text, the URL and the similar pages of the snippets. Five features are experimented as document features as shown in the Tables 1 and 2: "Title", "Summary Text", "URL Domain", "URL Folder" and "Similar pages". The feature "Title" comes from the title of the snippets which is the text *"Motorola RAZR V3 Reviews"* as in the following snippet example. The feature "Summary Text" comes from the plain text below the title which is the text *"User Reviews for the Motorola RAZR V3. Plus specs, features, discussion forum, photos, merchants, and accessories"*. The URL is divided into two parts, "URL Domain" which is *"www.phonescoop.com"* and "URL Folder" which is *"phones"*. Finally, the feature "Similar Pages" is extracted from the snippets of the similar pages by following the link "Similar pages", which is provided by the search engine to retrieve the similar pages which are related to the current snippet.

Motorola RAZR V3 Reviews
User Reviews for the Motorola RAZR V3. Plus specs, features, discussion forum, photos, merchants, and accessories.
www.phonescoop.com/phones/user_reviews.php?phone=547   Similar pages

---

The test results with the first 800 documents are shown in the Table 1. When unigrams are used, the best result comes from the feature selection option S6 (Table 1), which uses the Title and the URL domain and folder. When the summary text of the snippets is included as a document feature, it reduces the accuracy of the classification because the summary text can come from any part of the full text and, thus, it distracts the SVM model when it comes to classification. When the similar pages are included in the text, the accuracy of classifier also decreases significantly because the similar pages are mainly product information pages, but not product review documents.

The machine learning approach produces better results when using n-grams than when using just unigrams for model building and classification. In the study, the n-grams consist of unigrams, bigrams, and trigrams. For instance "review, "full review", and "unbiased review document" are valid terms in the n-grams. The best result comes from the S7 option, which uses n-grams of the Title and the URL domain and folder without any feature reduction. The options using the same features as S7 but with feature reduction, S8 and S9, do not perform better than S7 in terms of accuracies but the computation cost is significantly reduced since around 10,000 n-gram terms are reduced to 3,000 terms. For feature reduction, information gain and chi-square values are used [13].

Terms such as "unbiased review", "comparison" and "comprehensive", and domain names such as "reviews.cnet.com", "review.zdnet.com" and "www.pcmag.com" occur very frequently in the URL and text of the review snippets. For the non-review snippets, terms such as "price", "spec" and "shop" occur commonly.

Table 1. The SVM approach using 10-Fold Cross Validation with 800 documents

ID	Features					n-grams	Feature Reduction		Accuracy
	Title	Summary Text	URL Domain	URL Folder	Similar Pages		Chi Square	Information Gain	
S1	Y	Y	Y	Y	Y				74.25%
S2	Y	Y	Y	Y					79.14%
S3	Y	Y	Y						77.92%
S4	Y	Y							73.90%
S5	Y		Y						85.43%
S6	Y		Y	Y					86.07%
S7	Y		Y	Y		Y			**87.08%**
S8	Y		Y	Y		Y	Y		86.70%
S9	Y		Y	Y		Y		Y	86.58%

The test results of the machine learning approach when tested with the unseen 400 documents are shown in the Table 2. The SVM model which is built by training with the initial dataset of the 800 documents is tested for the unseen documents. The accuracies of the tests using n-gram terms are consistently better than the accuracies of the tests using unigram terms.

Table 2. The SVM approach with 400 unseen documents

ID	Features					n-grams	Feature Reduction		Accuracy
	Title	Summary Text	URL Domain	URL Folder	Similar Pages		Chi Square	Information Gain	
S1	Y	Y	Y	Y	Y				59.35%
S2	Y	Y	Y	Y					78.80%
S3	Y	Y	Y						75.81%
S4	Y	Y							71.32%
S5	Y		Y						81.55%
S6	Y		Y	Y					81.55%
S7	Y		Y	Y		Y			**83.04%**
S8	Y		Y	Y		Y	Y		82.79%
S9	Y		Y	Y		Y		Y	82.79%

3.2 Heuristic Approach

A heuristic approach is also developed to experiment if a simpler heuristic approach with semi-automatically constructed domain knowledge can perform as good as the machine learning approach. In contrast to the machine learning approach which uses thousands of terms, this approach uses only hundreds of terms for classification. It is based on the review and non-review lists of n-gram terms which are constructed by analyzing the 800 snippets. Through the analysis, meaningful terms with high information gain or chi-square values are taken into consideration. Also manually constructed terms are added to the lists. These lists then are used to distinguish the review and non-review documents using the title, the summary text and the URLs of the snippets. Some sample entries of the lists are shown in Table 3.

Table 3. Sample of n-gram terms for the heuristic approach

	Review	Non-review
Title	unbiased review * editor review * full review * review by * mobile review * guide exclusive comparison overview good ($N_{\text{Review-Title}}$ =25 entries)	shop * price * free * software download * best price * introduce service spec supply review image ($N_{\text{Non-Review-Title}}$ =25 entries)
Summary Text	unbiased review * editor review* cute * beauty * coverage * exclusive comprehensive footage guide compare compare editorial ($N_{\text{Review-Text}}$ =25 entries)	shop* share* sell* merchant * buyer * review tip photographic review article review write review image ($N_{\text{Non-Review-Text}}$ =25 entries)
URL	review.zdnet.com www.infosyncworld.com www.mobile-review.com www.trustedreviews.com asia.cnet.com www.pocket-lint.co.uk www.letsgodigital.org www.mobile-phones-uk.org.uk laptopmag.com cellphones.about.com ($N_{\text{Review-URL}}$ =25 entries)	www.amazon.com www.livingroom.org.au www.reviewcentre.com www.imobile.com.au mobilementalism.com www.dpreview.com www.steves-digicams.com www.letsgomobile.org www.pricerunner.com www.phonedog.com ($N_{\text{Non-Review-URL}}$ =25 entries)

*: indicates manually constructed terms

For the Title's review and non-review lists, n-gram terms with high information gain or chi-square values are collected first from the titles of the snippets, and the terms which appear mainly in review documents are added into the review list while those which appear more in non-review documents are added into the non-review list.

The distinguishing terms such as "editor review" for review titles and "software download" for non review titles are also included although they may not have high information gain or chi-square values, or may not appear in automatically generated n-grams. The review and non-review lists of the Summary Text are constructed in the same way. For the URL lists, only terms with high information gain or chi-square values are added into either the review or the non-review list.

The following mathematical formula is used for the heuristic approach. $W_{Heuristic}$ represents the classification output value where the positive or negative value indicates a review document or a non-review document respectively. The parameters α, β and γ are weights on the title, summary text and URL. Based on our trial and error analysis, their optimal values are 0.3, 0.1 and 0.5 respectively. It shows that the URL is given a higher weight than others. If a snippet comes from a known review site, it is most likely to be a review document regardless of the other terms in the snippet.

$$W_{Heuristic} = \alpha \cdot W_{H.Title} + \beta \cdot W_{H.Summary} + \gamma \cdot W_{H.URL}$$

$$W_{H.Title} = \sum_{i=1}^{N_{Review-Title}} TF_i - \sum_{j=1}^{N_{Non-Review-Title}} TF_j$$

$$W_{H.Summary} = \sum_{i=1}^{N_{Review-Text}} TF_i - \sum_{j=1}^{N_{Non-Review-Text}} TF_j$$

$$W_{H.URL} = \begin{cases} +1 & \text{If } URL \in \text{Review-URL List} \\ -1 & \text{If } URL \in \text{Non-Review-URL List} \\ 0 & \text{Else} \end{cases}$$

The heuristic approach is tested with the 800 snippets and the heuristic approach performs comparatively well as the machine learning approach, with only a small drop in accuracy rates. The best accuracy is achieved when the title, the summary text and the URL of the snippets are used together (H4 in Table 4).

Table 4. The heuristic approach with 800 documents

ID	Title	Summary Text	URL	Accuracy
H1	Y			65.37%
H2		Y		66.44%
H3			Y	77.41%
H4	Y	Y	Y	**84.08%**

Table 5 shows the results of the heuristic approach when tested with the unseen documents. The heuristic approach has lesser computation cost yet it performs quite close to the machine learning approach. The heuristic approach using only the URL shows significantly lower accuracy when it is tested with the 400 unseen documents (H3 in Table 5) than when it is tested with the initial 800 documents (H3 in Table 4). This is because some URLs from the 400 unseen documents do not match with URL terms in the URL lists collected from the 800 documents, and the URL lists alone are not enough to determine review documents.

Table 5. The heuristic approach with unseen 400 documents

ID	Title	Summary Text	URL	Accuracy
H1	Y			61.60%
H2		Y		64.59%
H3			Y	65.34%
H4	Y	Y	Y	**79.05%**

3.3 Hybrid Approach

This is to experiment and find out if the hybrid approach which is a combination of both the machine learning and the heuristic approach can be employed to improve the outcome of the classification. The results of the experiment show that the hybrid approach performs better than the machine learning and the heuristic approach though not very significantly. The outcome of the hybrid approach (W_{Hybrid}) is calculated by combining the outcomes of the SVM approach (W_{SVM}) and the heuristic approach ($W_{Heuristic}$). The parameters λ and μ are used to fine-tune the outcome. For this initial evaluation, the values are set as 1 to equally weigh the two approaches.

$$W_{Hybrid} \quad = \quad \lambda \cdot W_{SVM} + \lambda \cdot W_{Heuristic}$$

When testing with the 800 documents, the best options, HB2 and HB4 (Table 6), achieve an accuracy of 89.30%. HB2 is a combination of S7 (Table 1), a machine learning approach using n-grams without feature reduction, and H4 (Table 4), a heuristic approach using the title, summary text and the URL. On the other hand, HB4 is a combination of S9 (Table 1), a machine learning approach using n-grams with feature reduction using chi-square, and H4 (Table 4). It is also observed that the test results of hybrid approaches HB3 and HB4 which utilize feature reduction, perform slightly better than S7 (Table 1), a machine learning approach without any feature reduction.

Table 6. The hybrid approach with 10-fold cross validation

ID	SVM Option	Heuristic Option	Accuracy
HB1	S6	H4	87.81%
HB2	S7	H4	**89. 30%**
HB3	S8	H4	89.17%
HB4	S9	H4	**89.30%**

The same approaches are then tested with the 400 unseen documents. The hybrid approach generally performs better than the machine learning approach or the heuristic approach alone as shown in the Table 7. The best result comes from HB2, which is a combination of a machine learning approach using n-gram terms without feature reduction and a heuristic approach using the title, summary text and the URL.

Table 7. The hybrid approach with 400 unseen documents

ID	SVM Option	Heuristic Option	Accuracy
HB1	S6	H4	83.79%
HB2	S7	H4	**84.79%**
HB3	S8	H4	84.29%
HB4	S9	H4	84.04%

3.4 Error Analysis

When analyzing the errors encountered by the approaches, it is observed that some of the errors are inevitable mainly because the classification is done based on just snippets which are relatively very short and with incomplete sentences. In such a scenario, even human classifiers will not be able to distinguish snippets of the non-review documents from those of review documents without looking at the full texts.

The following is an example snippet of a review document with a rating which is wrongly classified by both approaches as a non-review document because the snippet does not have enough terms related to review documents.

URL: *www.vnunet.com/personal-computer-world/hardware/2187326/lexmark-c534dn*
Title: Review: Lexmark C534dn laser printer - vnunet.com
Summary Text: Fast monochrome and color printing in one compact device.

The following is an example snippet of a non-review document which is wrongly classified by both approaches as a review document because it has some terms related to review documents.

URL: *mobilereviews.o2.co.uk/userreview/home*
Title: Mobile reviews - Mobiles & Tariffs - O2
Summary Text: Welcome to O2. Read and write reviews on the latest mobile phones.

When testing the hybrid approach with the unseen 400 documents, it is observed that 28 out of 88 errors made by the heuristic approach are corrected by the machine learning approach. On the other hand, 7 out of 67 errors made by the machine learning approach are corrected by the heuristic approach. We believe that in the hybrid approach the machine learning and heuristic approaches use different logics and complement each other to give better performance.

4 Discussion and Conclusion

In conclusion, the machine learning approach using the SVM performs the best with just the title and URL (domain and folder names) of the snippets as phrase terms (n-grams) for classifying product review documents. When feature reduction techniques such as Information Gain and Chi-square statistics are applied, computation cost is reduced with only a slight drop in accuracies. The heuristic approach which mainly makes use of domain knowledge performs as well as the machine learning approach in our experiments. The heuristic approach with the title, URL and the summary text of the snippets gives the best performance. The hybrid approach which makes use of both machine learning techniques and domain knowledge performs slightly better than the machine learning approach alone.

The limitation of this study is that it is only conducted for the electronic product review documents through a search engine and it may not work consistently for other domains. For future work, more evaluations and experiments will be carried out with larger datasets and a wider range of products using various Web search engines. Furthermore, the heuristic approach can be improved by including more meaningful and distinguishing terms and enhancing the formula to achieve better performance for the unseen documents.

References

1. Boese, E.S., Howe, A.E.: Effects of Web Document Evolution on Genre Classification. In: Proceedings of the 14th ACM international conference on Information and knowledge management (CIKM 2005), Bremen, Germany, pp. 632–639 (2005)
2. Chen, H., Dumais, S.T.: Bringing Order to the Web: Automatically Categorizing Search Results. In: Proceedings of the ACM SIGCHI Conference on Human Factors in Computing Systems (CHI 2000), pp. 145–152 (2000)
3. Choi, B., Yao, Z.: Web Page Classification, Foundations and Advances in Data Mining, Studies in Fuzziness and Soft Computing, vol. 180, pp. 221–274. Springer, Berlin (2005)
4. Finn, A., Kushmerick, N., Smyth, B.: Genre classification and domain transfer for information filtering. In: Crestani, F., Girolami, M., van Rijsbergen, C.J.K. (eds.) Advances in Information Retrieval. LNCS, vol. 2291, pp. 353–362. Springer, Heidelberg (2002)
5. Joachims, T.: Text categorization with support vector machines: Learning with many relevant features. In: Proceedings of 10th European Conference on Machine-learning, Chemnitz, Germany, April 21-24, pp. 137–142 (1998)
6. Jones, K.S., Willet, P.: Readings in Information Retrieval. Morgan Kaufman, San Francisco (1997)
7. Kessler, B., Nunberg, G., Schutze, H.: Automatic detection of text genre. In: Proceedings of the Eighth Conference on European Chapter of the ACL (Association for Computational Linguistics), pp. 32–38 (1997)
8. Na, J.-C., Khoo, C., Chan, S., Hamzah, N.B.: A sentiment-based search in digital libraries. In: Proceedings of Joint Conference on Digital Libraries 2005 (JCDL 2005), Denver, pp. 143–144 (2005)
9. Pang, B., Lee, L., Vaithyanathan, S.: Thumbs up? Sentiment classification using machine-learning techniques. In: Proceedings of the 2002 Conference on Empirical Methods in Natural Language Processing, Philadelphia, PA, July 6-7, pp. 79–86 (2002)
10. Quinlan, R.: C4.5: Programs for Machine Learning. Morgan Kaufman, San Francisco (1993)
11. Sebastiani, F.: Machine-learning in automated text categorization. ACM Computing Surveys 34(1), 1–47 (2002)
12. Thet, T.T., Na, J.-C., Khoo, C.S.G.: Filtering Product Reviews from Web Search Results. In: Proceedings of ACM Symposium on Document Engineering (DocEng 2007), Winnipeg, Canada (August 28 - 31, 2007)
13. Yang, Y., Pedersen, J.O.: A Comparative Study on Feature Selection in Text Categorization. In: Proceedings of the fourteenth International Conference on Machine Learning, pp. 412–420. Morgan Kaufmann, San Francisco (1997)
14. Zeng, H.-J., He, Q.-C., Chen, Z., Ma, W.-Y., Ma, J.: Learning to Cluster Web Search Results. In: Proceedings of the 27th Annual International ACM SIGIR Conference, Sheffield, UK, pp. 210–217 (2004)

An Effective Algorithm for Dimensional Reduction in Collaborative Filtering

Fengrong Gao[1], Chunxiao Xing[1], and Yong Zhao[2]

[1] Research Institute of Information Technology,
Tsinghua University, Beijing 100084
`{gaofengrong,xingcx}@mail.tsinghua.edu.cn`
[2] Department of Computer Science and Technology,
Tsinghua University, Beijing 100084
`Zhaoyong04@mails.tsinghua.edu.cn`

Abstract. It is necessary to provide personalized information service for users through the enormous volume of information on the web. Collaborative filtering is the most successful recommender system technology to date and is used in many domains. Unfortunately collaborative filtering is limited by the high dimensionality and sparsity of user-item rating matrix. In this paper, we propose a new method for applying semantic classification to collaborative filtering. Experimental results show the high efficiency and performance of our approach, compared with tradition collaborative filtering algorithm and collaborative filtering using K-means clustering algorithm.

Keywords: Collaborative filtering, dimensionality reduction, semantic classification.

1 Introduction

With the rapid growth of the Web, people have to spend more and more time to search what they need. To deal with this difficulty, recommender systems have been developed to provide different services for different users. Collaborative filtering is one of the most successful recommender system technologies to date and is used in many domains. It works by collecting user feedback in the form of ratings for items in a given domain and seeks similarities between user rating histories. Collaborative filtering does not consider the content of items, so it supports for filtering items whose content is not easily analyzed with computers such as video, audio, restaurants, etc. For it provides recommendations based on user's ratings to items, most users seem less reluctant to provide item-rating information. So the user-item rating matrix is very sparse. Moreover, along with users and items increase, the matrix will be very high dimensional. Both the sparsity and the high dimensionality lower the accuracy and efficiency of recommendations.

In this paper, we propose an effective method to overcome the drawbacks in collaborative filtering. Our approach uses semantic classification to divide original user-item rating matrix into several low-dimensional dense user-item rating matrices, then use low-dimensional matrices to provide recommendations.

D.H.-L. Goh et al. (Eds.): ICADL 2007, LNCS 4822, pp. 75–84, 2007.

2 Related Work

The term "collaborative filtering" was first coined by Goldberg et al on the Tapestry email system[1]. A variety of collaborative filtering algorithms have been designed and deployed henceforth. The GroupLens[2~4] system is one of the first automated collaborative filtering systems to apply a statistical collaborative filtering to the problem of Usenet news overload. It identifies advisors based on the Pearson correlation of voting history between pairs of users. This method is usually called "correlation-based collaborative filtering". It is the most popular among all the algorithms and represents in many researches.

Besides correlation-based methods, some model-based algorithms [5~7] appeared which adopt data mining techniques such as clustering, classification, and Bayesian networks. These methods pre-compute a model based on a training data set and then use the model for predictions. Once the clustering process is complete, the efficiency of the recommendations can be very good, since the size of the data is much smaller than original user-item rating matrix. In section 5 besides comparing our approach to traditional collaborative filtering, we also experiment by comparing our approach to a collaborative filtering algorithm using clustering technique.

To reduce the dimensionality of data and tackle the sparsity problem, Singular value decomposition [8] is used to produce a low-dimensional representation of the original user-item space and a list of recommendations will be generated using low-dimensional space. Our method uses the idea "dimensionality reduction" of this paper. But our approach of dimensionality reduction is quite different from SVD.

Popescul et al [9] presented probabilistic mixture models for recommending items based on collaborative and content-based evidence merged in a unified manner. The model builds on two-way co-occurrence models and collaborative filtering. It incorporates three-way co-occurrence data (users, items and item content) presuming that users are interested in a set of latent topics which in turn generate both items and item content information. The approach of building probabilistic model mixing content data with collaborative data is also proposed in Ref. [10~11].

Breese et al [12] performed an empirical analysis of several collaborative filtering algorithms. Experiments show that the recommendations are correlative with the nature of the dataset, nature of the application, and the availability of votes with which to make predictions.

3 Semantic Classification-Based Collaborative Filtering

3.1 Problem Description

User ratings for items are described in traditional collaborative filtering algorithm as a matrix:

$$R_{m \times n} = \begin{bmatrix} r_{11} & \cdots & r_{1j} & \cdots & r_{1n} \\ \vdots & \ddots & \vdots & \vdots & \vdots \\ r_{a1} & \cdots & r_{aj} & \cdots & r_{an} \\ \vdots & \vdots & \vdots & \ddots & \vdots \\ r_{m1} & \cdots & r_{mj} & \cdots & r_{mn} \end{bmatrix}_{m \times n} \tag{1}$$

where r_{aj} denotes the score of item j rated by user a. If user a had not rated item j, $r_{aj} = 0$. m denotes the total number of users, and n denotes the total number of items. The prediction of user a to an unseen item j, i.e. P_{aj} is done based on the average ratings of user a and a weighted sum of co-rated items between user a and all his similar users based on $R_{m \times n}$.

$$P_{aj} = \overline{r}_a + k \sum_{i=1}^{m_a} w(a,i)(r_{ij} - \overline{r}_i) \tag{2}$$

where m_a is the number of users, similar to user a, who have rated item j. The weight $w(a,i)$ expresses the similarity between user a and user i. k is a normalizing factor such that the absolute values of the weights sum to unity.

For a target user (in subsequent sections, user a denotes the target user.), the collaborative system provides a list of unseen items descending ordered by predicted values.

We have known that both the total number of items and the total number of users are very large. And generally each user will only have rated a small percentage of the total number of items. So $R_{m \times n}$ in equation (1) is high dimensional sparse. The weakness of the original matrix led us to explore alternate methods for low dimensional representation.

3.2 Dimension Reduction Based on Semantic Classification

Domain-specific classification is used to organize web resources. Each item can be assigned to one or more classifications. For example, in the domain of movies, every movie can be classified according to the attribute "genre" of each item (the values of genre include Action, Adventure, Drama, and so on). In the domain of books, an attribute "category" of items is used to classify books. We use domain-specific classification information to partition original user-item rating matrix for dimensionality reduction. Each item belongs to one or more classes. Each class has at least one item. The following presents main process of dimensionality reduction method.

Step 1: Reduce the items in original matrix $R_{m \times n}$.

For each class p, find all the items belong to p from $R_{m \times n}$, $R_{m \times n}$ is converted to $R^p_{m \times n_p}$:

$$R^p_{m \times n_p} = \begin{bmatrix} r_{11} & \cdots & r_{1j} & \cdots & r_{1n_p} \\ \vdots & \ddots & \vdots & \vdots & \vdots \\ r_{a1} & \cdots & r_{aj} & \cdots & r_{an_p} \\ \vdots & \vdots & \vdots & \ddots & \vdots \\ r_{m1} & \cdots & r_{mj} & \cdots & r_{mn_p} \end{bmatrix}_{m \times n_p} \tag{3}$$

where n_p is the total number of items that belong to class p. Experiments show that $n_p \ll n$. Thus from $R_{m \times n}$ to $R^p_{m \times n_p}$ ($p = 1, \cdots, C$), the reduction of dimension "item" is successfully completed. If an item belongs to one more classes, it will be assigned to each of the classes it belongs to. For example, the genres of the movie "Toy Story" are "Animation", "Children's" and "Comedy". So the movie is assigned to each of the three genres.

Step 2: Reduce users in every $R_{m \times n_p}^{p}$.

In $R_{m \times n_p}^{p}$, if the number of items rated by user a is less than a threshold ϖ, the system considers user a is uninterested in this class of movies. So user a will be removed from the matrix $R_{m \times n_p}^{p}$. Then $R_{m \times n_p}^{p}$ is converted to $R_{m_p \times n_p}^{p}$:

$$R_{m_p \times n_p}^{p} = \begin{bmatrix} r_{11} & \cdots & r_{1j} & \cdots & r_{1n_p} \\ \vdots & \ddots & \vdots & \vdots & \vdots \\ r_{a1} & \cdots & r_{aj} & \cdots & r_{an_p} \\ \vdots & \vdots & \vdots & \ddots & \vdots \\ r_{m_p 1} & \cdots & r_{m_p j} & \cdots & r_{m_p n_p} \end{bmatrix}_{m_p \times n_p} \qquad (4)$$

where m_p is the total number of users within class p. Also our experiments show that $m_p \ll m$. We completed the reduction of dimension "user" from $R_{m \times n_p}^{p}$ to $R_{m_p \times n_p}^{p}$ for each p. In the next section we present how to generate recommendations with $R_{m_p \times n_p}^{p}$ ($p=1, \ldots, C$) for user a.

3.3 Modeling User Preferences

Traditional collaborative filtering algorithm represents user preferences as a list of <item, rating>. In semantic classification-based collaborative filtering, we model user preferences of a particular user (e.g. user a) as

$$U^{(a)} = \{R_{ap} | p = 1, \cdots, C\} \qquad (5)$$

where R_{ap} denotes the set of <item, score> whose items belong to class p, and they had been rated by user a. If user a rated no item within class p, R_{ap} will be NULL.

Example: Table 1 shows user 1 rated movies. Table 2 shows the relations between items and classes. In the cell of table 2, "1" represents the item belongs to the Class. From the data of table 1 and table 2, we model user preferences for user 1:

$U^{(1)} = \{R_{11}, R_{12}, R_{13}, R_{14}, R_{15}, R_{16}, R_{17}, R_{18}, R_{19}\}$, where

$R_{11} = \{<2,3>,<4,3>\}$, $R_{12} = \{<2,3>\}$, $R_{13} = \{<1,5>\}$, $R_{14} = \{<1,5>,<5,3>,<8,1>\}$,

$R_{15} = \{<1,5>,<4,3>,<8,1>\}$, $\qquad R_{16} = \{5,3\}$, $\qquad R_{17} = \phi$,

$R_{18} = \{<4,3>,<7,4>,<8,1>,<9,5>\}$, $R_{19} = \{<3,4>,<11,2>\}$.

3.4 Computing Similarity

There are many similarity measures including vector similarity, Pearson correlation coefficient, entropy-based uncertainty measure, and mean square difference to weight all users with respect to similarity with the target user. Researchers found that Pearson

Table 1. Examples of user rated items

User	Item	Rating
1	1	5
1	2	3
1	3	4
1	4	3
1	5	3
1	7	4
1	8	1
1	9	5
1	11	2

Table 2. Examples of item-class matrix

Item	Class								
	C1	C2	C3	C4	C5	C6	C7	C8	C9
1			1	1	1				
2	1	1							
3									1
4	1				1			1	
5			1			1			
6								1	
7			1	1				1	
8								1	
9						1			1
11	1				1			1	

correlation performs better than other similarity measures. In our algorithm, we use Pearson correlation coefficient to compute the similarity weight between user a and user i ($i=1,\cdots,m_p$) within class P ($p=1,\cdots,C$), i.e. $w^P(a,i)$:

$$w^P(a,i) = \frac{\sum_j (r_{aj} - \overline{r_{ap}})(r_{ij} - \overline{r_{ip}})}{\sqrt{\sum_j (r_{aj} - \overline{r_{ap}})^2 \sum_j (r_{ij} - \overline{r_{ip}})^2}} \tag{6}$$

where $j \in I_{ap} \cap I_{ip}$, I_{ap} is the set of items rated by user a within class P, I_{ip} is the set of items rated by user i within class P. Thus the summations over j are over the items for which both user a and user i have given ratings within class P. $\overline{r_{ap}}$ is the average rating of user a for items, which belong to class P.

$$\overline{r_{ap}} = \frac{1}{|I_{ap}|} \sum_{j \in I_{ap}} r_{aj} \tag{7}$$

If the total number of items rated by both user a and user i is under a threshold, it is considered that the two users have no common preferences for class p, i.e. $w^P(a,i) = 0$.

4   Recommendations Generation

4.1   Predicting Unseen Items

After the similarity weight is computed between user a and each user in class P, the prediction of unseen items can be computed using the following equation:

$$p_{aj}^P = \overline{r_{ap}} + k \sum_{i=1}^{m_{ap}} w^P(a,i)(r_{ij} - \overline{r_{ip}}) \tag{8}$$

where p_{aj}^p represents the prediction for the target user a for item j within class p. m_{ap} is the number of users, similar to user a, which have rated item j in class p. Synthetically, we compute the prediction for user a for unseen item j whenever the item belongs to more than one class or belongs to just one class.

If item j belongs to more than one class, then for user a the prediction for item j, i.e. p_{aj} is assigned to the maximum value among the classes that item j belongs to.

$$p_{aj} = \max_{j \in p} p_{aj}^p \qquad (9)$$

If item j belongs to just one class, then $p_{aj} = p_{aj}^p$.

4.2 Recommendation for a List of Items

In semantic classification-based collaborative filtering, unseen items for user a are descending sorted by the predicted value. Then the algorithm provides a list of highest predicted items for user a. This is commonly known as top-N recommendation. If user a prefers C semantic classes, for each class p about $\lceil N/C \rceil$ highest predicted items are selected for the recommendation. N is the total number of recommended items for user a, and C is the total number of semantic classes in user preference model of user a.

5 Experiments and Evaluation

In order to evaluate the quality of semantic classification-based collaborative filtering, we experimentally evaluate the performance among semantic classification-based collaborative filtering (SCF), traditional collaborative filtering (Tradtional CF)and collaborative filtering using K-means clustering (KCF). Some researchers proposed that data mining techniques could be applied to collaborative filtering systems. One of popular clustering algorithms used in model-based collaborative filtering is K-means. We use K-means clustering algorithm to partition the train data subset into K clusters. Then predict each item in test data subset, determining which cluster the item belongs to. In the algorithm, we assign the value K equals to C, i.e. the number of clusters equals to the number of genres.

5.1 Data Set

In our experiments, we use MovieLens data set contains 100,000 ratings of 1682 movies rated by 943 users. There are 18 genres in the data set, and each movie belongs to one or more genres. The data set is divided into 80% training set and 20% test set. In our approach, "genre" is used as semantic classification to partition user-item matrix of the training set.

5.2 Evaluation Metrics

The effectiveness of collaborative filtering algorithms has traditionally been measured by statistical accuracy and decision-support accuracy metrics. Statistical accuracy metrics evaluate the accuracy of a system by comparing the numerical recommendation scores against the actual user ratings for the user-item pairs in the test dataset. We use Mean Absolute Error (MAE), a statistical accuracy metrics, to report prediction experiments for it is most commonly used and easy to understand

$$MAE = \frac{\sum_{j=1}^{N}\left|p_j - r_j\right|}{N} \tag{10}$$

where $\{p_1, \dots, p_N\}$ are predicted values in the target set, and $\{r_1, \dots, r_N\}$ are all the real values for the same items. N is the total number of items in the target set.

Decision support accuracy metrics evaluate how effective a prediction engine is at helping a user select high-quality items from the set of all items. The ROC (receiver operating characteristic) sensitivity is an example of decision-support accuracy metrics. The metric indicates how effectively the system can steer users towards highly-rated items and away from low-rated ones. We use ROC-4 measure as the evaluation metric.

$$ROC - 4 = \sum_{j=1}^{N} w_j \Big/ \sum_{j=1}^{N} u_j \tag{11}$$

where
$$w_j = \begin{cases} 1 & r_j \geq 4 \text{ and } p_j \geq 4 \\ 0 & otherwise \end{cases}, \quad u_j = \begin{cases} 1 & p_j \geq 4 \\ 0 & otherwise \end{cases}$$

5.3 Experimental Results

5.3.1 Dimensionality reduction

Table 3 shows the number of items that each genre has. From Table 3 and Fig. 1, we can see the largest number is 725, 43.1 percent of original 1682 items, and the minimum number is 22, 1.3 percent of the original. The total number of items grouped by genres are larger than 1682 because many items belong to more than one genre.

Table 3. Number of items in semantic classification

Genre	Number of Items	Genre	Number of Items
1	251	10	24
2	135	11	92
3	42	12	56
4	122	13	61
5	505	14	247
6	109	15	101
7	50	16	251
8	725	17	71
9	22	18	27

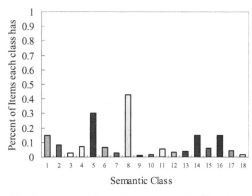

Fig. 1. Percent of items in semantic classification

As the same, users are partitioned into semantic classes. Table 4 shows the number of users that each genre has, and Fig. 2 shows the percent of users. We can see the difference of the number in each genre is very large. Some genres such as genre 8 interest almost all the users, and few people like genre 9. The percent averages at 0.38.

Table 4. Number of users in semantic classification

Genre	Number of Users	Genre	Number of Users
1	828	10	88
2	470	11	75
3	39	12	74
4	125	13	141
5	873	14	836
6	216	15	481
7	30	16	832
8	933	17	314
9	25	18	38

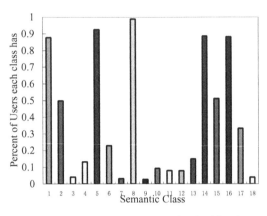

Fig. 2. Percent of users in semantic classification

5.3.2 Statistical Accuracy

When a user rates an item, the number of nearest neighbors affects the MAE for the algorithm. Figure 3 shows the dependence of the MAE on the number of nearest neighbors. SCF algorithm performs better than both traditional CF and KCF. On average, the SCF algorithm performs 7.3% better than the traditional CF algorithm, and it performs little difference between traditional CF algorithm and KCF algorithm.

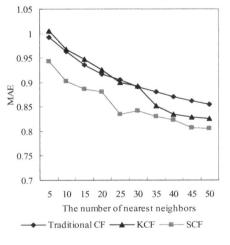

Fig. 3. MAE results

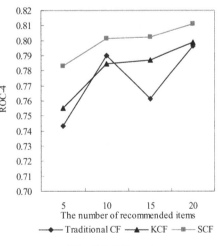

Fig. 4. ROC results

6 Conclusions and Future Work

In this paper, according to the attribute of items, we proposed a collaborative filtering based on semantic classification. Experiments shows it works well in dimensionality reduction and performs better quality of recommendations in the domain of movie recommendations. In the paper we suppose one genre is independent of another. But in many domain-specific classifications, the classes are correlative. For example, in the domain of book recommendations, the attribute "category" of books can be used as semantic classification. Different from movies, the category has multi-levels. Categories can be subdivided into sub-categories, and sub-categories are subdivided into sub-subcategories, and so on. We will apply other techniques such as association analysis to expand semantic classification-based collaborative filtering in the future research.

Acknowledgements

This work is supported by the National Natural Science Foundation of China under Grant No. 60473078 and National Science and Technology Support Plan of China (No. 2006BAH02A12). We also thank GroupLens research group for allowing us to use the data set they had developed.

References

1. Goldberg, D., Nichols, D., Oki, B., Terru, D.: Using collaborative filtering to weave an information tapestry. Communications of the ACM 35(12), 61–70 (1992)
2. Resnick, P., Iacovou, N., Sushak, M., et al.: GroupLens: An open architecture for collaborative filtering of netnews. In: Proceedings of the 1994 Computer Supported Cooperative Work Conference, ACM Press, New York (1994)
3. Sarwar, B.M., Konstan, J.A., Borchers, A., et al.: Using filtering agents to improve prediction quality in the grouplens research collaborative filtering system. In: Proceedings of 1998 Conference on Computer Supported Collaborative Work (November 1998)
4. Rashid, M., Lam, K., Karypis, G., Riedl, J.: ClustKNN: A Highly Scalable Hybrid Model- & Memory-Based CF Algorithm. In: proceedings of WEBKDD 2006, Philadelphia, Pennsylvania, USA (August 20, 2006)
5. Xue, G.R., Lin, C.X., Yang, Q., et al.: Scalable collaborative filtering using cluster-based smoothing. In: Proceedings of SIGIR 2005, Salvador, Brazil, August 15-19, pp. 114–121 (2005)
6. Aggarwal, C.C., Yu, P.S.: Data mining techniques for personalization. IEEE Bulletin of the Technical Committee on Data Engineering - Special Issue on Database Technology in E-Commerce 23(1), 4–9 (2000)
7. Kohrs, A., Merialdo, B.: Clustering for collaborative filtering applications. In: Computational Intelligence for Modeling, Control & Automation (CIMCA 1999), Vienna, IOS Press, Amsterdam (1999)
8. Zhang, S., Wang, W.H., Ford, J., et al.: Using singular value decomposition approximation for collaborative filtering. In: Proceedings of the Seventh IEEE International Conference on E-Commerce Technology (CEC 2005), pp. 257–264 (2005)

9. Popescul, A., Ungar, L.H., Pennock, D.M., et al.: Probabilistic models for unified collaborative and content-based recommendation in sparse-data environments. In: Proceedings of the Seventeenth Conference on Uncertainty in Artificial Intelligence (UAI-2001), pp. 437–444. Morgan Kaufmann, San Francisco (2001)
10. Basilico, J., Hofmann, T.: A joint framework for collaborative and content filtering. In: Proceedings of SIGIR, pp. 550–551 (2004)
11. Basilico, J., Hofmann, T.: Unifying collaborative and content-based filtering. In: Proceedings of the twenty-first international conference on Machine learning (ICML 2004) (2004)
12. Breese, J.S., Heckerman, D., Kadie, C.: Empirical analysis of predictive algorithm for collaborative filtering. In: Proceedings of the Fourteenth Conference on Uncertainty in Artificial Intelligence, Madison, WI, Morgan Kaufmann, San Francisco (1998)

Modeling and Learning User Profiles for Personalized Content Service

Heung-Nam Kim, Inay Ha, Seung-Hoon Lee, and Geun-Sik Jo

Intelligent E-Commerce Systems Laboratory,
Department of Computer Science & Information Engineering, Inha University
{nami,inay,shlee}@eslab.inha.ac.kr, gsjo@inha.ac.kr

Abstract. With the spread of the digital library and the web, users can obtain a wide variety of information, and also can access novel content. In this environment, finding useful information from a huge amount of available content becomes a time consuming process. In this paper, we focus on user modeling for personalization to recommend content relevant to user interests. We exploit the data mining techniques for identifying useful and meaningful patterns of users. Each user model, collectively called PTP (Personalized Term Pattern), is represented as both interest patterns and disinterest patterns. We present empirical experiments using *NSF research award* datasets to demonstrate our approach and evaluate performance compared with existing methods.

1 Introduction

Numerous technological developments related to the Internet and the World Wide Web provide anyone living in today's information society with accessing a variety of content and information on the web. Due to the nonstop growth of the internet information, users often face the challenges with huge amount of content, and need to waste plenty of time to find content relevant to their interest. Beside that, the advent of bolgs, DL (digital library), and RSS (Really Simple Syndication) generate millions of information overnight. As a result, such an information overload increases user's frustration to find out the content of their interest. Therefore, a user modeling for efficient personalization plays a significant role in modern information filtering system [7, 14].

One notable challenge in a user modeling is the ability to identify meaningful or useful patterns for users. In content-based personalization it is important to recognize meaningful patterns for representing items or contents [11]. For example, when content contains 'apple Macintosh computer', the semantic of 'apple' is discriminated from those of 'apple' in 'apple pie'. Likewise, mouse in 'optical mouse' implies not an animal but an input device of computers. Therefore, our aim is to build a user model that supports the identification of useful patterns of users, and thus can be used for personalized recommendation services. In our research, we exploit a data mining technique for identifying important pattern of user's preferences. Considering the contents of user interest (positive) and disinterest (negative), we mine the frequent term patterns residing in the user's positive contents and negative contents. And each

D.H.-L. Goh et al. (Eds.): ICADL 2007, LNCS 4822, pp. 85–94, 2007.

user model, collectively called PTP (Personalized Term Pattern), is represented as both interest patterns and disinterest patterns, which will boost recommendations of contents related to user interests. We also take advantage of content-based filtering approach to recommend content that is very close to not negative term patterns but positive term patterns.

The subsequent sections of this paper are organized as follows: The next section contains our approach for modeling user preference. In section 3, we describe a content-based filtering for a personalized recommendation. A performance evaluation is presented in section 4 and related work is discussed in section 5. Finally, we conclude with a discussion and future directions.

2 Modeling User Profiles from Positive and Negative Examples

In this section, we describe our approach to modeling user preference, which is mined from the user's interest contents (positive contents) and disinterest contents (negative contents). The proposed method is divided into three main types of tasks: (a) Observing relevance feedback of a given user, (b) Modeling user preference from observed contents, and (c) Generating content recommendations for a given user. Fig. 1 provides a brief overview of the proposed approach.

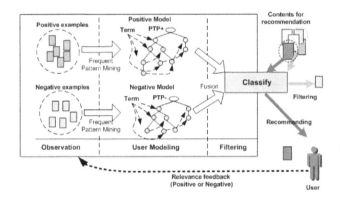

Fig. 1. Overview of the proposed method for personalized content recommendations

Since every user has different interests, feature selection for representing users' interests should be personalized and be performed individually for each user [7]. The first step in user modeling is the extraction of the terms from positive or negative contents that have been preprocessed by removing stop words and stemming words [12]. After extracting terms, each content C_j is represented as a vector of attribute-value pairs; $\vec{C_j} = (w_{1,j}, w_{2,j}, \ldots, w_{m,j})$, where $w_{i,j}$ is the weight of term T_i in C_j, which is computed by static TF-IDF term-weighting scheme [1] and defined as follows:

$$w_{i,j} = \frac{f_{i,j}}{\max_l f_{l,j}} \times \log \frac{n}{n_i}$$

where $f_{i,j}$ is the frequency of occurrence of term T_i in content C_j, n is the total number of contents in the collections, and n_i is the number of contents in which term T_i occurs.

Secondly, the frequent patterns that occur at least as frequently as a predetermined minimum support (*min_sup*), i.e., *PS* > *min_sup*, are mined from the positive examples and the negative examples, respectively [9]. If the pattern support of pattern P_k, that is composed of at least l different terms ($l \geq 2$), satisfies a pre-specified minimum support threshold, then pattern P_k is a frequent term pattern. Therefore, two types of a set of terms (pattern) for a user can be discovered; a set of positive frequent patterns, written as F_u^+, and a set of negative frequent patterns, written as F_u^-, $F_u^+ \cup F_u^- = F_u$. In our research, a content of a user corresponds to a transaction and terms extracted from the content are items in transaction.

Definition 1 (Pattern Support, PS). Let $T = \{T_1, T_2, \ldots, T_m\}$ be a set of terms, I_u a set of positive contents of user u where each content C^+ is a set of terms such that $C^+ \subseteq T$, and N_u a set of negative contents of user u where each content C^- is a set of terms such that $C^- \subseteq T$. Let pattern P_k be a set of terms. A content is said to contain a pattern if and only if $P_k \subseteq C^+$ or $P_k \subseteq C^-$. *Pattern support* for pattern P_k, $PS(P_k)$, in I_u or N_u is the ratio of contents in I_u or N_u that contain pattern P_k.

Definition 2 (Personalized Term Pattern, PTP). *Personalized term patterns* for user u, PTP_u, is defined as a set of frequent term patterns whose *pattern weights* are greater than a threshold value θ, i.e., $PTP_u \subseteq F_u$ and $PW(P_k) > \theta$. PTP_u can be divided into two groups, a set of positive patterns, written as PTP_u^+, and a set of negative patterns, written as PTP_u^-, such that $PTP_u^+ \subseteq F_u^+$, $PTP_u^- \subseteq F_u^-$, and $PTP_u^+ \cup PTP_u^- = PTP_u$.

Definition 3 (Pattern Weight, PW). Let $T(P_k) = \{T_1, T_2, \ldots, T_n\}$ be a set of terms contained in pattern P_k such that $P_k \in F_u$. *Pattern weight* of P_k, denoted as $PW(P_k)$, indicates the importance of each term in representing the pattern and is computed as follows:

$$PW(P_k) = \frac{1}{|T(P_k)|} \cdot \sum_{i \in T(P_k)} \left(\frac{1}{|E_u(i)|} \times \sum_{j \in E_u(i)} w_{i,j} \right)$$

where $E_u(i)$ is a set of positive (or negative) contents containing term T_i for user u and $w_{i,j}$ is the term weight of term T_i in content C_j.

Finally, we remove the patterns, which contain unnecessary terms, from F_u^+ and F_u^- of user u and model the user preference based on those patterns. A formal description of a model for user u, $\mathbf{M_u}$, follows: $\mathbf{M_u} = \langle (\mathbf{PTP_u^+}, \mathbf{PTP_u^-}) \rangle$, where PTP_u^+ models the interest patterns and PTP_u^- models the disinterest patterns (Definition 2). In other words, the PTP_u^+ is a set of personalized term patterns mined from positive contents of user u whereas the PTP^- is a set of personalized term patterns mined from negative contents of user u.

To save memory space and explore relationships of terms, the model is stored in a prefix tree structure, called Personalized Term Pattern tree [12]. For example, if four positive patterns are found after mining content of interest for user u as shown in Table 1(left), a tree structure of a model for user u is then constructed as follows.

Table 1. After mining positive and negative content of user u, four positive personalized term patterns (left table) and four negative personalized term patterns are found (right table)

Pattern	PTP$^+$	PS	Pattern	PTP$^-$	PS
P_1^+	$\{T_1, T_2, T_3\}$	0.56	P_1^-	$\{T_5, T_6, T_7\}$	0.52
P_2^+	$\{T_1, T_2, T_3, T_4\}$	0.51	P_2^-	$\{T_4, T_5, T_6\}$	0.41
P_3^+	$\{T_1, T_2, T_5\}$	0.47	P_3^-	$\{T_5, T_7\}$	0.37
P_4^+	$\{T_2, T_3, T_4\}$	0.32	P_4^-	$\{T_6, T_8\}$	0.31

All terms are stored in header table and sorted according to descending order of their frequency. First, create the root of the tree, labeled with "null". For the first term pattern, $\{T_1, T_2, T_3\}$ is insert into the tree as a path from root node where T_2 is linked as child of the root, T_1 is linked to T_2, and T_3 is linked to T_1. And *PS* and *length* of the pattern $(PS(P_1^+)=0.56, |T(P_1^+)| = 3)$ are then attached to the last node T_3. For the second pattern, since its term pattern, $\{T_1, T_2, T_3, T_4\}$, shares common prefix $\{T_2, T_1, T_3\}$ with the existing path for the first term pattern, a new node T_4 is created and linked as a child of node T_3. Thereafter, $PS(P_2^+)$ and $|T(P_2^+)|$ are attached to the last node T_4. The third, and fourth patterns are inserted in a manner similar to the first and second patterns. To facilitate tree traversal, header table is built in which each term points to its occurrence in the tree via a *Node-link*. Nodes with the same *term-name* are linked in sequence via such *node-links*. Finally, PTP$_u^+$ for user u is constructed as shown in Fig. 2(left). Likewise, PTP$_u^-$ in Table 1(right) is constructed as shown in Fig. 2(right).

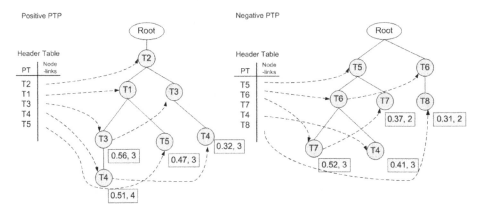

Fig. 2. A tree structure of M_u for personalized term patterns in Table 1

3   Content-Based Filtering for Personalized Service

In this paper, the filtering approach considers matched patterns between the new contents and **PTP** for a user. In addition, we consider both positive feedback and negative feedback for judging whether content are relevant or irrelevant to the user.

Definition 4 (Matched Pattern). Let $T(P_k)$ be a set of terms contained in pattern P_k such that $P_k \in PTP_u$. If all terms in contained P_k appear new content C_n, $T(P_k) \subseteq C_n$, then pattern P_k is deemed a *matched pattern* between P_k and content C_n.

The positive similarity, a measure of how positively the content is relevant to the user, between new content C_n and positive PTP of user u is defined in equation (1).

$$pos(u,C_n) = \frac{|MP^+|}{|PTP_u^+|} \times \sum_{P_k \in MP^+} |T(P_k)| \cdot PS^+(P_k) \tag{1}$$

where PTP_u^+ is a set of positive personalized patterns for user u, $T(P_k)$ is a set of terms contained in pattern P_k, and MP^+ is a set of matched patterns between PTP_u^+ and content C_n. $PS^+(P_k)$ refers to the positive support value of matched pattern P_k. The higher the similarity value, the more relevant the content is to the user.

Likewise, the negative similarity, a measure of how the content is irrelevant to the user, between content C_n and negative PTP of user u is defined in equation (2). However, as opposed to the positive similarity, the content which has the highest value of the negative similarity is the most irrelevant to the user.

$$neg(u,C_n) = \frac{|MP^-|}{|PTP_u^-|} \times \sum_{P_k \in MP^-} |T(P_k)| \cdot PS^-(P_k) \tag{2}$$

The main concept of the similarity schemes dictates that specific patterns (positive or negative) with numerous occurrences in user preference (positive or negative) present a greater contribution with regard to similarity than general patterns with a smaller number of occurrences. Finally, the combined similarity between user u and content C_n is obtained as the following.

$$Sim(u,C_n) = \alpha \cdot pos(u,C_n) + (\alpha - 1) \cdot neg(u,C_n) \tag{3}$$

where α is a parameter in [0,1] which specifies for adjusting the relative weighting between the positive similarity and the negative similarity. If $\alpha=0$ then $Sim(u,C_n)$ just takes $neg(u,C_n)$ into account whereas if $\alpha=1$ then $Sim(u,C_n)$ just coincides with $pos(u,C_n)$. Given two contents C_i and C_j, content C_i is of more interest to user u than content C_j if and only if a similarity between user u and content C_i is higher than that of content C_j, $sim(u,C_i) > sim(u, C_j)$.

Once the scores between user u and new contents are computed, the contents are sorted in order of descending score value. Thereafter, a set of N rank contents that have obtained higher similarity values are identified for user u, and then those contents are recommended to user u (Top-N recommendation) [8].

4   Experimental Evaluation

In this section, we present the quality evaluation of the proposed approach with experimental details. The experiment data is taken from NSF (National Science Foundation) research award abstracts [16]. The original data set contains 129,000 abstracts describing NSF awards for basic research from 1900 to 2003. However, the data is too large to be used for experiments, and thus we selected the award abstracts from

2000 to 2003, i.e. the selected data set contained 30,384 abstracts and 22,236 distinct terms as obtained from the abstracts. 10 users were participated in the experiments by providing a positive and negative feedback according to their interests from the total contents (30,384 contents). Whenever they found the content related to their preferences, they added that content into their positive list or negative list. To evaluate the performance, we divided the collected positive contents of the users into *a test set* with exactly 100 contents per user and *a training set* with the remaining contents. A model $\mathbf{M}_u$ of each user was constructed using only the *training set*. Thereafter, we computed the similarity scores of contents except the content list of a given user in the training set and subsequently identified a set of N rank contents that obtained the higher scores.

The performance was measured by looking at the number of *hits*, and their *ranking* within the *top*-N contents that were recommended by a particular scheme. We computed the quality measures that are defined as follows.

Hit Rate (HR). In the context of *top*-N recommendations, *hit-rate*, a measure of how often a list of recommendations contains contents that the user is actually interested in, was used for the evaluation metric [5, 8]. The *hit-rate* for user u is defined as:

$$HR(u) = \frac{|Test_u \cap TopN_u|}{|Test_u|}$$

where $Test_u$ is the content list of user u in the test data and $TopN_u$ is a *top*-N recommended content list for user u. Finally, the overall HR of the *top*-N recommendation for all users is computed by averaging these personal HR in test data.

Reciprocal Hit Rank (RHR). One limitation of the *hit-rate* measure is that it treats all hits equally regardless of the ranking of recommended contents. In other words, content that is recommended with a top ranking is treated equally with content that is recommended with an Nth ranking [8]. To address this limitation, therefore, we adopted *the reciprocal hit-rank* metric described in [8]. The *reciprocal hit-rank* for user u is defined as:

$$RHR(u) = \sum_{C_n \in (Test_u \cap TopN_u)} \frac{1}{rank(C_n)}$$

where $rank(C_n)$ refers to a recommended ranking of content C_n within the *hit set* of user u. That is, hit contents that appear earlier in the *top*-N list are given more weight than hit contents that occur later in the list. Finally, the overall RHR for all users is computed by averaging the personal $RHR(u)$ in test data. The higher the RHR, the more accurately the algorithm recommends contents.

4.1 Experimental Results

The performance evaluation is divided into two dimensions. In the first experiment, we determine the *minimum support* that controls the size of $\mathbf{M}_u$ and the parameter α that blends two similarity (positive and negative) measures. The second experiment presents successful performance of our method for a content relevant personalized recommendation in comparison with other approaches. In order to compare the performance of the proposed scheme, a probabilistic learning algorithm, which applies a *naïve Bayesian classifier* (denoted as *NB*) [3, 4], and a TF-IDF vector-based algorithm, which is employed in the *Webmate* system (denoted as *Webmate*) [2], were implemented. To make the comparison fair, both of the algorithms were designed to learn users' preferences from positive examples and negative examples. Because *Webmate* was not originally designed to learn from negative examples, the learning of negative examples is performed by subtracting the feature vector of a learned content from the profile [5]. For the content filtering process, in the case of *NB*, the probability of new content belonging to the "interest (positive)" class of a user divided by the probability of the content belonging to the "no interest" (negative) class is used. In the case of *Webmate*, contents are ranked using the calculated cosine similarity between contents and a user profile. The *top*-N recommendation of our strategy was then evaluated in comparison with the benchmark algorithms.

4.1.1 Experiments with α Value

First of all, we considered about two significant factors affecting the quality of our algorithm, which are minimum support (*min_sup*) and α value. A high *min_sup* discards more patterns, and thus remaining term patterns may not be sufficient to represent user preference. In contrast, a low *min_sup* may include many noise patterns. The other factor, parameter α, determines which will be given more weight, a positive similarity or a negative one. For our main comparisons with existing works, we empirically determined these two values which showed the most reasonable performance in both evaluation metrics, HR and RHR. *min_sup* values used for mining personalized term patterns were 5%, 8%, 10%, and 20%. In addition, we varied α value from 0 to 1 in an increment of 0.2. In this experiment we set the value of $N=100$ as the number of recommended contents and $\theta = 0.5$ as the pattern weight threshold.

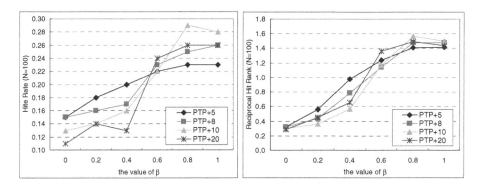

Fig. 3. Hit rate (HR) and reciprocal hit rank (RHR) according to variation of α value

Fig. 3 presents a variation of HR (left) and RHR (right), by changing the α value. We describe the lines as PTP + k where k means *min_sup* of 5%, 8%, 10%, and 20%. It can be observed from the graph that the parameter α affected the performance and overall performance was improved with the growth of α except for a few cases. Generally, with respect to HR, a low *min_sup* levels (i.e., 5%, 8%) showed better quality than a high *min_sup* levels (i.e., 10%, 20%) when α was close to 0 (negative similarity weighted).

On the contrary, high *min_sup* levels performed better (positive similarity weighted) when α was close to 1. When we compare the results of RHR, the four cases demonstrate similar types of charts and elevate RHR as the α value increases from 0.0 to 0.8; beyond this point, RHR deteriorates slightly. For example, when α is set to 0.8, *PTP+10* yields a RTR of 1.57, which is the best value, whereas it gives a RTR of 1.50 in the case of α=1. Roughly speaking, considering the positive similarity rather than the negative one between a user and contents might reflect user's preference better. We conclude from this experiment that the fusion of the positive and negative similarity for a content filtering is effective in terms of improving the performance, compared to the positive similarity or the negative similarity only.

4.1.2 Comparison with Other Methods

To experimentally compare the performance of our algorithm, we calculated the hit rate (HR) and the reciprocal hit rank (RHR) achieved by *PTP*, *Webmate*, and *NB* when the number of recommended contents N was 100 and 200. According to the results of the prior experiments in section 4.1.1, *min_sup* and α value were set to 10% and 0.8, respectively.

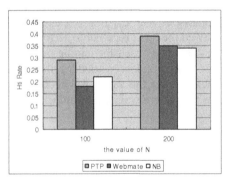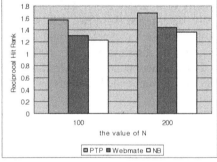

Fig. 4. Comparison of the hit rate (HR) and the reciprocal hit rank (RHR)

Fig. 4 depicts the HR (left) and the RHR (right). In general, with the growth of recommended contents N, HR, and RHR tend increase. However, overall HR performance of all three methods did not meet with good results throughout our experiments due to the huge size of total data set. Even though data set used for learning user preferences (a training set) was excluded from total data set, the number of recommended contents was less than 0.01% of total contents. Although HR for all algorithms is unsatisfactorily low, *PTP* provided considerably improved HR on all occasions compared to the benchmark algorithms. Similar conclusions can be made by looking at the RHR results as well. For example, when N is 100, *PTP* achieves 19% and 27% improvement in terms of RHR, compared to *Webmate* and *NB*, respectively. These results show that our algorithm can recommend contents at higher ranks for each user as well as it can recommend more accurate contents than the other two methods.

5   Related Work

This section briefly explains previous studies related to user modeling and personalized recommendation. Two approaches for recommender systems have been discussed in the literature, *i.e.*, a content-based filtering approach and a collaborative filtering approach [15]. Our research mainly focuses on the content-based filtering for personalized recommendations. Content-based approaches analyze information object of a user, usually textual contents, and build a model of personal preferences based on the features of the object. *Webmate* tracks user interests from his positive information only (i.e., documents that the user is interested in) and exploits the vector space model using TF-IDF method [2]. A classification approach has been explored to recommend articles relevant user profile, such as *NewsDude* and *ELFI* [3, 4]. In *NewsDude*, two types of the user interests are used: short-term interests and long-term interests. To avoid recommendations of very similar documents, short-term profile is used. For the long-term interests of a user, the probabilies of a document are calculated using Naïve Bayes to classify a document as interesting or not interesting. Instead of learning from users' explicit information, *PVA* learns a user profile implicitly without user intervention, such as relevance feedback, and represents it as keyword vector in the form of a hierarchical category structure [6] as similar to *Alipes* [5]. In *Newsjunkie*, novelty-analysis algorithm is employed to present novel information for users by identifying novelty of articles in the contexts of articles they have already reviewed [10]. Likewise our research, Lops et al. exploits user profiles consisting of two parts, the positive part for modeling user interests and the negative part for user disinterests [15]. Although these systems have their own method to build a user model, they do not deliberate on concurrence of terms and offer the ability to identify meaningful or useful patterns, which are important features for representing articles or contents [11].

6   Discussion and Conclusions

In the present work, we have presented a new method for modeling and learning user profiles that discriminates interesting information from uninteresting data. The major advantage of our proposed learning method is that it supports the identification of useful patterns of each user. In addition, mining from the contents of user interest (positive) and disinterest (negative), user models could identify disinterest patterns as well as interest patterns. In order to evaluate our work, we compare our experimental results with those of probabilistic learning model and vector space model. The experimental results demonstrate that the proposed method offers significant advantages in terms of improving recommendation quality.

Nevertheless, there remain some research questions. It remains to be evaluated, how different a threshold of pattern weight, θ, affects the learning result. Another research question is how to consider the changes of user interests efficiently. Once user models are built, it is difficult to reflect a new user feedback. Incremental learning is one of the interesting issues that we plan to consider for addressing this problem in the future.

References

1. Salton, G., Buckley, C.: Term Weighting Approaches in Automatic Text Retrieval. Information Processing and Management 24, 513–523 (1988)
2. Chen, L., Sycara, K.: WebMate: Personal Agent for Browsing and Searching. In: Proc. of the 2nd Int. Conf. on Autonomous Agents and Multi Agent Systems, pp. 132–139 (1998)
3. Billsus, D., Pazzani, M.J.: A hybrid user model for News story classification. In: Proc. of the 7th Int. Conf. on User Modeling, pp. 99–108 (1999)
4. Schwab, I., Pohl, W., Koychev, I.: Learning to Recommend from Positive Evidence. In: Proc. of Int. Conf. on Intelligent User Interfaces (2000)
5. Widyantoro, D.H., Ioerger, T., Yen, J.: Learning User Interest Dynamics with a Three-Descriptor Representation. Journal of the American Society for Information Science and Technology 52, 212–225 (2001)
6. Chen, C.C., Chen, M.C., Sun, Y.: PVA: A Self-Adaptive Personal View Agent. Journal of Intelligent Information Systems 18, 173–194 (2002)
7. Eirinaki, M., Vazirgiannis, M.: Web Mining for Web Personalization. ACM Transactions on Internet Technology 3, 1–27 (2003)
8. Deshpande, M., Karypis, G.: Item-based Top-N Recommendation Algorithms. ACM Transactions on Information Systems 22, 143–177 (2004)
9. Han, J., Pei, J., Yin, Y.: Mining Frequent Patterns without Candidate Generation: A Frequent-Pattern Tree Approach. Data Mining and Knowledge Discovery 8, 53–87 (2004)
10. Gabrilovich, E., Dumais, S., Horvitz, E.: Newsjunkie: Providing Personalized News-feeds via Analysis of Information Novelty. In: proc. of the 13th Int. Conf. on World Wide Web, pp. 482–490 (2004)
11. Chung, S., McLeod, D.: Dynamic Pattern Mining: An Incremental Data Clustering Approach. In: Spaccapietra, S., Bertino, E., Jajodia, S., King, R., McLeod, D., Orlowska, M.E., Strous, L. (eds.) Journal on Data Semantics II. LNCS, vol. 3360, pp. 85–112. Springer, Heidelberg (2005)
12. Kim, H.N., Kim, H.J., Jo, G.S.: Content-based Document Recommendation in collaborative Peer-to-Peer Network. In: Jin, H., Pan, Y., Xiao, N., Sun, J. (eds.) GCC 2004. LNCS, vol. 3251, pp. 575–582. Springer, Heidelberg (2004)
13. Flesca, S., Greco, S., Tagarelli, A., Zumpano, E.: Mining User Preferences, Page Content and Usage to Personalize Website Navigation. World Wide Web: Internet and Web Information Systems 8, 317–345 (2005)
14. Das, A., Datar, M., Garg, A., Rajaram, S.: Google News Personalization: Scalable Online Collaborative Filtering. In: Proc. of the 16th Int. Conf. on World Wide Web, pp. 271–280 (2007)
15. Lops, P., Degemmis, M., Semeraro, G.: Improving Social Filtering Techniques Through WordNet-Based User Profiles. In: Proc. of the 11th Int. Conf. on User Modeling, pp. 268–277 (2007)
16. Pazzani, M.J., Meyers, A.: NSF Research Awards Abstracts 1990-2003, http://kdd.ics.uci.edu/databases/nsfabs/nsfawards.html

Ontology-Based Fuzzy Retrieval for Digital Library

Tho Thanh Quan[1], Siu Cheung Hui[2], and Tru Hoang Cao[1]

[1] Faculty of Computer Science and Engineering
Hochiminh City University of Technology
Hochiminh City, Vietnam
qttho@cse.hcmut.edu.vn
[2] School of Computer Engineering
Nanyang Technological University
Singapore

Abstract. With the recent advancement of the Semantic Web, researchers are now considering developing ontology-based digital librarires for the sake of efficient information sharing, exchanging and retrieval. In addition, fuzzy queries have been also introduced to help readers to specify their queries more precisely when searching information in digital librarires. In this paper, we first propose an architecture that enables multiple digital libraries to collaborate in the Semantic Web environment. Then we discuss using fuzzy ontology to represent uncertain information in digital libraries and fuzzy queries for retrieving information from fuzzy ontology. An illustrative system is then developed for experiment purpose. Performance of our system is also evaluated and analyzed.

1 Introduction

Digital library is an organized repository of recorded knowledge which can be accessed in a digital and networked environment [1]. With the recent advancement of Semantic Web [2], there is much research considering developing digital library systems on the Semantic Web enviroment [3]. In addition, to improve the accuracy of information retrieval, some digital libraries also adopt fuzzy-based retrieval techniques [5]. In this paper, we propose an ontology-based technique for fuzzy document retrieval on digital libraries. The rest of this paper is organized as follows. Section 2 discusses a proposed Semantic Web-based architecture for exchanging and retrieving information cross multiple digital libraries. Section 3 presents a formal definitions of fuzzy ontology and fuzzy query. Section 4 provides performance evaluation of a experimental system. Finally, Section 5 concludes the paper.

2 Ontology-Based Digital Libraries on the Semantic Web

In this section, we propose an architecture supporting exchanging and retrieving information from multiple ontology-based digital libraries on the Semantic Web environment, given in Figure 1. In our architecture, the scholarly knowledge is not merely stored in a single scholarly database as typically designed in classic digital

D.H.-L. Goh et al. (Eds.): ICADL 2007, LNCS 4822, pp. 95–98, 2007.

libraries. Instead, the scholarly knowledge will be represented using ontological formalism as Scholarly Ontology. It enables scholarly knowledge stored in various Scholarly Ontologies can be shared among multiple digital libraries.

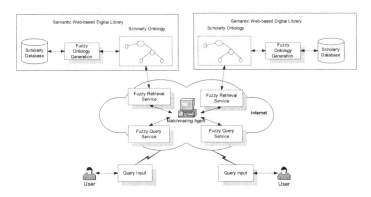

Fig. 1. Ontological-based Digital Libraries on the Semantic Web

3   Fuzzy Ontology and Fuzzy Query

The Semantic Web adopts ontology as standard for knowledge representation. Adopting from [7] , we formally define fuzzy ontology and fuzzy query as follows:

Definition 1 (Fuzzy Ontology). A *fuzzy ontology* F_O consists of 4 elements (C, A^C, R, X), where C represents a set of concepts; A^C represents a collection of attributes sets, one for each concept; $R = (R_T, R_N)$ represents a set of relationships, which consists of 2 elements: R_N is a set of *non-taxonomy relationships* and R_T is a set of *taxonomy* relationships. Each attribute value of an object or relationship instance is associated with a fuzzy *membership value* between [0,1] implying the uncertainty degree of this attribute value or relationship. X is a set of axioms.

Definition 2 (Fuzzy Query). A *fuzzy query* on a fuzzy ontology (C, A^C, R, X) is the fuzzy set $Q_f = \varphi(A')$ where $A' \subseteq A^C$.

To measure the relevance between an ontogical class and a query, we consider both of them as two fuzzy sets. Then we calculate the similarities between the fuzzy sets using fuzzy logic [8].

4   Performance Evaluation

In this section, we introduce an experimental system that makes use of Scholarly Ontology to support fuzzy scholarly retrieval. The Scholarly Ontology is generated from a citation database built on a set of 1400 scientific documents downloaded from the Institute for Scientific Information's (ISI) website[1], as depicted in Figure 2.

---

[1] http://www.isinet.com

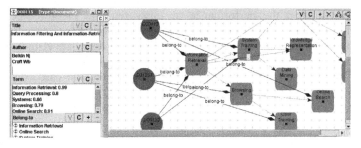

Fig. 2. Browsing documents in the Scholarly Ontology

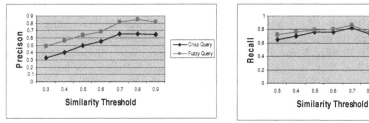

Fig. 3. Fuzzy retrieval result

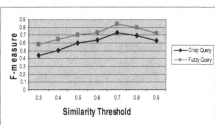

Fig. 4. (a) Performance evaluation based on precision. (b) Performance evaluation based on recall. (c) Performance evaluation based on F-measure.

We have evaluated the performance of fuzzy retrieval in our system. Fuzzy queries are formed based on an interface developed to allow user to specify the fuzzy membership of each query term. In addition, the retrieval resuls are also displayed the similarities between the queries and the retrieved documents as shown in Figure 3.

Figures 4(a), 4(b) and 4(c) give the performance evaluation based on precision, recall and F-measure on both the crisp queries and fuzzy queries. Similar to other keyword-based retrieval systems, crisp queries can obtain quite good performance in terms of recall. However, the precision on crisp queries is not that good. On the other hand, fuzzy queries have achieved better recall, but the improvement is not significant in comparison with crisp queries. However, the precision on fuzzy queries has achieved quite good improvement as compared to crisp queries. Finally, the F-measure on fuzzy queries is also much better than that of crisp queries as well.

5 Conclusion

This paper has proposed two extensions for the current generation of digital librarires. First, we propose to use ontology to represent scholarly information in digital libraries, thus making the libraries enable to share and exchange knowledge in the Semantic Web environment. Second, fuzzy theory is employed to process uncertain scholarly information as the forms of fuzzy ontology and fuzzy queries. A general architecture of digital libraries in the Semantic Web environment has been presented and an experimental system has also been developed to verify our ideas and techniques.

References

[1] Saracevic, T., Dalbello, M.: A Survey of Digital Library Education. Proceedings of the American Society for Information Science and Technology 38, 209–223 (2001)
[2] Berners-Lee T., Hendler J. and Lassila O., The Semantic Web, Scientific American, (2001), Available at: http://www.sciam.com/2001/0501issue/0501berners-lee.html
[3] Naing, M.-M, Lim, E.-P., Chiang, R.H.-L.: Core: A Search and Browsing Tool for Semantic Instances of Web Sites. In: Zhang, Y., Tanaka, K., Yu, J.X., Wang, S., Li, M. (eds.) APWeb 2005. LNCS, vol. 3399, Springer, Heidelberg (2005)
[4] Lim, E.P., Sun, A.: Web Mining - The Ontology Approach. In: Proceedings of The International Advanced Digital Library Conference (IADLC 2005), Nagoya, Japan (August 2005)
[5] IntraText Digital Library, available at: http://www.intratext.com/CERCA/Aiuto.htm
[6] Girill, T.R., Luk, C.H.: Fuzzy Matching as a Retrieval Enabling Technique for Digital Libraries. In: The Digital Revolution: Proceedings of the American Society for Information Science Mid-Year Meeting, San Diego, CA, pp. 139–145. Information Today, Inc. (May 1996)
[7] Quan, T.T., Hui, S.C, Fong, A.C.M., Cao, T.H.: Automatic fuzzy ontology generation for Semantic Web. IEEE Transactions on Knowledge and Data Engineering 18(6), 842–856 (2006)
[8] Zadeh, L.A.: Fuzzy Logic and Approximate Reasoning. Synthese 30, 407–428 (1975)

Feature Reinforcement Approach to Poly-lingual Text Categorization

Chih-Ping Wei[1], Huihua Shi[2], and Christopher C. Yang[3]

[1] Institute of Technology Management, National Tsing Hua University, Taiwan, ROC
[2] Department of Information Management, National Sun Yat-sen University, Taiwan, ROC
[3] Department of Systems Engineering and Engineering Management,
The Chinese University of Hong Kong, Shatin, N.T., Hong Kong

Abstract. With the rapid emergence and proliferation of Internet and the trend of globalization, a tremendous amount of textual documents written in different languages are electronically accessible online. Poly-lingual text categorization (PLTC) refers to the automatic learning of a text categorization model(s) from a set of preclassified training documents written in different languages and the subsequent assignment of unclassified poly-lingual documents to predefined categories on the basis of the induced text categorization model(s). Although PLTC can be approached as multiple independent monolingual text categorization problems, this naïve approach employs only the training documents of the same language to construct a monolingual classifier and fails to utilize the opportunity offered by poly-lingual training documents. In this study, we propose a feature reinforcement approach to PLTC that takes into account the training documents of all languages when constructing a monolingual classifier for a specific language. Using the independent monolingual text categorization (MnTC) technique as performance benchmarks, our empirical evaluation results show that the proposed PLTC technique achieves higher classification accuracy than the benchmark technique does in both English and Chinese corpora.

1 Introduction

With advances in information and networking technologies, organizations have been actively gathering competitive intelligence information from various online sources, and facilitating information and knowledge sharing within or beyond organizational boundaries. Such e-commerce and knowledge management applications generate and maintain a tremendous amount of textual documents in organizational repositories. To facilitate subsequent access to these documents, use of categories to manage this ever-increasing volume of documents is often observed at both organizational and individual levels. Text categorization deals with the assignment of documents to appropriate categories on the basis of their contents [1][5][6][17]. Central to text categorization is the automatic learning of a text categorization model from a training set of preclassified documents. The induced text categorization model then can classify (or predict) the particular category (or categories) to which a new document belongs.

Various text categorization techniques have been proposed [1][5][6][8][16][17]; however, most of them focus on monolingual documents (i.e., all documents are

D.H.-L. Goh et al. (Eds.): ICADL 2007, LNCS 4822, pp. 99–108, 2007.

written in the same language) in both the learning of a text categorization model and the category assignment of new documents. Because of the trend of globalization, an organization or individual often generates, acquires, and then archives documents written in different languages (i.e., poly-lingual documents). Besides, many countries adopt multiple languages as their official languages. Assume the languages involved in a repository include L_1, L_2, ..., L_s, where $s \geq 2$. That is, the set of ploy-lingual documents contains some documents in L_1, some in L_2, ..., and some in L_s. If an organization or individual has already organized these poly-lingual documents into existing categories and would like to use this set of precategorized documents as training documents for constructing text categorization models to classify into appropriate categories newly arrived poly-lingual documents, the organization and individual faces the poly-lingual text categorization (PLTC) problem.

PLTC can adopt a naïve approach by considering the problem as multiple independent monolingual text categorization problems. The naïve approach only employs the training documents of a language to construct a monolingual classifier of the same language and ignores all training documents of other languages. When a new document in a specific language arrives, we select the corresponding classifier to predict appropriate category(s) for the target document. However, this independent construction of each monolingual classifier fails to utilize the opportunity offered by poly-lingual training documents to improve the effectiveness of the classifier when the representativeness of the training documents of another language is higher.

For multilingual text categorization, some prior studies address the challenge of cross-lingual text categorization (i.e., learning from a set of training documents written in one language and then classifying new documents in a different language) [3][13]. However, prior research has not paid much attention to PLTC yet. This study is motivated by the importance of providing PLTC support to organizations and individuals in the increasingly globalized and multilingual environment. Specifically, we propose a PLTC technique that takes into account all training documents of all languages when constructing a monolingual classifier for a specific language. For purposes of the intended feasibility assessment and illustration, this study concentrates on only two languages involved in poly-lingual documents and deals with single-category documents rather than multi-category documents. To support linguistic interoperability between training documents in different languages, we rely on a statistical-based bilingual thesaurus that is constructed automatically from a collection of parallel documents. Experimentally, we evaluate the effectiveness of the proposed PLTC technique using independent monolingual classifiers built via the aforementioned naïve approach as performance benchmarks.

2 Literature Review

Text categorization refers to the assignment of documents, on the basis of their contents, to one or more predefined categories. Many text categorization techniques have been proposed [1][5][6][8][16][17], but most of them focus on monolingual documents. Central to text categorization is the automatic learning of a text categorization model from a set of preclassified documents that serve as training examples. The resulting categorization model will then be used to classify the particular category or categories to which an unclassified document belongs. The process of (monolingual) text categorization generally includes three main phases: feature extraction and selection, document representation, and induction [1][12].

Feature extraction extracts terms (or features) from the training documents. However, different languages exhibit different grammatical and lexical characteristics that significantly affect how the features in documents are segmented. Feature selection reduces the size of the feature space. Popular feature selection techniques include TF (term frequency), TF×IDF (IDF denotes inverse document frequency), correlation coefficient, χ^2 metric, and mutual information [6][7][10].

In the document representation phase, each document is represented by a vector space jointly defined by the top-k features selected in the previous phase and, in the meanwhile, is labeled to indicate its category membership. Binary (which indicates the presence or absence of a feature in a document), within-document TF, and TF×IDF are the most popular representation methods.

In the induction phase, a text categorization model(s) that distinguishes categories from one another, on the basis of the set of training documents is constructed. Prevalent learning techniques employed for text categorization include decision-tree induction, decision-rule induction [1][5], k-nearest neighbor (kNN) classification [8][16], neural network, Naïve Bayes probabilistic classifier [2][8], SVM [6], and statistical approach [17]. Sebastiani [9] offer empirical evaluations of different learning techniques for text categorization.

Cross-lingual text categorization (CLTC) deals with learning from a set of training documents (i.e., the training corpus) written in one language and then classifying unclassified documents (i.e., the prediction corpus) in a different language [3][4][13]. The major challenge facing CLTC is cross-lingual semantic interoperability that establishes the bridge between the representations of the training and prediction documents that are written in different languages. Although several studies have been conducted on CLTC, CLTC does not take poly-lingual preclassified documents as training examples as PLTC does. Therefore, CLTC is not able to take the advantage of the semantics embedded in poly-lingual training documents for text categorization model learning but relies on monolingual training documents and a translation mechanism to classify new documents in another language. In this work, we focus on PLTC with the support of automatic multilingual thesaurus.

3   Poly-lingual Text Categorization (PLTC) with Feature Reinforcement

In this study, we propose a feature reinforcement approach to PLTC with the support of an automatic multilingual (or bilingual when $s = 2$) thesaurus to address potential limitations of the naïve approach. Figure 1 shows the overall design of the proposed PLTC technique, which consists of three main tasks: 1) bilingual thesaurus construction for building a statistical bilingual thesaurus (in this study, English and Chinese) from a parallel corpus, 2) categorization learning for inducing a text categorization model for each language based on a set of preclassified documents in languages L_1 and L_2, and 3) category assignment for predicting appropriate categories for unclassified documents in either L_1 or L_2.

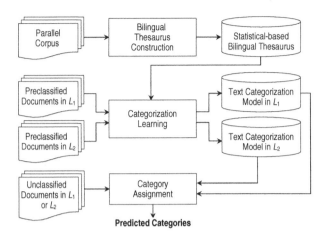

Fig. 1. Overall Process of Our Proposed PLTC Technique

3.1 Bilingual Thesaurus Construction

This task automatically constructs a statistical-based bilingual thesaurus using the co-occurrence analysis technique [15], commonly employed in cross-lingual information retrieval (CLIR) and CLTC research. Given a parallel corpus, the thesaurus construction process starts from term extraction and selection. In this study, we deal with only English and Chinese documents. We use the rule-based part-of-speech tagger [4] to tag each word in the English documents in the parallel corpus and then adopt the approach proposed by Voutilainen [11] to extract noun phrases from the syntactically tagged English documents. For the Chinese documents in the parallel corpus, we employ a hybrid of dictionary-based and statistical approaches to extract Chinese terms [14][15].

Subsequently, we adopt the TF×IDF scheme proposed by Yang and Luk [14] as the term selection metric. The term weight of a term f_j (English or Chinese) in a parallel document d_i (denoted as tw_{ij}) is calculated as:

$$tw_{ij} = tf_{ij} \times \log\left(\frac{N_P}{n_j} \times l_j\right)$$

where tf_{ij} is the term frequency of f_j in d_i, N_P is the total number of parallel documents in the corpus, n_j is the number of parallel documents in which f_j appears, and l_j is the length of f_j (where l_j denotes the number of English words if f_j is an English term or the number of Chinese characters if f_j is a Chinese term).

For each parallel document, the top k_{clt} English and k_{clt} Chinese terms with the highest TF×IDF values (i.e., tw_{ij}) and that simultaneously occur in more than δ_{DF} documents are selected for each parallel document. On the basis of the concept that relevant terms often co-occur in the same parallel documents, the co-occurrence analysis first measures the co-importance weight cw_{ijh} between terms f_j and f_h in the parallel document d_i as follows [14]:

$$cw_{ijh} = tf_{ijh} \times \log\left(\frac{N_P}{n_{jh}}\right)$$

where tf_{ijh} is the minimum of tf_{ij} and tf_{ih} in d_i, and n_{jh} is the number of parallel documents in which both f_j and f_h occur.

Finally, the relevance weights between f_j and f_h are computed asymmetrically as follows [14]:

$$rw_{jh} = \frac{\sum_{i=1}^{N_P} cw_{ijh}}{\sum_{i=1}^{N_P} tw_{ij}} \text{ and } rw_{hj} = \frac{\sum_{i=1}^{N_P} cw_{ijh}}{\sum_{i=1}^{N_P} tw_{ih}}$$

where rw_{jh} (or rw_{hj}) denotes the relevance weight from f_j to f_h (or from f_h to f_j).

After we estimate all directional statistical strengths between each pair of English and Chinese terms selected by the term extraction and selection phase, pruning of insignificant strengths is performed. Specifically, if the statistical strength from one term to another is less than a relevance threshold δ_{rw}, we remove the link. As a result, we construct a statistical-based bilingual thesaurus from the input parallel corpus.

3.2 Categorization Learning

The categorization learning task is an important component of our proposed PLTC technique. As we show in Figure 2, when training a monolingual classifier for language L_i (L_1 or L_2), our proposed categorization learning method takes into account not only the preclassified documents in L_i but also the preclassified documents in another language L_j as well as the statistical-based bilingual thesaurus. Specifically, to train a monolingual classifier for L_i, the categorization learning task involves four phases: feature extraction (for L_i and L_j), feature reinforcement and selection (for L_i), document representation (in L_i), and induction.

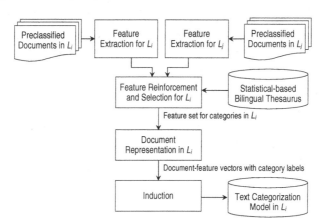

Fig. 2. Process of Categorization Learning (for L_i)

<u>Feature Extraction:</u> In this phase, we extract features from the preclassified documents in both languages. We employ the same feature extraction mechanisms as those in the bilingual thesaurus construction task to extract as features nouns and noun phrases

from preclassified English documents and Chinese terms from preclassified Chinese documents.

Feature Reinforcement and Selection: We assess the discriminating power of each feature in its respective training corpus and language. In this work, we adopt the χ^2 statistic metric, which measures the dependence between a feature f and a category C_i (denoted as $\chi^2(f, C_i)$). Using a two-way contingency table of f and C_i, let n_{r+} be the number of documents in C_i in which f occurs, n_{r-} be the number of documents in C_i in which f does not appear, n_{n+} be the number of documents in categories other than C_i in which f occurs, n_{n-} be the number of documents in categories other than C_i in which f does not appear, and n be the total number of documents under discussion. The χ^2 statistic of f relevant to C_i thus is defined as [18]:

$$\chi^2(f, C_i) = \frac{n \times (n_{r+}\,n_{n-} - n_{r-}\,n_{n+})^2}{(n_{r+} + n_{r-})(n_{n+} + n_{n-})(n_{r+} + n_{n+})(n_{r-} + n_{n-})}$$

Once the χ^2 statistic of the feature f relevant to each category C_i is derived, the overall χ^2 statistic of f for all categories T is calculated using the weighted average scheme [18]. That is,

$$\chi^2(f, T) = \sum_{C_i \in T} p(C_i) \times \chi^2(f, C_i)$$

where $p(C_i)$ is the number of documents in C_i divided by n.

After the χ^2 statistic scores for all features in both languages are obtained, we start to reassess the discriminating power of a feature in one language by considering the discriminating power of its related features in another language. The reason for such crosschecking between two languages is that if a feature in one language and its related features in another language are having high χ^2 statistic scores, it is likely that the feature has greater discriminatory power. However, inconsistent assessments between two languages (i.e., the χ^2 statistic score of a feature is high in one language but the χ^2 statistic scores of its related features are low in another language) will result in a lower confidence on the discriminatory power of the feature. In this work, we adopt this crosschecking process as feature reinforcement.

Assume a total of N_1 features in L_1 are extracted from the preclassified training documents (in L_1) and N_2 features in L_2 are extracted from the preclassified training documents (in L_2). Given a feature f_i in L_1, let $R(f_i)$ be the set of features in L_2 that have direct cross-lingual associations to f_i according to the statistical-based bilingual thesaurus derived previously. The alignment weight for f_i in L_1 (denoted as $aw(f_i)$) from its related features (i.e., $R(f_i)$) in L_2 is defined as follows:

$$aw(f_i) = \frac{\sum_{\forall g_j \in R(f_i)} \chi^2(g_j) \times rw_{g_j f_i}}{|R(f_i)|} \times \log\frac{N_2}{|R(f_i)|}$$

where $\chi^2(g_j)$ is the χ^2 statistic score of feature g_j, and $rw_{g_j f_i}$ is the relevance weight from g_j to f_i as specified in the statistical-based bilingual thesaurus.

Subsequently, we use the following formula to arrive at the overall weight of a feature f_i by combining the weights of f_i derived from the training documents in both languages:

$$w(f_i) = \alpha \times \chi^2(f_i) + (1-\alpha) \times aw(f_i)$$

where α denotes the tradeoff between the χ^2 statistic score of f_i in its original language and the alignment weight of f_i derived from the other language (where $0 \le \alpha \le 1$).

Once the overall weights of all features are derived for both languages, we then perform feature selection. For each language (L_1 or L_2), we select the k features with the highest overall weights as features to represent each training document of the respective language.

Document Representation: In this phase, the training documents of each language are represented using the corresponding feature set selected previously. In this study, we consider three prevalent document representation schemes that include binary, within-document TF and TF×IDF and empirically evaluate their effects on classification effectiveness. That is, each training document d_i forms a document-feature vector $\vec{d_i}$.

Induction: The induction phase is to induce two monolingual text categorization models from the preclassified documents in L_1 and L_2, respectively. We adopt the Naïve Bayes probabilistic classifier and Support Vector Machine (SVM) as alternative learning algorithms because of their popularity in prior research on text categorization. The Naïve Bayes classifier uses the joint probabilities of words and categories to estimate the probabilities of categories fitting a particular document. In contrast, SVM is based on Structural Risk Minimization principle and defined over a vector space where the classification or categorization problem is to find a decision surface that best separates the positive and negative training examples with the maximum margin.

3.3 Category Assignment

In the category assignment task, we categorize each unclassified document in L_1 or L_2 using the corresponding text categorization model induced previously. According to the language used in the unclassified document, we use the respective feature extraction method (described in Section 3.1) to extract features from the unclassified document and employ binary, within-document TF, or TF×IDF representation scheme to represent the target unclassified document. Finally, the feature vector of the document is used to determine an appropriate category on the basis of the corresponding text categorization model.

4 Empirical Evaluation

In this section, we report our empirical evaluation of the proposed PLTC approach. In the following subsections, we detail the design of our empirical experiments, including data collection, evaluation procedure and criteria, and our benchmark technique. Subsequently, we discuss important evaluation results.

4.1 Data Collection

To construct a statistical-based bilingual thesaurus requires the parallel documents in two languages. News presses from Government Information Center, Hong Kong

Special Administrative Region of The People's Republic of China (accessible at http://www.info.gov.hk/) were collected for constructing a statistical-based bilingual thesaurus. Specifically, the parallel corpus collected for our experimental purpose contains 2074 pairs of Chinese and English news presses.

Two additional monolingual document corpora were collected for evaluating the effectiveness of our proposed PLTC technique. The English and Chinese corpora are news presses collected from Government Information Center, Hong Kong. Both the English and Chinese corpora consist of 278 news presses related to eight categories. We merge these two monolingual corpora into a poly-lingual corpus for our evaluation purpose.

4.2 Evaluation Procedure and Criteria

To evaluate the effectiveness of PLTC, we randomly select 50% of the documents in the English and the Chinese corpora respectively as our training dataset and the remainder 50% of the documents in these two corpora as the testing dataset. To avoid the bias caused by random sampling, we repeat the sampling and train-and-test process 10 times and evaluate the effectiveness of the PLTC technique under investigation by averaging the performance obtained from these 10 individual trials. We measure the effectiveness of PLTC on the basis of classification accuracy, which is defined as the percentage of documents in the testing dataset that the PLTC technique under investigation correctly classifies into the predefined categories.

4.3 Performance Benchmark

As mentioned previously, the PLTC problem can be simply approached as multiple independent monolingual text categorization problems. That is, we construct for each language a monolingual text categorization model (i.e., classifier) on the basis of the training examples of the respective language only. For an unclassified document, we select the corresponding classifier to predict the appropriate category for the target document. In this study, we adopt this naïve approach as our benchmark technique and refer it as the MnTC technique.

4.4 Comparative Evaluation

We first conduct a series of tuning experiments to determine appropriate values for the parameters involved in bilingual thesaurus construction. Our experimental results suggest that setting δ_{DF} as 3, k_{clt} as 30, and δ_{rw} as 0.15 would be appropriate. Thus, these values are adopted for our subsequent experiments. Moreover, we also perform tuning experiments to determine the value for α (required by the PLTC technique). Our results show the best classification accuracy is achieved when α equals to 0.1. Thus, we employ this value for the subsequent comparative evaluation.

As we summarize in Tables 1 and 2, across all representation schemes and learning algorithms examined, our proposed PLTC outperforms the benchmark technique (i.e., MnTC) in both document corpora (i.e., English and Chinese). In addition, the PLTC technique using the Naïve Bayes classifier and binary representation achieves the best classification accuracy (i.e., 72.42% and 71.49%) across two different corpora.

Table 1. Comparison of Effectiveness of PLTC and MnTC on English Corpus

	Representation	Classification Accuracy		Δ
		MnTC	PLTC	
Naïve Bayes	Binary	67.63%	72.42%	4.79%
	TF	68.25%	70.26%	2.01%
	TF×IDF	67.24%	68.54%	1.30%
SVM	Binary	66.14%	68.87%	2.73%
	TF	62.45%	68.97%	6.52%
	TF×IDF	62.59%	68.54%	5.95%

Note: Δ denotes the improvement, calculated as (Classification Accuracy of PLTC – Classification Accuracy of MnTC), in Tables 1–2.

Table 2. Comparison of Effectiveness of PLTC and MnTC on Chinese Corpus

	Representation	Classification Accuracy		Δ
		MnTC	PLTC	
Naïve Bayes	Binary	65.61%	71.49%	5.88%
	TF	64.29%	67.63%	3.34%
	TF×IDF	63.48%	67.34%	3.86%
SVM	Binary	62.33%	65.64%	3.31%
	TF	58.92%	64.12%	5.20%
	TF×IDF	58.68%	63.57%	4.89%

5   Conclusion and Future Research Directions

In this work, we have investigated poly-lingual text categorization (PLTC). Many text categorization techniques have been proposed in the literature; however, most of them deal with monolingual documents. In response, we propose a feature-reinforcement-based PLTC technique that takes into account all training documents of all languages when constructing a monolingual classifier for a specific language. Using the independent monolingual text categorization (MnTC) technique as performance benchmarks, our empirical evaluation results show that our proposed PLTC technique achieves higher classification accuracy than the benchmark technique does in both English and Chinese corpora.

Some future research works related to this study include the following: Our proposed PLTC technique focuses only on two languages. It would be interesting to extend our proposed PLTC technique when the preclassified poly-lingual documents are written in more than two languages. In addition to PLTC, other poly-lingual document management issues (e.g., poly-lingual event detection) require further research attention.

Acknowledgments

This work was supported by the National Science Council of the Republic of China under the grants NSC 93-2416-H-110-021 and NSC 94-2416-H-110-002.

References

[1] Apte, C., Damerau, F., Weiss, S.: Automated Learning of Decision Rules for Text Categorization. ACM Transactions of Information Systems 12(3), 233–251 (1994)

[2] Baker, L.D., Mccallum, A.K.: Distributional Clustering of Words for Text Classification. In: Proceedings of the 21st International ACM SIGIR Conference on Research and Development in Information Retrieval (SIGIR 1998), pp. 96–103 (1998)

[3] Bel, N., Koster, C.H.A., Villegas, M.: Cross-Lingual Text Categorization. In: Koch, T., Sølvberg, I.T. (eds.) ECDL 2003. LNCS, vol. 2769, pp. 126–139. Springer, Heidelberg (2003)

[4] Brill, E.: Some Advances in Rule-Based Part of Speech Tagging. In: Proceedings of the 12th National Conference on Artificial Intelligence (AAAI-94), Seattle, WA, pp. 722–727 (1994)

[5] Cohen, W.W., Singer, Y.: Context-sensitive Learning Methods for Text Categorization. ACM Transactions on Information Systems, 17(2), 141–173 (1999)

[6] Dumais, S., Platt, J., Heckerman, D., Sahami, M.: Inductive Learning Algorithms and Representation for Text Categorization. In: Proceedings of the 1998 ACM 7th International Conference on Information and Knowledge Management (CIKM 1998), pp. 148–155 (1998)

[7] Lam, W., Ho, C.Y.: Using A Generalized Instance Set for Automatic Text Categorization. Proceedings of the 21st International ACM SIGIR Conference on Research and Development in Information Retrieval (SIGIR 1998), 81–89 (1998)

[8] Larkey, L., Croft, W.: Combining Classifiers in Text Categorization. In: Proceedings of the 19th Annual International ACM SIGIR Conference on Research and Development in Information Retrieval (SIGIR 1996), Zurich, Switzerland, pp. 289–297 (August 1996)

[9] Sebastiani, F.: Machine Learning in Automated Text Categorization. ACM Computing Surveys, 34(1), 1–47 (2002)

[10] Schutze, H., Hull, D.A., Pedersen, J.O.: A Comparison of Classifiers and Document Representations for the Routing Problem. In: Proceedings of the 18th International ACM SIGIR Conference on Research and Development in Information Retrieval, pp. 229–237 (1995)

[11] Voutilainen, A.: Nptool: A Detector of English Noun Phrases. In: Proceedings of Workshop on Very Large Corpora, Ohio, pp. 48–57 (June 1993)

[12] Wei, C., Hu, P., Dong, Y.X.: Managing Document Categories in E-Commerce Environments: An Evolution-Based Approach. European Journal of Information Systems 11(3), 208–222 (2002)

[13] Wei, C., Lin, Y. T., Yang, C. C.: Cross-Lingual Text Categorization for Global Knowledge Management, Working Paper, Institute of Technology Management, National Tsing Hua University, Hsinchu, Taiwan, R.O.C. (June 2005)

[14] Yang, C.C., Luk, J.: Automatic Generation of English/Chinese Thesaurus Based on a Parallel Corpus in Laws. Journal of the American Society for Information Science and Technology 54(7), 671–682 (2003)

[15] Yang, C.C., Luk, J., Yung, S., Yen, J.: Combination and Boundary Detection Approach for Chinese Indexing. Journal of the American Society for Information Science 51(4), 340–351 (2000)

[16] Yang, Y.: Expert Network: Effective and Efficient Learning from Human Decisions in Text Categorization and Retrieval. In: Proceedings of the 17th International ACM SIGIR Conference on Research and Development in Information Retrieval, Dublin, Ireland, pp. 13–22 (July 1994)

[17] Yang, Y., Chute, C.G.: An Example-Based Mapping Method for Text Categorization and Retrieval. ACM Transaction on Information Systems 12(3), 252–277 (1994)

[18] Yang, Y., Pedersen, J.O.: A Comparative Study on Feature Selection in Text Categorization. In: Proceedings of 14th International Conference on Machine Learning, pp. 412–420 (1997)

Development of Prototype Morphological Analyzer for the South Indian Language of Kannada

T.N. Vikram and Shalini R. Urs

International School of Information Management, University of Mysore, Manasagangotri,
Mysore-570006, Karnataka, India
{shalini,vikram}@isim.ac.in

Abstract. A prototype morphological analyzer for the south Indian language of Kannada is presented in this work. The analyzer is based on Finite state machines and can handle 500 distinct Noun and Verb stems of Kannada. The morphological analyzer can simultaneously serve as a stemmer, part of speech tagger and spell checker and hence it becomes a very efficient tool for content management.

Keywords: Kannada Morphology, Finite State Machine, Kannada Content Management, Natural Language Processing.

1 Introduction

The onset of localization of the content has capacitated the penetration of internet into those regions which do not speak English, particularly Asia. People can read and post things in their own native languages now. However, the current capabilities of the localized edition of internet is very limited. Key word based searching for the local languages is yet to be developed. Text categorization, summarization and retrieval has not been achieved in most of the Asian languages due to the lack of the essential stemming algorithms which are language specific. Similarly automatic translation of the pages to English or any other language is facilitated only if there is an efficient Part of Speech tagger(POS) [22]. As in the case of a stemming algorithm, most of the Asian languages also lack POS taggers for their respective languages. This can be addressed by developing a morph analyzer for that given language. A morph analyzer outputs the stem, the POS tag and affix for any given word. As a result the morph analyzer can be used for both stemming and part of speech tagging simultaneously.

In view of this, we have attempted to develop a prototype Kannada Morph Analyzer. Kannada is the official language of the south Indian state of Karnataka, with about 44 million speakers. Though a language of rich literary history, it is resource poor when viewed through the prism of computational linguistics. There are hardly any attempts apart from the work of Sahoo and Vidyasagar [5] where a Kannada WordNet is attempted and a Kannada Indexing software prototype by Settar [6]. Both of them are highly constrained by the lack of a morphology analyzer. Unlike English where most the morphotactic changes do not bring about change in spellings, Kannada words change spellings when the stems are inflected, which adds to the complexity of developing the morph analyzer. The analyzer is based on Finite state machines and can handle 500 distinct Noun and Verb

D.H.-L. Goh et al. (Eds.): ICADL 2007, LNCS 4822, pp. 109–116, 2007.

stems of Kannada. The morphological analyzer can simultaneously serve as a stemmer, part of speech tagger and spell checker simultaneously, and hence it becomes a very efficient tool for content management.

The paper is organized as follows. In Section 2 we briefly describe the state of the art in morphology analysis of various languages. Language specific morphology for Kannada is explained is Section 3. In Section 4 we explain the development of the proposed morph analyzer for Kannada. Finally we conclude this work with some discussion in Section 5.

2   The Current State of the Art in Morphological Analysis

The morphological analysis for English is far more advanced than any other contemporary languages [1]. Some recent advances in stemming for Germainc and other European languages can be found in Braschler and Ripplinger [7]. A comparative analysis of the various stemming algorithms for nine European Languages is presented in the survey report by Tomlinson [8]. A few stemmers for Asian languages are also proposed in the literature. Lee [9] has proposed a lexical analyzer and stemmer for Korean. It is implemented using finite state machine. A Chinese-English cross language information retrieval based on Chinese stemming is proposed in Min et al [10]. But stemming does not play a very crucial role in Chinese and Japanese because noun phrases never undergo morphotactic changes [2]. A few attempts for Malay language morph analysis is also seen [3].

Relatively little literature is available for Indic languages. An unsupervised morphology learner for Assamese language is proposed by Sharma et al. [11]. An automatic spell check for Assamese is proposed by Das et al. [12]. A Morph analyzer for Oriya has been proposed in Mohanty et al. [13], which works on the paradigm of finite state machines. A few prototype morph analyzers for Tamil, Bangla, Oriya, Assamese and Manipuri has been attempted [14].

The future of any content management activities in Kannada relies on the language technologies like spell checker development, POS tagger and stemmer. A morph analyzer serves as a spell checker, POS tagger and stem identifier simultaneously and hence this works assumes importance. To the best of our knowledge there is no research literature available with regard to the development of a morphology analyzer in Kannada, and hence this work assumes importance. Kannada has 38 basic character. Also 330 conjuncts are formed due to combination of vowels and consonants. Kannada has 100,000 basic stems and more than a million morphed variants formed due to more than 5000 distinct character variants. We report here the development of a finite state Kannada morphology for Nouns and Verbs. It has been implemented on the AT & T FSM Toolkit [17]. The nuances of verb and noun morphology of Kannada are explained in the section to follow.

3   Kannada Morphology

In Kannada, the derivation of words is either by combining two distinct words or by affixes. During the combination of two words spelling changes might occur. Eg: mugilu (Sky) + eVttara (High)= mugileVttara (Sky High). Word combination occurs in two ways in Kannada, namely *Sandhi* and *Samasa*[15]. However we do not handle compound formation morphology in this work. The other method in which word formation happens is through affixes.

Kannada inherits many of the affixes from Sanskrit. Most of the Kannada affixes are inflectional suffixes called *vibhakatis*[15]. Spelling changes occur in a large number of cases with the application of suffixes. Kannada inherits 20 prefixes (*upasarga*s) from Sanskrit. Prefixes in many cases change the meaning of the words in a way that the derived words may be a treated as root words themselves.

3.1 Nouns

Nouns represents the gender, rationality, case and number in Kannada. The nouns ending with –anu and –aLu are identified as masculine and feminine respectively. Kannada nouns in their singular for do not have any markers attached to them. –gaLu, –a, –aru are generally considered as the plural markers. Eg: The root *bAlaka* (Boy) is pluralized as *bAlakaru* (Boys), *maneV* (House) is pluralized as *maneVgaLu* (Houses).

Similar case marking exists in Kannada as in other Dravidian languages. The case markers for the corresponding cases is given in Table 1. However they cannot be merely concatenated to the roots.

Table 1. Case markers for nouns [18]

Case	Marker
Nominative	-0(u,nu,lu,ru)
Accusative	-annu, -vanna,
Genetive	-a
Dative	-ge, -ige, -akke, -kke
Locative	-alli, -yalli,
Instrumental/Ablative	-inda, YiMda
Vocative	-ee / vowel length

Table 2. Inflected noun and its meaning when different markers are concatenated

Inflected Nouns	Meaning in English	Type of Inflection made on the Noun by markers
hani	Drop	-
hani-gaLu	Drops	Plural
hani-yiMda	From the Drop	Singular + Ablative
hani-geV	To the Drop	Singular + Dative
hani-ya	Of the Drop	Singular + Genitive
hani-yalli	In the Drop	Singular + Locative
hani-yannu	The Drop	Singular + Accusative
hani-galYiMda	Because of Drops	Plural + Ablative
hani-galYigeV	To the Drops	Plural + Dative
hani-galYa	Of the Drops	Plural + Genitive
hani-galYalli	In the Drops	Plural + Locative
hani-galYannu	The Drops	Plural + Accusative

Nouns that terminate with vowels like (eV, i, u, A) are appended with an *a*, preceded by morphophonomically inserted y or eV. These case markers are strictly for singular cases. For plural cases the markers themselves undergo certain morphotactic changes. An example is considered in Table 2, which illustrates the inflection of the noun *hani*(Drop), with singular and plural case markers.

For convenience the stems are separated from the suffix with a hyphen in Table 2. It shall be observed in Table 2, that the morphophonemic *y* is inserted in the Singular Locative and Accusative cases.

3.2 Verbs

Kannada verbs occur in both Finite and non-Finite form. Most of the verb stems are in non-finite form for which tense, markers and grammatical forms are added. The singular polite form of a verb is generally obtained by adding +i, and the plural polite form is obtained by adding +ri. For example consider the word *noVdu* (See), the singular and plural polite forms thus obtained are *noVdi* (Please See) and *noVdiri* (Please be so kind enough to see) [24].

Table 3. Inflected verb and its meaning when different markers are inflected

Inflected Verbs	Meaning in English	Tense	PNG
nadeV- yuttA-ne	He will walk	Future Progressive	3SM
nadeV - yuttA-lYeV	She will walk	Future Progressive	3SF
nadeV - yuttA-re	They will walk	Future Progressive	3P-
nadeV - yuttidda-ne	He is walking	Present Progressive	3SM
nadeV - yuttidda-lYeV	She is walking	Present Progressive	3SF
nadeV - yuttidda-re	They are walking	Present Progressive	3P-
nadeV - yuttidda-nu	He was walking	Past Progressive	3SM
nadeV - yuttidda-lYu	She was walking	Past Progressive	3SF
nadeV - yuttidda-ru	They were walking	Past Progressive	3P-
nadeV - yuttiddI-ya	You are walking	Present Progressive	2S-
nadeV - yuttidde-ne	I am walking	Present Progressive	1S-

The important aspect of verb morphology is the Person-Number-Gender (PNG) and the tense marker concatenated to the verb stems. Unlike English, Indic languages add Gender information to verbs also. The general syntax for this is Verb stem + Tense Marker + PNG Marker. The morphotactic changes that occur in a verb when tense information is added is highly subjective. For example the morphotactic changes in the verb *nadeV* (To Walk) is given in Table 3. Note that the PNG marker in Table 3 is as follows, (1/2/3)(S/P)(M/F/-), where 1/2/3 denote 1st , 2nd , 3rd Person, S/P indicate Singular/Plural and M/F/- indicates Masculine, Feminine or Neutral Gender.

The development of the proposed prototype morph analyzer for Kannada is illustrated in the next section.

4 Construction of Morph Analyzer for Kannada

We employ finite state machines (FSM) for the development of Kannada morph analyzer. The primary attraction for using a FSM, for the purpose of developing a morph analyzer is its speed and efficiency. Many natural language processing techniques routinely employ FSM for shallow parsing, syllabification, tokenization and spell checking[16]. When compared to many unsupervised methods of learning morphology, an FSM based morphology development is a more tedious process because all the rules and the morphophonemic changes have to be hard coded. But the major advantage is that once a verb or noun paradigm is identified it is just a matter of identifying the stems with which it can be concatenated. The formal grammar for Kannada nouns and verbs thus identified and are given in Fig. 1 and Fig. 2 respectively.

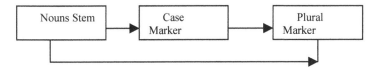

Fig. 1. A Formal Grammar for Kannada Nouns

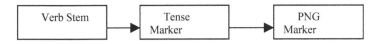

Fig. 2. A Formal Grammar for Kannada Verbs

Morphological analysis with FSM is based on the assumption that the mapping of the words to their underlying analysis forms a regular set, and there is a regular relation between these sets. In languages where morphotactics is morph concatenation only, FSM's are straight forward to apply. Handling non-concatenative or partially concatenative languages is highly challenging [19].

The development of the Morph analyzer for Kannada is hindered by the lack of publicly available dataset. Hence we have created a dataset of 500 distinct noun and verb stems. The dataset is in the Roman transliterated form and we have used the ITRANS [23] prescribed Kannada to Roman character mapping. One of the major difficulties in developing any language analysis tools for Indic languages are that the number of diacritics and compound character symbol totals to about 80,000 in number [21]. Unlike in English it is just 52 distinct symbols, the upper and lower cases of the 26 alphabets.

The dataset that we have created has 1014 distinct Kannada character symbols. With this we have implemented a Finite State Transducer(FST) on the AT&T [17] Toolkit. Transducer is a kind of machine, which translates a given input into a specified output [20]. For a given input the designed transducer outputs the stem, part of speech and case marker. A transducer is created for stem with all its morphotactic changes. The transducer for the noun *hani*, illustrated in Table 2 is given in Fig. 3.

The circles in the Fig. 3 indicate the states and the arrows indicate the transitions. Each transition is assigned a symbol of form $x:y$, where x is the input symbol and y being the corresponding output symbol. In order to maintain time efficiency, the transducer is

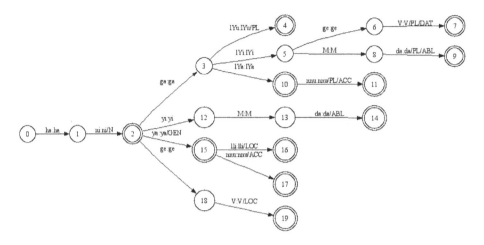

Fig. 3. FST for the noun "hani". Legend-> N: Noun; PL: Plural; DAT: Dative; ABL: Ablative; ACC: Accusative; LOC: Locative; GEN: Genative.

designed as a deterministic finite automata (DFA). DFA has minimal number of redundant transitions and hence the complexity of the network is reduced [20].

Consider the word '*haniyiMda*' of the stem *hani* illustrated in Table 2, as the input given to the transducer given in Fig. 3. The transducer produces the output as '*hani*/N *yiMda*/ABL'. N stands for noun and ABL stands for ablative. The stem *hani* is thus concatenated with the POS tag: N and *yiMda* is concatenated with the case marker: ABL. The stem *hani*/N is output from the transitions 0, 1 and 2, and the case marker *yiMda*/ABL is output from the transitions 2, 12, 13 and 14.

Likewise an FST is written for each of the individual Noun and Verb stems to accept them. An FST accepts a query only if the word is contained within its transitions and not otherwise. During query the word has to be subjected to parsing by all the developed FSTs, and the query will be accepted by only one of the FST which contains the word in its transitions. However passing a query input string to obtain its part of speech and stem for all the developed FSTs is unwieldy. This is overcome by merging all the developed FSTs and making it a single unified FST. AT & T toolkit provides the necessary commands to merge individual FSTs.

The developed prototype analyzer has the capability to handle around 7000 distinct words from 500 distinct noun and verb stems. But it is far from a being full fledged morph analyzer as pronoun and adjective morphology have not been included in this work.

5   Contributions and Conclusion

We have developed a prototype morph analyzer for Kannada for the very first time in the literature. The developed morph analyzer can be used as a spell checker, POS tagger and stemmer simultaneously. This serves as an efficient tool for the preprocessing activities of

Kannada document digitization and content management. The performance of the existing OCRs for Kannada can be improved by modifying the morph analyzer to a spell checker, thereby correcting the mistakes, which the OCR has incurred [4]. As it also serves as a stemmer, Kannada document summarization and classification is made possible, which has not been attempted yet. It also serves as a tool for language translation because it identifies the POS tag. POS tag identification is the first pre-processing for machine translation. Our future goal is to develop a morph analyzer, which can handle words from 15,000 distinct stems from the current capability of 500 stems.

Acknowledgement

This paper is in continuation of the project carried out by T. N. Vikram at the Microsoft NLP Summer School (May 3-17, 2007) at the Indian Institute of Science (IISc), Bangalore, India. He would like to thank his team mates V.N. Manjunath Aradhya, S. Noushath and team mentor S. Baskaran for the support rendered during the project days at IISc. Support from MSR India is appreciated.

References

1. van Rijsbergen, C.J., Robertson, S.E., Porter, M F : New models in probabilistic information retrieval, British Library, London (1980)
2. Zhou, Y., Qin, J., Chen, H., Nunamaker, J.F.: Multilingual Web Retrieval: An Experiment on a Multilingual Business Intelligence Portal. Digital Object Identifier (2005), doi:10.1109/HICSS.2005.450
3. Idris, N., Syed, S.M.F.D.: Stemming for Term Conflation in Malay Texts. International Conference on Artificial Intelligence (IC-AI 2001) (2001)
4. Ma, Q.: Natural language processing with neural networks. Language Engineering Conference, pp. 45–56 (2002)
5. Sahoo, K., Vidyasagar, E.V.: Kannada WordNet - A Lexical Database. TENCON Asia Pacific, pp. 1352–1356 (2003)
6. Setter, S., Goswami, S., Abhishek, H K.: Indexing software for Ancient Kannada Books. Language Engineering Conference (2002)
7. Braschler, M., Ripplinger, B.: How Effective is Stemming and Decompounding for German Text Retrieval? Information Retrieval, 291–306 (2004)
8. Tomlinson, S.: Lexical and Algorithmic Stemming Compared for 9 European Languages with Hummingbird SearchServer^{TM} at CLEF 2003. pp. 286–300 (2003)
9. Lee, C.Y.: Local grammar based lexical analyzer for Korean language. In: Proceedings of VEXTEL (1999)
10. Min, J., Sun, L., Zhang, J.: ISCAS in English-Chinese CLIR at NTCIR-5. In: Proceedings of NTCIR (2005)
11. Sharma, U., Kalita, J., Das, R.: Unsupervised learning of morphology for building lexicon for a highly inflectional language. ACL SIGPHON, 1–10 (2002)
12. Das, M., Borgohain, S., Gogoi, J., Nair, S.B.: Design and implementation of spell checker for Assamese (2002)

13. Mohanty, S., Santi, P.K., Adhikary, K.P.D.: Analysis and Design of Oriya Morphological Analyser: Some Tests with OriNet. In: Proceeding of symposium on Indian Morphology, phonology and Language Engineering, IIT Kharagpur (2004)
14. http://tdil.mit.gov.in/TDIL-OCT-2003/morph%20analyzer.pdf]
15. Hiremath, R.C.: The Structure of Kannada. PhD Thesis. Karnatak University (1961)
16. Amsalu, S., Gibbon, D.: Finite state morphology of Amharic. Workshop on RNLAP (2005)
17. http://www.research.att.com/ fsmtools/fsm/
18. Sharada, B.A.: Transformation of Natural language into an indexing language: Kannada- A case study. PhD Thesis. University of Mysore (2002)
19. Kay. Nonconcatenative Finite State Morphology. EACL. pp. 2–10 (1985)
20. Aho, A.V., Sethi, R., Ulmann, J.D.: Compilers: Principles, Techniques and Tools. Addison wesley, Reading (1985)
21. Pal, U., Chaudhuri, B.B.: Indian script character recognition. Pattern Recognition 37, 1887–1899 (2004)
22. Cao, H.-L., Zhao, T.-J., Li, S., Sun, J., Zhang, C.-X.: Chinese POS tagging based on bilexical co-occurrences. Machine Learning and Cybernetics Conf. (2005)
23. http://www.indictrans.in
24. http://ccat.sas.upenn.edu/plc/kannada/

Semantic Similarity Measures for Malay Sentences

Shahrul Azman Noah, Amru Yusrin Amruddin, and Nazlia Omar

Faculty of Information Science & Technology
Universiti Kebangaan Malaysia
Bangi Selangor
samn@ftsm.ukm.my, amruyusrin@yahoo.com, no@ftsm.ukm.my

Abstract. The concept of semantic similarity is an important element in many applications such as information extraction, information retrieval, document clustering and ontology learning. Most of the previous works regarding semantic similarity measures have been traditionally defined between words or concepts (i.e. word-to-word similarity), thus ignoring the text or sentence that the concepts participate. Semantic text similarity was made possible with the availability of resources in the form of semantic lexicon such as the WordNet for English and GermaNet for German. However, for languages such as Malay, text similarity proved to be difficult due to the unavailability of similar resources. This paper, however, describe our approach for text similarity in Malay language. We used a preprocessed Malay dictionary and the overlap edge counting based method to first calculate the word-to-word semantic similarity. The word-to-word semantic similarity measure is then used to identify the semantic sentence similarity using a modified approach for English language. Results of the experiments are very encouraging, and indicate the potential of semantic similarity measure for Malay sentences.

Keywords: Sentence similarity, semantic similarity measures, information retrieval.

1 Introduction

Most of the previous work in information retrieval regarding similarity has been focused primarily on text similarity whereby input query is compared with collection of documents and some ranking results will be obtained. The vector space model is perhaps the most popular approach still employed in text similarity [1]. Text similarity has also been used in relevance feedback [2], document clustering [3], information extraction [4] and ontology learning [5]. Semantic text similarity on the other hand is a concept whereby a set of sentences or terms within term lists are assigned a metric based on the likeness of their meaning content [6].

Measures of semantic similarity have been traditionally defined between words or concepts, and much less between text segments of two or more words. The emphasis on word-to-word similarity metrics is probably due to availability of resources that explicitly specify the relations among words such as the WordNet [7]. Although the method to measure the similarity between pair of texts can be done by measuring similarity of co-occurring words, the chances to get good measures are very slim and therefore few other aspects need to be considered such as word ordering and semantic word meanings.

D.H.-L. Goh et al. (Eds.): ICADL 2007, LNCS 4822, pp. 117–126, 2007.
© Springer-Verlag Berlin Heidelberg 2007

In the case of other non-dominant languages such as the Malay language, the measures for text similarity proved to be difficult due to the non-availability of a lexical database similar to the Wordnet. In this paper, we describe the sentence similarity measure for Malay language. We based our approach from the work of Li et al. [8] with some modifications particularly on measuring the word-to-word similarity. The next section provides a brief review on related work in this area particularly in Malay language. Section 3 describes the proposed approach and section 4 discuss our initial experiments findings. Section 5 concludes our work and provides future work directions.

2 Background

Works relating to measuring similarity between sentences and documents in English are extensive [8, 9, 10], but there have been very little or any work which relate to semantic sentence similarity for Malay language. The nearest to our knowledge would be the work of [11] which exploit word-to-word semantic similarity to enhance Malay documents retrieval. In their work, the similarity between words is defined by direct translation of English WordNet.

Most of the sentence similarity measures mainly concern with 'calculating' the availability or non-availability of words in the compared sentences [9, 10]. Therefore, the word overlap measures, TF-IDF measures, relative frequency measures and probabilistic models have been the popular method for evaluating similarity.

In semantic sentence similarity measure, the first task is to get the word-to-word semantic measures of the participating sentences. There is a relatively large number of word-to-word similarity measures previously proposed in the literature, which according to [7] can be clustered into two groups: corpus based measures and knowledge based measures. Corpus-based measures of word semantic similarity seek to identify the similarity between words using information derived from large corpora [12, 13]. Turney [12] proposed Pointwise Mutual Information measures which was based from the term co-occurrence method using counts over large corpora. Another popular approach is the Latent Semantic Analysis (LSA) whereby the term co-occurrences are captured by means of dimensionality reduction operated by a singular value decomposition (SVD).

Knowledge-based measures on the other hand identify the semantic similarity between words by calculating the degree of relatedness among words using information from dictionary or thesaurus [14, 15]. For example the Leacock and Chodorow method [14] count the number of nodes of the shortest path between two concepts in WordNet. The work by Resnik [15] and Li et al. [8] also use the Wordnet to calculate the semantic measures. The Lesk method [16] defined semantic similarity between two words based on overlap measures between the corresponding definitions as provided by a dictionary.

As can be seen work focusing on Malay semantic sentence similarity is little or none. Most of the established and proven works were in English. In this work we experiment the use of semantic sentence similarity for Malay. The proposed method is typically comparable with other methods for English sentences [7, 8] of which pair of texts is compared base on the derived syntactic and semantic information. In this method, we follow the approach proposed in [8] with some modifications.

3   The Proposed Text Similarity Method

Fig. 1 illustrates the procedure for measuring the sentence similarity between two Malay sentences. In adopting the method proposed by [8], a joint distinct word set is formed for the two sentences. For each submitted sentence, a raw semantic vector is obtained by exploiting an open Malay dictionary database. Unlike other English text similarity methods which rely on the WordNet in calculating the word semantic similarity measures, our approach uses the overlap edge counting-based method which was originally proposed by Lesk [16]. In this case the semantic similarity between words is based on the counting of overlaps between dictionary definitions of the compare words.

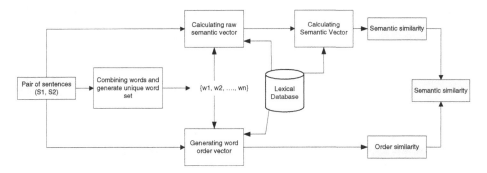

Fig. 1. The process for semantic similarity between sentences

A word order vector is formed for each sentence, again using information from the open dictionary. Since each word in a sentence contributes differently to the meaning of the whole sentence, the significance of a word is weighted by using information content derived from the open dictionary. This is another limitation of Malay language which does not yet contain any document corpus such as the Brown corpus for English. By combining the values of raw semantic vector with information content from the open dictionary, a semantic vector is obtained for each of the two sentences. Semantic similarity is then computed based on the two semantic vectors, whereas an order similarity is calculated using the two order vector. Finally, the sentence similarity is derived by combining semantic similarity and order similarity.

We describe in detail the aforementioned procedure in the following sections.

3.1   Semantic Similarity of Words

As previously mentioned, the semantic similarity measures between words can be grouped into corpus-based measures and knowledge measures. We chose to focus on the knowledge-based measure as a large corpus for Malay language related sources is not currently in existence. Furthermore as lexical database for Malay language similar to WordNet is not yet available, we chose to use an open dictionary. The open dictionary contains 69,344 rows of data with 48,177 Malay words which is based from the Kamus Dewan 3rd Edition. The dictionary, however, is still not yet in a Machine Readable Dictionary (MRD) format (i.e the dictionary is still in a human readable format), therefore a few pre processing are required. The dictionary was parsed by filtering and eliminating symbols, short form words, verbs, and other words not found in the dictionary.

After investigating a number of methods for knowledge-based measures, the only suitable method to use is the Lesk's method [16]. This is due to the nature of the generated MRD dictionary which only contains meanings of words and not the hierarchical structure of words that models the human common sense knowledge of general language usage similar to WordNet [17].

Therefore, we proposed that the similarity $sim(w_1, w_2)$ between words w_1 and w_2 is the multiplication of ratios for the meanings of words w_1 and w_2 as follows:

$$sim(w_1, w_2) = r(C, M_{w_1}) \cdot r(C, M_{w_2}) \tag{1}$$

where C is the set of unique overlap words found in the meanings of w_1 and w_2 and M refers to the meanings of the respective words. Therefore, $r(C, M_{w1})$ refers to the ratio between the counts of meanings that contains any of the words in C with all the meaning associated with w_1.

3.2 Semantic Similarity Between Sentences

Sentences are aggregation of words, therefore, it is common to use words in the sentences to represent the sentences. Using the method proposed by [8], the semantic vector of words is dynamically formed solely based on the compared sentences. This approach is slightly different with the conventional vector space model which requires the comparison of all words existed on the document corpus.

So, assuming we are comparing between sentences, S_1 and S_2, a joint distinct word set S, is formed between S_1 and S_2:

$$S = S_1 \cup S_2$$
$$= \{w_1, w_2, \ldots\ldots, w_n\}; w_i \text{ are distincts}$$

We don't consider morphological variants among words. Therefore, the words *makan* (eat), *makanan* (food) and *pemakanan* (nutrition) are all considered as three distinct words and forms part of the set S. For example if we have the sentences: S_1: *Saya berjalan ke sekolah* (I walked to school); S_2: *Dia berkereta ke bandar* (He drived to town), then we will have $S = \{Saya\ berjalan\ ke\ sekolah\ Dia\ berkereta\ bandar\}$. The joint word set S, is viewed as the semantic information for the compared sentences. In other words the semantic information for sentences S_1 and S_2 are derived from the joint set S. To derive the semantic information content of S_1 and S_2, a term-term matrix is constructed as follows:

$$S_i = \begin{matrix} & S = w_1 & w_2 & \ldots & \ldots & \ldots & \ldots & w_n \\ & \begin{matrix} q_1 \\ q_2 \\ . \\ . \\ . \\ q_m \end{matrix} & \begin{bmatrix} x_{1,1} & x_{1,2} & .. & .. & .. & .. & x_{1,n} \\ x_{2,1} & x_{2,2} & .. & .. & .. & .. & x_{2,n} \\ . & & .. & .. & .. & .. & . \\ . & & .. & .. & .. & .. & . \\ x_{m,1} & x_{m,2} & .. & .. & .. & .. & x_{m,n} \end{bmatrix} \end{matrix}$$

whereby $x_{i,j}$ represents the similarity measure between the i-th word in the compared sentence and the j-th word of the joint set. The value of $x_{i,j} = 1$, if q_i and w_i are the same words, whereas if $q_i \neq w_i$, the similarity measure is computed using the word-to-word semantic similarity method previously described. In the case of $q_i \neq w_i$, the similarity value of $x_{i,j}$ is only considered if $x_{i,j} > \xi$, whereby ξ is a specified threshold value. Anything less than ξ, is assumed not semantically similar.

From our experiment of 200 pairs of antonyms, the $\xi = 0.18$ has been selected due to its dominance as shown in Fig. 2.

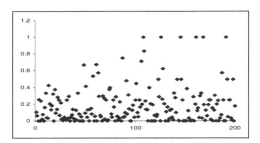

Fig. 2. Distribution of the word-to-word similarity measures for antonyms

The raw semantic vector of S_i ($i = 1,2$), i.e. $\check{s}$; can then be computed, whereby $\check{s} = \{(max\ (x_{1,1}...x_{m,1})),......,\ (max(x_{1,n}...x_{m,n}))\}$. For example if we compared between, the joint set $S = \{negara,\ Malaysia,\ aman,\ sentosa,\ jepun,\ maju\}$ with the compared sentence $S_1 = \{negara,\ Malaysia,\ aman,\ sentosa\}$, we will get the following term-term matrix:

$$\begin{bmatrix} 1 & 0 & 0 & 0 & 0 & 0 \\ 0 & 1 & 0 & 0 & 0 & 0.327 \\ 0 & 0 & 1 & 0 & 0 & 0 \\ 0 & 0 & 0 & 1 & 0 & 0 \end{bmatrix}$$

therefore, the raw semantic vector $\check{s}$ for S_1 will be $\{1\ 1\ 1\ 1\ 0\ 0.327\}$.

For the calculation of the semantic vector s_i, the following formula is used:

$$s_i = \check{s} \cdot I(w_i) \cdot I(\tilde{w_i}) \tag{2}$$

where w_i is a word in the joint word set S, and w_i is its associated word in the sentence. The value of $I(w)$ is calculated by referring to the MRD dictionary, using the following formula:

$$I(w) = 1 - \frac{\log(n+1)}{\log(N+1)} \tag{3}$$

where n is the number of rows of meaning containing the word w and N is the total number rows (meaning) in the dictionary. Then, the semantic similarity between the two compared sentences is simply the cosine coefficient between the two semantic vectors.

$$S_s = \cos(s_1, s_2) = \frac{s_1 \cdot s_2}{|s_1| \times |s_2|} \tag{4}$$

3.3 Word Order Similarity Between Sentences

Measuring the word similarity is rather a straightforward process and used the similar joint word set as discussed in the previous section. Assuming that we have a pair of sentences, L_1 and L_2 of which:

L_1: *Negara Malaysia aman sentosa*
L_2: *Jepun negara maju*

therefore, we will have a join set $L = \{Negara, Malaysia, aman, sentosa, Jepun, maju\}$. Similarly with the semantic similarity, the vector of the word order is derived from the joint set L. A term-term matrix is constructed and the word-to-word similarity measure is calculated using the method discuss in section 3.1. The resulting matrix for the sentence L_1 and the joint set L is similar to the one presented in section 3.2.

The word order vector for L_1, i.e. u_1 is constructed based on the existence or the highest word-to-word similarity between the joint set L and L_1. Therefore we will have $u_1 = (1\ 2\ 3\ 4\ 0\ 2)$, the last value of u_1 is equal to 2 because the word *maju* in L is strongly similar with the word Malaysia, which is the second position in L_1. Similarly we will get $u_2 = (2\ 2\ 3\ 3\ 1\ 3)$, derived from the following matrix.

L_2 \ L	negara	Malaysia	aman	sentosa	Jepun	Maju
Jepun	0	0	0	0	1	0
negara	1	0.58	0	0	0	0
maju	0	0	0.282	0.163	0	1
u_2	(2	2	3	3	1	3)

Using the word order similarity as follows:

$$S_r = 1 - \frac{|u_1 - u_2|}{|u_1 + u_2|} \tag{5}$$

we will get $S_r(L_1, L_2) = 0.828$. The word order similarity in (5) is determined by the normalized difference of word. Li et al. [8] has demonstrated that the formula is an efficient metric for measuring word order similarity.

3.4 Combine Sentences Similarity

The combine sentences similarity represents the overall sentences similarity, which is the summed of the semantic similarity and the word order similarity as follows:

$$Sim(S_1, S_2) = \delta S_s + (1 - \delta)S_r \tag{6}$$

whereby δ is a damping factor, which decides the contribution of the involved similarity measures (i.e. S_s and S_r). Li et al. [8] suggested that δ should be greater than 0.5 due to the importance of lexical elements presented in semantic similarity [18].

4   Initial Experimental Testing

We have conducted an initial testing in order to evaluate of the proposed modified approach. The result of the testing is as illustrated in Table 1. Due to brevity, only portion of the result is shown.

Table 1 separates the result into semantic similarity, order similarity and sentence similarity for $\partial = 0.5$. The testing as illustrated in Table 1 compares the first sentence of the list with the remaining six sentences. To assist discussion, we called the first sentence of the list as the 'target sentence' and the remaining six sentences as the 'compared sentences'. Each of the six compared sentence is being weight against the target sentence. Human ranking of similarity is just our (human) opinion about the relevancy ranking of the compared sentences with the target sentence.

Result in Table 1 shows consistent outcome between our ranking of similarity and the approach sentence similarity measures, with very minimal differences.

Table 1. Initial testing result*

Sentence Tested	Sentences Compared	Human Ranking of Similarity	Semantic Similarity	Order Similarity	Sentence Similarity
Target sentence 1					
Saya pergi ke sekolah.					
	Saya pergi ke sekolah	1	1	1	1
	Saya berjalan ke sekolah	2	0.95	1	0.98
	Saya pergi ke madrasah	3	0.90	1	0.95
	Saya pergi ke kedai	4	0.61	0.89	0.74
	Dia pergi ke kedai	5	0.59	0.89	0.74
	Saya makan nasi di kedai	6	0.46	0.67	0.57
Target sentence 2					
Saya membaca buku sambil minum air kopi.					
	Saya membaca buku sambil minum air kopi.	1	1	1	1
	Saya membaca buku sambil minum air teh.	2	0.87	0.62	0.74
	Saya membelek majalah sambil minum air teh.	3	0.54	0.60	0.57
	Saya menonton televisyen sambil minum air teh	4	0.62	0.60	0.61
	Ahmad menonton televisyen sambil minum air teh	5	0.55	0.64	0.60
	Saya menonton televisyen sambil baring.	6	0.44	0.61	0.53

* For brevity, we don't provide the English translation of the tested sentences presented in Table 1

As mentioned earlier we do not consider morphological variant among words. However, further testing and analysis of the approach, found that morphological variants do play a significant role. For example, consider the following compared sentences and their respective similarity measures. The underlined words are the morphological variants in Malay although they referred to the same words when translated into English. In this case 'kahwin' (married) is the root word for 'berkahwin' (got married).

S_1 = Saya suka lelaki bujang itu.   => I like that bachelor man
S_2 = Saya suka lelaki belum berkahwin itu. => I like that *unmarried* man
S_3 = Saya suka lelaki belum kahwin itu.   => I like that *unmarried* man

$S(S_1, S_2) = 0.579;$ $S(S_1, S_3) = 0.902$

As we can see, words that were stemmed to their root word give higher similarity measures. Therefore, aspects of morphological variants should be considered. The machine processing however might be the drawback if the morphological variants are to be considered.

We have also conducted some random testing by selecting pair of sentences of which the relevancies are known. Table 2 shows portion of the testing result. If the paired sentences are assumed to be relevant if the similarity measures > 0.5; then we have consistent result except for pair number 3. For pair number 3, the high similarity measures value is due to the high word-to-word similarity between '*tidur*' (sleep) and '*katil*' (bed).

Table 2. Similarities between selected sentences

	Sentences Pair	Sentences Pair (English Translation)	Rel.	Similarity Measure
1.	*Saya hendak tidur*	I want to go to sleep	Y	0.772
	Saya sangat mengantuk	I am very sleepy		
2.	*Ali sangat lapar*	Ali is very hungry	Y	0.643
	Ali hendak makan	Ali wants to eat		
3.	*Saya hendak tidur*	I want to go to sleep	N	0.667
	Saya bermain atas katil	I am playing on the bed		
4.	*Ahmad ke kuliah*	Ahmad went to a lecture	Y	0. 547
	Ali belajar di kelas	Ali study in class		
5.	*Tayar kereta pancit*	Punctured car tyre	Y	0.939
	Tayar motosikal pancit	Punctured motorcycle tyre		
6.	*Saya bermain di padang*	I am playing at the field	N	0.133
	Ibu memasak kari ikan	Mother is cooking fish curry		
7.	*Saya ada tukul*	I have a hammer	N	0.314
	Beri limau itu	Give that lemon		

5   Conclusions and Near Future Works

This semantic sentence similarity is important in many applications such as information retrieval, information extraction and ontology learning. While research in this area has been dominated for English language, little or no work has been focus for Malay language. This paper has presented an approach based on the work of [8] to provide a semantic measure for Malay sentences. This approach compares pair of sentences by first finding the similarity measures among words. The word-to-word similarity measure is derived from an on-online Malay dictionary using the overlap edge counting based method. The obtained word-to-word similarity measures are then used to construct the semantic vector and the word order vector. Lastly the sentence similarity is derived from the sum of the aforementioned vectors.

Initial experiments have shown consistent and encouraging results which indicate the potential use of this modified approach to practical applications previously mentioned. However more testing and evaluation works need to be conducted particularly involving real test data and human experts. Therefore further testing has been our main near future work.

As previously mentioned, morphological variants do provide significant similarity measures. Few stemming algorithms for Malay words are currently in existence such as [19, 20]. However, we need to further investigate on how significant morphological variants have in terms of sentence similarity. This is quite crucial due to the limitation of stemming algorithm relating to understemming and overstemming [21].

Our other future works include applying the sentence similarity measures to information retrieval activities of Malay documents. Apart from that the evaluation of word-to-word similarity should be extended to other method such as the term co-occurrence corpus base method and the semantic network which requires the construction of a linguistic ontology similar to WordNet.

Acknowledgments. The authors wish to thank the Ministry of Higher Education for the fund provided for this project and also the anonymous referees for their helpful and constructive comments of this paper.

References

1. Salton, G., Lesk.: Computer evaluation of indexing and text processing. Prentice Hall, Englewood Cliffs (1971)
2. Smucker, M.D., Allan, J.: Find-similar: similarity browsing as a search tool. In: Proceedings of the 29th annual international ACM SIGIR conference on Research and development in information retrieval, pp. 461–468. ACM Press, New York (2006)
3. Zeng, H.-J., He, Q.-C., Chen, Z., Ma, W.-Y., Ma, J.: Learning to cluster web search results. In: SIGIR 2004 (2004)
4. Mooney, R.J., Bunescu, R.: Mining Knowledge from Text Using Information Extraction. SIGKDD Explorations 7(1), 3–10 (2005)
5. Buitelaar, P., Cimiano, P.: Bernardo Magnini Ontology Learning from Text: An Overview. In: Buitelaar, P., Cimiano, P., Magnini, B. (eds.) Ontology Learning from Text: Methods, Evaluation and Applications Frontiers in Artificial Intelligence and Applications Series, vol. 123, IOS Press, Amsterdam, Trento, Italy (2005)
6. Cilibrasi, R., Vitanyi, P.M.B.: Similarity of objects and the meaning of words. In: Cai, J.-Y., Cooper, S.B., Li, A. (eds.) TAMC 2006. LNCS, vol. 3959, Springer, Heidelberg (2006)
7. Mihalcea, R., Corley, C., Strapparave, C.: Corpus based and knowledge based measures of text semantic similarity. In: Proceedings of the American Association for Artificial Intelligence (AAAI 2006) (2006)
8. Li, Y., McLean, D., Bandar, Z.A., O'Shea, J.D., Crockett, K.: Sentence similarity based on semantic nets and corpus statistics. IEEE Transactions on Knowledge and Data Engineering 18(8), 1138–1150 (2006)
9. Metzler, D., Bernstein, Y., Croft, W.B., Moffat, A., Zobel, J.: Similarity Measures for Tracking Information Flow. In: Proceedings of the CIKM 2005, pp. 571–524 (2005)
10. Tatu, M., Moldovan, D.: A semantic approach to recognizing textual entailment. In: Proceedings of the conference on Human Language Technology and Empirical Methods in Natural Language Processing, pp. 371–378 (2005)
11. Hamzah, M.P., Sembok, T.M.: Enhance retrieval of Malay documents by exploiting implicit semantic relationship between words. Enformatika 10, 89–94 (2005)
12. Turney, P.: Mining the web for synonyms: PMI-IR versus LSA on TOEFL. In: Proceedings of the 12th European Conference on Machine Learning (2001)
13. Karov, Edement.: Similarity-based Word Sense Disambiguation. Computational Linguitics 24(1), 41–59 (1998)

14. Leacock, C., Chodorow, M.: Combining local context and WordNet sense similarity for word sense identification. WordNet, An Electronic Lexical Database. The MIT Press, Cambridge (1998)
15. Resnik, P.: Using information content to evaluate the semantic similarity. In: Proceedings of the 14th International Joint Conference on Artificial Intelligence (1995)
16. Lesk, Automatic Sense Disambiguation Using Machine Readable Dictionaries: How to Tell a Pine Cone from an Ice Cream Cone (1986)
17. Miller, G.A.: WordNet: a lexical database for English. Communication of the ACM 38(11), 39–41 (1995)
18. Wiemer-Hastings, P.: Adding syntactic information to LSA. In: Proceedings of the 2nd Annual Conference on Cognitive Science, pp. 989–993 (2000)
19. Ahmad, F., Yusoff, M., Sembok, T.M.T.: Experiments with a Stemming Algorithm for Malay Words. JASIS 47(12), 909–918 (1996)
20. Othman, A.: Pengakar perkataan melayu untuk sistem capaian dokumen. MSc Thesis. National University of Malaysia (1993)
21. Xu, J., Croft, W.B.: Corpus-based stemming using coocurrence of word variants. ACM Transactions on Information Systems 16(1), 61–81 (1998)

Enabling Resource Selection Based on Written English and Intellectual Competencies

Ayako Morozumi[1], Liddy Nevile[2], and Shigeo Sugimoto[1]

[1] University of Tsukuba, 1-2, Kasuga,
Tsukuba City, Ibaraki, Japan
{moro,sugimoto}@slis.tsukuba.ac.jp
[2] La Trobe University, Kingsbury Drive,
Bundoora, Australia
liddy@sunriseresearch.org

Abstract. A growing number of people are using the Web to access English-language resources, among other things. In Asian countries, for example, many people want access to English texts. Many Asians are not as competent reading English as they may be in the intellectual content of their domain. The problem of accessibility to English texts is significant simply because of the number of people involved. The problems for second language English readers are similar to those for many dyslexic first language readers. We propose a descriptive model that supports adaptability of texts for the benefit of such people based on FRBR and AccessForAll standards.

Keywords: FRBR, AccessForAll, resource descovery, selection, adaptation, accessibility, English second language, ESL, dyslexia.

1 Introduction

There are at least three major groups of readers with language-skill problems who want access to intellectually stimulating and specialist English texts:

- people with domain expertise who lacking sufficient English reading skills to access the English literature in their field of interest;
- people with domain expertise who need translations of English literature, and
- people with dyslexia.

Although there are texts in many languages, there is sufficient interest in English literature for it to be the focus of this paper. Previously we described a metadata schema model for users seeking appropriately 'accessible' resources [1]. In this paper, we focus on the selecting a resource depending both on the knowledge level of the user and the (sometimes second-language) reading skills of the reader.

We consider the problem for second-language readers, translators (particularly automated ones) and people with dyslexia to be similar: in all cases it is important to have plain English without distracting or confusing metaphors, or complicated language constructions such as the subjunctive mood or passive voice.

D.H.-L. Goh et al. (Eds.): ICADL 2007, LNCS 4822, pp. 127–130, 2007.

2 Research Base

We follow the AccessForAll model of accessibility [2] in which a user specifies their needs and preferences and a resource is discovered and, where necessary modified, to match these requirements. Accessibility, in this sense, means that the display, control and content of the resource is suited to the user, regardless of any disabilities they may have or circumstances causing a lack of access. In this paper, the lack of access of concern is to content that is originally English.

We consider that unless resources are suitably described for both their content level and their reading level and these two properties are related, users will not have be able to find suitable resources for their personal use. Further, we are concerned that the reading level of a text in a second-language is not the same as the reading level for a first-language reader, and that this should be described differently.

2.1 Functional Requirements for Bibliographic Records

Although the Functional Requirements for Bibliographic Records (FRBR) [3] were originally developed for books, they are increasingly proving useful for digital resources. In this paper, we want a resource that contains the intellectual content that is in the original *work* that resulted in an original *expression*. Having found the right *expression*, the user will need to discover a *manifestation* of that expression to access as a suitable item when delivered. These are the four levels of abstraction, called entities, in the FRBR model.

2.2 AccessForAll

The AccessForAll metadata model depends upon a description of a user's access needs and preferences being specified and matched by the characteristics of a resource, as described in the resource's metadata. This, of course, requires the availability of a matching service.

The innovation in the AccessForAll approach to accessibility is that the matching service can enable cumulative, distributed components to be combined to make a resource more accessible, even after the resource has been published in its original form. In some cases, this involves the de-construction of a resource into components and the re-construction including alternative, adapted or augmented components.

It is the just-in-time accessibility that is to be exploited in this paper. This means that a Web service, for instance SWAP [4], could render a resource accessible by providing a translation of it, or a plain English version.

The AccessForAll approach advocates description of both needs and preferences because for some people a need is crucial, and if not satisfied the resource will not be useful at all, while for others the stated need is a preference, and if not satisfied, may make for difficulties that will be tolerated by that user.

3 Metadata Standards and the Needs of the User

The purpose of our study is to provide useful models for resource characteristic description and possibly for complementary needs and preferences descriptions. What is needed is a way of making it easy for users to have their discovery results matched to their needs and preferences. The following paragraphs discuss metadata standards for resource discovery and functional augmentation for resource access from the viewpoint of accessibility.

3.1 Metadata Standards

The AccessForAll specifications have been formalised by the International Standards Organization [5]. They provide for two descriptions: the description of the user's needs and preferences and the potentially matching properties of the resource. This model can be extended simply as it depends simply upon the presence of the descriptions and the matching service.

We consider that a user first needs to discover a resource based on the intellectual content of the resource. FRBR provides a standard framework for such a description and it has been partially or completely implemented in many available standards, such as Dublin Core, MARC-21, MODS, etc.

Having found the work and the expression, the problem is to find a suitable manifestation to deliver the right item. This means a manifestation that has suitable characteristics such as that it is in plain English, which can be easily and most accurately translated, read, or interacted with by a person with dyslexia.

DC Audience [6] is a standard term used to describe the class of person to whom the resource is aimed, or for whom it is considered suitable. The class of people is usually expressed as citizens, or Grade 4 children. AccessForAll avoids such judgments in favour of descriptions of properties of the resource so that individual users can assess the suitability of the resource for them as individuals. This is significant because users do not have the same needs and gross classifications can eliminate resources a particular user could enjoy.

In MARC-21, there are elements such as *Reading Grade Level* and *Interest Grade Level* [7]. We foresee no problem with using such established standard descriptions but note that it is not the author or publisher's target audience that should determine the access but rather the user's needs and preferences.

We are arguing therefore, for a standard way of expressing these in combination that can be accessed by AccessForAll services. The AccessForAll model provides a way of doing the matching and applications such as The Inclusive Learning Exchange (TILE [8]) already show how implementations might work.

3.2 Functions to Augment Accessibility

Web services are emerging that are capable of automating such processes. There are automatic translators. There are online services that offer immediate translation by humans. There are services that simplify texts in appropriate ways for some readers, transform them, we might say. SWAP is an example of such a service [4].

This paper does not call for a new model so much as an additional set of elements developed according to the established model. We argue that once the relevant characteristics are ascertained and standardized, determining the relevant values is not necessarily a human task. Reading levels have been determined automatically in the past. Potentially, second-language reading levels could similarly be determined.

'Clutter' that causes serious problems for dyslexic readers may be an example of what is relevant, as would be the use of metaphors, passive voice, subjunctive tenses, etc.

4 Conclusion

In this paper, we have drawn attention to what, we believe, is a significant problem in quantitative as well as qualitative terms. We adopt the position that the World Wide Web and associated technologies and practices have opened the way for wide international participation in intellectual endeavour but that currently there is unacceptable effective discrimination against non-English readers. We propose some requirements for consideration if this situation is to be averted. We argue that by describing the English expression within the text in appropriate, standard metadata, a significant benefit would be derived from a more accessible internationalized knowledge base.

References

1. Morozumi, A., Nevile, L., et al.: Using FRBR for the Selection and Adaptation of Accessible Resources, International Conference of Dublin Core and Metadata Applications. In: Proceedings of DCMI Conference 2006, Mexico, University of Colima (2006)
2. Nevile, L., Treviranus, J.: Interoperability for Individual Learner Centred Accessibility for Web-based Educational Systems. IEEE TCLT's Journal of Educational Technology & Society 9(4) (2006)
3. International Federation of Library Associations and Institutions: Functional Requirements for Bibliographic Records (FRBR), http://www.ifla.org/VII/s13/frbr/frbr.pdf
4. Semantic Web Accessibility Platform, http://www.ubaccess.com/swap.html
5. Heath, A., Nevile, L., Treviranus, J.: Individualized Adaptability and Accessibility in E-learning, Education and Training Parts 1, 2 and 3, http://jtc1sc36.org/doc/36N1139.pdf, http://jtc1sc36.org/doc/36N1140.pdf, http://jtc1sc36.org/doc/36N1141.pdf
6. Dublin Core Metadata Initiative, http://purl.org/dc/terms/
7. http://www.loc.gov/marc/bibliographic/ecbdnot1.html#mrcb521
8. The Inclusive Learning Exchange (TILE), http://www.barrierfree.ca/tile/index.htm

QRselect: A User-Driven System for Collecting Translation Document Pairs from the Web

Kyo Kageura[1], Takeshi Abekawa[1], and Satoshi Sekine[2]

[1] Graduate School of Education, University of Tokyo,
7–3–1 Hongo, Bunkyo-ku, Tokyo 113–0033, Japan
{kyo,abekawa}@p.u-tokyo.ac.jp
[2] Computer Science Department, New York University,
715 Broadway, New York, NY 10003 USA
sekine@cs.nyu.edu

Abstract. In this paper we introduce a system that collects English-Japanese translation document pairs from the Web that are relevant to subject keywords specified by the user. The system, QRselect, is specifically designed to meet the needs of online volunteer translators who, in the process of translation, want to refer to a small and specific set of translation document pairs which are relevant to what they are translating. A system which collects relevant existing translated documents and makes them available for reference in the translation process will therefore greatly help these translators. Against this backdrop, we developed a prototype translated document collection system and evaluated its performance. We also examined the users' role in improving the system.

1 Introduction

The ever-expanding breadth of global communication on the Internet has recently been accompanied by an increase in the number of online volunteer translators [1], who translate online documents in a variety of fields such as politics, culture, area studies, sports, computers, etc. and publish their translations on the Web. Loose networks of translators dealing with similar or related subjects have been and continue to be formed.

These translators often play vital roles in distributing important information that would otherwise not appear in mainstream media and in promoting critical media literacy in the age of Internet. Despite this, and despite the fact that there are many translation-aid systems, there has been no system to date that specifically aims at aiding online volunteer translators.

Against this backdrop, we are currently developing a system that aids online volunteer translators. As a part of this, we developed QRselect, a system that collects translation document pairs from the Web, based on translators' requests or specifications. This is, in a sense, a system that constructs each translator's private digital library of relevant translation document pairs, and enables translators to refer to relevant existing translated documents systematically.

D.H.-L. Goh et al. (Eds.): ICADL 2007, LNCS 4822, pp. 131–140, 2007.

In this paper we will first explain the basic needs of translators that led us to the development of QRselect. We will then introduce the basic structure of QRselect and how it works, and present the results of an evaluation of the system's basic performance, with a diagnosis. We will also discuss the response of translators to the system and the extent to which they are willing to cooperate with us to improve the system's performance and to make it fully effective.

2   Translators' Need for Existing Translated Documents

One of translators' key needs – we interviewed eight online volunteer translators and also obtained opinions by e-mail from twelve other translators – is the ability to refer to and recycle bilingual translations of various language units, such as proper names, repetitive quotations, domain-dependent expressions, etc., from existing translated texts that deal with the same or similar topics and which are judged to have sufficiently high quality. Two things characterise this need:

(i) What translators are looking for within existing relevant translations are not *linguistically similar examples*, but concrete information showing *translation conventions* relevant to the group of texts to which the text that the translator is translating will belong, a group characterised by such basic traits as subject topics, register, etc. In other words, what translators would like to be able to refer to in the process of translation is an *archive of relevant texts*, not an unanchored *corpus* that represents *language* in general [7]. This information need is different from and complementary to the need to check a broader range of reference sources.

(ii) Translators want a system that helps them refer to what they want to refer to. When they look for existing translation document pairs that are relevant to what they are translating, the documents translators refer to tend to be very dense and few in number – in the order of tens or often less. Size cannot compensate for the relevance to their requirements; it is the fact that they checked the documents that they thought they needed to check, which may be few in number, that enables them to finalise the translation. This incidentally corresponds to the claim made in the field of natural language processing that the usefulness and effectiveness of a corpus depends qualitatively on the aim and that a larger corpus may not necessarily perform better [8,10].

Currently, many translators take several steps in order to refer to existing relevant translation document pairs, including checking pages they know as well as using Google to find new translations. What is desirable here, therefore, is to automatise this process by developing a mechanism to collect a set of translation text pairs which are relevant to the text that the translator is translating, according to the translator's request. As such a group of texts is defined vis-à-vis the translator's need which is mostly determined by the particular text which the translator is translating or the particular subject area with which the translator is mainly concerned, a system which collects such translation document pairs should function in a user-driven manner. This requirement contrasts with the need to collect large bilingual corpora from the Web for use as a basic resource for natural language processing [4,12] or to obtain a broad-coverage list of bilingual word pairs from a large corpora [2,5,6,9,11,16].

Translators regard these two types of information as qualitatively different. This can be understood in analogy to the behaviour of patent translators, who *always need to check* existing translations in the archive of patent documents. They must do so as much to be able to make their own decisions with confidence as to look for translation expressions which they do not know. In order for translators to make their documents authentic and acceptable to readers, the process of situating the translated document within a set of relevant existing documents is an essential process of translation. Irrespective of whether translators adopt existing expressions and phrases or not, and irrespective of whether they have basic information on expressions from other reference sources, this process is therefore a *sin qua non* for translation and cannot be compensated for by other information sources, however large they may be.

Taking the above into account, we have developed QRselect, a user-driven system to collect from the Web a specific set of translation document pairs relevant to the document that the translator deals with.

3   The QRselect Prototype System

3.1   Basic Structure of the System

The QRselect system operates in two different modes, i.e. dynamic mode and batch mode. In the following, we focus primarily on an explanation and evaluation of the dynamic mode. Figure 1 shows the overall framework of the QRselect prototype system.

1. The user inputs Japanese keywords relevant to the topic of the document that the user is translating. In batch mode, the user registers a list of URLs under which translated documents relevant to the translator's interests are published frequently.
2. The system retrieves a specified number of Japanese Web documents relevant to the Japanese keywords, using an existing search engine. When the retrieved documents are evaluated as a translation in steps 3 to 6, the search is expanded to "similar pages", as the site that includes a translated document may well include many translated documents. In batch mode, the system checks the update log of the registered sites, and collects the newly published documents.
3. For each retrieved page, the system detects the anchor link given by the `<a http="">` `</a>` tag, traverses the anchor link, and obtains the target page of the link. The system only traverses the anchor links in close proximity to the reserved words, which should indicate that the target page is the original document. After having analysed scores of translated documents, we adopted seven reserved words, i.e. "   " (original document), "   " (source), "   " (English), "   " (source article), "   " (original), "   " (source text), "   " (source language).

4. For each Japanese document retrieved in step 2 and for each document detected in step 3 as a source document candidate, the system applies a simplified version of Webstemmer [15] and extracts the textual area.
5. For the pair of textual areas extracted in step 4, the system calculates the similarity of the texts. This is done by transforming the words in the English text into Japanese using a publicly available large-scale English-Japanese dictionary [3]. Currently, simple content words are used. The similarity is calculated by the ratio of the number of matched word tokens to the total number of Japanese word tokens in the text.
6. The system identifies the pairs whose similarity score given in step 5 is above a given threshold as translation document pairs.

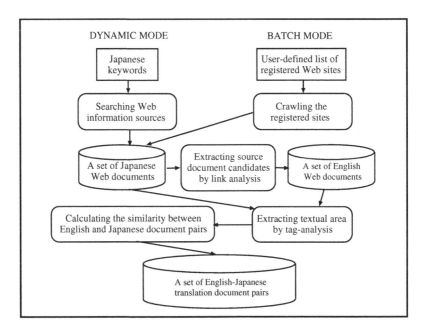

Fig. 1. The Overall Framework of the QRselect Prototype System

The prototype is implemented in Java and operates on Tomcat. Note that the search proceeds from Japanese documents to English documents, which is more efficient because the number of Japanese documents translated from English is much larger than the vice-versa.

3.2   Quantitative Evaluation

We evaluated the dynamic version of the QRselect prototype using 33 keyword sets provided by two translators and two evaluators. The keywords were roughly categorised into five groups, i.e. (a) geographical areas or countries; (b) current affairs; (c) information technology; (d) law, culture and sports; and (e) others.

The keyword sets are given in Table 1. The experimental settings were as follows: (i) the target number of Japanese pages to be retrieved in the experiment was set to 100, which means that for each keyword set, we retrieved 100 Japanese Web pages and detected the translation pages among them; (ii) Google was used as a search engine; (iii) the similarity threshold was set to 0.1, on the basis of a preliminary analysis of the performance. We decided on 100 Web pages because all the translators we consulted scan less than 100 snippets when they check related translations. Due to the system configuration, the total number of retrieved pages may not be exactly 100. Note also that, in servicing the system, we may use the Yahoo Japan search engine because the API provided by Yahoo Japan allows more searches per day than the Google API. There is not much difference in the performance of these two search engines as far as the Japanese pages are concerned. The choice of search engine is external to the system.

Table 1 shows the results of the evaluation. We only give basic figures for precision and recall, and do not give other IR-like performance measures applied to the ordered list of outputs [13], even though it is possible to order the output by means of similarity scores. This is because the user requirement for QRselect is to provide a sufficient amount of information with a manageable level of precision, and the concept of a "trade-off" between precision and recall is not relevant. Each row indicates a keyword set (33 in total). The meaning of the signs in the column are as follows:

- A: The total number of Japanese pages retrieved.
- T: The number of pairs consisting of Japanese translations and their English originals which are accessible through the link (= MH + Y). This is the target that the QRselect prototype should cover.
- C: Translation pair candidates output by QRselect. This is divided into:
  ○ CY: Correct output.
  ○ CE: Error, i.e. they are not a translation pair.
- M: Miss, i.e. when QRselect does not output the translation pair but the Japanese page is a translation of some English documents available online. This is further divided into[1]:
  ○ MH: Original page exists in HTML or related tagged forms.
  ○ MI: No tagged links, erroneous links or original pages.
- N: The number of non-translation Japanese pages which are correctly identified by QRselect as non-translations.
- P: Precision = CY/C.
- R: Recall = CY/T.

All in all, the system gave a modest performance, with the overall precision being 0.74 and recall 0.35.

---

[1] We also checked for misses caused by the fact that the document was in pdf format, but there were none in the data we evaluated.

Table 1. Evaluation of the QRselect Prototype System for 33 Keywords

Keyword set	A	T	C	CY	CE	M	MH	MI	N	P	R
(a) Colombia	92	1	1	0	1	2	1	1	89	0	0
(a) Colombia, drug	58	25	17	17	0	11	8	3	30	1	0.68
(a) Colombia, Uribe	32	20	19	19	0	1	1	0	12	1	0.95
(a) Venezuela	95	3	1	1	0	3	2	1	91	1	0.33
(a) Venezuela, Chavez	81	3	3	3	0	0	0	0	78	1	1
(a) Falluja	97	14	3	3	0	19	11	8	75	1	0.21
(a) Falluja, Aljazeera	96	31	7	4	3	29	27	2	60	0.57	0.13
(a) Baghdad, resistance	98	36	17	17	0	20	19	1	61	1	0.47
(b) Abu Graib, human rights	97	26	14	7	7	19	19	0	64	0.5	0.27
(b) separation wall	93	11	6	3	3	12	8	4	75	0.5	0.27
(b) Chomsky, Iraq, invasion	96	14	9	9	0	7	5	2	80	1	0.64
(b) Katrina, Hispanic	93	4	21	1	20	3	3	0	69	0.05	0.25
(b) China, censorship	98	30	12	12	0	20	18	2	66	1	0.4
(b) Sellafield, BNG	93	8	1	1	0	12	7	5	80	1	0.125
(b) Said, Arafat, Zionism	51	9	8	7	1	5	2	3	38	0.88	0.78
(b) Catholic, contraception	94	9	6	4	2	6	5	1	82	0.67	0.44
(b) veterans, suicide	91	3	4	1	3	4	2	2	83	0.25	0.33
(c) Torvalds	97	40	4	4	0	36	36	0	57	1	0.1
(c) Stallman	94	17	13	10	3	7	7	0	74	0.77	0.59
(c) Napster, file exchange	97	52	31	31	0	22	21	1	44	1	0.60
(c) Halloween document	79	2	5	0	5	7	2	5	67	0	0
(c) Linux, developing countries	93	13	15	10	5	4	3	0	1	74	0.67
(c) Google, library, scan	96	16	2	1	1	15	15	0	79	0.5	0.06
(d) Free culture	94	25	6	4	2	21	21	0	67	0.67	0.16
(d) Krugman, column	97	15	22	7	15	10	8	2	65	0.32	0.47
(d) Seattle Post, Mariners	94	36	1	1	0	40	35	5	53	1	0.028
(d) China, football	99	11	1	1	0	17	10	7	81	1	0.09
(d) Shunsuke, local, media	87	0	0	0	0	0	0	0	87	—	—
(d) F1, interview, driver	99	62	1	1	0	64	61	3	34	1	0.02
(d) Ghibli, export	63	1	0	0	0	2	1	1	61	—	0
(d) Hollywood, star, article	97	1	1	1	0	0	0	0	96	1	1
(e) Nablus report	100	23	11	10	1	23	13	10	66	0.91	0.43
(e) John Pilger	95	26	19	18	1	10	8	2	66	0.95	0.69
Total	2936	587	281	208	73	451	379	72	2204	0.74	0.35

3.3   Diagnosis

The overall figure, however, means little because the user is concerned only with the performance for a specific subject topic. Of much greater importance are the causes of errors (CE) and misses (MH).

Errors (CE) can be divided into the following patterns:

1. The Japanese pages are not translations, but refer to English documents as an information source or as related information. This pattern accounted for 67 cases, 41 of which were caused by a single Web site.

2. The Japanese pages are translations, but the system traversed the wrong link and detected false pages. This pattern accounted for six cases, four of which did not have correct links to the original. In two cases the system traversed the wrong link because many translated texts were contained in a single Japanese page.

Misses (MH) can be divided into the following patterns:

1. The link was not detected by the QRselect system, because the anchor link to the original document was not accompanied by the reserved words assumed by the QRselect system. This type of miss accounted for 187 cases.
2. The system properly detected the original text by traversing the link but identified that the pages were not translations at the stage of similarity calculations. This type of miss accounted for 192 cases. This was caused by (i) poor performance in the extraction of the textual area by tag analysis, and/or (ii) the limitations of the simple English-to-Japanese word transformations in the dictionary-based similarity calculation. Although these two causes are interdependent and it is difficult therefore to specify which is the main cause, in at least 25 cases the improvement of tag analysis is essential because in these cases the tag analysis failed to identify the main textual area. On the other hand, in 60 cases the similarity score was above 0.08 (note that the threshold was set to 0.1). For these cases it can reasonably be predicted that the improvement of dictionary-based matching methods will lead to a reduction in misses.

In summary, errors and misses were caused by three main factors, i.e. (a) mistakes in detecting anchor links to the original English pages, (b) errors or insufficiencies in textual area extraction by tag analysis, and (c) insufficiencies in dictionary-based similarity evaluation between Japanese and English pages.

4 The Social Model: Translators' Potential Contributions

Among the factors that caused the errors and misses just summarised, the latter two ((b) and (c)) are problems that should be technically solved. On the other hand, the first issue, i.e. mistakes in detecting anchor links to the original English pages, can and should be solved socially, although technical refinements are needed. The social solution can be achieved by promoting the involvement of and contributions by translators: If translators realise the merit of recycling and referring to information made by translators working in similar fields through QRselect, it is expected that they will agree to provide anchor links to the original English texts in a controlled and consistent manner. This is essential for such systems as QRselect, because QRselect has specific target users (online volunteer translators) to whose tasks it aims to contribute, and any successful system of this nature should evolve via interaction between the system and its users.

We consulted eight online volunteer translators about the possibility of adding extra tags or keywords to improve the performance of QRselect. Although most translators we consulted refused to use extra HTML- or XML-based meta-tags, most of them were at the same time happy to provide explicit anchor links to English originals by `<a></a>` tags on the translated document page, and a basic word near the anchor link to indicate that the link is to the original English page. Only one translator we consulted was positive about the use of meta-tags. This is probably partly due to the fact that most translators are reluctant to concern themselves with the technical aspects of the publication of their translated documents, and partly due to the fact that most online volunteer translators publish their own essays and comments as well as translations on a single page and do not manage sites specialised for translations.

Five translators also said that they would modify existing translations which they have published online to make them conform to a format that can be dealt with by QRselect.

If that sort of cooperation can be assumed by online volunteer translators, the 187 misses caused by a lack of reserved words near the anchor links would be avoided, as well as a certain number of misses categorised under MI. In addition, such cooperation may well contribute to reducing errors as well, because it would help reduce the erroneous identification of incorrect links to some referred pages. As the development of QRselect was triggered by requests from translators working online, there is a good chance of extending translators' cooperation in improving the system performance, which in turn would contribute to making a wider range of reference functions available to the translation community. Among the eight translators we consulted about this issue, four started adding specific keywords systematically to their anchor links. We are hoping that this cycle will not only enhance the performance of QRselect but also activate further online translators' activities through the use of QRselect.

5   Conclusions and Outlook

In this paper, we have introduced QRselect, a user-driven system for collecting subject-specific translation document pairs from the Web, and evaluated its performance. We also discussed the social aspect of the system, in which translators would contribute to the improvement of the system to their own potential benefit.

Although there is a general trend in the realm of Web-based information systems towards dealing with a huge, ever-increasing amount of data, there are areas where users require a limited but relevant range of data from the Web which is specific to their concerns. This can be understood in analogy to the relation between the quest for the universal library and the necessity for a personal library. These requirements are independent and complementary, and one cannot compensate for the other. Collecting and recycling existing translation document pairs relevant to the document that the translator is translating is one such area where relevance to the user rather than largeness of scale is required. Although the current performance of the QRselect prototype is moderate, there is a good chance of improving the system performance to make the system fully serviceable, especially given that we can assume translators' cooperation in making the system more effective.

In the evaluation, we focused on the dynamic mode of the QRselect system. In the real-world setting, however, we envisage that these two modes, i.e. the dynamic mode based on keywords and the batch mode based on registered Web sites, will be used in a mutually complementary manner. Although the current system performance is moderate, there is a good chance of improving the performance to a realistically useful level by improving technical aspects and by obtaining translators' cooperation.

Currently, we are enhancing the QRselect prototype in three directions:

1. Improving the performance of the module that extracts the textual area from tagged Web documents by tag-analysis.
2. Improving the granularity in calculating the similarity between Japanese and English documents using dictionaries.
3. Incorporating the non-linguistic clues to enhance the performance of similarity calculation.
4. Expanding the system so that it can deal with language pairs other than English and Japanese. We are currently modifying the system to cover English-French translations. We are also developing an interface through which users can modify and adapt the system to language pairs that the user wants.

In the fully operational system into which QRselect is incorporated, recyclable bilingual linguistic units, such as proper names, technical terms, fixed phrases and quotations [14], will be automatically extracted from translation document pairs and these units will be made available to users. In addition, we are also planning to extend the QRselect system by adding a module which automatically generates Japanese keywords when the translator specifies an English document to translate, instead of asking the translator to specify keywords to activate the QRselect dynamic module.

Acknowledgements

This research is partly supported by grant-in-aid (A) 17200018 "Construction of online multilingual reference tools for aiding translators" by the Japan Society for the Promotion of Sciences (JSPS) and the National Institute of Information and Communication Technology (NiCT). The authors would like to thank the anonymous reviewers for their valuable comments.

References

1. Boitet, C., Bey, Y., Kageura, K.: Main research issues in building web services for mutualized, non-commercial translation. In: Proceedings of the 6th Symposium on Natural Language Processing (2005)
2. Cao, Y., Li, H.: Base noun phrase translation using web data and the EM algorithm. In: Proceedings of COLING 2002, pp. 127–133 (2002)

3. Eijiro (2006), `http://www.eijiro.jp/`
4. Fukushima, K., Taura, K., Chikayama, T.: Fast and accurate method for detecting English-Japanese parallel texts. In: Proceedings of the COLING/ACL Workshop on Multilingual Language Resources and Interoperability, pp. 60–67 (2006)
5. Fung, P.: A statistical view on bilingual lexicon extraction. In: Proceedings of AMTA 1998, pp. 1–16 (1998)
6. Huang, F., Zhang, Y., Vogel, S.: Mining key phrase translations from web corpora. In: Proceedings of HLT/EMNLP 2005, pp. 483–490 (2005)
7. Kageura, K.: The status of "corpus" in human translation. In: Proceedings of the 12th Annual Meeting of the Japan Society of Natural Language Processing, pp. 452–455 (2006)
8. Morin, E., Daille, B., Takeuchi, K., Kageura, K.: Bilingual terminology mining – using brain, not brawn comparable corpora. In: Proceedings of ACL 2007, pp. 664–671 (2007)
9. Nagata, M., Saito, T., Suzuki, K.: Using the web as a bilingual dictionary. In: Proceedings of the Workshop on Data-driven Methods in Machine Translation, pp. 95–102 (2001)
10. Péry-Woodley, M.-P.: Quels corpus pour quels traitements automatiques? Traitement Automatique des Langues 36, 213–232 (1995)
11. Rapp, R.: Automatic identification of word translations from unrelated English and German corpora. In: Proceedings of ACL 1999, pp. 519–526 (1999)
12. Resnik, P., Smith, N.A.: The web as a parallel corpus. Computational Linguistics 29, 349–380 (2003)
13. Sakai, T.: For the realisation of better IR systems. IPSJ Magazine 47, 147–158 (2006)
14. Shinagawa, T., Mori, T., Kageura, K.: Extraction and alignment of textual blocks from online translation document pairs. In: Proceedings of the 12th Annual Meeting of the Japan Society of Natural Language Processing, pp. 520–523 (2006)
15. Shinyama, Y.: Webstemmer (2006), `http://www.unixuser.org/~euske/python/webstemmer/index.html`
16. Utsuro, T., Kida, M., Tonoike, M., Sato, S.: Collecting novel technical term from the Web by estimating domain specificity of a term. In: Matsumoto, Y., Sproat, R.W., Wong, K.-F., Zhang, M. (eds.) ICCPOL 2006. LNCS (LNAI), vol. 4285, pp. 173–180. Springer, Heidelberg (2006)

Humanities Graduate Students' Use Behavior on Full-Text Databases for Ancient Chinese Books

Ming-der Wu and Shih-chuan Chen

Department of Library & Information Science,
National Taiwan University, Taipei, Taiwan
{mdwu,r92126002}@ntu.edu.tw

Abstract. Digitizing ancient books, especially those related to the humanities, is practiced in many countries. The number of full-text databases in the humanities is increasing. Studies have shown that ancient books are important resources for humanities scholars and researchers. However, comparatively little research has been done concerning the use of those databases. Thirty graduate students majoring in Chinese Literature or History were interviewed in this study. This study attempts to answer the following questions: How do interviewees use the databases? Do they encounter any problems? What do they have to say concerning ancient books in digital or paper form? The results show that humanities graduate students use ancient books databases to locate information concerning their research interests. Most of them are satisfied with the search functions and feel that the databases are convenient to use. However, they comment that the coverage, quality, and search interface could be improved upon. As well, a few graduate students suggest that links to related resources should be added. They state that they do not totally rely upon the databases and continue to use paper sources.

Keywords: Full-text databases, Ancient Chinese books, Humanities graduate students, User behavior.

1 Introduction

A great deal of research has revealed that different forms of information resources hold different degrees of significance for researchers in different fields. In the field of science and technology, journals are more important than books, but in the humanities the reverse is true. Moreover, older books and documents are among the most important resources for research in the humanities. Due to developments in information technology, many resources are now available in the digital form. Some resources' original forms are digital to begin with. Additionally, there are many digitization projects being implemented around the world to transform paper resources into digital ones. Digitization of ancient books concerning the humanities is especially valued by those in academia.

Although the amount of the digital resources is increasing, there has been little research done concerning humanities researchers' use behavior of those resources. Graduate students are much more likely than humanities scholars to use digital resources. Barrett [1]

D.H.-L. Goh et al. (Eds.): ICADL 2007, LNCS 4822, pp. 141–149, 2007.

states that "graduate students are passing through some extremely formative years in their academic careers, where evolving research habits will influence a lifetime of scholarship to follow." Therefore, it is important to understand how graduate students use digital resources. Even though full-text ancient books databases have advantages such as being accessible remotely, time-saving, and having convenient search functions, how graduate students use these resources remains unknown. How do they search the databases? Do they encounter any problems? What do they have to say concerning ancient books in digital or paper form? These issues are worth taking a closer look at.

2 Literature Review

Stone [14] describes the diversity of research materials that humanities scholars require, most frequently in book and journal formats. She mentions that books play a greater role than journals. Talja [15] points out that humanities scholars typically use old or established theories to make sense of new topics. Reynolds [13] depicts that "humanists use books and older materials, rely heavily on their own personal collections, use a variety of material rather than a well-defined core of material."

The digitization of ancient humanities documents is regarded as important in academic circles. For example, looking abroad, a wide variety of Western classics figure heavily in Project Gutenberg [12]. The University of Virginia's Electronic Text Center has collected circa 70,000 digitalized humanities texts, including works by significant authors in the UK and the US, classical British and American novels, Bibles, and much more. Texts in 15 languages, which include *300 Tang Poems*, *Shi Jing*, *The Dream of the Red Chamber*, and many more Chinese classics, are available on the Center's website [5].

Digitization of ancient documents is also valued in Taiwan. The Academia Sinica's Electronic Chinese Ancient Document Project includes the *Twenty-five Dynastic Histories*, the *Thirteen Confucian Classics*, *Taisho Tripitaka,* Taiwanese historical materials, and other ancient books and records [7]. The National Central Library made efforts to digitize rare books and create a database that is searchable by metadata and offers links to the images of the documents [10]. In addition, the National Palace Museum and other institutions have projects to digitize ancient documents [11].

The number of humanities databases is increasing. However, relatively few studies have been done concerning humanities researchers' use of databases or online information retrieval systems. Some studies were done when information technology was in its infancy; the information technology of that time naturally cannot be compared to today's. Katzen [8] explains why humanities users rarely conducted online searches and indicates that there are relatively few humanities databases available and that retrospective resources are not adequately covered. Watson-Boone [17] indicates that if online databases were easier to use and if humanities scholars were to receive training and change their research habits, they could make better use of existing tools and services. Hoogcarspel [6] states that electronic texts in the humanities have problems such as a lack of information that is vital to humanities scholars such as title screens and a reference to the original print version. Bates [2] and her colleagues published a series of articles sponsored by Getty Information Institute. The two-year study concludes that most humanities scholars make little use of the online database. Scholars appreciate that the database covers many topics, but complain about the difficulty of its search language and the lack of availability of desired resources. It is interesting to note that scholars regard themselves as experts in their subjects and did not expect to learn anything new from the database. Massey-Burzio [9] investigates to what extent humanities faculty value information technology and view its relation to their research and teaching. She concludes that they "definitely feel the pressure to use and deal with technology." Humanities faculty appreciate the advantages of computer searches, but they feel it is uncomfortable and inconvenient to read off a computer screen for a long time. In a study comparing scholars in different disciplines use of e-journals and databases, Talja and Maula [16] classify humanities scholars as "low level users."

Chen [3] investigates students majoring in Chinese language who had used the ancient Chinese books' full-text databases. The study shows that students had a positive response to the Web version of the database concerning its screen prompts, search functions, response time, search results, results display, and the system when taken as a whole, but were less than impressed with the online help and error messages. In other words, users praised the convenience in using the databases, yet felt frustrated that the system did not provide immediate assistance when they encountered problems during the search process. Wu, Huang and Chen [18] survey the use of full-text databases for ancient Chinese books among humanities faculty. Their results show that professors affirmed the convenience of the databases, but pointed out several issues they encountered while using the system, such as the version of the paper source, accuracy, special characters, layout design, and search and browse functions. It is also noted that there is room for improvement in the coverage, quality of content, and search interface. During the period the research was conducted, all professors used full-text databases; however, they verified their findings with paper versions. Some professors felt that the ancient books databases were helpful and convenient in doing academic research but expressed concern with regard to their graduate students' over-dependency on digital resources.

Graduate students are more apt to use databases than humanities scholars. The study by Delgadillo and Lynch [4] shows that graduate students majoring in history are better than their professors at finding information on the Internet. Yet students' research methods and habits mirrored those of their professors. They conclude that future historians will "still have to adhere to principles of interpretation and creativity" which remain unchanged by information technology. Barrett [1] indicates that humanities graduate students frequently use a variety of electronic resources. Most students learn about electronic resources through their supervisors and colleagues or discover the resources on their own. Students praise several advantages of the electronic resources, including efficiency, the saving of time, the speed of word processing, and the convenience of remotely accessing full-text journals.

3   Methods

Thirty graduate students from the Department of Chinese Literature (15 students) and Department of History (15 students) of National Taiwan University were interviewed to investigate their use behavior of full-text databases for ancient Chinese books.

Participants consisted of 18 male and 12 female students. Selection of participants was based on recommendation by professors in these two departments. Most participants are either working on their master's theses or doctoral dissertations or preparing their research topics. The interview schedule was similar to the one being used in an earlier research conducted by the authors [18] with minor revisions. The questions in the schedule were divided into three categories. The first category was designed to analyze the participants' use behavior on the full-text databases. The second category investigated participants' views on paper versions and databases. The third category explored participants' suggestions for improving the databases. All participants have prior experience in using the databases. In order to help participants better recall their experience of using those databases, at the beginning of each interview, the list of the University Library's full-text databases for ancient Chinese books was presented to each participant. Interviews were recorded and averaged 45 minutes in length.

4   Results

4.1   Use of Full-Text Databases for Ancient Chinese Books

4.1.1   Learning About the Databases
All except one participant learned about the databases from their peers and instructors. Some instructors who commonly used the databases introduced their students to them or taught what to pay special attention to while using them. Six participants mentioned that they also learned of the databases during courses in library instruction.

Most databases provide user manuals or help messages to give information about the usage of the databases. Four participants feel that these manuals or help messages are either too long or too complex and therefore rarely use them. Participants explained that after using the databases a few times they would master it themselves. One participant mentioned that she once asked for a librarian's assistance and learned how to use the databases thereafter.

4.1.2   Installation of Software
A few databases for ancient Chinese books come in an Internet-based CD-ROM version. Users need to borrow the CD-ROM from the library and install the software onto their own computers in order to use the databases. Participants in this study seldom install the software as they feel it to be troublesome, and it is seen as more convenient to use Web-based databases.

In order to display some variant form of Chinese characters, old characters, or rarely-used characters on the computer screen, a few databases provide character creation software for users to install. Without the software, random codes or spaces will appear on the screen instead of the appropriate characters. Six participants mentioned that they do not install the software because they are incapable of doing so.

4.1.3   Methods of Retrieval
While accessing the databases, participants use more search functions than browse functions. Search functions are utilized when participants have keywords (such as

personal names, place names, proper nouns) on hand, whereas browse functions are utilized when participants know or roughly know the location of the sources they are looking for. If participants do not have access to a paper version and need to read the texts quickly, they would browse the full text on the screen.

When conducting a search, most participants used basic searches rather than combined searches (i.e. Boolean logic search). In the basic search, participants mainly used two or more characters as opposed to a single character. They explained that using a single character would retrieve too many results, making it difficult to find relevant items. On the other hand, a few participants preferred to use a single character and then to filter through the long list of results. Participants did not make use of combined searches because they were either unaware that it could be used to perform searches within the databases or because they felt that this kind of search would retrieve fewer results. Although most of the participants used a basic search method only, twenty-one of them responded that they were satisfied with search results done in these databases.

4.1.4 Reading Methods
Twenty-four participants said that they usually read full-text ancient books directly on the screen. Only a few printed out the documents to read them. Some participants mentioned that they copy sentences or paragraphs and save them to their computer for reading at a later time. Those who read directly on the screen mainly did so because their purpose was to locate information. After locating what they were looking for, they would turn to the paper version

4.1.5 Difficulties Encountered
Ancient Chinese books are printed vertically while texts are displayed horizontally in the databases. The horizontal layout was not favored by humanities graduate students because they are not accustomed to reading that way. Five participants mentioned that the divide between the main body of text and its annotations was not clear in certain databases, which make them difficult to read. One participant mentioned that it was difficult to read the annotation because its font was smaller than the main text. Some databases do not state the source from which they were digitized, which made it troublesome for students to determine the appropriate counterpart paper version. Three participants said that they disliked the search interface of some databases and felt that to search in paper versions was easier. "I don't know how to start my search and it is slow to flip the pages on the screen," said one participant. Some participants complained that they needed to click back several pages if they wanted to start a new search. The complexity of the multi-layer structure forced users to click through several pages to locate what they were looking for. If the speed of their home computer's Internet connection was slow, then much time was wasted in using the databases.

4.2 Comments on Paper Versions and Databases

4.2.1 Choice Between Paper Version and Database
When searching for information, most participants would first use the databases to locate the source of the information and then confirm their findings using paper

versions. Participants agreed that searching the databases was comparatively faster and more convenient than searching paper versions. They found that they could gather more results which may otherwise have been missed had paper versions been searched. Also, there is no time and location limit in using the databases, which makes it a better choice. Texts in some ancient books in the databases are punctuated. Four participants said this made the texts much more readable than those in paper versions that did not contain punctuation.

However, participants mentioned that errors showed up in the ancient Chinese books databases, such as incorrect characters and punctuation. They found the accuracy of the databases to be questionable. Therefore, when writing term papers, theses, dissertations, or other important articles, all participants mentioned that they would verify quotations in corresponding paper versions. Most of the time, they used the full-text databases as a tool to find information, but they preferred to use paper versions for further reading.

Participants complained that texts were displayed fragmentally on the screen, meaning that the contextual structure was lost. They preferred to read the paper version to avoid misunderstanding the texts. Moreover, it was found to take more energy to read text on-screen. Participants explained that it was uncomfortable to read text on-screen for long periods of time.

4.2.2 Effect on Theses or Dissertations

Participants responded that full-text databases for ancient books had a noticeable effect on their theses or dissertations, especially with regard to the amount of information they could collect. Participants mentioned that the time spent collecting information would be much longer if the databases did not exist. Thirteen participants   generally felt that using databases broadens the scope of their theses or dissertations. Many of them mentioned that copying and pasting sentences or paragraphs from the full-text databases saves time when they write their theses or dissertations.

4.2.3 Dependence on the Databases

Among the 30 participants, 12 indicated that the existence of full-text databases would not cause them to depend less on paper versions. However, the other 18 participants stated that being able to use the databases has indeed caused them to refer less frequently to paper versions, particularly while in the information gathering stage. They explained that when conducting research, they could not solely depend on a single source, and that the full-text databases helped them gain access to a number of comprehensive resources from related fields.   Although some students used paper versions less frequently, this does not mean that the databases can completely replace paper sources. "Paper versions are irreplaceable. They are much comfortable to read and they provide contextual structure," said one participant.

The study by Wu, Huang and Chen [18] indicated that some professors, especially senior professors, are concerned about the effect full-text databases would have on graduate students in their research. Professors worried that graduate students will totally depend on the databases and refer less frequently to paper versions. Most participants remarked that they understood their professors' concern, but they recognized that the full-text database was an increasingly indispensable tool. In the past, researchers would need to go through the traditional research process, which required extensive reading. Participants said that they spend less time searching for information but still need to build on their ability to comprehend and analyze. They understood that research could still be conducted without the full-text databases but that more time would be needed. Some participants were not concerned, as they felt if their professors had any concerns about the information students gleaned from full-text databases, they could simply request that their students confirm the information in paper versions.

4.3 Suggestions to Full-Text Databases for Ancient Chinese Books

4.3.1 Coverage
Participants felt that the coverage of ancient Chinese books databases should be as exhaustive as possible. Annals, notes, novels, and plays were mentioned as deserving representation in the database in addition to the standard classics. A few participants suggested that important annotations to ancient books should also be covered.

4.3.2 Quality
Participants suggested that errors such as incorrect characters and punctuation should be avoided. They also suggested that images should also be included in the database for the purpose of verifying the errors found in the text file. The databases should give a clear indication of the version of its paper source. Participants questioned the accuracy of the databases and recommended that databases use authorized versions or popular versions in order to improve their credibility. "It is unnecessary for a database to collect all versions, but the version covered should be the best one," said one participant.

4.3.3 Search Interface
Ancient Chinese people usually had several names. A place name may also change over time. Participants suggested that biographical dictionaries, gazetteers or authority files for personal and place names be included in the databases, such that any form of a name used in a search can retrieve all documents containing different evolutions of names. To make a precise search, some participants suggested that the databases should allow users to specify their searches by personal name or place name. A search limited by dates should also be provided. As for browsing, a detailed table of contents would help users to browse any level or part of the book. Participants said databases would be easier to use if there were a unified search interface for all databases. A cross-database search function was also recommended. Generally speaking, participants expected a much more user-friendly search interface.

4.3.4 Display of Search Result
The length of search results is usually short: only the sentence or paragraph that includes the keyword was displayed. Sixteen participants suggested that more sentences or paragraphs should be displayed in order to help the user understand the context.

4.3.5 File Saving
When copying an entire web page to a Word file, participants sometimes encountered problems such as the typeset failing to display correctly or annotations being lost. To

solve such problems, they would need to re-organize the typesetting. They suggest users would benefit if different file formats, such as Word, .txt, or PDF, were provided within the databases.

4.3.6 Other Functions

Participants suggested that hyperlinks to related documents, such as commentaries or interpretations about a given book, be included. This kind of snowball technique links related materials and thus broadens the scope of searches.

Three participants also spoke of including a space for discussion within the databases for users to share their opinions or comments. Personalized settings, such as layout settings or pre-selected databases, could be provided in accordance with individual needs.

5 Conclusions

Researchers in the humanities are stereotypically seen as less likely to access the Internet and use digital resources and to be less familiar with information technology. However, all humanities graduate students interviewed in this study accessed the Internet and ancient books databases. They appreciated the advantages of the databases and have no major problem in searching the databases. Humanities graduate students took a serious approach to utilizing the databases in that they always verified their findings in paper versions.

In comparison with the study made by Wu, Huang and Chen [18], there is no significant difference concerning the use of full-text databases for ancient Chinese books by humanities scholars and graduate students. However, professors relied less on the databases and emphasized the importance of paper versions, while harshly criticizing the flaws found in the databases. Graduate students, on the other hand, were found to be more familiar with information technology. They had more proposed changes, including concrete suggestions concerning the design of the database's search interface. For the most part, professors are familiar with the contents of ancient books. They basically use the database as a tool to locate or cross-check information. Just like the findings of Bates [2], professors do not expect to learn new things from the database. However, graduate students may use the database to discover information to broaden their knowledge on the subjects.

Full-text databases for ancient books are important research resources in the humanities. In the future, the number of databases is certain to rise. Digitization of paper versions is, however, merely the first step. Creators of databases should understand user behavior and utilize the newest information technology to enhance database functions such that databases may become more important and relevant to the humanities.

References

1. Barrett, A.: The Information-seeking Habits of Graduate Student Researchers in the Humanities. Journal of Academic Librarianship 31, 324–331 (2005)
2. Bates, M.J.: The Getty End-User Online Searching Project in the Humanities: Report No. 6: Overview and Conclusions. College and Research Libraries 57(6), 514–523 (1996)

3. Chen, T.C.: A Case Study on Human-Machine Interaction for the Handy Text Retrieval System. Journal of Library and Information Science 24(2), 65–85 (1998)
4. Delgadillo, R., Lynch, B.P.: Future Historians: Their Quest for Information. College & Research Libraries 60, 245–259 (1999)
5. Electronic Text Center: Electronic Text Center Collections, http://etext.lib.virginia.edu/collections/
6. Hoogcarspel, A.: The Rutgers Inventory of Machine Readable Text in the Humanities: Cataloging and Access. Information Technology and Libraries 13, 27–34 (1994)
7. Hsieh, C.C., Lin, H.: A Survey of Full-text Data Bases and Related Techniques for Chinese Ancient Documents in Academia Sinica, http://www.sinica.edu.tw/tdbproj/handy1/
8. Katzen, M.: Application of Computers in the Humanities. Information Processing and Management 22, 259–267 (1986)
9. Massey-Burzio, V.: The Rush to Technology: A View from the Humanists. Library Trends 47, 620–639 (1999)
10. National Central Library: Guide to Rare Books Image System, http://rarebook.ncl.edu.tw/rbook.cgi/store/frameset.htm
11. National Digital Archives Program, Taiwan: Outcomes of the Program: Rare Books, http://www.ndap.org.tw/2_catalog/visit_folder.php?id=386
12. Project Gutenberg Literary Archive Foundation: Online Book Catalog: Top 100, http://www.gutenberg.org/browse/scores/top
13. Reynolds, J.: A Brave New World: User Studies in the Humanities Enter the Electronic Age. Reference Librarian 49(50), 61–81 (1995)
14. Stone, S.: Humanities Scholars: Information Needs and Uses. Journal of Documentation 38, 292–313 (1982)
15. Talja, S.: Information Sharing in Academic Communities: Types and Levels of Collaboration in Information Seeking and Use. New Review of Information Behavior Research 3, 143–160 (2002)
16. Talja, S., Maula, H.: Reasons for the Use and Non-use of Electronic Journals and Databases: A Domain Analytic Study in Four Scholarly and Disciplines. Journal of Documentation 59, 673–691 (2003)
17. Watson-Boone, R.: The Information Needs and Habits of Humanities Scholars. RQ 34, 203–216 (1994)
18. Wu, M.D., Huang, W.C., Chen, S.C.: Humanities Scholars and Chinese Ancient Books Databases. Journal of Library and Information Studies 4 (2006 in press)

Annotations and Digital Libraries: Designing Adequate Test-Beds

Maristella Agosti, Tullio Coppotelli*, Nicola Ferro, and Luca Pretto

Department of Information Engineering, University of Padua, Italy
{agosti,coppotel,ferro,pretto}@dei.unipd.it

Abstract. The increasing number of users and the diffusion of *Digital Libraries (DLs)* has increased the demand for newer and improved systems to give better assistance to the user during the search of resources in collections managed by *Digital Library Systems (DLSs)*. In this perspective, the annotations made on documents offer an interesting possibility for improving both the user experience of the DLS and the retrieval performance of the system itself. However, while different approaches based on annotations have been proposed, they still lack a full experimental evaluation, mainly because an experimental collection with annotation is missing. Therefore, this paper addresses the problem of setting an adequate experimental test-bed for DL search algorithms which exploit annotations, and discusses a flexible strategy for creating test collections with annotated documents.

1 Introduction

When users search a *Digital Library (DL)* they usually need answers to their information needs. They interact with the *Digital Library System (DLS)* to materialize, to the best of their abilities, their need for information. After this step, the system, interacting with the DL content, tries to retrieve the maximum possible number of documents relevant to user queries. End-users hope that the DLS can help them meet their needs by providing the documents that they are searching for. This, however, turns out to be a hard and twofold problem, because on the one hand users find it difficult to correctly explain their needs (i.e. to materialize the query to the system) and on the other DLSs have problems finding the correct resources (i.e. retrieving documents useful to the user).

Several studies have been performed to identify newer and better algorithms which aim to improve retrieval effectiveness and better satisfy end-user information needs. In this perspective, the annotations made on documents offer an interesting possibility for improving information access performance. The additional information contained in the annotations and the hypertext which connects annotations to documents are exploited to define search strategies which

---

* The work of Tullio Coppotelli has been partially supported by a grant of MicroWave Network S.p.A., Padua, Italy.

D.H.-L. Goh et al. (Eds.): ICADL 2007, LNCS 4822, pp. 150–159, 2007.

merge multiple sources of evidence, thus increasing system effectiveness and helping users to meet their needs. DLSs which implement these new techniques require the correct evaluation of both the user interaction with the system and
system effectiveness. However, while different approaches based on annotations
have been proposed, they still lack a full experimental evaluation; this is because
an adequate test collection is missing. Without a test collection with annotated
documents setting up a correct evaluation test-bed and comparing DLSs that
use annotations with DLSs that do not is impossible. It is then hard to decide if
these approaches introduce improvements and which of them work better. This
paper focuses on the effectiveness aspect and addresses the problem of designing
an adequate experimental test-bed to evaluate DL search functionality which exploit annotations. The next section overviews related work. Section 3 describes
the main characteristics of our approach. Section 4 presents two algorithms that
cooperate to create the annotated test-collection. Finally, Section 5 concludes
the paper.

2   Annotations and Digital Libraries

The concept of annotation is complex and multifaceted and covers a wide range
of different areas. Annotations can be considered as metadata: additional data
which concern an existing content and clarify the properties and the semantics of
the annotated content. In this sense, annotations have to conform to some specifications which define the structure, the semantics, the syntax, and, maybe, the
values annotations can assume. On the other hand, annotations can be regarded
as an additional content which concerns an existing content: they increase the
existing content by providing an additional layer of elucidation and explanation.

Full advantage of annotations can be taken by providing a DLS with annotation capability [1]. The primary effect of introducing annotations is to enrich the
DL content; for example, by using annotations the content of a document can
be broadened with personal considerations to propose different points of view
or to underline text passages that need further discussion. In addition, annotations allow users to actively integrate DLs into their way of working to create
a cooperative environment where annotations become the medium for users to
communicate with each other.

Another important characteristic of annotations is their heterogeneity. Annotations in DLs are created by different authors with different backgrounds
and at different times: the user who annotates a document may know more recent information then the author about the topic; he or she may disagree with
the document content and would like to communicate this different opinion to
the readers of the document. This heterogeneity is a key-point that allows for
dynamic improvement in the content of the document, and by using this new
information it is possible to better estimate the relationship between documents
and queries, a feature which is so important in document retrieval.

Finally, different media can be annotated such as text, video or images and annotations themselves can be multimedia objects. However, this study focuses on

the use of textual annotations to annotate textual documents. For an extensive study on annotations and their formal definition refer to [2].

Golovchinsky et al. [4] proposed the use of highlight annotations as a way to implement query expansion and relevance feedback. The results showed how this approach increases the effectiveness of the system with respect to the simple use of relevance feedback, but it limits annotation to only one facet: their use as a relevance feedback. Frommholz et al. [5] proposed a system that implements annotations for collaboration among scholars. Annotations were used to provide advanced content and content-based access to the underlying digital repository. This work adopted a broader view on annotations and enables the creation of a collaborative experience over the DL (increasing the user experience of the DLS) but it does not present any evaluation of the system effectiveness. Agosti and Ferro [3] proposed an algorithm that allows the concurrent search of documents over multiple DLs. Annotations were used to naturally merge and link personal contents with the information resources provided by the DLSs and were exploited during the research not only to rank documents better but also to retrieve more relevant documents. This study lacks an extended evaluation but it is the first which calls for a test collection that could enable the evaluation of effectiveness of these kinds of systems.

3   A New Approach to Test-Bed Design

The use of test collections to evaluate the effectiveness of DLSs is a commonly accepted practice. Existing evaluation campaigns (TREC[1], CLEF[2], NTCIR[3]) have provided a wide range of test collections. They are reusable, produce reproducible results, encourage collaboration among researchers and cross comparison of system performance. Although these collections are general purpose and are suitable for a wide range of systems, when it comes to the need to evaluate new techniques it is possible that an adequate test collection is still lacking and consequently needs to be created. Nevertheless, creating a new collection could be itself a hard task requiring resources and time. Therefore, it can be useful to accomplish an intermediate step that still allows a reliable evaluation of the effectiveness using an alternative technique as proposed for example in [6]. Carterette et al. [7] observed that this can be the case when a researcher is performing a preliminary investigation of a new retrieval task. Hence, when it comes to the evaluation of DLSs with annotated documents, the use of an alternative technique for collection creation is a viable option.

The usual approach to test collection creation, the TREC approach [8], requires: 1) finding and acquiring a suitable set of documents; 2) manually creating annotations and topics; and 3) evaluating the relevance of documents to each topic, i.e. deciding which documents are relevant to those topics. In the case of an annotated collection: 1) assessors cannot be used to manually create the set

---

[1] http://trec.nist.gov/

[2] http://www.clef-campaign.org/

[3] http://research.nii.ac.jp/ntcir/index-en.html

of annotations because a wide range of annotations written by different authors in different periods of time is needed to maintain their heterogeneous nature; 2) the pooling method is used to reduce the number of documents which human assessors need to assess for each topic. This method relies on the existence of a certain number of experiments but, in the case of annotated documents, the lack of these experiments prevents us from applying this method; hence it would be necessary to judge the relevance of each document to each topic. All these additional issues make the creation of an annotated test collection particularly difficult and, once again, confirm the need for new strategies.

This paper deals with these problems and presents a new strategy that enables the fast creation of a reliable test collection with annotated documents. The proposed technique requires starting from an already existing test collection and then creating a parallel collection of related annotations over it. These annotations are human written documents themselves that are matched to other documents on the basis of objective features, thus trying to simulate the behaviour of a human annotator that does not simply underline some passages but annotates passages of document with extensive annotations. The only constraint for the starting collection is that the documents have to be objectively divisible in more than one set (the motivation will be clarified later in this section). This strategy is not limited to the creation of a single collection: by using as a starting point collections with different characteristics, monolingual or multilingual, general or specialized, it enables the multiple creation of new collections that inherit the characteristics of the original ones. This strategy reduces the overall effort of the collection creation and has the following advantages: 1) the results of systems evaluation with the new collection are directly comparable with the results previously obtained with the original one; the testing of the systems with both collections enables the direct comparison between systems that use annotations and systems that do not, allowing the evaluation of improvements and in general the impact that new algorithms or their refinements have in DLSs; 2) it exploits existing pools to deal with a sufficient number of experiments without the expensive need for a new assessment; 3) it allows the fast creation of multiple collections with different characteristics and the evaluation of the algorithms in different contexts; and 4) it respects the heterogeneous nature of annotations.

The starting test collection can be represented as a triple $C = (D, T, J)$ where D is the set of documents, T is the set of topics and J is the set of relevance assessments defined as $J = D \times T \times \{0, 1\}$ (binary relevance). The documents D of the chosen test collection must be divisible in two disjoint sets, D_1 and $\hat{A}$, where $D = D_1 \cup \hat{A}$ and $D_1 \cap \hat{A} = \varnothing$. We have conducted preliminary experiments where D_1 were newspaper articles and $\hat{A}$ were agency news of the same year [9]. The annotated collection is $C' = (D_1', T, J)$, where D_1' contains exactly the same document as D_1 with the addition of annotations over these documents. Topics and relevance assessments are exactly the same. In C' we use a subset A of $\hat{A}$ to annotate the documents in D_1, thus $\hat{A}$ is the set of candidate annotations and A is the set of actual annotations. The strategy goal is then to find which candidate annotations can be used to correctly annotate documents in D_1 and create the annotation hypertext over these documents. To identify these relationships we take advantage of the fact that in C the topics are made over both D_1 and $\hat{A}$ (thus their relevance to each topic has been judged): if in C both a candidate annotation and a document have been judged relevant to the same topic then we infer that it is possible to annotate that document with that candidate annotation. Referring to Figure 1, these couples (document, annotation) are those connected by a two-edge path in the undirected graph $G_1 = (V_1, E_1)$ where $V_1 = D_1 \cup T \cup \hat{A}$ and $E_1 = (D_1 \cup \hat{A}) \times T$. In G_1 each edge represents a human assessment i.e. a path between annotation $\hat{a}$ and document d passing through topic t means that a person assessed both $\hat{a}$ and d relevant to t. This relevance property creates a path between documents and candidate annotations that is used in Section 4 to introduce annotations in C'. The intuition is that the strength of these paths allows the use of candidate annotations as real annotations for connected documents and that these annotations reflect human annotative behaviour.

4   The Two Cooperating Algorithms

In this Section we introduce the two cooperating algorithms; in 4.1, we shortly outline a method that uses the human information contained in the original collection to introduce annotation in C, because full details where given in [10]. In 4.2, a new automatic technique is proposed which can be partnered to the one of 4.1 to discover an additional set of annotations.

4.1   Exploiting Assessor Assessments

Once graph $G_1 = (V_1, E_1)$ is given, the problem of matching a candidate annotation with a suitable document can be addressed. The proposed algorithm makes use of the human relevance assessments in C for matching candidate annotations with documents. The first aim of the algorithm is to match each candidate annotation $\hat{a}$ with the most suitable document d, bringing to the surface the relationship between documents. These matches respect the annotation constraint proposed in [3], i.e. one annotation can annotate only one document, and when more than one match is possible, the algorithm heuristically tends to choose matches which maximize the number of annotated documents—indeed, maximizing the number of annotated documents is the second aim of the algorithm. If a match is possible then $\hat{a} \in A$ otherwise $\hat{a} \in (\hat{A} - A)$ and, at this point, cannot be used as a real annotation.

The algorithm works in two phases. In the first phase it constructs a weighted bipartite graph G_b on the basis of G_1, i.e. the graph whose edges represent positive relevance assessments. In the second phase the algorithm works on the weighted bipartite graph G_b to properly match a candidate annotation with a document. The construction of the weighted bipartite graph $G_b = (V_b, E_b)$ is immediate (see Figure 1): the vertices of G_b are all the vertices of G_1 which represent documents or candidate annotations, that is $V_b = D_1 \cup \hat{A}$, and an edge

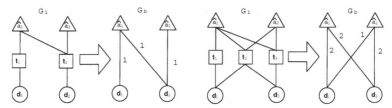

Fig. 1. Examples of the construction of graph G_b, starting from graph G_1

between candidate annotation $\hat{a}$ and document d exists if and only if $\hat{a}$ and d have been judged relevant to at least one common topic, that is $t \in T$ exists such that edges $\hat{a}$-t and t-d are in E_1. Moreover, a weight is assigned to each edge $\hat{a}$-d in E_b, which gives the number of common topics between $\hat{a}$ and d. These weights take account of the fact that when $\hat{a}$ and d are assessed as relevant to more than one common topic at the same time, it is reasonable to suppose that the bond between the candidate annotation $\hat{a}$ and the document d will be strengthened. Once G_b is constructed, the second phase of the algorithm works on G_b to reach the two aims described above (a detailed presentation has been presented in [10]).

4.2   Exploiting the Whole Pool

The method suggested in 4.1 has the great advantage of bringing to the surface hidden human work, matching annotations with documents using the judgements that human assessors gave during relevance assessments. With this approach only the candidate annotations that have been judged relevant to at least one topic can be matched; it is important to note that it cannot find any match for all the other candidate annotations. Although in our preliminary experiments the number of couples (document, annotation) that this method could find was promising, in the original collection there were still a certain number of good couples that were not matched. Hence we propose an automatic technique that tries to bring to light these couples.

So far only the information about the relevance of documents to topics has been used and other information contained in C was discarded. Now we focus on the pool of documents and particularly on the reasons that caused the documents to be inserted in the pool for a certain topic (and at a later time assessed). Since we utilized one of the CLEF test collections for our initial experiments we are in the condition to exemplify by referring to that collection. The following topic is useful to illustrate how information previously discarded now can be used: "Alberto Tomba's skiing victories". As usually happens with the pooling method, the pool corresponding to that topic contains both relevant and not relevant documents; it includes not only documents about Alberto Tomba's skiing victories but also documents about skiing competitions where Alberto Tomba did not participate, those where he participated without winning and documents about his social life. While these documents have been judged not relevant to the topic and then are useless for the previous algorithm, they still contain useful information that can be used to introduce new annotations in C' coupling documents and annotations about the same subject matter, e.g. the social life of Alberto Tomba. With this aim, a complementary algorithm is proposed that first creates more paths between candidate annotations and documents building the graph using the whole pool and then brings to light the most interesting.

We define E as the set of all edges that can be created using the whole pool: for each topic t an edge e is added to the graph between t and each document d_i or annotation a_j such that $\forall i, j \ d_i$ and a_j are in the pool of topic t (the previous set E_1 is a subset of E). A_2 is the set of actual annotations matched with the previous algorithm and E_2 is the subset of E incident with A_2. A new graph $G_2 = (V - A_2, E - E_2)$ is then obtained using the whole pool and removing, due to the annotation constraint, all the candidate annotations already matched by the previous algorithm. Starting from G_2 a bipartite graph G_{b2} is built using the topics as connections. The main difference with the previous method is that these edges no longer reflect human assessments. The drawback of the choice to include all the documents of the pool is that there are paths in the graph that are not suitable for use as annotations for any documents. As a consequence good annotations with the previous algorithm can no longer be identified and a new strategy is required to evaluate the quality of the relationships between candidate annotations and documents and to decide which edges can be used to annotate a document. With this goal in mind, four evaluation parameters are introduced and their score is merged to compute a unique weight for each edge of G_{b2}: the affinity between topics, the score obtained using an *Information Retrieval Tool (IRT)*, the annotation generality and their temporal nearness. Each parameter measures a different aspect of the relationship between documents and candidate annotations, and their union permits an objective measure of annotation suitability. This algorithm can no longer match all annotations with document, but it does aim to annotate the greatest possible number of documents with good quality annotations; the very poor quality of some candidate annotations prevents their use as annotations, even if some of them could annotate non-annotated documents.

The affinity between documents and candidate annotations is a score P_a, ranging, like the other parameters, between 0 and 1. It uses G_{b2} structure to measure the superimposition in the content of two or more topics and the similarity of documents involved. The probability that two documents cover the same subject matter increases when the affinity increases, while edges incident with vertices with very low affinity lead to bad annotations. The formal definition of affinity between topics T_i and T_j $(i \neq j)$ is:

$$P_a^{(ij)} = \frac{|T_{ij}|}{max(|T_{ij}|)}, \quad \texttt{where} \quad T_{ij} = \{(\hat{a}, d) \subseteq \hat{A} \times D_1 | \hat{a} \in \hat{A}_i \cap \hat{A}_j, \texttt{ and } d \in D_i \cap D_j\}$$

where $D_i, D_j \subseteq D_1$ are the sets of documents that are in the pool for these topics and $\hat{A}_i, \hat{A}_j \subseteq \hat{A}$ are the sets of annotations in the same pool.

The pooling method used to create the pool of documents in the original collection, and then to create the graph G_{b2}, selects the documents that enter the pool using the experiments run by different systems. These experiments contain information about both the ranking of the documents and their score. Although this information is discarded after the creation of the pool, it could be used to weight each edge in the graph G_{b2}. It is then useful to recompute these scores using an IRT, focusing directly on the relevance of a document to an annotation without considering their relevance to a topic. The idea is to bring to the surface documents that cover the same subject matter but that are not relevant to the same topic. High scores along not relevant paths do not contradict the assessor assessments because assessors only judged the relevance to a given topic while documents can still be related to some other topic not considered in the original collection, like in the previous VIP example. This score is computed by first creating an index over all the candidate annotations and then querying the system using the content of each document as query. In this way we obtain an ordered list of possible annotations for each document. The first K annotations of that list are considered valid while the edges between documents and annotations not in the list are deleted from G_{b2} because their relationship is too weak, i.e. the superimposition on their content is low. The use of an IRT obtained the twofold result of eliminating from G_{b2} weak edges and weighting those remaining with the score P_{ir} assigned by the IRT.

The generality score P_g is computed based on the inverse number of edges incident to the annotation's vertex, that is, the number of topics per annotation. In G_{b2} it is no longer true that increasing the number of topics in which a couple (document, annotation) belongs also increases the quality of that couple; it is only the generality of the annotation which increases. An annotation included in a lot of topic pools necessarily has to be a generic one.

The last parameter, P_t, measures the temporal nearness between documents and annotations, regardless of their order (it does not matter which comes first). The probability that documents and annotations cover the same subject matter increases when the temporal nearness increases. It is more probable to find good matches considering documents and candidate annotations temporally near.

Once these parameters have been defined, it is convenient to compute a unique score to evaluate the strength of each (document, annotation) couple: Score $S = \alpha_a * P_a + \alpha_{ir} * P_{ir} + \alpha_g * P_g + \alpha_t * P_t$ with $\alpha_a + \alpha_{ir} + \alpha_g + \alpha_t = 1$. The discriminating power of these parameters is not equal so α_a, α_{ir}, α_g and α_t have been introduced to correctly weight their importance. These weights have to be set depending on the original test collection, but it seems convenient to use a high weight for the IRT score (the most important) and a lower weight for the generality score (the less important one). Once a unique weight S is computed the maximum number of couples of vertices from the graph G_{b2} needs to be selected taking into account their quality. The algorithm presented in Section 4.1 cannot be applied because now the matching problem is more complex. The main difficulty is that there is a trade-off between the number of documents that can be annotated and the quality of these annotations: if the algorithm simply selects all possible annotations ignoring their weight, the result would be a collection with annotations of poor quality while, on the other hand, selecting only the

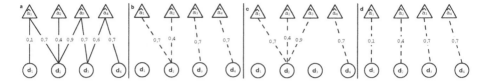

Fig. 2. From left: a is the starting graph, b is obtained with the suggested algorithm, c is obtained maximizing the score and d maximizing the documents coverage

best matches, very few documents could be annotated, reducing the advantage of this new approach.

A new greedy algorithm is proposed that resolves this trade off by proceeding in phases and trying to maximize both the number and the quality of the annotations. The examples in Figure 2 help to understand the goals of the algorithm. Starting from the input graph in Figure 2a it is possible to maximize the score of the edges (Figure 2c), obtaining a total score of 2.7 with the drawback of annotating only 2 documents over 4. The choice to maximize the number of annotated documents (Figure 2d) instead obtains a score of 1.9 with 4 documents annotated. The proposed strategy deals with the trade-off by obtaining the comprehensive score of 2.5, annotating 3 documents. The result is the annotation of an average number of documents over the two presented choices, paying only a negligible loss in the overall quality of the selected annotations. To obtain this result, the annotations that can annotate only one document are selected first. Because these annotations cannot annotate other documents, only in this case is it sufficient to apply a threshold to their quality, selecting all the edges over this threshold. In a second phase the annotations of good quality are preferred. In this phase it is important not to make counterproductive choices and only those couples $(\hat{a}_i, d_j)$ are accepted where the document d_j could not be annotated by other good annotations. The third phase is the most complex. In this phase the algorithm selects only those edges that allow it to annotate documents that are not already annotated. For each d_i not annotated a search is made for the best annotation $\hat{a}_j$ and then for the best document d_k that this annotation could annotate. If $i = k$ then the following statements are true: d_i is not already annotated, a couple $(\hat{a}_n, d_i)$ or $(\hat{a}_j, d_n)$ does not exist with a score better than that of $(\hat{a}_j, d_i)$. When these statements are verified the couple $(\hat{a}_j, d_i)$ is the best choice and if the weight of that arc is over a threshold, the document can be annotated. After these phases, to annotate more documents the algorithm relaxes the constraint $i = k$, allowing the selection of those couples where $|P(\hat{a}_j, d_k) - P(\hat{a}_j, d_i)| < \epsilon$, defining $P(\hat{a}_j, d_k)$ and $P(\hat{a}_j, d_i)$ as the weights of edges $(\hat{a}_j, d_k)$ and $(\hat{a}_j, d_i)$. With this relaxed constraint other good annotations can be introduced into the collection, thus permitting a small increase in the probability of making bad choices, that is, moving away from the optimum solution. This phase is iterated increasing the value of ϵ although at a certain point it would no longer be possible to annotate new documents. The last phase then relaxes the other constraint allowing the annotation of already annotated documents. In this phase another threshold is used to avoid the presentation of bad annotations to the assessor. By correctly choosing the parameters and thresholds of this algorithm the best fitting mixture of good quality annotation and annotated documents can be obtained, and even if the solution is not the optimum one, it is adequate for the problem we need to solve. Once the algorithm has produced a set of annotations, a human assessor has the possibility of evaluating these annotations to ensure collection reliability.

5   Conclusions

An approach to automatically create a test collection with annotated documents has been proposed, using two innovative algorithms. The preliminary results and the manual inspection of the created annotation test collection have confirmed its quality. Future work intends to complete the evaluation of the proposed approach also taking into account some of the comments of the reviewers, which need further investigation to be fully answered.

References

1. Agosti, M., Ferro, N.: Annotations: Enriching a Digital Library. In: Koch, T., Sølvberg, I.T. (eds.) ECDL 2003. LNCS, vol. 2769, pp. 88–100. Springer, Heidelberg (2003)
2. Agosti, M., Ferro, N.: A Formal Model of Annotations of Digital Content. ACM Trans. on Information Systems (TOIS) 26(1), 1–55 (2008)
3. Agosti, M., Ferro, N.: Annotations as Context for Searching Documents. In: Crestani, F., Ruthven, I. (eds.) CoLIS 2005. LNCS, vol. 3507, pp. 155–170. Springer, Heidelberg (2005)
4. Golovchinsky, G., Price, M.N., Schilit, B.N.: From Reading to Retrieval: Freeform Ink Annotations as Queries. In: Proc. 22nd Annual Int. ACM SIGIR Conf. on Research and Development in IR, pp. 19–25. ACM Press, New York (1999)
5. Frommholz, I., Brocks, H., Thiel, U., Neuhold, E., Iannone, L., Semeraro, G., Berardi, M., Ceci, M.: Document-Centered Collaboration for Scholars in the Humanities–The COLLATE System. In: Koch, T., Sølvberg, I.T. (eds.) ECDL 2003. LNCS, vol. 2769, pp. 434–445. Springer, Heidelberg (2003)
6. Sanderson, M., Joho, H.: Forming test collections with no system pooling. In: SIGIR 2004: Proc. of the 27th Annual Int. ACM SIGIR Conf. on Research and Development in IR, pp. 33–40. ACM Press, New York (2004)
7. Carterette, B., Allan, J., Sitaraman, R.: Minimal test collections for retrieval evaluation. In: SIGIR 2006: Proc. of the 29th Annual Int. ACM SIGIR Conf. on Research and development in IR, pp. 268–275. ACM Press, New York (2006)
8. Voorhees, E.M., Harman, D.K.: TREC: Experiment and Evaluation in Information Retrieval (Digital Libraries and Electronic Publishing). MIT Press, Cambridge (2005)
9. Coppotelli, T.: Creazione di una collezione sperimentale per la valutazione di sistemi di reperimento dell'informazione che utilizzino le annotazioni (in Italian). Master's thesis, Dept of Information Engineering, Univ. Padua (2006)
10. Agosti, M., Coppotelli, T., Ferro, N., Pretto, L.: Exploiting Relevance Assessment for the Creation of an Experimental Test Collection to Evaluate Systems that Use Annotations. In: Thanos, C., Borri, F., Candela, L. (eds.) DELOS 2007. LNCS, vol. 4877, pp. 195–202. Springer, Heidelberg (2007)

A Method of Fair Use in Digital Rights Management*

Yong Zhong[1,2], Zhu Zhen[2], Dong-mei Lin[2], and Xiao-lin Qin[3]

[1] Department of Computer Science and Technology,
Nanjing University of Science and Technology,
210094 Nanjing, China
[2] Information and Education Technology Center, Foshan University,
528000 Foshan, China
[3] Institute of Information Security,
Nanjing University of Aeronautics and Astronautics,
210016 Nanjing, China
zhongyong@fosu.edu.cn

Abstract. Fair use is a difficult problem to implement in DRM systems due to its vagueness and uncertainty. We propose a fair use mechanism based on *rights assertion* without limitation, *audit logging* and *misuses trigger*, which brings a fair use mechanism nearer to offline world than that of existing DRM systems.

Keywords: Intellectual property rights, fair use, digital rights management.

1 Introduction

Fair use is a difficult problem to implement in Digital Rights Management (DRM). Mulligan et al.[1] show that current Rights Expression Languages (RELs) could not express or even approximate most of the limitations posed in copyright law. Cheun et al.[2] propose a two-part approach to approximate fair use rights in LicenseScript by *Rights assertion* and *Audit logging*. But in their method, content owners still retain strong control over the rights asserted by users. We propose a fair use mechanism based on *rights assertion* without limitation, *audit logging* and *misuses trigger*, which is the first method to permit users to assert any new fair use rights without restrictions but still can prevent users abusing their rights, which is closest to the offline world.

2 Fair Use in LucScript

LucScript[3] is a new Rights Expression Languages developed by us, which is based on Active-U-Datalog[4] language that is a logical update language integrating active rules. An Active-U-Datalog program $P=IDB \cup EDB \cup AR$ consists of an extensional database EDB, of an intensional database IDB and of a set of active rules AR. A license of LucScript is a 5-tuple (D, IDB, AR, EDB, BV) where D is a unique identifier

---

* This paper is supported by the National Natural Science Foundation of China under Grant No. 60673127.

D.H.-L. Goh et al. (Eds.): ICADL 2007, LNCS 4822, pp. 160–164, 2007.

representing the data that the license refers to, *IDB* is the deductive rules set of the license, *AR* is the active rules set of the license, *EDB* is the set of ground extensional atoms of the license, *BV* is a set of bindings, i.e., a set containing elements of the form *name≡value* such as {*expires≡'06/12/31'*}. Predicate *binding(name, value)* is used to manage binding set. We call $\Delta=IDB \cup EDB \cup AR \cup BV$ the program of the license and show a license as *lic(D, Δ)*.

In our fair use mechanism showed as Fig.1, the content owner can't limit the fair use rights assertion of consumers directly, but they can set some trigger rules in *assertion license* or *audit license* to alert the third party authority if the consumers have some misuses matching the rules. Consumers have three methods to execute their fair use rights: (1) a range of preauthorized fair uses modeled on existing fair uses in offline world are encoded into the license, (2) a consumer who wants to engage in a nonpreauthorized use could submit a request to the third party authority to acquire an authorization, and (3) if the third party authority denies the request, or if the consumer judges the burden of applying to the authority to be too great, the consumer can simply assert the right and executes it by his own way, and the tradeoff is that the assertion and some of information of his will be recorded in an *audit license*.

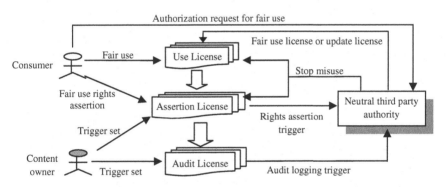

Fig. 1. Mechanism of fair use in LucScript

Suppose Alice wants to use the book *e-book_stars*. Firstly, She must download the encrypted book from the publisher. Then she must purchase a *use license* in order to read the book. Assume the license purchased by Alice is as follows. The license permits the read privilege to Alice before expire time (line 2), where the predicate *display(D, n, m)* means the consumer can read the book *m* pages from page *n*.

(1) *License (e-book_stars, //D*
(2)   { *display(D, n, m)←current_date(d), binding('expire', d'), d≤d'* },   *//IDB*
(3)   { },   *//AR*
(4)   { }, *// EDB*
(5)   { *owner≡'Alice'; id≡'12346'; version≡'11.6.1'; expire≡'07/12/31' ; type≡'use'* }, *//BV*
(6))

Alice thinks she should have the privilege to print the book, but there unexpectedly is no print function in the license. So she decides to request the right by sending the *request license* as follows, whose *IDB* rule (line 2) is the right that Alice is requesting.

(1) *License (e-book_stars,   //D*
(2)   *{   print(D, n, m)←current_date(d), binding('expire', d'), d≤d' },   //IDB*
(3)   *{ },   //AR*
(4)   *{ },   // EDB*
(5)   *{   owner='Alice'; id='12347'; expire='07/12/31' ; type='request' },   //BV*
(6) *)*

After receiving the request, the third party authority thinks the request is reasonable, so an *update license* as follows is sent back to the consumer. The update rule of the license (line 2) will insert an *EDB* rule to the *use license*.

(1) *License (e-book_stars,   //D*
(2)   *{   update()←+EDB("print(D, n, m)←current_date(d),binding ('expire', d'), d≤d' ") },   //IDB*
(3)   *{ },   //AR*
(4)   *{ },   // EDB*
(5)   *{   owner='Alice'; id='12348'; expire='07/12/31' ; type='update' },   //BV*
(6) *)*

After the *update license* comes, the update will be added to Alice's *use license* and the license will become the following license, which grants the print privilege to Alice.

(1) *License (e-book_stars,   //D*
(2)   *{   display(D, n, m)←current_date(d), binding('expire', d'), d≤d';*
(3)   *print(D, n, m)←current_date(d), binding('expire', d'), d≤d'   },   //IDB*
(4)   *{ },   //AR*
(5)   *{ },   // EDB*
(6)   *{   owner='Alice'; id='12346'; version='11.6.1'; expire='07/12/31' ; type='use' },   //BV*
(7) *)*

For excerpt privilege, Alice decides to assert the rights simply. However, the content owner thinks there are no any reasons that a consumer can excerpt his books more than a half, so he decides to set a trigger to alert third party authority if the consumer excerpts his books more than a half. After Alice assert her rights by submitting the transaction ?-(+*IDB*("*modified(e-book_stars, n, m)←.*"), the *assertion license* is as follows, where *modified(e-book_stars, n, m)* means the consumer can excerpt the books *m* pages from page *n* and since the assertion has no limit on *n* and *m*, which grants Alice excerpts any page. Extensible predicate *modified_pages (p)* keeps the pages that have been modified by consumers, where *p* is the type of set.

(1) *License (e-book_stars,   //D*
(2)   *{   **modify(e-book_stars, n, m)←.** ;   // the asserted right*
(3)   *alert(D, p)←create(lic(D, Δ)), insert(lic(D, Δ), p), binding('agentIP', I), send(I, lic(D, Δ));*
(4)   *insert(lic(D, Δ), p)←Δ?+(+BV(b)), BV_name(b, name), name<>'type';*
(5)   *insert(lic(D, Δ), p)←Δ?+(+binding('type', 'alert'));*
(6)   *insert(lic(D, Δ), p)←Δ?+(+ modified_pages(p))*

(7)   },   //IDB

(8)   { -modified_pages(p), +modified_pages(p')← +modify_a(D, n, m), modified_pages(p), append(p,
               n, m, p');

(9)   +()←+modify_a(D, n, m), modified_pages(p), binding('max_pages', max), sizeof(p, k), plus(k,
               max, max'), max'>max, append(p, n, m, p'), alert(D, p')

(10)   },   //AR

(11)   { modified_pages ({}) },   // EDB

(12)   { owner≡'Alice'; id≡'12360'; type≡'assert'; max_pages≡'100', agentIP≡'202.192.168.38' }, //BV

(13))

The rule of line (3) explains how to create the *alert license*, where *create(lic(D, Δ))* create a license in memory and *send(I, lic(D, Δ))* sends the license to the third party authority. The rule of line (4) inserts the *BV* data of *assertion license* except *type* binding to the *alert license*. The rule of line (5) sets the *type* of the license. The rule of line (6) set the pages that have been modified by consumers. The *AR* rule of line (8) appends the page modified by consumers to *modified_pages(p)* when consumers modify a page. The *AR* rule of line (9) activates the creation and send of *alert license* when the pages modified by the consumers are more than half of the book, where +() is a no update operation for rule syntax. The *alert license* will be as follows.

(1) License (e-book_stars,   //D

(2)   { },   //IDB

(3)   { },   //AR

(4)   { modified_pages ({1,3,7,...}) },   // EDB

(5)   { owner≡'Alice'; id≡'12360'; type≡'alert'; max_pages≡'100', agentIP≡'202.192.168.38' }, //BV

(6))

At last, when Alice assert her rights by submitting the transaction ?-(+IDB("modified(e-book_stars, n, m)←."), a transaction -(+log(User, Action, time, purpose)) will also be submitted to the *audit license*, where *User* is the identifier of the consumer, *action* is the asserted right, *purpose* is the purpose of asserting the right. The *audit license* is as follows after above right assertion.

(1) License (e-book_stars,   //D

(2)   { },   //IDB

(3)   { },   //AR

(4)   { log('Alice', 'modify(e-book_stars, n, m)←.', '1/7/2007 13:30:25', 'comment') },   // EDB

(5)   { owner≡'Alice'; type≡'audit'; max_pages≡'100', agentIP≡'202.192.168.38' }, //BV

(6))

3 Conclusions

In this paper we propose a fair use mechanism based on LucSciprt Language, which we show is a unique solution to fair use in DRM and is nearer to offline world than that of existing DRM systems.

References

1. Erickson, J.S.: Fair Use. DRM, and Trusted Computing. Communications of the ACM 46(4), 34–39 (2003)
2. Chong, C.N., Etalle, D.S., et al.: Approximating Fair Use in LicenseScript. In: 6th Int. Conf. of Asian Digital Libraries, Kuala Lumpur, Malaysia, pp. 432–443 (2003)
3. Zhong, Y., Lin, D.M.: Report of LucScript Rights Expression Languages. Technical report, Information and Education Technology Center, Foshan University, China (2007), http://www.fosu.edu.cn/zhongyong/LucScript.pdf
4. Bertino, E., Catania, B., Gori, R.: Active-U-Datalog: Integrating Active Rules in a Logical Update Language. In: Kifer, M., Voronkov, A., Freitag, B., Decker, H. (eds.) Dagstuhl Seminar 1997, DYNAMICS 1997, and ILPS-WS 1997. LNCS, vol. 1472, pp. 107–133. Springer, Heidelberg (1998)

Ontology-Based Metadata Integration in the Cultural Heritage Domain*

Thomais Stasinopoulou[1], Lina Bountouri[1], Constantia Kakali[1], Irene Lourdi[1], Christos Papatheodorou[1], Martin Doerr[2], and Manolis Gergatsoulis[1]

[1] Department of Archive and Library Sciences, Ionian University,
Palea Anaktora, Plateia Eleftherias, 49100 Corfu, Greece
{boudouri,papatheodor,manolis}@ionio.gr, stasthomais@yahoo.gr,
nkakal@panteion.gr, elourdi@lib.uoa.gr
[2] Institute of Computer Science Foundation for Research and Technology, Vassilika
Vouton P.O.Box 1385 GR 711 10 Heraklion, Crete Greece
martin@ics.forth.gr

Abstract. In this paper, we propose an ontology-based metadata integration methodology for the cultural heritage domain. The proposed real - world approach considers an integration architecture in which CIDOC/CRM ontology acts as a mediating scheme. In this context, we present a mapping methodology from Encoded Archival Description (EAD) and Dublin Core (DC) metadata to CIDOC/CRM, and discuss the faced difficulties.

Keywords: Ontology-based Integration, Metadata interoperability, Cultural Information, CIDOC/CRM, EAD, DC.

1   Introduction

Making cultural resources accessible requires rich metadata structures, able to cover the variety of material held in memory institutions (such as archives, bibliographic and electronic material). Nowadays, it is common to find metadata sources, which may differ in various aspects, even if the resources they describe originate from the same application domain. This phenomenon is especially observed for the cultural resources, for which several metadata standards have been developed in order to cover the documentation needs and the peculiarities of every type of material.

Taking into account the metadata variety and the increasing demand for targeted global search, unified access and data exchange between heterogeneous

* The research work of the authors I. Lourdi, L. Bountouri, and M. Gergatsoulis was partially co-funded by the European Social Fund (75%) and National Resources (25%) -Operational Program for Educational and Vocational Training (EPEAEK II) and particularly by the Research Program "PYTHAGORAS II". The research work of the authors C. Kakali, Th. Stasinopoulou, C. Papatheodorou and M. Doerr was supported by DELOS Network of Excellence on Digital Libraries funded by European Commission.

D.H.-L. Goh et al. (Eds.): ICADL 2007, LNCS 4822, pp. 165–175, 2007.

cultural sources, emphasis is given to matters of interoperability and integration at various levels, such as syntactic, schematic, system and also at the more complex semantic level. Data Integration has been a dynamic and challenging research area for many years. However, nowadays, research interests are moving from Data Integration to Semantic Integration in many communities and disciplines, such as e-government and cultural heritage. This movement is being seriously influenced by the new notion and philosophy that the Internet tends to acquire, the Semantic Web. Semantic Integration is the process of using a conceptual representation of the data and of their relationships to eliminate possible heterogeneities [4]. One of the main Semantic Web infrastructure elements, which are an important means in semantic integration scenarios, are ontologies. Their nature allows the sophisticated, extended and rich expression of meanings, and - at the same time - the ability of reasoning.

In this context, ontologies can be considered as an important building block for integration architectures [9], into which metadata originating from diverse sources can be semantically mapped and integrated [5,13]. They are preferred in comparison to other schemas, because of their ability to conceptualize particular domains of interest and express their rich semantics.

In our approach, we use CIDOC/CRM [2] ontology as a conceptual representation of cultural heritage domain to promote semantic integration between different metadata schemas, such as Encoded Archival Description [14] and Dublin Core [18], and eliminate their possible semantic heterogeneities. We address the problems that arise when creating semantic mappings from metadata schemas to ontological models, with the intention of achieving semantic interoperability. We document the use of those mappings in an architecture integrating different cultural metadata sources. We also present the methodology followed to create the mappings and we give an overview of the EAD elements to CIDOC mapping, extending the mappings presented in [17]. EAD is the most well known standard for archival description. It is an XML-based descriptive schema, intended to create electronic finding aids, which include the necessary information for the identification, management and interpretation of an archive. EAD has been used in order to provide archival metadata in digital libraries.

2   Problem Definition and Related Work

2.1   Mapping Metadata to Ontologies

Mapping metadata schemas to ontologies is a complicated procedure, once those two forms have many differences between them in various levels.

Scope and Function: Metadata have a completely different scope and function in comparison with ontologies. Metadata are used to describe, identify, facilitate the access, usage and management of (digital) resources. Ontologies define entities in a more abstract level, with the intention of conceptualizing a domain of interest. They do not provide specific elements for the description of a resource, but a general definition of the basic notions of a field and the relations between them.

Expression of Semantics: Metadata schemas are created for resources' identification and description and - most of the times - they do not express rich semantics. Even though the meaning of the metadata information can be processed by humans and its relationship to the described resource can be understood, for machine processing the actual relationships are frequently not obvious. In contrast to metadata schemas, ontologies provide rich constructs to express the meaning of data. For example, in DC we write that "a specific poet is the creator of a poem" by assigning a value to DC.creator. On the contrary, in CIDOC we can express general statements about the creation of poems denoting that an Actor (poet) participates in a Creation Event which produces a Linguistic Object (poem). In this way, the knowledge concerning the poem creation becomes explicit and machine "understandable"[8].

Moreover a plethora of conceptual expressions should be aligned for mapping a metadata schema to an ontology. For example, EAD carries two main semantic structures: (a) the metadata of a finding aid and (b) the encoding of a finding aid itself. On the other side, the combination of the CIDOC entities and properties generates a large number of conceptual expressions that should be studied in order to select the semantically closest of them to map the metadata elements.

2.2   Related Work

Works related to ontology-based integration, usually emphasize on element and structure level mappings and transformations (i.e. elements to classes, attributes to properties etc.). In [5], authors map the XML data of every local source to an RDFS local ontology, created by transforming the XML elements and attributes to RDFS classes and properties. An additional characteristic of their method is that they preserve the structure of an XML local source inside the local RDFS ontology. Then, local ontologies are merged to a global ontology for unified access and semantic integration of local data sources. In [13], an XML data integration approach is presented based on the Web Ontology Language (OWL). More specifically, the proposed architecture maps XML structures (such as elements and attributes) to OWL structural components (such classes, properties, etc.) and thus they convert the XML data to an OWL global ontology. In order to define mappings, mapping languages have been proposed (see [11] for example).

In [1] the intention of the work is to propose a mechanism for the cultural information sources integration. The authors map pieces of information contained in XML fragments to domain specific ontologies, such as CIDOC, defining (1) a mapping language that describes the resources by a set of rules relating XPath location paths to the concepts and roles of an ontology and (2) a query rewriting algorithm for translating user queries into queries expressed in an XML query language, which are send for evaluation to XML sources.

Even though these approaches define semantic integration formulas, they are strongly oriented to integrate XML data to RDFS and OWL ontology languages, giving emphasis to define structure mappings or model mappings between them.

However, their effectiveness in mapping really complex semantically data structures, such as metadata schemas, has not yet been tested.

3   An Ontology-Based Mediator

In our approach, we focus on the need to develop information systems able to provide access to heterogeneous data sources. We consider the existence of various cultural sources described with different metadata schemas and there is the demand that our users retrieve information from them. For this purpose, we propose to employ a mediator able to semantically integrate the various schemas. Specifically, we consider CIDOC ontology as the global schema and we define mappings from the metadata schemas to CIDOC and vice versa.

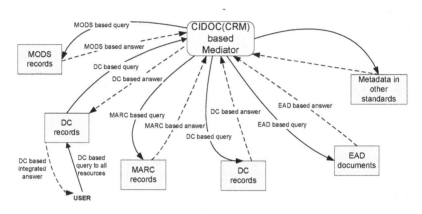

Fig. 1. An Ontology-Based Mediator

We selected CIDOC as the mediating ontology because it is a core ontology designed to be applied for the documentation, integration, mediation and exchange of heterogeneous cultural information. It is a conceptual model, composed of *entities*, which are organized into a hierarchy and semantically related to each other with *properties*. In detail, the CIDOC defines the complex interrelationships that exist between objects, actors, events, places and other concepts in the cultural heritage field [10].

According to [3] the value of CIDOC/CRM becomes apparent when it is used as the basis for data transfer and exchange between different systems, schemas and semantics. In such a scenario, CIDOC acts as a mediated schema to which different metadata can be mapped. Given that it is a core ontology, it allows gathering all necessary cultural information in a suitable form for further reasoning [7]. Figure 1 presents an architecture in which a set of data sources exists, each of them following a possibly different metadata schema. All these schemas are mapped to CIDOC. Users can pose their queries to a local data source following the restrictions of the local metadata schema. The local query engine returns the results from its source and promotes the query to the mediator which translates the query to suitable forms, using the appropriate mappings, and forwards them to be answered by the other sources. Finally, the results from each source are collected and returned to the user. Note that the queries in the DC sources might be written in a query language such as SPARQL while the queries in EAD sources might be in XQUERY.

For example, suppose that users wish to find metadata records (archival finding aids, Dublin Core records, etc.) describing documents published by a person whose name is "John Smith". In terms of Dublin Core (DC), the author is looking for records for which DC.publisher= "John Smith" and DC.type= "text". Suppose that users pose their (appropriately formed) query to a DC local data source. Then, the DC records from the local source matching the query are returned to the users. The query is then propagated to the mediator and transformed, using a set of mapping rules from DC to CIDOC [12], into an equivalent query in terms of CIDOC/CRM. In this query, the conditions (corresponding to the conditions of the initial query) locate the values that should be checked through appropriately formed CIDOC paths such as[1]: E33(Linguistic Object)-*P94 (has created/was created)*-E65 (Creation Event)-*P14(performed)[with subproperty P14.1 (in the role of)*-*E55(Type)= "Publisher"]*-E39(Actor)-*P131(is identified by/identifies)*-E82(Actor Appellation)= "John Smith". The CIDOC query is then transformed to other formats and propagated to the corresponding sources. For example, a local source keeping EAD data receives a query whose condition applies on the values returned by the path /ead/eadheader/filedesc/publicationstmt/publisher/name evaluated on the EAD data. This condition compares the returned value with the string value "John Smith". If they match, the whole finding aid is returned to the mediator and then to the users (through the local client) after being transformed into DC format.

4 Mapping Metadata to CIDOC

In this section we introduce the basic methodology steps followed in order to define the mappings between metadata schemas and the CIDOC ontology. Additionally, given that mappings from different metadata schemas (DC, EAD etc.) to CIDOC are required in order to develop the ontology-based mediator described in 3, we define mappings from Dublin Core and EAD to CIDOC. Due to space reasons, we only give an overview of part of the EAD mapping. For a complete reference of DC and EAD mapping to CIDOC see [16,12].

4.1 Methodology

Path-Oriented Approach: In our methodology, a mapping from a source schema to a target schema transforms each instance of the source schema into a valid instance of the target schema [11]. For metadata to CIDOC mapping, we interpret the metadata paths to semantic equivalent CIDOC paths. We define a

---

[1] The notation E*nn*, *Pnn* corresponds to CIDOC entities and properties respectively.

CIDOC path as a chain of the form entity-property-entity, such that the entities associated by a property correspond to the property's domain and range. For example, a CIDOC path is E33(Linguistic Object)-*P94 (has created)*-E65 (Creation Event)-*P14(performed)*-E39(Actor) denoting that an author (Actor, E39) during a Creation Event (E65) generated a poem (Linguistic Object, E33). A metadata path is defined as a sequence of elements, subelements (or element refinements), encoding schemes and vocabulary terms, starting from the metadata schema root element separated by the slash symbol (/). For instance, the path /ead/eadheader/filedesc/titlestmt/author/name, is a part of the metadata of an archival description encoded in EAD and denotes the name of the author of an archival description. Moreover the path /DC/DC.Date/DC.Date.Created denotes the creation date of a resource.

It is worth mentioning that this approach is appropriate since both the metadata and the ontology participating in the integration scenario encode information via paths. An indicative example that confirms the need to follow the "path" approach is that most of the metadata schemas provide elements which, even if they have the same element name, depending on their path position declare different semantics. For instance, the EAD element <corpname> declares the organization responsible for the creation, accumulation, or assembly of the described materials before their incorporation into an archival repository, when included in the path /ead/archdesc/did/originator/corpname, and the institution or agency responsible for providing intellectual access to the archival materials being described, when included in the path /ead/archdesc/did/repository/corpname.

Event-Orientation: An issue we are facing while mapping metadata schemas to CIDOC core ontology is the event orientation of the specific ontology. Metadata, such as EAD and DC, are data structures oriented to describe objects and, as most of the metadata schemas, focus on the described object. On the other hand, CIDOC is event based. Its main notions are the temporal entities and events, and the presence of CIDOC entities, such as Actors, Dates, Places, Objects, etc. implies their participation to an event or an activity [7]. For instance, in order to map the DC.Creator element of a physical object in CIDOC terms, we should map this element to the CIDOC entity Actor (E39). However, in CIDOC persons perform particular roles through events, such as Period (E4), Event (E5) and any other incident valid for a certain time. As a result, the entity Actor (E39) could be interlinked to the described object only with the intermediation of an event or an activity, which indicates that the event - taking place in a particular date - resulted in the described object.

Wrapper Elements: An additional issue evoked during the mapping is that there exist metadata schemas, such as EAD, TEI and MODS, composed of many wrapper elements, which group relative information. For example, EAD Header and Archival Description - which are two basic elements of EAD - are wrapper elements. However, most wrapper elements do not have any semantics by their own, but are used to group the elements that belong to them. In fact their semantics are expressed through the semantics of the elements that they contain.

For example, the wrapper element Descriptive Identification (<did>) contains all the elements that provide the basic identification information for an archive, such as the originator (<originator>), the title (<unittitle>) and the physical location (<physloc>) of the described archive. Therefore in our approach we do not define any mappings for the "semantic - free" wrapper elements. Similarly, we do not map any formatting elements, such as tables, list of items etc.

4.2   Mapping EAD to CIDOC

EAD schema is composed of three basic elements:

- The EAD Header (<eadheader>), which is a mandatory element that includes information about the EAD finding aid (i.e. includes the metadata for the archival description and not the archival description itself).
- The Front Matter (<frontmatter>), which is an optional element that contains publication information, such as the title page information of the printed finding aid etc.
- The Archival Description (<archdesc>), which is a mandatory element that incorporates information about the archival description itself.

In our effort, we define mappings from the EAD Header and the Archival Description elements to CIDOC. We ignore Front Matter since it is extremely rarely used. Furthermore, those two groups of elements are mandatory and they constitute the core descriptive part of a finding aid.

An EAD path is a sequence of EAD elements and subelements, starting from the schema root element <ead> separated by the slash symbol (/). For instance, the path /ead/eadheader/filedesc/titlestmt/author/name, denotes the name of the author of an archival description. Specifically, this path is a part of the metadata of an archival description, since it includes the element <eadheader> and information about the name of the person who created the archival description. Therefore, we have to map the EAD paths to CIDOC paths in a way that satisfies the semantic equivalence taking into account the points mentioned in 4.1.

The Archival Description (<archdesc>) is an element that identifies the archive itself, describing its content and context of creation. From this element, we can derive the following information for an archive: (a) its description, (b) its material substance and (c) the information that it carries. In CIDOC terms this information is mapped to the following classes:

- E31 (Document) and E33 (Linguistic Object), denoting that the Archival Description is a text which describes (documents) an archive.
- E22 (Man-Made Object), declaring that the archive is a physical object created by human activity.
- E73 (Information Object) and E33 (Linguistic Object), since these classes refer to immaterial items that include human memory and do not depend on any specific physical carrier.

Given that the <ead> element is equivalent to the entities (E31 Document) and E33 (Linguistic Object), the corresponding CIDOC path to the EAD path

/ead/archdesc is: {E31 (Document), E33(Linguistic Object)}-*P106 (is composed of/forms part of)*-{E31 (Document), E33(Linguistic Object)}-*P70 (documents/is documented in)*-E22 (Man-Made Object)-*P128 (carries/is carried by)*-{E73 (Information Object), E33 (Linguistic Object)} (see Figure 2).

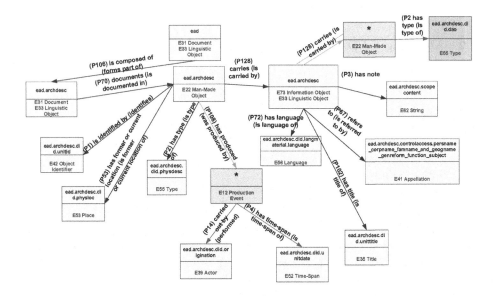

Fig. 2. An indicative part of the Archival Description mapping

According to all the above, the subelements of Archival Description are linked either to E22 (Man-Made Object), when they provide information about the archive as a physical object, or to E73 (Information Object) and E33 (Linguistic Object), when they provide information that refer to the archive as an informational carrier, or to E31 (Document) and E33 (Linguistic Object), when the provide descriptive information about the archive.

Most of the information contained in the subelements of Descriptive Identification wrapper element (<did>) identifies core information about the described materials and it is linked to the E22 (Man-Made Object) entity dealing with material substance of the archive. For instance consider the path (See Figure 2): /ead/archdesc/did/origination which corresponds to the originator of the archive. The <origination> is an Actor (E39) and we have to link the archive production event with this person. Since CIDOC does not provide a direct link from an object to the person that created it, we consider the archive Production Event (E12) and we link it with the archive using the property P108 (has produced/was produced by). The originator is linked with the production event using the property P14 (carried out/performed). Hence, the CIDOC path is:

{E31 (Document), E33(Linguistic Object)}-*P106 (is composed of/forms part of)*-{E31 (Document), E33(Linguistic Object)}-*P70 (documents/is documented in)*-E22 (Man Made Object)-*P108 (has produced/was produced by)*-E12 (Production Event)-*P14 (carried out by/performed)*-E39 (Actor).

Similar mappings could be generated for the elements Physical Location (<physloc>), Physical Description (<physdesc>), etc. Moreover, there are many EAD elements which are related to the archive as an information carrier. For instance, the element <controlaccess> contains the thematic metadata of an archive. The mapping for the <controlaccess> and its subelements denoting access points, such as /ead/archdesc/controlaccess/persname, to CIDOC is (See Figure 2): {E31 (Document), E33(Linguistic Object)}-*P106 (is composed of/forms part of)*-{E31 (Document), E33(Linguistic Object)}-*P70 (documents/is documented in)*-E22 (Man Made Object)-*P128(carries/is carried by)*-{E73 (Information Object), E33(Linguistic Object)}-*P67 (refers to/is referred to by)*-E41 (Appellation). A similar mapping is followed for the element Title of the Unit (<unittitle>), Scope and Content (<scopecontent>), Abstract (), etc.

For the development of digital libraries consisting of archival material, elements referring to the digital version of the archive are significant. Those elements are Digital Archival Object (<dao>), Digital Archival Object Group (<daogrp>), Digital Archival Object Location (<daoloc>) and Digital Archival Object Description <daodesc>. All of them are linked with the entity Information Object (E73) since they carry information about the digitized form of the archive. For instance, the element Digital Archival Object (<dao>) provides information about the digital representation of an archive and its components parts (e.g. its URI).

5   Conclusion

Metadata semantic interoperability in the cultural heritage domain is one of the main issues in the digital environment. In our attempt to accomplish that goal, we proposed a semantic integration mechanism, so as to provide unified access to collections of heterogeneous material. In this context, we described an ontology-based integration architecture and addressed the issues of mapping metadata schemas to ontologies. What is more, we presented part of the necessary mappings from cultural heritage metadata to CIDOC mediated ontology.

The mapping definition between the metadata schemas and CIDOC was complex enough. One of the difficulties encountered was the absence of ontology concepts semantically equivalent to metadata fields. In this case, our research team proposed the creation of new classes and properties [12]. An additional issue was the event-based logic that CIDOC implements. Due to that fact, we had to make use of intermediate CIDOC activity and event entities to represent the relationships expressed in metadata between objects (i.e. the archive) and persons (i.e. the creator of the archive). In case of EAD, the mapping difficulties were empowered because of the two different - but related - semantic structures it includes: the metadata of the finding aid (<eadheader>) and the finding aid itself (<archdesc>). In order to evaluate the handling of the specific difficulties, our future work is to define the inverse mapping from CIDOC to metadata schemas and implement the metadata-CIDOC-metadata query engine.

To conclude, the mapping defined can be encoded using automated tools, such as OWL editors and XML technologies [11]. However, human intervention is necessary in order to define the semantic mapping, given that it is a deep conceptual work.

References

1. Amann, B., Fundulaki, I., Scholl, M., Beeri, C., Vercoustre, A.M.: Mapping XML Fragments to Community Web Ontologies. In: WebDB, pp. 97–102 (2001)
2. CIDOC Documentation Standards Working Group and CIDOC CRM SIG. The CIDOC Conceptual Reference Model, http://cidoc.ics.forth.gr/
3. Crofts, N., Doerr, M., Gill, T.: The CIDOC Conceptual Reference Model: a Standard for Communicating Cultural Contents. In: Cultivate Interactive, vol. (9) (February 2003)
4. Cruz, I.F., Xiao, H.: The Role of Ontologies in Data Integration. Journal of Engineering Intelligent Systems 13(4), 245–252 (2005)
5. Cruz, I.F., Xiao, H., Hsu, F.: An Ontology-Based Framework for XML Semantic Integration. In: Proceedings of the 8th International Database Engineering and Applications Symposium (IDEAS 2004), Coimbra, Portugal, July 7-9 (2004)
6. Doerr, M.: Mapping of the Dublin Core Metadata Element Set to the CIDOC CRM. Technical Report 274 (July 2000)
7. Doerr, M.: The CIDOC CRM An Ontological Approach to Semantic Interoperability of Metadata. AI Magazine 24, 75–92 (2003)
8. Doerr, M., Hunter, J., Lagoze, C.: Towards a Core Ontology for Information Integration. Journal of Digital Information 4(1) (2003)
9. Doerr, M., Lagoze, C., Hunter, J., Baker, T.: Building Core Ontologies: a White Paper of the DELOS Working Group on Ontology Harmonization. White paper, DELOS Network of Excellence on Digital Libraries (2002)
10. ISO. Information and documentation: A reference ontology for the interchange of cultural heritage information, ISO 21127 (2006)
11. Kondylakis, H., Doerr, M., Plexousakis, D.: Mapping Language for Information Integration. Technical Report 385 (December 2006)
12. Kakali, C., Lourdi, I., Stasinopoulou, T., Bountouri, L., Papatheodorou, C., Doerr, M., Gergatsoulis, M.: Integrating Dublin Core metadata for cultural heritage collections using ontologies. In: Proc. Int'l Conf. on Dublin Core and Metadata Applications (DC 2007), Singapore, 27 - 31 August, pp. 128–139 (2007)
13. Lehti, P., Fankhauser, P.: XML Data Integration with OWL: experiences and challenges. In: Proc. of SAINT 2004, pp. 160–170. IEEE Computer Society Press, Los Alamitos (2004)
14. Library of Congress. Encoded Archival Description (2002), http://www.loc.gov/ead/
15. Partridge, C.: The Role of Ontology in Integrating Semantically Heterogeneous Databases. Technical Report 05/02, LADSEB-CNR (June 2002)
16. Stasinopoulou, T., Doerr, M., Papatheodorou, C., Kakali, K.: WP5 - Task 5.5.: EAD mapping to CIDOC/CRM. Report, DELOS-WP5 - Task 5.5 Ontology-driven Interoperability (2007)

17. Theodoridou, M., Doerr, M.: Mapping of the Encoded Archival Description DTD Element Set to the CIDOC CRM. Technical Report 289 (June 2001), http://cidoc.ics.forth.gr/docs/ead.rtf
18. DCMI Usage Board. DCMI Metadata Terms (2006), http://dublincore.org/documents/dcmi-terms/
19. DCMI Usage Board. DCMI Type Vocabulary 2006, http://dublincore.org/documents/dcmi-type-vocabulary/
20. Uschold, M., Gruninger, M.: Ontologies: principles, methods and applications. Knowledge Engineering Review 11(2), 93–155 (1996)

It Is the Time for the Digital Library to Meet the Enterprise Architecture

José Borbinha

IST – Instituto Superior Técnico / INESC-ID, Av. Rovisco Pais, 1049-001 Lisboa, Portugal
jlb@ist.utl.pt

Abstract. The purpose of this paper is to raise arguments to support the proposal that we should promote the discussion of the Digital Library in a structured way, aligned with the emerging perspective of the Enterprise Architecture. In this sense, the Digital Library practitioners should be motivated to give more emphasis to the need to better integrate its efforts and body of knowledge with the more generic area of Information Systems, where important concepts, regulations and good practices have been emerging, defined by authorities, the industry and the multiple stockholders of each specific scenario. Concluding, it is time for the Digital Library to mature by recognizing that it is, simply, a case of an Information System, which is specific only in what concerns the requirements derived of its specific business goals.

1 Introduction

The title and motivation for this paper was inspired by [4]. The content was also inspired by [1]. In his paper Michael Lesk was himself inspired by the seven ages of man, described by Shakespeare, giving us that way a very interesting description of the evolution of the area of Information Retrieval. However, after a careful reading we can recognize that the scope of this description covers much more than the traditional area of Information Retrieval, also comprising the area of the Digital Library (DL).

Lesk's paper was written in 1995, on the same time the D-Lib magazine was debuting[1], and was precisely in the first issue of D-Lib that William Arms expressed his eight key general principles for a generic DL architecture.

I propose now to revisit these two works, twelve years after their first publication, with two main purposes in mind: to review their contents at the light of our actual knowledge; to use that effort as a process to try to characterize the actual thinking of the DL as a problem and the main emerging related challenges. The ultimate goal is to raise arguments to prove that, from now, we should not continue promoting the DL by mainly raising generic goals and addressing the technological related issues. Alternatively, the DL community should be motivated to better structure its goals and give more attention to the need to integrate its efforts and body of knowledge with the more generic area of Information Systems, where important concepts have been emerging recently that must not be ignored. Specifically, those are the cases of the concepts of Enterprise Architecture and Enterprise Architecture Framework.

---

[1] http://www.dlib.org

D.H.-L. Goh et al. (Eds.): ICADL 2007, LNCS 4822, pp. 176–185, 2007.

But why is this really important? First, let us develop a simple analysis…

One can conceive "DL deployments" in mainly two scenarios: as a purpose in itself (the DL as the main business goal); or as a contribution to other purposes (technology and processes created from a "DL perspective" in order to be used to support more generic goals). The first scenario will continue sustaining the DL has a relevant concept, where it might be not too difficult to acknowledge the right credits to the right communities contributing for that. It also might be possible to assure that relevance and credits in the second scenario (making the acronym DL^2 equivalent to others such as ERP, CRM, SCM, etc.), but in any of the cases the DL community has to make it happen.

The need to rationalize resources, to apply standard governance's models and business processes, as also the need to accomplish with strict legal and auditing requirements, have been pushing governments and private organizations to promote and impose Enterprise Architecture Frameworks to central administration services, public services and enterprises in general[3,4]. Assuming that DL's technology has reached a maturity for formal deployments at these levels, than those specific requirements concerning management, legal and business issues, and especially concerning accountability, can not be ignored.

2 "Key Concepts in the Architecture of the Digital Library"

Arms' presents eight general principles representing concepts and requirements for the DL architecture. Quoting them in short:

1. **The technical framework exists within a legal and social framework:** "Early networked information systems were developed by technical and professional communities, concentrating on their own needs. The emphasis was on making information available (…) without charge. The digital library of the future will exist within a much larger economic, social and legal framework. (…)"
2. **Understanding of digital library concepts is hampered by terminology:** "(…) Certain words cause such misunderstandings that they are best expunged from any precise discussion of the digital library. The list includes "copy", "publish", "document", and "work". Other words have to be used very carefully and their exact meaning made clear whenever they are used. An example is "content". (…)"
3. **The underlying architecture should be separate from the content stored in the library:** "Separating general functions from those specific to the type of content has other benefits. It encourages different markets to emerge, and allows a legal framework in which storage, transmission and delivery of digital objects is separate from activities to create and manage the intellectual content."

---

[2] DL – Digital Library; ERP – Enterprise Resource Planning; CRM – Costumer Relationship Management; SCM – Supply Chain Management.

[3] "Congress is enforcing its mandate that the Defense Department develop systems compatible with the DOD Business Enterprise Architecture - with the threat of jail time and hefty fines for the department's comptroller." - http://www.gcn.com/print/23_33/27950-1.html?topic= enterprise-architecture

[4] http://www.dmreview.com/article_sub.cfm?articleId=1038091 (Zachman, Basel II and Sarbanes-Oxley).

4. **Names and identifiers are the basic building block for the digital library:** "Names are a vital building block for the digital library. Names are needed to identify digital objects, to register intellectual property in digital objects, and to record changes of ownership. They are required for citations, for information retrieval, and are used for links between objects."

5. **Digital library objects are more than collections of bits:** "A primitive idea of a digital object is that it is just a set of bits, but this idea is too simple. The content of even the most basic digital object has some structure, and information, such as intellectual property rights (…)."

6. **The digital library object that is used is different from the stored object:** "The architecture must distinguish carefully between digital objects as they are created by an originator, digital objects stored in a repository, and digital objects as disseminated to a user."

7. **Repositories must look after the information they hold:** "Since digital objects contain valuable intellectual property, the stored form of a digital object within the repository includes information that allows for it to be managed within economic and social frameworks."

8. **Users want intellectual works, not digital objects:** "Which digital objects should be grouped together can not be specified in a few dogmatic rules. (…) The underlying architecture (…) must provide methods for grouping digital library objects and must provide means for retrieval."

3 "The Seven Ages of Information Retrieval"

Lesk provides an historical description and a vision of the future of the area of Information Retrieval that makes it clearly coincident with the DL.

According to Lesk, **Childhood** (1945-1955) is described as the time when Vannevar Bush proposed is vision for the Memex [2]. The **Schoolboy** (1960s) "…were a time of great experimentation in information retrieval systems". **Adulthood** (1970s) was when "…retrieval began to mature into real systems". **Maturity** (1980s) was reached with "…the steady increase in word processing and the steady decrease in the price of disk space... The use of online information retrieval expanded". Lesk wrote his paper during the **Mid-Life Crisis** (1990s), when "Things seemed to be progressing well: more and more text was available online, it was retrieved by full-text search algorithms, and end-users were using OPACs. (…) Nevertheless it was still an area primarily of interest to specialists in libraries".

After this, it was supposed to come the time for **Fulfillment** (2000s): "Which will it be? I believe that in this decade we will see not just Bush's goal of a 1M book library, so that most ordinary questions can be answered by reference to online materials rather than paper materials, but also the routine offering of new books online, and the routine retrospective conversion of library collections. We will also have enough guidance companies on the Web to satisfy anyone, so that the lack of any fundamental advances in knowledge organization will not matter". Accordingly, **Retirement** (2010) is the age when "…central library buildings on campus have been reclaimed for other uses, as students access all the works they need from dormitory room computer. (…) Most students, faced with a choice between reading a book and watching a

TV program on a subject, will watch the TV program. (…) Educators will probably bemoan this process. (…). As for the researchers, there will be engineering work in improving the systems, and there will be applications research as we learn new ways to use our new systems."

4   The Age of the Digital Library

At a first glance one might be tempted to consider the DL not as a continuum or a specialization of the area of Information Retrieval, but a child of it. This might be an argument for those willing to "reset" Lesk's scale of time, probably in order to give a "second live", or a "second chance" for the DL. I must stress that I disagree of that!

In my opinion, Lesk uses a description of the area of Information Retrieval that really makes it overlap the DL, and his vision is correct. Also, this includes not only the direct references to goals and processes easily identified with that, but also the multiple references to border areas, such as Artificial Intelligence. Lesk is rally talking about the same body of motivations and goals than we have been using as a reference for the DL! In this sense, the DL should be now in its fulfillment age! We have the "Million Books Project"[5]; reference works are common to find as e-books; Yahoo, Google, del.icio.us are fairly well guiding us in the labyrinth of the Web, etc. Z39.50[6], once a specific answer to specific requirements for technical interoperability from specific DL business goals has become irrelevant after the emerging of the web-OPAC, which in itself is disappearing, integrated in the "enterprise portal" or replaced by new processes based on the OAI-PMH[7]. Concerning semantic interoperability, one other common issue in Digital Libraries, it also is a common issue in most of the attempts to integrate businesses and processes among any different organizations. The concept of metadata registries, also usually raised by the DL, started in fact in the industry, due to very practical and generic needs. In fact, since the emerging of HTTP, XML, web-services (whenever they are based on SOAP or simply on REST), etc., that we can not claim anymore any key challenges for technical or semantic interoperability to be specific to the DL. They are simply generic issues in any kind of Information System!

Also automatic indexing, metadata extraction and "knowledge organization" in general are meeting the "traditional" corporate information systems area, trough the vital role played nowadays in any organization by document management systems, enterprise content management (the digital content as asset), and the dematerialization of the processes in general. In those scenarios, the "digital object" is not the exception anymore, but the rule, so even once DL (and archives) very specific issues such as digital preservation have been emerging as a normal concern in any organization, as historical information is making less sense, since all the information available is now critical for any good business governance ("archives" are now "repositories").

Aligned with this tendency, even the roles are changing. And in fact Lesk closes his paper with this very interesting paragraph:

---

[5] http://www.archive.org/details/millionbooks
[6] http://www.loc.gov/z3950/agency/
[7] http://www.openarchives.org/OAI/openarchivesprotocol.html

"Will, in a future world of online information, the job of organizing information have higher status, whatever it is called? I am optimistic about this, by analogy with accountancy. Once upon a time accountants were thought of as people who were good at arithmetic. Nowadays calculators and computers have made arithmetical skill irrelevant; does this mean that accountants are unimportant? As we all know, the answer is the reverse and financial types are more likely to run corporations than before. So if computers make alphabetizing an irrelevant skill, this may well make librarians or their successors more important than before. If we think of information as a sea, the job of the librarian in the future will no longer be to provide the water, but to navigate the ship."

Accordingly, we can finish this point by concluding that even if there are areas of competence that we can claim as specific to a vision of the DL, we should differentiate its relevance as discipline, with a specific body of knowledge, from the possible applications of that body of knowledge to solve problems in specific scenarios. I mean, I believe that from now the DL community will be not requested anymore to provide technology, but expertise and services. In fact, reviewing Arms' key concepts, we can claim that none of them are really specific to the DL, but instead generic goals, constraints, requirements or good practices that we can find in multiple other areas:

About business goals and business environment:
1. The technical framework exists within a legal and social framework...
7. Repositories must look after the information they hold...
8. Users want intellectual works, not digital objects...

About business concepts and business domain
2. Understanding of digital library concepts is hampered by terminology...
5. Digital library objects are more than collections of bits...

About information systems design and good practices
3. The underlying architecture should be separate from the content stored in the library...
4. Names and identifiers are the basic building block for the digital library...
6. The digital library object that is used is different from the stored object...

---

5 Enterprise Architecture

The ANSI/IEEE 1471-2000 standard [3] defines architecture as "the fundamental organization of a system, embodied in its components, their relationships to each other and the environment, and the principles governing its design and evolution." According to this, the Enterprise Architecture emerges to help organizations to understand and express their business, structure and processes. The term Enterprise Architecture has, on the same time, two meanings: on one side it is the term given to the map of and organization and the plan for its business and technology continuous change; on the other side it is also the term given to the process to govern all of that.

The ability to have detailed views, planning and analytical knowledge of a system are vital tools to address new unavoidable requirements associated with the Web,

Table 1. The Zachman Framework

View	What (Data)	How (Function)	Where (Network)	Who (People)	When (Time)	Why (Motivation)
Scope	Things important to the business	Processes the business performs	Locations the business operates	Organizations important to the business	Events significant to the business	Business goals/strategies
Business Model	e.g., Semantic Model	e.g., Business Process Model	e.g., Business Logistics System	e.g., Work Flow Model	e.g., Master Schedule	e.g., Business Plan
System Model	e.g., Logical Data Model	e.g., Application Architecture	e.g., Distributed System Architecture	e.g., Human Interface Architecture	e.g., Processing Structure	e.g., Business Rule Model
Technology Model	e.g., Physical Data Model	e.g., System Design	e.g., Technology Architecture	e.g., Presentation Architecture	e.g., Control Structure	e.g., Rule Design
Components	e.g., Data Definition	e.g., Program	e.g., Network Architecture	e.g., Security Architecture	e.g., Timing Definition	e.g., Rule Specification
Instances	e.g., Data	e.g., Function	e.g., Network	e.g., Organization	e.g., Schedule	e.g., Strategy

XML and the concept of Service Oriented Architecture (SOA) [8]. The most important keyword associated with this new scenario is "flexibility"! Under this, the design and development of information systems builds on a global view of the world in which services are assembled and reused to quickly adapt to new goals, business needs and tasks. This means that the configuration of a system might have to change at any moment, removing, adding or replacing services on the fly, in alignment with the new business requirements. This is what Enterprise Architecture provides.

5.1   Enterprise Architecture Framework

Considering that the ultimate goal of the DL is to be able to offer solutions to address problems properly, than we must recognize that such solutions must be always a mix of an organizational structure with the related set of activities and services. Therefore, we'll have an enterprise, in the sense of a business activity. Accepting that, than we should ask now how organizations (enterprises) in other business areas address their issues related to information, processes and technology. That is the scope of the area of Information Systems[8]. The purpose of an information system in an organization is to support processes, and not surprisingly, professionals dealing with that use methodologies, models and frameworks to address their activities.

An Enterprise Architecture framework is a communication tool to support the Enterprise Architecture process. It consists in a set of concepts that must be used to guide during that process. The first Enterprise Architecture framework, also the most

---

[8] We should remember that the ACM – Association for Computer Machinery, identifies the area of "Digital Libraries" in its classification system with the coding H.3.7, under "Information Storage and Retrieval" (class H.3) and "Information Systems" (class H), as it can be seen at http://www.acm.org/class/

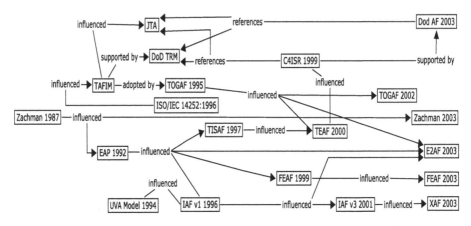

Fig. 1. The Enterprise Architecture Frameworks History Overview[9]

comprehensive and famous of them, is the Zachman framework[10], defined as "…a formal, highly structured, way of defining an enterprise's systems architecture. (…) to give a holistic view of the enterprise which is being modelled." the Zachman framework is resumed in simple terms in Table 1, where each cell can be related with a set of models, principles, services, standards, etc., whatever is needed to register and communicate its purpose. The meanings of the lines in this table are:

Scope (Contextual view; Planner) defined the business purpose and strategy; Business Model (Conceptual view; Owner) describes the organization, revealing which parts can be automated; System Model (Logical view; Designer) outline of how the system will satisfy the organization's information needs, independently of any specific technology or production constraints; Technology Model (Physical view; Builder) tells the system will be implemented, with the specific technology and ways to address production constraints; Components (Detailed view; Implementer) details each of the system elements that need clarification before production; Instances (Operational view; Worker) gives a view of the functioning system in its operational environment.

Concerning the meanings of the columns, What refers to the system's content, or data; How to the usage and functioning of the system, including processes and flows of control; Where to the spatial elements and their relationships; Who to the actors interacting with the system; When represents the timings of the processes; Why represents the overall motivation, with the option to express rules for constraints where important for the final purpose.

From this Framework many other Enterprise Architecture frameworks for specific areas have been developed. Those have been developed by research entities (such as E2A[11]), governmental bodies[12] (such as FEAF, TEAF, TOGAF, etc.) and private

---

[9] Redrawn from [6] (more details can be found in this reference).

[10] Originally conceived by John Zachman at IBM [9], this framework is now in the public domain, through the The Zachman Institute for Framework Advancement. For more details see http://www.zifa.com

[11] http://www.enterprise-architecture.info/ (Institute for Enterprise Architecture developments).

[12] http://www.eagov.com; http://www.eaframeworks.com/frameworks.htm; http://www.whitehouse.gov/omb/egov/a-1-fea.html

companies (such as the IAF[13], from Cap Gemini). The process has been also influenced by other related activities, as illustrated in the conceptual map in the Figure 1.

5.2 Enterprise Architecture and Governance

Enterprise Architecture is an instrument to manage the operations and future development in an organization. In this sense, in order to practice a correct Enterprise Architecture, planning and development must take in consideration the overall context of corporate and IT governance. This list of references expresses very well the complexity of the Enterprise Architecture process: Strategic Management: Balanced Scorecard[14]; Strategy Execution with EFQM[15]; Quality Management with ISO 9001[16]; IT Governance with COBIT[17]; IT Service Delivery and Support with ITIL[18]; IT Implementation with CMM[19] and CMMI[20].

6   The Goal of the Digital Library

How could we now define the goal of the DL? In my view, this simple statement might be enough to express that: **The goal of the DL is to provide access to selected intellectual works.** This goal comprises this way the three more generic (first level) business processes of the DL: Collection building; Discovery; Access.

We could express this goal with more words, but quite for sure that those would be redundant. We could also express this goal with more details, but quite for sure that such would be only a matter of specialization.

In fact, for a specific case second and other lower level processes must be identified, but these will depend of the specific context (the details of the "Scope" line in the Zachman Framework). For example, storage will be a requirement derived from access. Also the goal to provide access at any moment produces the requirement of preservation. In the same sense, registration is a requirement derived from discovery (to make it to be possible to find or be aware of a resource we produce requirements for cataloguing, indexing, descriptive metadata, etc.). Selectivity can be seen as a goal in itself, from which we can express relevant functional requirements (policies of collection building can be important in educational and professional libraries, in order to promote efficiency for the users), or it can be simply a consequence of a non-functional requirement associated to the fact that it might still be impossible, for a specific system, to provide discovery and access to everything produced by the focused organization (at least for now…).

---

[13] http://www.capgemini.com/services/soa/ent_architecture/iaf/

[14] http://www.balancedscorecard.org/ (The balanced scorecard management system).

[15] http://www.efqm.org/ (European Foundation for Quality Management excellence model).

[16] http://www.iso.org/iso/en/CatalogueDetailPage.CatalogueDetail?CSNUMBER=21823 (ISO 9001: Quality management systems – Requirements).

[17] http://www.isaca.org/cobit/ (Control Objectives for Information and related Technology standard).

[18] http://www.itsmf.org/ (IT Infrastructure Library best practices).

[19] http://www.sei.cmu.edu/cmm/ (Capability Maturity Model for Software).

[20] http://www.sei.cmu.edu/cmmi/ (Capability Maturity Model Integration).

7 Conclusions

Concluding, the DL community must prepare itself for a dignified retirement age by moving its established knowledge from research to engineering, in order to take part in more generic goals[21].

A framework can be described as "a set of assumptions, concepts, values, and practices that constitutes a way of viewing the current environment" [5]. Frameworks can be used as basic conceptual structures to solve complex issues. Concluding, and in alignment with the vision already expressed by the DLF Service Framework Working Group[22], I think that the DL community should "get out of the box" and give more attention to the development of conceptual frameworks giving preference to scopes, goals requirements and processes, in the sense as those concepts are already common in Enterprise Architecture processes ([7] is a classic and stills one of the most cited reference for that purpose) and Enterprise Architecture Frameworks ([6] can be a very simple comprehensive reference for this).

What should it be the process for that and what kind or level of frameworks should we envisage for this work? As also described in [5], "a reference model is an abstract framework for understanding significant relationships among the entities of some environment that enables the development of specific architectures using consistent standards or specifications supporting that environment (…) and is independent of specific standards, technologies, implementations, or other concrete details". Still in [5], "a reference architecture is an architectural design pattern that indicates how an abstract set of mechanisms and relationships realizes a predetermined set of requirements".

Should we have reference models and reference architectures for the DL?

Maybe yes. Maybe it makes sense to develop such references for specific goals and processes, such as Digital Preservation, Institutional Repositories, etc.!

But also maybe not, or at least as some of us have been trying to do it, especially if we give credit to this external observer that wrote one[23]:

"A framework should be developed at a particularly high level, encompassing only the common and agreed upon elements of library processes. Whilst you may need to dig deep to collect and confirm processes, the framework itself, I suggest, should remain fairly high -providing individual enterprises the ability to compare, contrast and build upon that framework in their own context. That said, libraries have been around for a very long time, I'm certain that libraries have many business processes that they commonly share.

---

[21] Off course that the retirement age for the Digital Library will occur naturally, when its children and grandchildren will emerge with new issues and challenges, on the top of its shoulders. Our "intellectual youngest cousin", the Semantic Web, could be one of those descendents, but in spite of the "good schools" where it has been breed and educated, it remains uncertain if it will be able to provide practical value. The Web 2.0, like the "new kid on the block", is bringing new and fresh fascinating ideas, but its informality makes us nervous; it is not clear yet if its actual effectiveness is not only a transient property resulting from the enthusiasm of the schoolboys.

[22] http://www.diglib.org/architectures/serviceframe/

[23] http://ea.typepad.com/enterprise_abstraction/2006/11/dlf_services_wo.html (this entire blog, from Stephen Anthony, deserves a close reading by any Digital Library practitioner).

What am I really saying? I'm saying there are at least 2 levels of architecture here. The high level meta-architecture (framework) that's generally agreed upon amongst libraries, and then there's a true enterprise-level architecture that's needed within an institution to meet specific needs. The enterprise-level architecture should, ideally, use the framework to guide their architecture development and implementations... but a framework can never fully accommodate the specific business needs, planning and implementation required within an organization."

Concluding, maybe it is time to recognise that the focus of the DL should move from the perspective of the engineer to the perspective of the architect[24].

References

[1] Arms, W.: Key Concepts in the Architecture of the Digital Library. D-Lib Magazine, (July 1995), http://www.dlib.org/dlib/July95/07arms.html

[2] Bush, V.: As We May Think. Atlantic Monthly 176(1), 101–108 (1945), http://www.theatlantic.com/doc/194507/bush

[3] IEEE: IEEE Std 1471-2000 IEEE Recommended Practice for Architectural Description of Software-Intensive Systems –Description (October 9, 2000)

[4] Lesk, M.: The Seven Ages of Information Retrieval. In: Proceedings of the Conference for the 50th anniversary of As We May Think, pp. 12–14 (1995), http://www.lesk.com/mlesk/ages/ages.html

[5] OASIS - Organization for the Advancement of Structured Information Standards. Reference Model for Service Oriented Architecture. Committee Specification 1. 2 (August 2006), http://www.oasis-open.org/committees/download.php/19679/soa-rm-cs.pdf

[6] Schekkerman, J.: How to survive in the jungle of Enterprise Architecture Frameworks. Trafford Publishing, ISBN 1-4120-1607-X (2004)

[7] Spewak, S.: Enterprise Architecture Planning – Developing a Blueprint for Data, Applications and Technology. John Wiley & Sons Inc, Chichester (1993)

[8] Thomas, E.: Service-Oriented Architecture: Concepts, Technology, and Design. Prentice Hall PTR, Englewood Cliffs (2005)

[9] Zachman, J.: A Framework for Information Systems Architecture. IBM Systems Journal 26(3) (1987) IBM Publication G321-5298

---

[24] http://answers.google.com/answers/main?cmd=threadview&id=233551 (Google Answer entry).

Multimedia in Cultural Heritage Collections: A Model and Applications[*]

Cristina Ribeiro[1,2], Gabriel David[1,2], and Catalin Calistru[1,2]

[1] Faculdade de Engenharia, Universidade do Porto
[2] INESC—Porto
Rua Dr. Roberto Frias s/n, 4200-465 Porto, Portugal
{mcr,gtd,catalin}@fe.up.pt

Abstract. The paper presents a multimedia database model accounting for the representation of documents, collections and the associated metadata. Appropriate structures are provided for descriptive metadata and for metadata resulting from automatic content analysis. The model is based on the identification and unification of the main concepts in the archival standards and the audiovisual area.

The main features of the model, designed to support multimedia database applications, are the integration of descriptive and content analysis metadata, the association of metadata to collections as well as to items, the extensibility with respect to the inclusion of new descriptors and the support to several retrieval modes. The MetaMedia application development platform, based on the model, has been used to support the construction of a historic documentation collection where a common web interface provides collection administrators, metadata creators and visitors a multi-faceted view of the repository.

Keywords: Models for multimedia repositories, information retrieval, cultural heritage applications.

1   Introduction

The organization of large digital repositories has required specialists from diverse areas to explore aspects that range from the conceptual aspects of digital collections as information systems [1] to the development of operational platforms to support the organization and access to digital collections [2,3,4]. The success of text-based retrieval has raised the expectations of users concerning the possibilities of search on multimedia collections. Many existing cultural heritage repositories are better viewed as multimedia, and current digital objects are more and more so.

To deal with multimedia collections, it is necessary to handle the content itself, which may require specific storage and presentation devices, and to manage the associated metadata that may be of different nature and be generated according

---

[*] Supported by FCT under project POSC/EIA/61109/2004 (DOMIR).

D.H.-L. Goh et al. (Eds.): ICADL 2007, LNCS 4822, pp. 186–195, 2007.

to a variety of standards. In cultural heritage collections, where objects are typically subject to some kind of expert analysis, an operational system must satisfy two goals. The first is to allow rigorous descriptive metadata to be handled and associated to items. The second is to offer content-based search using state-of-the-art technologies.

The paper presents a multimedia database model accounting for the representation of documents, collections and the associated metadata. The model is based on the identification of the main concepts in the archival standards as well as those from the audiovisual area. The main features of the model are the integration of descriptive and content analysis metadata, the association of metadata to collections as well as to items, the extensibility with respect to the inclusion of new descriptors and the support to several retrieval modes.

The model has been designed to support multimedia database applications, and its scope is distinct from the models underlying current standards and from the reference models adopted in several communities. The former [5,6,7], due to their specificity, prescribe strict definitions and formats for the elements of a description. The latter [8,9,10] are more complex models encompassing the objects and their processes in an organization. A model designed to support a multimedia database can be restricted to a set of core features and still allow the incorporation of data from different standards and support the storage and retrieval of individual objects and collections.

The paper is organized as follows. Section 2 introduces the concepts adopted to describe and manage multimedia items. Section 3 outlines the proposed model. The following sections present the MetaMedia application development platform, the case study and the retrieval methods, along with a preliminary evaluation of the proposed framework.

2 Metadata and Standards

The generalized production of digital documents brings new requirements to the activity of librarians and archivists coming from the nature of the documents—increasingly structured and in different media, and the target audiences—ranging from scientists and professionals to unspecialized users.

Metadata covers aspects of media such as its description, content analysis, technical details, terms of use and administrative aspects and it can be automatically generated or manually associated to the documents. There is currently a great interest in automating the metadata extraction process. Automatic analysis of text such as performed by indexing tools can be regarded as an automatic metadata generation process. With visual content, it is also possible to automatically generate descriptors for features such as color or texture.

2.1 Standards

Standards tend to address different aspects of the digital content, the ones most relevant for the intended use of the items in the respective community. While it may be essential for a library to have detailed information on a scientific journal, title and author descriptors may be enough for documents in a web site; a film distribution corporation may require media- and genre-specific descriptors to provide search on the movie catalog, but the same items when used by a broadcaster require information on the days and times when they will be programmed.

In the library area, there are well-established standards that support applications and metadata sharing [5,11]. Recent work concerns conformance to the XML language syntax, inclusion of sound and image documents, stronger networking.

ISAD and ISAAR [6] are standards for archival description. Their basic principles are multilevel structure and uniform description, and they handle both the records and the people and organizations involved in their creation.

Dublin Core (DC) [12] has appeared to solve the problem of the lack of description of documents on the web. It consists of a set of basic descriptors such as title, creator and date, intentionally kept at a basic non-specialist level. DC is being widely adopted as part of other web-related standards, such as those for the Semantic Web initiatives [13].

MPEG-7 [7] comes from the audiovisual signal processing community and aims at creating metadata for complex multimedia items. The emphasis is on descriptors that can be automatically extracted from audiovisual content, leaving descriptive metadata for other standards such as DC. MPEG-21 [14] originated in the same community and concerns metadata for handling the multimedia delivery chain rather than item or collection descriptions.

3   Multimedia Database Model

The goal of adopting a multimedia database model is twofold. A model can abstract the underlying technologies and therefore allow an application to be maintained as technology evolves. On the other hand, metadata production is expensive and repository managers want to import it whenever available, requiring a model for representing the items.

One thing to consider is whether it is actually necessary to design a model, as most initiatives on standards already rely on their own models and metadata reference models are also available in several domains [8,9]. The fact is a model can only be useful if it treats data at the convenient abstraction level. Following the model for one standard would lead to a specialization in its application domain, whereas reference models have a much broader scope than required for the basic representation and retrieval tasks. We have therefore chosen to design a compact model capturing concepts from several relevant standards and amenable for implementation. The model is centered on the integration of descriptive and content analysis metadata of multimedia objects, to support automatic indexing and retrieval tasks.

In the sequel we will first present the core concepts of the MetaMedia model and then an outline of the data model with the relevant relationships.

3.1   The Concepts

The model is organized around four main principles. The first one is that multimedia objects are usually organized in a *part-of hierarchy*. To each level one can associate a set of attributes characterizing the corresponding set of items. These attributes typically cover aspects related to the creation context of the multimedia object, its format, support and access conditions.

The second principle is that of *uniform description*, whereby the same set of attributes is used for an individual object, for composite objects and for sets of related objects. This principle has been followed in the standards for archival description such as ISAD [6] and is very useful in the representation of large collections: metadata is frequently available for sets of items rather than items, and inheritance can make it useful further down the hierarchy.

The third principle applies to actual multimedia data. It is concerned with the internal structure of individual multimedia objects. The actual multimedia object is stored in one or more *segments* which are parts of multimedia items.

The contents of each segment can be analysed by appropriate tools, which generate specialized *descriptors*. For example, the video track segment of a specific multimedia item may have associated motion activity and color descriptors, while the audio track segment of the same item may be connected to a melody contour descriptor. The fourth principle states that the descriptor resulting from the analysis of some feature of a segment is expressed as an XML file of an appropriate format.

According to the principles just outlined, four main concepts were selected. The first one, the Description Unit (DU), is already present in archival description [6], corresponds to the concept of Digital Item in the audiovisual standards and captures the notion of an object or collection of objects with an associated context.

DUs are organized in hierarchies that may have various topologies and different semantics for their levels. Such hierarchies can be created for new collections and can be extracted from existing ones. In both cases they capture the nature of the datasets. A Scheme that defines the possible levels, their semantics and their interconnections is required. Figure 1 shows a sample from a part-of hierarchy and the corresponding scheme.

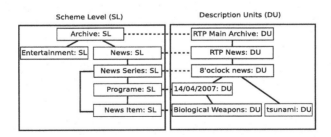

Fig. 1. Scheme Instance Example

The third concept is the Segment, following the MPEG-7 vocabulary, and captures the notion of some part of an actual multimedia item, such as a video sequence that is reused in a new documentary work. A segment has no context of its own, getting it from the DU of the object it belongs to.

The fourth concept is that of a Descriptor. The sense in which Descriptor is used is the one established by the MPEG-7 standard—a representation of a feature [7]. In the model, a Descriptor is regarded as the result of analysing a Segment. An image Segment, for example, can be associated to its corresponding instances of the *DominantColor* and *NumberOfFaces* descriptors, a video Segment can be associated to its *MotionActivity* descriptor and an audio Segment to its *MelodyContour*.

3.2   The Data Model

Figure 2 shows a simplified version of the data model. The concepts in the model are associated with the main classes. Control of the hierarchy is provided by the Scheme Level class. Each Description Unit is of a specified level; the structure of the levels, omitted in the simplified model, allows the application development platform to enforce creation and structure of the DUs according to the repository schema. In Figure 1, for instance, a Description Unit at the level of News Item can be a direct descendant of instances of Program or News Series, but not of News or Archive. Attributes such as title, author, date and copyright apply at the fine-grained level of the document as well as at the coarser grain of the whole collection. They are therefore captured as attributes in the Description Unit and appear uniformly at all levels.

The Segment class embodies the corresponding concept, modeling object fragments that belong to items in the Description Units. The association between Description Units and Segments is visible as the Contents class. Segments are further specialized as text, image, video and audio.

Descriptors such as *MotionActivity*, *DominantColor* and *MelodyContour* are likely to be used only for specific kinds of Segments. The Descriptor class represents them and the association of segments to descriptors is modeled with the Descriptor Instance class.

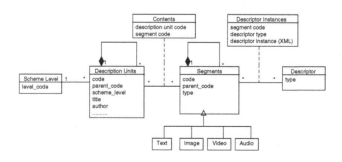

Fig. 2. The MetaMedia Model

In the model, the DU and Segment classes are very similar in structure: both offer a hierarchical organization of their objects. The main distinction lies in the nature of the objects and in the kind of associated metadata. An instance of Description Unit captures a document for which a well-established description is available, and which has been related to other documents according to the repository hierarchy. An instance of segment is appropriate for representing a video frame for an object whose description is in the associated DU and for which some automatic low-level descriptors have been produced. The part-of structure for the segments has no predefined structure and is intended to follow the granularity of the existing analysis tools.

4   MetaMedia — A Multimedia Database Platform

The current MetaMedia model has evolved from an early version of the model [15,16]. Representation of descriptors in XML allows the incremental addition of descriptors with arbitrary structure. This feature is important both for representing text annotations according to specified vocabularies and for dealing with multimedia items. Processing visual and audio information is a very active area of research and new descriptors become available frequently [17].

Having a model with a compact set of concepts has allowed the development of a software platform where object collections, object relations, high-level metadata and diverse low-level descriptors can be managed. The platform has been designed as a web portal supported on a database system, and is currently under test using several kinds of collections.

A collection interface built on the MetaMedia platform gives access to the structure of Description Units, but also to diverse associated segments. A collection of text documents may have extracted text—a segment for each document, and automatically extracted descriptors for the text—content annotations based on a controlled vocabulary. A video collection may have multidimensional descriptors such as *Color* or *Motion Activity*, automatically obtained for some of the video segments. Each kind of descriptor will be used for content-based access to the objects in the collection.

Content-based retrieval of textual documents has succeeded in providing useful answers to common queries based on automatic indexing. Content-based retrieval of audiovisual materials has to deal with the "semantic gap" [18] separating the low semantic level of automatically extracted descriptors from the high-level concepts people use in search. Many experimental systems explore content-based retrieval of multimedia objects [17,18]. The approach in MetaMedia is to allow search using the high-level concepts to be combined with search based on object similarity. There are no systems with these combined features in current production.

5   The "Terra de Santa Maria" Documentation Center

The "Terra de Santa Maria" historic documentation center [19] has been a case study for the MetaMedia platform. The collection is a virtual archive of medieval

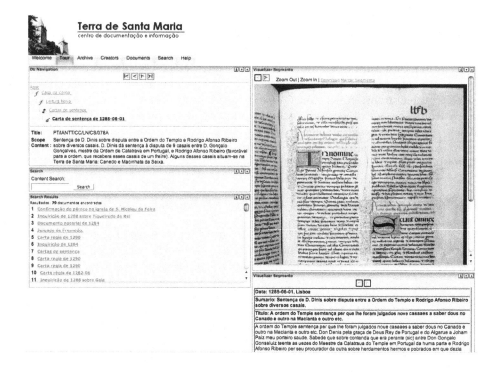

Fig. 3. Interface of the Historic Archive

documents, for which there are transcriptions in either Latin or archaic Portuguese. Information for a document in the repository includes three parts: the digitized image, the document transcription and the archival description according to the ISAD/ISAAR standards.

The interface is available in English and in Portuguese, and the archival descriptions have been generated in Portuguese. There are several views on the document, intended for different kinds of users.

On the *Archive* tab, it is possible to browse the structure of the archive, designed by archivists, and to edit the descriptions. The *Creators* tab has detailed information on the creators for the current unit. Under the *Document* tab users may explore the contents of the documents. A medievalist studying one of the parchments uses this mode to observe the digitized image and its transcription side by side and to possibly upload his own analysis of the text.

Specialists need to search on specific items, while casual users require a simple keyword-based retrieval mode. Both are offered, and it is also possible to use the concepts marked up on the document transcriptions for search purposes. Textual segments corresponding to the digital content have been marked using an XML tool. Markup identifies key concepts that would be hard to spot on the original latin documents and is mainly intended for improving retrieval.

6   Collection Building and Management

Interesting documents are essential when we want to offer access to a digital repository, and the available metadata may allow rich views on their contents. A tradeoff exists between the simplicity of the view which is offered on documents and their relations and the detailed access to the collection building tools. Having a compact set of concepts underlying the model helps to put the collection manager in control of the relevant aspects of the collection and to automate the building process.

In the MetaMedia platform there is one task strictly reserved for the collection administrator: the design of the repository hierarchic structure. The part-of relationship exists in traditional archives due to their provenance-based structure, but can also be found in most large collections, where it is useful to account for the organization of collections and sub-collections and deal with the increasing complexity of digital items.

Based on a chosen hierarchy, items and their metadata are added. Descriptive metadata is available both at document and at collection level, and it is possible to add it at any point in the collection production. Segments in the form of text, image or video may also be associated to items at any point. Taking advantage of this information in search requires a rebuild of the low-level indexes and automatic descriptors. A collection is ready for presentation when the automatic addition of extracted descriptors and the indexing of the textual and non-textual content have been performed.

7   Search and Retrieval

A multimedia repository must offer several retrieval modes. The user may search on textual content for textual segments of the documents, on the structured contextual metadata available as descriptors in the Description Units, and on the visual features for image/video segments of the items.

Query by keyword is the most straightforward and requires a full-text indexing system. The platform uses the Apache Lucene search technology [20], configured to index selected parts of the available textual information: the textual content itself and parts of the descriptive information. A structured query interface is intended for specialized users, who are aware of the meaning of the descriptive metadata in the Description Units. Search on the contextual metadata is handled by the built-in indexes of the relational database management system.

Query by visual features is integrated with the keyword search. In a collection where documents have textual segments, keyword search provides an initial answer which is expanded by visual and audio similarity. A collection with just image or video objects can also be searched with textual queries, provided that some minimal concept annotation is present. Concepts in the query are extracted and matched against the annotations, resulting in an initial set of items. Image and audio similarity are then used to expand the answer and a relevance feedback interface is offered to refine the query. The "BitMatrix" multidimensional indexing technique [21] is used to ease the computation of image similarity and to allow the tuning of the set of descriptors to the nature of the collection.

When collections have substantial structure, browsing can be an effective way of guiding a retrieval process. The MetaMedia platform keeps the context of items being visited, and it is therefore possible to browse the hierarchy, locate an interesting sub-collection, view its description, travel down to a document and analyze its visual or textual content. From a selected image or video segment it is also possible to start a new search using low-level similarity.

8   Conclusions

The representations used by libraries and archives are being extended to deal with audiovisual content. Object descriptions must capture their features and also support the navigation and search by unspecialized users. Many standards have emerged, addressing common requirements and specificities of some content types and application domains.

In this work we have identified the main concepts underlying the standards and built a compact model suitable for implementation in a relational database system with associated repositories for multimedia items. The model allows structured descriptors to deal with the fine-grained features of digital items on various domains. A web platform is used to build repositories, to associate metadata to the items and to search the collections.

The contributions or the paper are: the analysis of existing standards from the point of view of an operational multimedia database system; a multimedia database model; a platform for managing, building and accessing digital repositories; and a search interface integrating descriptive and content metadata where high- and low-level descriptors are interleaved.

A case study shows the appropriateness of the model in a historic documentation center, an application domain where diverse description and content analysis metadata is available. The model has proved robust and the configurable hierarchy has been useful both for representing the structure of the repository and for making it accessible to the public.

The MetaMedia platform is currently being used for searching video collections with no descriptive metadata. The video retrieval process is being prepared for a retrieval evaluation process. Several open issues remain, namely in the search with automatically extracted descriptors, the collaborative environment for uploading segment descriptors and the user interfaces for complex items.

References

1. Gonçalves, M.A., Fox, E.A., Watson, L.T., Kipp, N.A.: Streams, Structures, Spaces, Scenarios, Societies (5S): A Formal Model for Digital Libraries. ACM Trans. Inf. Syst. 22, 270–312 (2004)
2. DSpace: The DSpace digital repository system (2007), http://dspace.org/
3. Lagoze, C., Payette, S., Shin, E., Wilper, C.: Fedora: an architecture for complex objects and their relationships. Int. J. on Digital Libraries 6, 124–138 (2006)

4. Witten, I.H., Bainbridge, D.: Building digital library collections with Greenstone.. In: JCDL, p. 425 (2005)
5. Library of Congress: MAchine-Readable Cataloguing (MARC) (2007), http://www.loc.gov/marc/
6. ISAD(G): General International Standard Archival Description, Second edition (1999), http://www.ica.org/
7. Sanchez, J.M.M., Koenen, R., Pereira, F.: MPEG-7: The Generic Multimedia Content Description Standard, Part 1. IEEE MultiMedia 9, 78–87 (2002)
8. International Council of Museums: CIDOC Conceptual Reference Model (2006), http://cidoc.ics.forth.gr/
9. DELOS Network of Excellence on Digital Libraries: A Reference Model for Digital Library Management Systems (2007), http://www.delos.info/
10. Hunter, J.: Combining the CIDOC CRM and MPEG-7 to Describe Multimedia in Museums. In: Proceedings of MW 2002: Museums and the Web, Archives and Museum Informatics (2002)
11. Library of Congress: Metadata Encoding and Transmission Standard (METS) (2007), http://www.loc.gov/standards/mets/
12. Dublin Core Requirements Group: Dublin Core Metadata Initiative (2007), http://dublincore.org/
13. W3C Consortium: Semantic Web (2007), http://www.w3.org/2001/sw/
14. Burnett, I., Van de Walle, R., Hill, K., Bormans, J., Pereira, F.: MPEG-21: Goals and Achievements. IEEE MultiMedia 10, 60–70 (2003)
15. Ribeiro, C., David, G.: A Metadata Model for Multimedia Databases. In: Proceedings of ICHIM 2001: International Cultural Heritage Informatics Meeting, Archives and Museum Informatics (2001)
16. Ribeiro, C., David, G., Calistru, C.: A Multimedia Database Workbench for Content and Context Retrieval. In: MMSP IEEE Workshop, IEEE Computer Society Press, Los Alamitos (2004)
17. Lew, M.S., Sebe, N., Djeraba, C., Jain, R.: Content-based multimedia information retrieval: State of the art and challenges. ACM Trans. Multimedia Comput. Commun. Appl. 2, 1–19 (2006)
18. Zhao, R., Grosky, W.I.: Narrowing the Semantic Gap—Improved Text-Based Web Document Retrieval Using Visual Features. IEEE Transactions on Multimedia 4, 189–200 (2002)
19. Comissão de Vigilância do Castelo de Santa Maria da Feira.: Centro de Documentação da Terra de Santa Maria (2007), http://www.castelodafeira.pt/
20. Apache Software Foundation: Apache Lucene 2.2 (2007), http://lucene.apache.org/
21. Calistru, C., Ribeiro, C., David, G.: Multidimensional Descriptor Indexing: Exploring the BitMatrix.. In: Sundaram, H., Naphade, M., Smith, J.R., Rui, Y. (eds.) CIVR 2006. LNCS, vol. 4071, pp. 401–410. Springer, Heidelberg (2006)

Graph-Based Indexing and Querying on Image Corpora with Unified Visual Semantic and Relational Descriptions

Mohammed Belkhatir

School of Information Technology, Monash University
Belkhatir.mohammed@infotech.monash.edu

Abstract. We propose in this paper to integrate the semantic description of the image and the relational characterization of its components through an architecture which follows a sharp process for generating image index and query representations and computing their correspondence. This architecture relies on an expressive representation formalism handling high-level image descriptions and a conceptual query framework in an attempt to operate image indexing and retrieval operations beyond keyword-based and loosely-coupled state-of-the-art systems. At the experimental level, we evaluate its retrieval performance through recall and precision indicators on a test collection of 2500 color photographs.

1 Introduction and Related Work

State-of-the-art image indexing and retrieval systems are mainly based on characterizing the image content through an automatic process mapping low-level extracted features (such as color histograms or Gabor matrices for respectively color and texture extraction) to semantic-based keywords (among them [6,7,10]). The major disadvantage of this class of frameworks relies on the specification of restrained and fixed sets of semantic-based keywords which are moreover not sufficient to accurately represent non-textual documents, such as images. Regarding the fact that several artificial objects have high degrees of variability with respect to signal properties, an interesting solution is to extend the extracted visual semantics with signal characterizations in order to enrich the image indexing vocabulary and query language. Therefore, a new generation of systems integrating semantics and signal descriptions has emerged and the first solutions [8,13] are based on the association of textual annotations to characterize semantics with a relevance feedback (RF) scheme operating on low-level signal features. These approaches have three major drawbacks: first, they fail to exhibit a single framework unifying low-level data and semantics, which penalizes the performance of the system in terms of retrieval efficiency. Then, as far as the query process is concerned, the user is to query both textually in order to express high-level concepts and through several and time-consuming RF loops to complement his initial query. Therefore, this solution for integrating semantics and signal features, relying on a cumbersome query process, does not enforce facilitated and efficient user interaction. Finally, these systems do not take into account the relational spatial information between visual entities, which affects the quality of the retrieval results. Indeed, the need of an expressive index and query language for manipulating multimedia documents, and in particular one supporting relational characterization of the content, has been highlighted in [11].

D.H.-L. Goh et al. (Eds.): ICADL 2007, LNCS 4822, pp. 196–205, 2007.
© Springer-Verlag Berlin Heidelberg 2007

We propose a unified multi-facetted framework unifying visual semantics and relational spatial characterization for automatic image retrieval that enforces expressivity through the use of symbolic descriptors to characterize the image content. After specifying a fully-automatic framework extracting the visual semantics, we enrich the description of images through the specification of processes establishing a correspondence between extracted low-level features and high-level spatial concepts. For example, with the semantic concepts "huts" and "grass" one might assign relations such as "above", "disconnected", "below" and "near" characterizing the fact that huts are above and disconnected from the grass, which is itself below and near huts. Therefore, not only do we characterize visual semantics, but also spatial relations linking them. For this, we consider an efficient operational model that allows relational indexing and is adaptable to symbolic image retrieval: **conceptual graphs** (CGs) [12]. However, contrarily to the EMIR[2] system [9] which was one of the early attempts at using CGs for image retrieval and limited its descriptive power to the basic semantics associated with these graphs (i.e. the conjunction of concepts and relations), we extend their operational semantics to handle a rich image query language consisting of the 3 major boolean operators (conjunction, disjunction and negation). Indeed, we are interested in dealing with non trivial queries involving the combination of visual semantics and spatial relations and the possibility to associate boolean operators to these queries. This would allow the user to retrieve images with "huts above **and** disconnected from the grass", "people at the left **or** at the right of buildings" or "houses **not** covered by vegetation"...

In the remainder, we first present the general organization of our image retrieval architecture. We deal in sections 3 and 4 with the visual semantics and spatial characterizations. Section 5 will specify the query framework. We finally present in section 6 the validation experiments conducted on a test collection of 2500 photographs.

2   An Architecture for Integrating Semantic and Spatial Descriptions

We propose an image retrieval architecture illustrated in fig. 1 which consists of five processing modules to integrate semantic and relational descriptions:

- The first provides the extraction of the image visual semantics through a statistical joint probability distribution tagging framework. Starting from a physical image (seen as a matrix of pixels), this framework allows to highlight the perceptually-meaningful visual entities with their associated semantic characterization in the form of a vector of semantic concepts with their recognition probabilities (further details are provided in section 3.1).
- The second module handles the image content representation and is based on a *multi-facetted* image model unifying visual semantics and spatial features. The **object facet** describes an image as a set of **image objects** (**IOs**) abstract structures representing visual entities within an image. The **visual semantics facet**, formally specified in section 3, describes the image semantic content and is based on labeling IOs with a semantic concept using the outcome of the semantic extraction module. E.g., in fig. 1, the first IO (Io1) is tagged by the semantic concept *People*. The **spatial facet**, detailed and formalized in section 4, describes the relational characterizations between pairs of IOs in terms of symbolic spatial relations. E.g. the first IO (Io1) is **inside** the second IO (Io2).

- The third module consists in applying the image model to specify an image index representation and therefore provides a representation of an image document in the corpus with respect to the multi-facetted image model. It is a structure called image index representation. In fig. 1, an image belonging to the corpus is characterized by a multi-facetted representation.
- At the other end of the architecture spectrum, the fourth module allows to translate a query with semantic and relational descriptions into a high-level structure with respect to the multi-facetted image model. It is a structure called image query representation. In fig. 1, a user full-text query "Images composed of people inside water" is translated into an image query representation closely following the multi-facetted image model.
- Finally, the fifth module deals with the correspondence process and the definition of a matching function between index and query representations. The image query representation is compared to all index representations of image documents in the corpus and a relevance value, estimating their degree of similarity is computed in order to rank all image documents relevant to a query. The search results are then displayed through the interface of the image retrieval system.

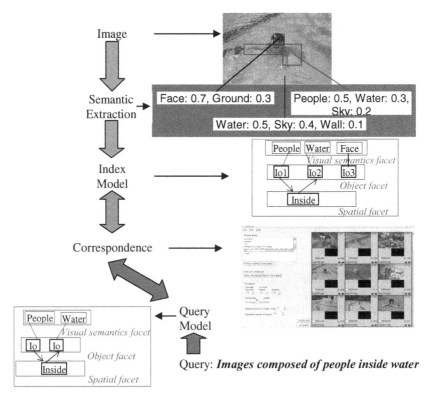

Fig. 1. Image retrieval architecture coupling semantic and relational descriptions

In order to instantiate the image model within an image retrieval framework, we choose a representation formalism capable to represent IOs, the visual semantics they convey and their relational characterizations: CGs. They have indeed proven to adapt to the symbolic approach of image retrieval [1,2,9,11] and allow to represent components of our image retrieval architecture and specify expressive index and query representations. Formally, a CG is a finite, bipartite, connex and oriented graph. It features two types of nodes: concept and relations. In the graph [2007]←(Year)←[ICADL]→(Location)→[Hanoi], concepts are between brackets and relations between parenthesis. This graph is semantically interpreted as: the ICADL conference of year 2007 is held in Hanoi. Concepts and relations are organized within lattice structures ordered by the IS-A relation.

3 The Visual Semantics Facet

3.1 Extracting the Semantics

Semantic concepts are learned and then automatically extracted given a visual ontology. Its specification is strongly constrained by the application domain [9]. Indeed dealing with corpus of medical images would entail the elaboration of a visual ontology that would be different from an ontology considering computer-generated images. In this paper, our experiments in section 6 are based on collections of general-purpose color photographs.

Several experimental studies presented in [10] have led to the specification of twenty categories or picture scenes describing the image content at a global level. Web-based image search engines (google, altavista) are queried by textual keywords corresponding to these picture scenes and 100 images are gathered for each query. These images are used to establish a list of semantic concepts characterizing objects that can be encountered in these scenes. A total of 72 semantic concepts to be learnt and automatically extracted are specified. Fig. 2 shows their typical appearance.

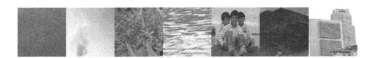

Fig. 2. Semantic concepts: ground, sky, vegetation, water, people, mountain, building

The indexing process is characterized by a statistical model which takes into account the joint distribution of semantic concepts on the one hand and symbolic signal features (color and texture) on the other hand. Starting from a learning set which includes IOs corresponding to visual entities, this model is instantiated by considering color and texture features of sets of connected rectangular regions used to generate the semantic concepts and their associated probabilities from this joint distribution. This process allows to highlight perceptually-meaningful visual entities with their associated semantic characterization in the form of a vector of semantic concepts with their recognition probabilities (further details can be found in [2]).

E.g., three visual entities linked to three IOs are highlighted from the example image in fig. 1. The first IO (*Io1*) is linked to a vector of semantic concepts with the highest recognition probability corresponding to the concept *people*, the second IO (*Io2*) to a vector of semantic concepts with the highest recognition probability corresponding to the concept *water* and the third IO (*Io3*) to a vector of semantic concepts with the highest recognition probability corresponding to the concept *face*.

3.2 Model of the Visual Semantics Facet

IOs are represented by *Io* concepts and the semantic concepts are organized within a multi-layered lattice ordered by a specific/generic order. An instance of the visual semantics facet is represented by a set of CGs, each one containing an *Io* type linked through the conceptual relation *is_a* to a semantic concept. Let us note that only the semantic concept with the highest recognition probability is considered as far as the CG representation of the facet is concerned. The graph controlling the generation of all visual semantics facet graphs, called visual semantics graph, is: $[Io] \rightarrow (is\_a) \rightarrow [SC]$. E.g., graphs $[Io1] \rightarrow (is\_a) \rightarrow [people]$, $[Io2] \rightarrow (is\_a) \rightarrow [water]$ and $[Io3] \rightarrow (is\_a) \rightarrow [face]$ are the representation of the visual semantics facet in fig. 1 and can be translated as: the first, second and third IOs are respectively associated to semantic concepts *people*, *water* and *face*.

4 The Spatial Facet: From Low-Level Spatial Features to High-Level Relational Description

Taking into account spatial relations between visual entities is crucial in the framework of an image retrieval system since it enriches the index structures and expands the query language. It is indeed shown in the study published in [5] that people frequently describe images by formulating spatial descriptions such as «...**at the left of**...» or «...**below**...». Also, dealing with relational information between image components allows to enhance the quality of the results of an information retrieval system [11]. We study in this part methods used to represent spatial data and deal with the automatic generation of high-level spatial relations following a first process of low-level extraction.

4.1 The Relation-Oriented Approach

In order to model the spatial data, we consider the «relation-oriented» approach which allows to explicitly represent the relevant spatial relations between IOs without taking into account their basic geometrical features. Our study features the four modeling and representation spaces:

- The Euclidean space gathers the coordinates of image pixels. Starting with this information, all knowledge related to the other representation spaces can be deduced.
- We consider in the topological space five relations inspired from [3] and justify this choice by the fact that they are exhaustive and relevant in the framework of an image retrieval system. Let io1 and io2 two IOs, these relations are (s_1=**P**,io1,io2) : 'io1 is a part of io2', (s_2=**T**,io1,io2) : 'io1 touches io2 (is externally connected)',

(s_3=**D**,io1,io2) : 'io1 is disconnected from io2', (s_4=**C**,io1,io2) : 'io1 partially covers (in front of) io2' and (s_5=**C_B**,io1,io2) : 'io1 is covered by (behind) io2'. Let us note that these relations are mutually exclusive and characterized by the the important property that each pair of IOs is linked by only one of these relations.
– The Vectorial space gathers the directional relations: Right (s_6=**R**), Left (s_7=**L**), Above (s_8=**A**) and Below (s_9=**B**). These relations are invariant to basic geometrical transformations such as translation and scaling.
– In the metric space, we consider the fuzzy relations Near (s_{10}=**N**) and Far (s_{11}=**F**).

4.2 Automatic Spatial Characterization

4.2.1 Topological Relations

In our spatial modeling, an IO *io* is characterized by its center of gravity io_c and by two pixel sets: its interior, noted io_i an dits border io_b. We define for an image an orthonormal axis with its origin being the image left superior border and the basic measure unity the pixel. All spatial characterizations of an object such as its border, interior and center of gravity are defined with respect to this axis.

In order to highlight topological relations between IOs, we consider the intersections of their interior and border pixel sets through a process adapted from [4]. Let io1 and io2 be two IOs, the four intersections are: io1_i $\cap$ io2_i, io1_i $\cap$ io2_b, io1_b $\cap$ io2_i and io1_b $\cap$ io2_b.

Each topological relation is linked to the results of these intersections as follows:

- (P, io1, io2) iff. io1_b$\cap$io2_b = $\varnothing$, io1_i$\cap$io2_b $\neq \varnothing$, io1_b$\cap$io2_i=$\varnothing$ & io1_i$\cap$io2_i $\neq \varnothing$
- (T, io1, io2) iff. io1_b$\cap$io2_b $\neq \varnothing$, io1_i$\cap$io2_b= $\varnothing$, io1_b$\cap$io2_i=$\varnothing$ & io1_i$\cap$io2_i=$\varnothing$
- (D, io1, io2) iff. io1_b$\cap$io2_b=$\varnothing$, io1_i$\cap$io2_b=$\varnothing$, io1_b$\cap$io2_i=$\varnothing$ & io1_i$\cap$io2_i=$\varnothing$
- (C,io1,io2) iff. io1_b$\cap$io2_b=$\varnothing$, io1_i$\cap$io2_b=$\varnothing$, io1_b$\cap$io2_i $\neq \varnothing$ & io1_i$\cap$io2_i $\neq \varnothing$
- (E_C, io1, io2) iff. io1_b$\cap$io2_b=$\varnothing$, io1_i$\cap$io2_b $\neq \varnothing$, io1_b$\cap$io2_i=$\varnothing$ & io1_i$\cap$io2_i $\neq \varnothing$

The strength of this computation method relies on associating topological relations to a range of necessary and sufficient conditions linked to spatial attributes of IOs (i.e. their interior and border pixel sets).

4.2.2 Directional Relations

The computation of directional relations between io1 and io2 is based on their centers of gravity io1_c($x1_c$, $y1_c$) and io2_c($x2_c$, $y2_c$), the minimal and maximal coordinates along x axis ($x1_{min}$, $x2_{min}$ et $x1_{max}$, $x2_{max}$) as well as the minimal and maximal coordinates along y axis ($y1_{min}$, $y2_{min}$ et $y1_{max}$, $y2_{max}$) of their four extremities.

We will say that io1 is at the left of io2, noted (L,io1,io2) iff. $x1_c < x2_c \wedge x1_{min} < x2_{min} \wedge x1_{max} < x2_{max}$. Also, io1 is at the right of io2, noted (R,io1,io2) iff. $x1_c > x2_c \wedge x1_{min} > x2_{min} \wedge x1_{max} > x2_{max}$.

We will say that io1 is above io2, noted (A,io1,io2) iff. $y1_c > y2_c \wedge y1_{min} > y2_{min} \wedge y1_{max} > y2_{max}$. Also, io1 is below io2, noted (B,io1,io2) iff. $y1_c < y2_c \wedge y1_{min} < y2_{min} \wedge y1_{max} < y2_{max}$.

We illustrate these definitions in fig. 3 where the IO corresponding to huts (io1) is above the IO corresponding to the grass (io2). It is however not at the left of the latter since $x1_c < x2_c$ but $x1_{min} > x2_{min}$.

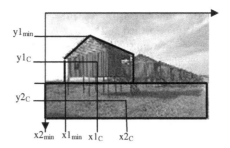

Fig. 3. Characterization of directional relations

4.2.3 Metric Relations

In order to distinguish between the Near and Far relations, we use the constant $D_{sp}=$ $d(\vec{0},0.5*[\sigma_1,\sigma_2]^T)$ where d is the Euclidean distance between the null vector $\vec{0}$ and $[\sigma_1,\sigma_2]^T$ is the vector of standard deviations of the localization of centers of gravity for each IO in each dimension from the overall spatial distribution of all IOs in the corpus. D_{sp} is therefore a measure of the spread of the distribution of centers of gravity of IOs. This distance agrees with results from psychophysics and can be interpreted as the bigger the spread, the larger the distances between centers of gravity are. Two IOs are **near** if the Euclidean distance between their centers of gravity is inferior to D_{sp}, **far** otherwise.

4.3 Conceptual Index and Query Structures for the Spatial Facet

4.3.1 Spatial Index Structures

IOs are related pairwise through an index spatial meta-relation (ISM), compact structure summarizing spatial relationships between these IOs. ISMs are supported by a vector structure **sp** with eleven elements corresponding to the previously explicited spatial relations. Values sp[i], $i \in [1,11]$ are booleans stressing that the spatial relation s_i links the two considered IOs. E.g., Io1 is related to Io2 through the ISM <**P:1**, T:0, D:0, C:0, C_B:0, R:0, L:0, A:0, B:0, N:0, F:0>, translated as Io2 being part of (inside) Io3.

4.3.2 Spatial Query Structures

Our framework proposes an expressive query language which integrates visual semantics and symbolic spatial characterization through boolean operators. A user shall be able to link visual entities with a boolean conjunction of spatial relations such as in Q1: "huts **above** AND **disconnected** from grass", a boolean disjunction of spatial relations such as in Q2: "people **at the left** OR **at the right** of buildings" and a negation of spatial relations such as in Q3: "houses NOT **covered by** vegetation".

Three types of conceptual structures are specified to support the previously defined query types. *And* spatial meta-relations (ASMs) represent the signal distribution of an IO by a conjunction of spatial relations; *Or* spatial meta-relations (OSMs) by a disjunction of spatial relations and *Not* spatial meta-relations (NSMs) by a negation of spatial relations. The ASM <P:0, T:0, **D:1**, C:0, C_B:0, R:,0 L:0, **A:1**, B:0, N:0, F:0>$_{AND}$, the OSM <P:0, T:0, D:0, C:0, C_B:0, **R:1**, **L:1**, T:0, B:0, N:0, F:0>$_{OR}$ and the NSM <P:0,

T:0, D:0, C:0, **C_B:1**, R:0, L:0, T:0, B:0, N:0, F:0>$_{NOT}$ respectively correspond to the spatial characterizations featured in queries Q1, Q2 and Q3.

4.4 Graph Representation of the Spatial Facet

Spatial meta-relations are elements of partially-ordered lattices organized with respect to the type of the query processed (we will not detail this organization here). There are 2 types of basic graphs controlling the generation of all the spatial facet graphs. **Index spatial graphs** link two IOs through an ISM: **[Io1]→(ISM)→[Io2]**. **Query spatial graphs** link two IOs through *And*, *Or* or *Not* spatial meta-relations: **[Io1]→(ASM)→[Io2]**; **[Io1]→(OSM)→[Io2]** and **[Io1]→(NSM)→[Io2]**.

Eg, the index spatial graph [Io1]→[<**P:1**, T:0, D:0, C:0, C_B:0, R:0, L:0, A:0, B:0, N:0, F:0>]→[Io2] is a graph of the index representation of the spatial facet in figure 1 and is interpreted as: io1 is linked to io2 through the index spatial meta-relation <**P:1**, T:0, D:0, C:0, C_B:0, R:0, L:0, A:0, B:0, N:0, F:0> (i.e. io1 being part of io2).

5   The Query Module

Our conceptual architecture is based on a unified framework allowing a user to query with both semantic and relational descriptions. The representation of a query is, like image index representations, obtained through the combination (join operation [12]) of CGs over the visual semantics and spatial facets. E.g., the query Q1 is represented by the CG:

[Io1]→(sct)→[Huts]   **[Io2]→(sct)→[Grass]**
     ⇘ [<P:0, T:0, **D:1**, C:0, C_B:0, R:,0 L:0, **A:1**, B:0, N:0, F:0>$_{AND}$]   ⇗

Also, the query Q2 is represented by the CG:

[Io1]→(sct)→[People]   **[Io2]→(sct)→[Buildings]**
     ⇘ [<P:0, T:0, D:0, C:0, C_B:0, **R:1, L:1**, T:0, B:0, N:0, F:0>$_{OR}$]   ⇗

Finally, the query Q3: "houses not covered by vegetation" is represented by the CG:

[Io1]→(sct)→[Houses]   **[Io2]→(sct)→[Vegetation]**
     ⇘ [<P:0, T:0, D:0, C:0, **C_B:1**, R:0, L:0, T:0, B:0, N:0, F:0>$_{NOT}$]⇗

The evaluation of similarity between index and query representations is achieved through a correspondence function: the CG projection operator. This operator allows to identify within the index CG sub-graphs with the same structure as the query CG, with nodes being possibly restricted (i.e. they are specializations of the query CG nodes).

6   Validation Experiments: An Application to Home Photographs

We implement the theoretical framework presented in this paper and validation experiments are carried out on a corpus of 2500 color photographs used as a validation corpus in [1,2,7]. IOs within the 2500 photographs are automatically assigned a vector of semantic concepts with their corresponding recognition probabilities as shown in section 3.1 and characterized with a visual semantics facet CG as shown in section 3.2. Also, pairs of IO are characterized with spatial index structures (section 4.3.1) and linked through an index spatial graph as shown in section 4.4.

As opposed to state-of-the-art keyword-based frameworks [6,7,10], we wish to retrieve photographs that represent elaborate image scenes and propose 12 queries characterizing relative location of visual entities such as *people near vegetation* along with their ground truths among the 2500 photographs. The evaluation of our formalism is based on the notion of *image relevance* which consists in quantifying the correspondence between index and query images. We compare our system with a system based on a semantic keyword-based approach: the *Visual Keyword* system S_1 [7] and a state-of-the-art loosely-coupled system S_2 combining a textual framework for querying on semantics and a RF process operating on low-level signal features.

For each proposed query in table 1, we construct relevant textual query terms using corresponding visual semantics and spatial characterizations as input to our system (e.g. *people near vegetation*). The retrieval results for this query are given in fig. 4. S_1 processes three series of three random relevant photographs for each query (they correspond to people near vegetation as far as our example query is concerned). Also each query in table 1 is translated in relevant textual data to be processed by the semantic framework of S_2 ('people, vegetation' for *people near foliage*). Then to refine the results, three random relevant photographs are selected as input to the RF framework.

Fig. 4. Results for "People near vegetation"

Table 1. Queries

People touch pool
Buildings left <u>and</u> right of people
People in (part of) water
Foliage left and right of people
People near buildings
People in front of mountains
Close-up of people (people not related to any IO)
Close-up of buildings (buildings not related to any IO)
People in front of buildings
People near foliage
Cityscape (view from far)
Mountain (view from far)

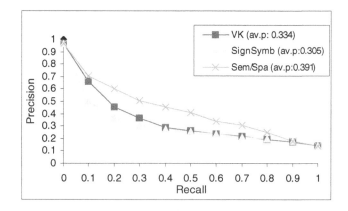

Fig. 5. Recall/Precision curves

Recall/precision curves of fig. 5 illustrate the average results obtained for all queries considering the corpus of 2500 images: the curve associated with the *Sem/Spa*
legend illustrates the results in recall and precision obtained by our system, the curve
associated with the *VK* legend by S_1 and the curve associated with the *SignSymb* legend by S_2. The average precision of our system (0.391) is approximately 17,1% higher
over the average precision of the VK system (0.334) and approximately 28,2% higher
over the average precision of the loosely-coupled state-of-the-art system (0.305). We
notice that improvements of the precision values are significant at all recall values.
This shows that when dealing with elaborate queries which combine multiple sources
of information (here visual semantics and spatial characterizations) and thus require a
higher level of abstraction, the use of an "intelligent" and expressive representation
formalism (here the CG formalism within our framework) is crucial. As a matter of
fact, our system complements automatic keyword-based approaches (in this case the
VKs) through the enrichment of their query frameworks with spatial characterization.
Moreover, it outperforms state-of-the-art loosely-coupled solutions by proposing a
unified high-level and expressive framework optimizing user interaction and allowing
to query with precision over visual semantics and symbolic spatial relations.

References

1. Belkhatir, M.: A Full-Text Framework for the Image Retrieval Signal/Semantic Integration. In: Andersen, K.V., Debenham, J., Wagner, R. (eds.) DEXA 2005. LNCS, vol. 3588,
   pp. 113–123. Springer, Heidelberg (2005)
2. Belkhatir, M.: On the Signal/Semantic Integration for Symbolic Image Indexing and Retrieval. PhD Thesis of the Joseph Fourier University (2005)
3. Cohn, A., et al.: Qualitative Spatial Representation and Reasoning with the Region Connection Calculus. Geoinformatica 1, 1–44 (1997)
4. Egenhofer, M.: Reasoning about binary topological relations. In: Proceedings of SSD, pp.
   143–160 (1991)
5. Hollink, L., et al.: Classification of user image descriptions. Int. Journal of Human Computer Studies 61(5), 601–626 (2004)
6. Jeon, J., et al.: Automatic image annotation and retrieval using cross-media relevance
   models. In: Proceedings of ACM SIGIR, pp. 119–126 (2003)
7. Lim, J.H., Jin, J.S.: A structured learning framework for content-based image indexing and
   visual query. Multimedia Systems 10(4), 317–331 (2005)
8. Lu, Y., et al.: A unified framework for semantics and feature based relevance feedback in
   image retrieval systems. In: Proceedings of ACM Multimedia, pp. 31–37 (2000)
9. Mechkour, M.: EMIR2: An Extended Model for Image Representation and Retrieval. In:
   Revell, N., Tjoa, A.M. (eds.) DEXA 1995. LNCS, vol. 978, pp. 395–404. Springer, Heidelberg (1995)
10. Mojsilovic, A., Rogowitz, B.: Capturing image semantics with low-level descriptors. In:
    Proceedings of IEEE ICIP, pp. 18–21 (2001)
11. Ounis, I., Pasca, M.: RELIEF: Combining expressiveness and rapidity into a single system.
    In: Proceedings of ACM SIGIR, pp. 266–274 (1998)
12. Sowa, J.F.: Conceptual structures: information processing in mind and machine. Addison-
    Wesley publishing company, Reading (1984)
13. Zhou, X.S., Huang, T.S.: Unifying Keywords and Visual Contents in Image Retrieval.
    IEEE Multimedia 9(2), 23–33 (2002)

Copyright, Patent and Trade Secret on Digital Libraries: Current Issues and Future Trends

Hideyasu Sasaki[1,2]

[1] Ritsumeikan University, Department of Information Science and Engineering
6-4-10 Wakakusa, Kusatsu, Shiga, 525-0045 Japan
[2] Attorney-at-Law, New York State Bar
hsasaki@alumni.uchicago.edu

Abstract. In this paper, we discuss current issues and future trends on intellectual properties of digital libraries by interpreting legal concepts in engineering manner as a reference to Asia-Pacific DL researchers and practitioners. First, we discuss problems on copyright entities in digital libraries and patent objects in their retrieval mechanisms. Second, we formulate the conditions of copyrightability on the multimedia databases as digital libraries and the patentability on the parameter setting components in retrieval mechanisms. Third, we discuss a new direction for protecting numerical parametric information as trade secret embedded in the patentable parameter setting components.

1 Introduction

Digital library is the global information infrastructure in the networked society [1]. The intellectual property protection of digital libraries is a critical issue in the digital library community, which demands frameworks for recouping their investment in database design and system implementation. A digital library, as an information system, consists of digital contents stored in databases and their retrieval mechanisms. Intellectual property law gives incentive to advance appropriate investment in database design and implementation with two conventional types of intellectual property protection: copyright and patent [2,3]. Nevertheless, present legal studies are not satisfactory as the source of technical interpretation of the intellectual properties regarding digital libraries. The intellectual property protection of the digital libraries demands clear and concise frameworks.

In this paper, we would describe the technical and legal issues on digital library as the objects of copyright, patent and trade secret that have not been discussed with sufficient attention at the present. The principal concern of this paper is to present the conditions of copyrightability on the multimedia databases and the patentability on the parameter setting components in retrieval mechanisms with the directions for protecting numerical parametric information in the parameter setting components as trade secret. Our secondary concern is to provide researchers and practitioners in the DL community with legal references on the concepts, issues, trends and frameworks of intellectual property protection regarding digital libraries in engineering manner.

The scope of this paper is restricted within the current standard of laws and cases in transnational transaction and licensing of intellectual properties regarding digital library. Cultural diversity in the Asia-Pacific region is a source of legislative differences

D.H.-L. Goh et al. (Eds.): ICADL 2007, LNCS 4822, pp. 206–215, 2007.

in intellectual property laws, though those countries join international trade agreements for intellectual property rights. We discuss the harmonized IP law standard regarding digital library with which the Asian-Pacific countries are able to keep up with the foregoing countries.

2 Background

In this section, we discuss three issues on the intellectual property protection regarding digital libraries. The first issue is the copyright protection of databases into which the digital contents are stored as form digital libraries. The second issue is the patent protection of the retrieval mechanisms of database systems or digital library systems. The third issue is the trade secret on the numerical parametric values for retrieval operations in the parameter setting components.

2.1 Copyright on Digital Libraries

U.S. Copyright Act [4] defines that a compilation or assembling of individual contents, *i.e.*, preexisting materials or data, is a copyrightable entity as an original work of authorship. Gorman and Ginsburg [5], and Nimmer, et al. [6] state that a compilation is copyrightable as far as it is an "original work of authorship that is fixed in tangible form".

Digital library systems consist of digital contents which are indexed and stored in databases for appropriate retrieval operations and the retrieval mechanisms which are optimized and applied to object domains of those databases. The entire database is copyrightable in the form of a component of "contents-plus-indexes" while static indexes or metadata are fixed to digital contents in a tangible medium of repository, *i.e.*, database. Static indexes or metadata represent a certain kind of categorization of the entire content of each database (See Fig. 1).

The originality on the categorization makes each database copyrightable as is different from the mere collection of its individual contents. What kind of categorization should be original to constitute a copyrightable compilation on the database? The Court of American Dental Association v. Delta Dental Plan Association [7] determined that minimal creativity in compilation sufficed this requirement of originality on databases. Any standard or framework on the requirement is not clear in the technical or engineering meanings. A uniform framework on the categorization regarding indexes or metadata of databases must be formulated in engineering manner.

The European Union has legislated and executed a scheme for protecting a database including its content *per se*, known as the sui generis right of database protection [8,9,10]. That European scheme shares the same issue on the originality regarding the categorization of digital contents in databases.

2.2 Patent on Digital Library Systems

U.S. Patent Act [11] defines that a data-processing process or method is patentable subject matter in the form of a computer-related invention, *i.e.*, a computer program.

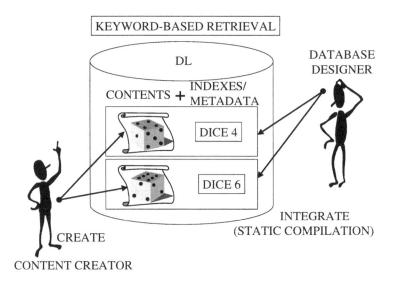

Fig. 1. Formulation for copyrighting digital libraries

The computer program is patentable as far as the "specific machine . . . . . . produce(s) a useful, concrete, and tangible result . . . for transforming . . . " physical data ("physical transformation") [12].

The computer-related inventions often combine means for data-processing, some of which are prior disclosed inventions. A retrieval mechanism in a digital library system consists of a number of "processes", *i.e.*, methods or means for data processing in the form of combination of computer programs. A set of programs focuses on image processing, while another set of programs operates text mining, for example.

Meanwhile, the processes in a retrieval mechanism of a digital library system comprise means or components for parameter setting which is adjusted to retrieve specific kinds of digital contents, for example, images in certain domains. The problem is that which process is to realize technical advancement (nonobviousness) on its combination of the prior arts and is to be specific/enable on its parameter setting. These two issues are emerging problems in the advent of digital library systems. Uniform frameworks on the novel combination and the specific parameter setting must be formulated in engineering manner, respectively.

2.3 Trade Secret in Parameter Setting

Another emerging problem is discussed on the parameter setting of retrieval mechanisms. Patent application on the parameter setting components demands applicants as developers to make public the detailed know-how on the best range of parametric values in practice.

The discovery of those parametric values needs considerable pecuniary investment in research and development. That kind of knowledge should be kept covered in the form of trade secret but not be open in public via patent application. The DL community demands a framework that determines which parameter setting component should be patentable and kept secret regarding digital library systems.

3   Frameworks for Intellectual Property Protection

In this section, we outline the frameworks for intellectual property protection regarding digital library systems: copyrightable database, patentable retrieval mechanism and embedded trade secret on numerical parametric values for retrieval.

3.1   Digital Library as Copyrightable Entity

Our framework for copyrighting the digital library determines which type of database should be copyrightable in the form of a component of contents-plus-indexes [13,14,15]. The collection of static indexes and individual contents forms a component of contents-plus-indexes. That component identifies the entire content of each database, as is a static and copyrightable compilation. Copyrightable compilation is to be of sufficient creativity, *i.e.*, originality in the form of a component of contents-plus-indexes.

The set of conditions on the original categorization regarding indexes or metadata is formulated as below [14,15]: A categorization regarding indexes or metadata is original only when

1. The type of indexes or metadata accepts discretionary selection in the domain of a problem database; otherwise,
2. The type of taxonomy regarding indexes or metadata accepts discretionary selection in the domain of a problem database.

A typical case of non-original categorization is a photo film album database which has indexes of consecutive numbers. That case does not accept any discretion in the selection of the type of indexes or metadata, or the type of taxonomy. The photo film album database uses its respective film numbers as indexes for its retrieval operations. The taxonomy of the indexes is only based on the consecutive numbering without any discretion in its selection of the type of indexes or taxonomy regarding a digital library.

Meanwhile, the discretionary selection of the type of indexes or metadata, or taxonomy constitutes copyrightable compilation of minimal creativity, *i.e.*, originality on the categorization regarding indexes or metadata. A typical case of discretionary selection of the type of indexes or metadata is the web document encyclopedia as a digital library. Suppose that a database restores pictures of starfish which are manually and numerically numbered by day/hour-chronicle interval that is based on their significant life stages from birth to death. That database is to be an original work of authorship as a copyrightable compilation in the form of a component of contents-plus-indexes. That database of discretionary type of numbering or indexing is an original, *i.e.*, copyrightable database.

3.2   Digital Library System as Patentable Mechanism

Our framework for patenting the retrieval mechanisms of digital library system determines which type of retrieval mechanism should be patentable in the form of a

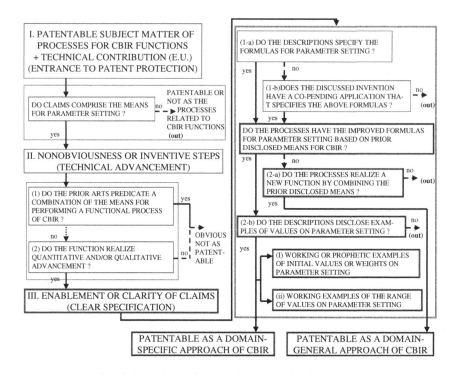

Fig. 2. Formulation for patenting the retrieval processes

component of novel combination of prior disclosed processes and/or a component of specific parameter setting (See Fig. 2) [16,17,18,15]. The frameworks focus on the following three requirements for patentability: "patentable subject matter" (entrance to patent protection), "nonobviousness" (technical advancement) and "enablement" (specification) [19].

The requirement for nonobviousness on the combination of the processes for data-processing as the retrieval mechanism in a digital library system is listed as below [18]:

1. The processes for performing a retrieval mechanism must comprise the combination of prior disclosed means to perform certain mechanism which is not predicated from any combination of the prior arts; in addition,
2. The processes for performing a retrieval mechanism must realize quantitative and/or qualitative advancement.

Otherwise, the discussed processes are obvious so that they are not patentable as the processes for performing a retrieval mechanism.

First, a combination of prior disclosed means should not be "suggested" from any disclosed means "with the reasonable expectation of success" [20]. Second, its asserted function on the discussed mechanism must be superior to the conventional functions which are realized in the prior disclosed or patented means in the field of the retrieval mechanism of digital library system. On the latter issue, several solutions for performance evaluation are proposed and applicable. Another general strategy is restriction of the scope of problem claims into a certain narrow field to which no prior arts have been applied. This claiming strategy is known as the local optimization of application scope.

A component for parameter setting realizes thresholding operations in the form of a computer program with a set of ranges of parametric values. In retrieval mechanisms, parametric values determine, as thresholds, which candidate image is similar to an exemplary requested image by computation of similarity of visual features [21,22,23,24]. That parameter setting component is to be a computer-related invention in the form of computer program as far as that parameter setting is sufficiently specified to enable a claimed invention or retrieval mechanism [25].

The requirement for enablement on the parameter setting component of the retrieval mechanism in a digital library system is listed as below [18]:

(**1-a**) The descriptions of the processes for performing a retrieval mechanism must specify the formulas for parameter setting; otherwise,
(**1-b**) the disclosed invention of the processes should have its co-pending application that describes the formulas in detail; in addition,
(**2-a**) the processes must perform a new mechanism by a combination of the prior disclosed means; otherwise,
(**2-b**) the processes should have improved formulas for parameter setting which is based on the prior disclosed means for performing a retrieval mechanism, and also should give examples of parametric values on parameter setting in descriptions.

For 2-b, the processes must specify the means for parameter setting by "giving a specific example of preparing an" application to enable those skilled in the arts to implement their best mode of the processes without undue experiment [26,27]. U.S. Patent and Trademark Office [25,28] suggested that the processes comprising the means, *i.e.*, the components for parameter setting must disclose at least one of the following examples of parametric values on parameter setting:

(**i**) Working or prophetic examples of initial values or weights on parameter setting;
(**ii**) Working examples of the ranges of parametric values on parameter setting.

The "working examples" are parametric values that are confirmed to work at actual laboratory or as prototype testing results. The "prophetic examples" are given without actual work by one skilled in the art.

3.3 A Simulation Example for the Formulated Procedural Diagram

The proposed formulation in Fig. 2 should be clear with its application to an exemplary digital library system. We apply it to "Virage Image Retrieval"(VIR), which was developed in the early 1990s as a typical content-based retrieval of visual objects stored in digital image database systems. VIR is an indexing method for an image search engine with "primitives", which compute similarity of visual features extracted out of typical visual objects, e.g., color, shape and texture of images. VIR evaluates similarity of images with ad hoc weights, *i.e.*, parametric values, which are given to the parameter setting components for correlation-computation, by user-preference. Its claims consist of "function containers" as means-plus-functions for feature extraction and similarity computation. Its first claim, as described below, constitutes the primitives as the means-plus-functions. Those primitives realize a domain-general approach of CBIR by the formulas on parameter setting.

VIR Claim # 1.
A search engine, comprising: a function container capable of storing primitive functions; ... a primitive supplying primitive functions ......, wherein the primitive functions include an analysis function ...... of extracting features from an object .... .

First in Fig. 2, on its patentable subject matter, its retrieval processes consisting of the formula for parameter setting are to be determined as patentable subject matter in the form of computer programs. Those data-processing processes generate physical transformation on a specific machine, *i.e.*, a computer memory with certain classification results. Second, on its nonobviousness, those data-processing processes are inventive steps that consist of combinations of the prior arts on thresholding functions as implemented in the integration of classification based on similarity computation, visual feature extraction and automatic indexing techniques. Those combinations are not predicated from any conventional keyword-based retrieval technique. Third, on its enablement, VIR's description of preferred embodiments gives its clear specification on the formulas for parameter setting that realizes a domain-general approach of CBIR that was a brand new technology at the time.

VIR Description ......
For primitives having multiple dimensions, ..., An equation for an exemplary Euclidean metric is as follows.
Primitive design. A primitive encompasses a given feature's representation, extraction, and comparison function. ...... The constraints are as follows: Primitives, in general, map to cognitively relevant image properties of the given domain. The formulation should take advantage of a threshold parameter (when available),....... .

The retrieval mechanisms of digital library systems are patentable in the form of components of novel combinations of prior disclosed processes and/or components of specific parameter settings while they are to satisfy the above conditions.

3.4 Embedded Trade Secret in Parameters

It is necessary to prepare a framework that determines how and which part of parameter setting components should take the form of trade secret. The problem is how to interpret the "working examples" of initial values or weights on parameter setting and the ranges of parametric values.

The requirement for patenting parameter setting components as computer-related inventions demands inventors to make public their discovered "working examples" on those parameter values: initial values or ranges. The practice in patent application, nonetheless, does not always force applicants to disclose to examiners complete and perfect evidences on those initial values or ranges of parametric values, but those values as should work in their best mode at the present art.

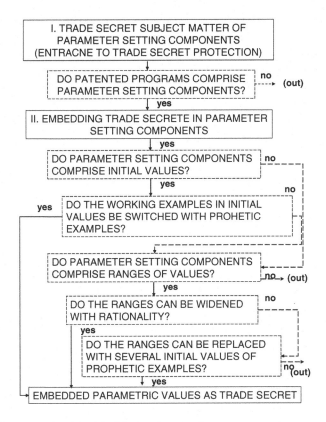

Fig. 3. Formulation for embedding trade secret in parameter setting components

In the reality of application practice, inventors have three choices for embedding trade secrets on their know-how of parametric values in the forms of patentable parameter components:

1. On the initial values, their prophetic examples should be disclosed in patent application, instead of working examples;
2. On the ranges of parametric values, those ranges should be widened as possible at the best but not complete mode;
3. Otherwise, the ranges of parametric values should be replaced with several initial values of prophetic examples.

Fig. 3 shows when a certain set of numerical parametric values in parameter setting components should be embedded as trade secret even those components have been patented as claimed inventions.

4   Conclusions

In this article, we have discussed issues on intellectual property protection regarding digital library systems which consist of indexed multimedia digital contents in databases and retrieval mechanisms with numerical parametric values. We have presented the frameworks for copyrighting the databases of digital library systems in the forms of components of contents-plus-indexes, patenting the retrieval mechanisms of digital library systems in the forms of combinations of processes and/or components of parameter settings and for embedding the trade secret of numerical parametric values in parameter setting components in retrieval mechanisms.

Acknowledgements

This study is supported financially in part by the Grant-in-Aid for Scientific Research ("KAKENHI") of the Japanese Government: No. 18,700,250 (FY 2006-2009).

References

1. Borgman, C.L.: From Gutenberg to the Global Information Infrastructure: Access to Information in the Networked World. Digital Libraries and Electronic Publishing. MIT Press, Cambridge, MA (2000)
2. Jakes, J.M., Yoches, E.R.: Legally Speaking: Basic Principles of Patent Protection for Computer Science. Communications of the ACM 32(8), 922–924 (1989)
3. Junghans, C., Levy, A.: Intellectual Property Management: A Guide for Scientists, Engineers, Financiers, and Managers. John Wiley & Sons, Hoboken, NJ (2006)
4. U.S. Copyright Act.: 17 U.S.C. Sec. 101, & 103 (2005)
5. Gorman, R.A., Ginsburg, J.C.: Copyright: Cases and Materials, 6th edn. University casebook series. The Michie Company, Charlottesville, NC (2002)
6. Nimmer, M.B., Marcus, P., Myers, D.A., Nimmer, D.: Cases and Materials on Copyright & Other Aspects of Entertainment Litigation Including Unfair Competition, 7th edn. Lexis-Nexis, Dayton, OH (2006)
7. American Dental Association v. Delta Dental Plan Association.: 126 F.3d 977 (7th Cir. 1997)
8. Reinbothe, J.: The Legal Protection of Non-creative Databases. In: Proc. of the Database Workshop of the International Conference of Electronic Commerce and Intellectual Property. WIPO. Geneva, Switzerland (September 14–16 1999)
9. Samuelson, P.: Legally Speaking: Legal Protection for Database Content. Communications of the ACM 39(12), 17–23 (1996)
10. Aplin, T.: Copyright Law in the Digital Society: The Challenges of Multimedia. Hart Publishing, Oxford, U.K (2005)
11. U.S. Patent Act.: 35 U.S.C. Sec. 101, 103, & 112 (2005)
12. In re Alappat.: 33 F.3d 1526, 31 U.S.P.Q.2d 1545 (en banc) (Fed. Cir. 1994)
13. Sasaki, H., Kiyoki, Y.: A Proposal for Digital Library Protection. In: Proc. of the 3rd ACM/IEEE-CS Joint Conference on Digital Libraries, Houston, TX, May 27–31, p. 392. IEEE Computer Society Press, Los Alamitos (2003)
14. Sasaki, H., Kiyoki, Y.: Copyrighting Digital Libraries from Database Designer Perspective. In: Chen, Z., Chen, H., Miao, Q., Fu, Y., Fox, E., Lim, E.-p. (eds.) ICADL 2004. LNCS, vol. 3334, pp. 11–14. Springer, Heidelberg (2004)
15. Sasaki, H., Kiyoki, Y.: Multimedia Digital Library as Intellectual Property. In: Design and Usability of Digital Libraries: Case Studies in the Asia Pacific, pp. 238–253. Idea Group Press (2005)

16. Sasaki, H., Kiyoki, Y.: Patenting Advanced Search Engines of Multimedia Databases. In: Lesavich, S. (ed.) Proc. of the 3rd International Conference on Law and Technology. International Society of Law and Technology (ISLAT), Cambridge, MA, November 6–7, pp. 34–39. Acta Press, Anaheim, Calgary, Zurich (2002)
17. Sasaki, H., Kiyoki, Y.: Patenting the Processes for Content-based Retrieval in Digital Libraries. In: Lim, E.-p., Foo, S.S.-B., Khoo, C., Chen, H., Fox, E., Urs, S.R., Costantino, T. (eds.) ICADL 2002. LNCS, vol. 2555, pp. 471–482. Springer, Heidelberg (2002)
18. Sasaki, H., Kiyoki, Y.: A Formulation for Patenting Content-based Retrieval Processes in Digital Libraries. Journal of Information Processing and Management 41(1), 57–74 (2005)
19. Merges, R.P., Duffy, J.F.: Patent Law and Policy: Cases and Materials, 3rd edn. LexisNexis, Dayton, OH (2002)
20. In re Dow Chemical Co.: 837 F.2d 469, 473, 5 U.S.P.Q.2d 1529, 1531 (Fed. Cir. 1988)
21. Rui, Y., Huang, T.S., Chang, S.F.: Image Retrieval: Current Techniques, Promising Directions and Open Issues. Journal of Visual Communication and Image Representation 10(4), 39–62 (1999)
22. Smeulders, A.W.M., Worring, M., Santini, S., Gupta, A., Jain, R.: Content-based Image Retrieval at the End of the Early Years. IEEE Trans. on Pattern Analysis and Machine Intelligence 22(12), 1349–1380 (2000)
23. Yoshitaka, A., Ichikawa, T.: A Survey on Content-based Retrieval for Multimedia Databases. IEEE Trans. on Knowledge and Data Engineering 11(1), 81–93 (1999)
24. Deb, S.: Multimedia Systems and Content-based Retrieval. Idea Group Inc., Hershey, PA (2004)
25. U.S. Patent and Trademark Office.: Examination Guidelines for Computer-related Inventions, 61 Fed. Reg. 7478 (Feb. 28, 1996) ("Guidelines") (1996), Available:
http://www.uspto.gov/web/offices/pac/dapp/oppd/patoc.htm
26. Autogiro Co. of America v. United States.: 384 F.2d 391, 155 U.S.P.Q. 697 (Ct. Cl. 1967)
27. Unique Concepts, Inc. v. Brown. 939 F.2d 1558, 19 U.S.P.Q.2d 1500 (Fed. Cir. 1991)
28. U.S. Patent and Trademark Office.: Examination Guidelines for Computer-related Inventions Training Materials Directed to Business, Artificial Intelligence, and Mathematical Processing Applications ("Training Materials") (1996), Available:
http://www.uspto.gov/web/offices/pac/compexam/examcomp.htm

A Query-Free Retrieving Method Based on Content Elements' Order for Multimedia News Archives

Daisuke Kitayama and Kazutoshi Sumiya

School of Human Science and Environment, University of Hyogo
1-1-12 Shinzaike-honcho, Himeji, Hyogo 670-0092, Japan
ne07p001@stshse.u-hyogo.ac.jp, sumiya@shse.u-hyogo.ac.jp

Abstract. Video and text-news content have recently been broadcast on TV, newspapers, and the Internet. Although video content on out-of-date news is of little value for viewing, it can be considered to have value by comparing it to related content. Repeated news should especially be compared, e.g., the Olympic games and international expositions. We propose a method of retrieving comparison content based on the order of news elements. It is composed of two parts. The first is an analysis of news content that someone is browsing. The second is the automatic generation of queries for retrieving content on comparison news.

1 Introduction

Information is generally distributed by the news not only on TV and by newspapers but also by the Internet. However, the news is only reported on these sites short term (about a week or a month). The currency of the news from these sites is generally assumed to be important. Articles that compare past Olympic events with present ones, on the other hand, are composed of feature articles. Old news that is not browsed is not considered to be of any value. However, we considered the value by finding the relation between past and present news.

The method we propose has two processes. First, objects or things that have been reported and news behaviors (actions) are extracted based on the order of content elements, which differs depending on the media. Second, news articles are retrieved that can effectively be compared with the news article that the user is browsing. A user can automatically obtain content to understand the news with our method simply by browsing news articles and selecting the comparison query. Figure 1 outlines the concept underlying the method we propose.

2 Related Work

Shin et al. [1] proposed generating query from natural language questions. Their proposed systems automatically analyzed a user's question using the 5-W 1-H keywords. Our proposed method needs neither complex grammatical analysis

D.H.-L. Goh et al. (Eds.): ICADL 2007, LNCS 4822, pp. 216–219, 2007.

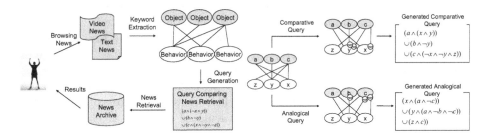

Fig. 1. Concept underlying query-free comparative news retrieval

Fig. 2. Generation of comparison queries

nor dictionary building because it does not depend on specific keywords. Instead the composition keywords of the news are simply extracted.

Ohshima et al. [2] proposed methods of extracting the sibling page. Yumoto et al. [3] proposed methods of extracting relational page sets. They proposed a method of detecting the relations between content using a vector space model. The relations with our method are detected based on keywords without using a vector space model.

3   Keyword Extraction Using Order of Content Elements

We defined the order of content elements as elemental units in the order of news content. They have different features based on media[4][5]. We consider that the objects of subjects described in news are often expressed as nouns, and behaviors are expressed as sets of verbs. One news item can be expressed by using the noun for the object and the verb for the behavior.

The method we propose extracts objects from subjects described in the news using elemental units. We consider that the object of a subject in video news accurately describes the object spotted at the scene. In this way, the degree of importance of the object keyword can be calculated from the word density in the transcription of video news. The degree of importance of the object keyword, a, in video news can be calculated as

$$obj\_val = \frac{i}{dist(a_1, a_i)} \tag{1}$$

where a_i is the ith noun a in the news. Function $dist$ calculates the distance between sentences. The distance between sentences is represented by a number and means how many sentences there are between two keywords. The distance between sentences is 1 when they appear in the same sentence. We consider that the positions where the objects of subjects described in text news are dispersed. The degree of importance of the object keyword, a, in text news is calculated as

$$obj\_val = min\left(\frac{\sum_{i=1}^{n} dist(s_1, a_i)}{n}, ..., \frac{\sum_{i=1}^{n} dist(s_j, a_i)}{n}, ..., \frac{\sum_{i=1}^{n} dist(s_m, a_i)}{n}\right) \tag{2}$$

where s_j is the jth sentence in text news. The minimum value of the element is extracted using function min because the expectation is unknown.

The method we propose extracts news behaviors using the order in which content is presented. We considered the conclusion to be described at the end of video news, and the verb that shows action in the conclusion expressed the news behavior. The degree of importance of the behavior keyword is calculated by the position it appears in the transcription of video news. The degree of importance of the behavior keyword in video news is calculated as

$$beh\_val = \sum_{i=1}^{S}(\frac{i}{S} \times count(V_i))$$ (3)

where i is the ith sentence in all S sentences, and V_i is a verb set that appears in the ith sentence. Function $count$ calculates the number of verbs to be calculated in V_i. We considered that a news-behavior keyword in text news would appear at the beginning where details on the conclusion are described. The degree of importance of the behavior keyword in text news is calculated as

$$beh\_val = \sum_{i=1}^{S}(\frac{S-i+1}{S} \times count(V_i)).$$ (4)

4   Queries Generation for Retrieving Comparison Articles

Comparison articles are those which can be compared by focusing attention on the browsing news. We defined the news where the focus of attention was an object as a comparative article, and that where the focus was a behavior as an analogical article. The query is generated based on a graph where the content element is described. The content element graph is a bipartite composed of the object keyword and the behavior keyword. The link shows the relation between the object and the behavior. We consider that the relation between the object keyword and the behavior keyword in video news is determined by a range where the word density of an object keyword is high. However, the relation between the object keyword and behavior keyword in text news was determined using the same paragraph.

Comparative queries are automatically generated to retrieve comparative articles related to what the user is currently browsing. The user can confirm whether the situation has been consistent over time. The upper part of Figure 2 shows how a comparative query is generated.

Analogical queries are automatically generated to retrieve analogical articles related to what the user is now browsing. The user can understand behavior in detail. The lower part of Figure 2 shows the generation of an analogical query.

5   Evaluation

We did an experiment to evaluate our proposed method by assessing the retrieved results for generating queries using a news elements' graph. The data

Table 1. Experimental result of retrieving comparison article

Video News No.	Comparative			Analogical			Text News No.	Comparative			Analogical		
	Precision	Recall	F-measure	Precision	Recall	F-measure		Precision	Recall	F-measure	Precision	Recall	F-measure
1 Text	0.50	0.25	0.33	0.80	0.33	0.47	3 Text	0.00	0.00	0.00	0.50	0.25	0.33
1 Video	1.00	0.25	0.40	0.64	0.39	0.49	3 Video	0.33	0.40	0.36	0.00	0.00	0.00
1 All	0.67	0.25	0.37	0.68	0.37	0.48	3 All	0.22	0.18	0.20	0.27	0.14	0.19
2 Text	0.00	0.00	0.00	0.43	0.33	0.38	4 Text	0.29	0.40	0.33	0.05	0.33	0.08
2 Video	0.00	0.00	0.00	0.44	0.71	0.54	4 Video	0.17	0.50	0.25	0.33	0.27	0.30
2 All	0.00	0.00	0.00	0.43	0.57	0.49	4 All	0.20	0.43	0.27	0.13	0.29	0.18

sets for each generated query in the experiment were about 180 news items in the news archive[1]. Video news and text news were included in these data sets. Test subjects extracted correct-answer sets from data when browsing news that generated queries and then compared queries. There were three test subjects. A correct-answer set was a set of articles that two or more test subjects extracted. We evaluated the proposed method by means of precision, recall, and F-measure calculated using correct-answer sets.

The results are listed in Table 1. F-measure of query generated from video news is higher than F-measure of query generated from text news. We considered that different media are not equally processable in our proposed method. Therefore, we should improve the algorithm.

6   Concluding Remarks

A content element graph with the degree of importance based on the order of content elements was presented, and the generation of queries to compare articles that had been retrieved using this graph was proposed. We also evaluated the retrieved results using comparative queries generated with our proposed method. In future work, we plan to: compare methods of calculating keywords in experiments with conventional degrees of importance, and improve generation of queries based on the relations between individual keywords.

References

1. Shin, S.E., Seo, Y.H.: Query Generation Using Semantic Features. In: Sugimoto, S., Hunter, J., Rauber, A., Morishima, A. (eds.) ICADL 2006. LNCS, vol. 4312, pp. 234–243. Springer, Heidelberg (2006)
2. Ohshima, H., Oyama, S., Tanaka, K.: Sibling Page Search by Page Examples. In: Sugimoto, S., Hunter, J., Rauber, A., Morishima, A. (eds.) ICADL 2006. LNCS, vol. 4312, pp. 244–253. Springer, Heidelberg (2006)
3. Yumoto, T., Tanaka, K.: Page Sets as Web Search Answers. In: Sugimoto, S., Hunter, J., Rauber, A., Morishima, A. (eds.) ICADL 2006. LNCS, vol. 4312, pp. 244–253. Springer, Heidelberg (2006)
4. Wikinews: Style guide, http://en.wikinews.org/wiki/Wikinews:Style_guide
5. Analizing News, http://akasaka.cool.ne.jp/kakeru3/bs3.html

---

[1] About 180 news items were assumed because 8 news items were used every month for 18 months and about 40 news items were selected by subjects as correct answers.

Adaptive Search Suggestions for Digital Libraries

Sascha Kriewel and Norbert Fuhr

University of Duisburg-Essen

Abstract. In this paper, an adaptive tool for providing suggestions during the information search process is presented. The tool uses case-based reasoning techniques to find the most useful suggestions for a given situation by comparing them to a case base of previous situations and adapting the solution. The tool can learn from user participation.

A small, preliminary evaluation showed a high acceptance of the tool, even if improvements are still needed.

1 Introduction

A common and well known problem in the design of information retrieval (IR) and digital library systems is how to support end users in finding good strategies for satisfying their information need. Despite many advances in making information search technology available to the larger public instead of just search professionals, the effective use of these information retrieval technologies remains a challenge [8,11,19].

While all digital library and information retrieval systems provide low-level search actions, in [4] Bates identifies three higher levels of abstraction for categorizing search functionalities: tactics, stratagems, and strategies. There has been extensive work on supporting users in executing moves, tactics, and even stratagems, but there is a definite lack of support on the highest level [9,10,14], i.e. in helping searchers to choose the most appropriate or useful action for their specific situation to form an overall search strategy or plan.

In [17] we suggest that an ideal search system would provide useful strategies for completing specific tasks – either upon user request or pro-actively – and help users improve their search experience by raising awareness of the strategic aspect of searching. In this way, users would acquire procedural knowledge to create better search strategies for themselves.

In this work, a new tool is introduced that suggests appropriate search tactics and stratagems based on the current situation of the user. The tool, which has been implemented as part of the DAFFODIL framework [13], uses case-based reasoning (CBR) techniques to find and rank suggestions according to the similarity of the user's situation to previous situations. It also learns which of the several presented suggestions were successfully employed by users and which not.

D.H.-L. Goh et al. (Eds.): ICADL 2007, LNCS 4822, pp. 220–229, 2007.

2   Providing Strategic Help

For the purpose of this paper, we will follow the definition of a *search strategy* used in [4]. A strategy is seen as a complete search plan encompassing possibly many tactics and stratagems used in the process of an information search. To form such a search strategy, users need to be able to select the appropriate stratagems or tactics supported by the information system. It should be noted however, that searching is an opportunistic process and search goals can be shifting throughout the task. It is often more fruitful for a searcher to follow promising opportunities that arise from previous results, instead of sticking to a straight path towards the perfect result set. In [3] Bates describes the *berry picking* model of information seeking.

From the view of the user recognizing these strategic opportunities remains a problem, as does using the strategic options available to maneuver out of a perceived dead-end during a search. In fact, users rarely exploit the advanced capabilities and features of modern search systems, even if these would improve their search. They might not be aware of their existence, might not understand them, or don't know in which situations they could be effectively employed. Search systems providing strategic assistance could improve search effectiveness by suggesting the use of these advanced features or options automatically [15].

In [10] Brajnik et al. describe a strategic help system based on collabora- tive coaching, which tries to assist users by providing them with suggestions and hints during so-called critical or enhanceable situations. The system uses a hand-crafted knowledge base of 94 production rules to provide suggestions based on the tactics and stratagems proposed in [2,4]. The strategic help module was integrated into FIRE, a user-interface to a Boolean IR system. Only six peo- ple participated in the user evaluation, but the results showed promise for the usefulness of strategic suggestions.

Belkin et al. describe information retrieval in terms of information-seeking be- haviors or strategies (ISSs) [5,7]. They present a characterization of such behav- iors using a small set of dimensions. In the course of a single information-seeking episode, users will engage in several such ISSs, moving from one to the next. In a feature-rich search system like DAFFODIL[13] many of these ISSs correspond to specific supported stratagems or available tools, and the movement of the user between the various tools is similar to the movement from one ISS to another. Belkin et al. suggest a mixed-initiative system supporting specific procedures during information seeking episodes, which change and branch as the user in- teracts with the system. In [6] the MERIT system based on these concepts is presented. This system uses scripts derived by case-based reasoning techniques to guide users through an information-seeking episode.

Case-based reasoning emerged in the early 1980s as an Artificial Intelligence (AI) approach to solving a new problem based on the known solutions of similar problems. A stored case is a previously experienced problem situation which has been captured and retained together with its solution. New cases can be "solved" by comparing them to the case base of previous situations, finding the most similar ones, adapting them to the current situation, and then re-using them. Learning in CBR occurs as a by-product of problem solving. Whenever a specific case is successfully solved, this experience is retained in order to solve similar problems in the future (and accordingly for failures) [1].

3   The ASDL Suggestion Tool

To help users in their particular search situation with useful strategic search advice, the ASDL (Adaptive Support for Digital Libraries) Suggestion Tool was developed and integrated into DAFFODIL, an existing search system developed at the University of Duisburg-Essen [18]. DAFFODIL (pictured in Fig. 1) offers a rich set of tools with a correspondingly large number of possible user actions from which to draw suggestions. It also provides facilities for easily logging and gathering information about the search progress of users. DAFFODIL consists of an agent-based back-end and a graphical user client, which presents search, browse and extraction services in the form of desktop tools [12].

The ASDL module is composed of three main components:

- the *Observing Agent*, which collects information about the current search activities of the user and the results sent by the information sources;
- the *Reasoning Agent*, which retrieves appropriate suggestions fitting the user's current situation, and ranks them according to similarity;
- the actual *Suggestion Tool*, which adapts and presents the search suggestions within the Daffodil user interface, and allows for automatic execution and user judgments on suggestions.

3.1   Observing the User

For the purpose of providing search suggestions each completed user query with its corresponding set of result documents is considered as a single situation. The description for the situation is gathered from the original query of the user, the results sent by the different information sources, and additional information extracted by the Observing Agent. Thus a complete description of a search situation within the ASDL module contains

- the search terms, search fields, Boolean operators, and information sources used;
- the number of results returned and system response time;
- a list of the most frequent terms, authors, journals, or conferences extracted from the results.

3.2   Finding Suggestions

The ASDL module uses case-based reasoning techniques to find situationally appropriate suggestions, where a search situation is considered as a case description and suggestions are possible solutions. Users can judge suggestions as useful for their current situation, thereby adding to the case base. The initial case base was build by creating one or two iconic cases for each suggestion.

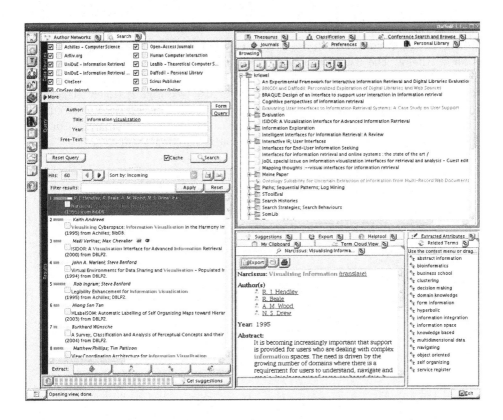

Fig. 1. The DAFFODIL desktop: Search Tool (left), Personal Library (top right), Detail View and Related Terms (bottom right)

For a given situation, the reasoning component tries to find the most similar cases from the database and returns the solutions, i.e. suggestions, ranked according to the descreasing similarity of the corresponding case to the current situation. If the same solution is suggested more than once, only the highest similarity is currently considered.

The similarity $sim_T(a, c)$ of two situations a and c is computed as the *weighted mean* (using weights w_k) of the individual similarity scores sim_k between the various aspects of a situation (see 3.1 above).

$$sim_T(a_i, c) := \frac{\sum_{k=0}^{N} w_k \cdot sim_k(a_{ik}, c_k)}{\sum_{k=0}^{N} w_k} \tag{1}$$

For determining the similarity between two vectors of term weights (e.g. the extracted terms or authors from results), the normalized inner product is used.

3.3   Presenting the Help

When the results of a search are presented to the user, an unobtrusive button is shown below the list of results (visible in Fig. 1 to the right of the search progressbar). Upon clicking this button, the search suggestions are presented in form of a ranked list to the right of the result list (see Fig. 2). Each suggestion consists of a short descriptive title, a longer explanation and a small colored bar representing the estimated applicability of the suggestion to the current situation (i.e. the similarity score computed by the reasoning component).

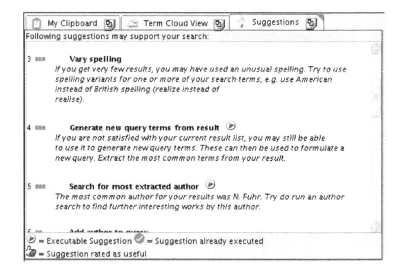

Fig. 2. The Suggestion Tool

Where possible, suggestions are adapted to the current search. E.g., a suggestion to perform an author search will suggested the most frequently extracted author name from the current results (if the query leading to the result was already an author search for that author, the second most frequently extracted author will be substituted).

A user can automatically execute most of the suggestions by double clicking on them, or by using the context menu. The list of suggestions remains visible, and additional suggestions can be tried. If a suggestion proves to be useful for furthering the search, the user can easily judge it as appropriate by using the context menu. Small icons are used to mark suggestions that can be automatically executed, that have already been executed in the current situation, or that have been judged as useful by the user.

3.4   Suggestions

A total number of sixteen suggestions were implemented for a first test, ranging from terminological hints (*vary spelling*) to suggestions for using different tools from the DAFFODIL toolbox (*show co-author network*). Wherever possible, the suggestions were made executable.

1. *Browse conference proceedings* using the most common conference from the current result set.
2. *Browse journals* using the most common journal from the current result set.
3. *Browse a classification* to find better terms to describe the information need.
4. Use the Thesaurus Tool to *Find a narrower term.*
5. Use the Thesaurus Tool to *Find a more general term.*
6. Use the Thesaurus Tool to *Find a related term.*
7. Use the Network Tool to generate a visual *co-author graph* for the most common author from the current result set.
8. *Perform an author search* using the most common author from the current result set (who was not already part of the current query).
9. *Restrict the query* by adding the most common *author* from the current result set (who was not already part of the current query).
10. *Restrict the query* by adding the most common *term* from the current result set (that was not already part of the current query).
11. *Restrict the query* by showing only the *most recent publications.*
12. *Restrict the query* by using the *phrase* operator for phrases.
13. *Broaden the query* by removing a search term from a conjunction.
14. *Broaden the query* by replacing the implicit conjunction with a disjunction.
15. *Vary spelling*, e.g. to compensate for differences in spelling between American and British English.
16. *Extract new query terms* from the current result documents and show them in form of a weighted list (also called *tag cloud*).

4   Evaluation

A pilot study was conducted to evaluate the ASDL prototype. It was a light-weight study with two main goals: gaining a first understanding of user acceptance of strategic suggestions, and evaluating the appropriateness of ranking suggestions according to previous users' situations and actions.

In preparation for the experiment the system was trained (using a search topic unrelated to that of the evaluation task), so that the case base contained at least one situation for each suggestion where it would be an appropriate advice. Additionally, no cut-off value for the similarity value was specified, so that all suggestions that could be applied to a situation were presented each time, even if their score was low.

A total of twelve participants were asked to perform a comprehensive search task with the help of the Suggestion Tool (ST): five graduate and seven under-graduate students of computer science or communication and media science from the University of Duisburg-Essen. They worked with a simplified version of the DAFFODIL system containing a sub-selection of tools, and performed their searches over several collections of computer science articles. The ST's purpose was described to them, and they were given an introduction to the DAFFODIL system where necessary.

The search task itself was only loosely defined, but comprehensive enough that even experienced searchers were expected to issue several queries and use a number of different tools from DAFFODIL's toolbox to gather a satisfying number of results. The participants of the evaluation were asked to "collect articles and additional information such as central authors, search terms, conferences or journal issues to prepare a course paper on methods for visualization of result lists in information retrieval systems". This search task was chosen to provide the searchers with a simulated work task and was only slightly modified from a real work task.

Each searcher was given between 40 and 50 minutes of time to work on the task, during which they were asked to use the suggestion tool for help and for generating new search ideas. During this time, one experimenter was available to give technical help and intervene on system errors. No strategic or terminological help was provided by the experimenter, and the searchers were referred to the Suggestion Tool instead.

Each session was logged using the logging framework of the Daffodil system [16]. While no searcher spent less than 40 minutes on the task, all searchers were asked to stop at the 50 minutes mark. On average they spend about 47 minutes searching, examining documents, and storing results. A combined total of 198 queries were issued by the searchers, and 143 items stored as relevant (an average of 16.5 queries and 11.9 stored items per user).

During their tasks the participants requested search advice a total of 94 times (about 7.8 times per searcher or about once every two queries). After reading the search advice, 62 suggestions were directly followed (i.e. executed semi-automatically) by the participants.

The experiment was accompanied by a two-part questionnaire. The first part of the questionnaire was designed to gather information about the domain and search knowledge of the participants. The second part was given after the participant had concluded the task, and contained four question blocks in which the searchers were asked to judge their experiences with the system.

4.1   Results

Because of the low number of participants the results from the pilot study can only be taken as first indicators. However, some of the results were promising and merit further exploration of the Suggestion Tool. The questions used a seven-point Likert scale from 'fully disagree' (coded as 1) to 'fully agree' (coded as 7). Users were invited to elaborate on their answers, and were asked to give specific examples of problems.

Clarity. Nearly all participants found the suggestions to be easy to understand. The explanations why a specific suggestion might be useful were generally found to be clear (modal score of 7, median score of 6.5). Participants rarely encountered suggestions that they interpreted differently from the intended meaning and where they were surprised by the results of the advice (modal and median score of 6).

A particular problem that was mentioned by several users was the description of suggestion #16, which didn't explain clearly how the list of terms was derived.

Appropriateness. Since the suggestions were selected and ranked according to their similarity to previous users' situations, it was interesting to see if this ranking method arranged the suggestions in a contextually appropriate manner. While the results were only marginally positive (modal score of 4, median score of 5), it turned out that this was mainly the result of the experimental setup. Nearly all problems reported were with very low scoring suggestions, which had been included in the presentation to give the participants a broader range of suggestions to choose from. A threshold to keep out low scoring suggestions is clearly necessary. Suggestions with scores of 0.5 and higher were universally deemed appropriate.

Usefulness. The high scoring suggestions in each situation were found to be generally useful and helpful (median and modal score of 6). Ten of the twelve participants reported that thanks to the suggestions, they had employed tactics and stratagems that they normally wouldn't have used or thought of. All ten stated that they found those helpful and would try to use them independently in future search tasks.

Particular suggestions that were mentioned as helpful were the hints about the different tools from the DAFFODIL toolbox (thesaurus, classification browser, social network tool), the purpose or even existence of which were not apparent to novice or casual users of the system. The experienced users liked the extraction of common authors, conferences or terms from the result set that was provided by the suggestions. Several mentioned that while they might have eventually used some of the suggested tactics on their own, the advice provided by the ST helped them to realize some of the available opportunities more quickly and to avoid lengthy and ultimately frustrating trial-and-error.

Interface. The visual presentation of the suggestions was generally well-liked, and the participants easily understood the ranking, as well as the icons used by the tool. 75% of the users also preferred that the suggestions were only shown on specific request instead of automatically. The availability indicator proved to be sufficient to notify the user about the existence of new suggestions, while still being unobtrusive enough not to interfere with the original search.

Interaction with the tool proved to be more problematic. Two major flaws were discovered which impaired the initial usability:

1. The icon marking a specific suggestion as "automatically executable" was readily understood. However, because of its similarity to the play button used in many media players, a large number of users tried to execute the suggestion by clicking on the icon (instead of using double-click activation as expected).

2. Although most of the users were willing to judge good suggestion, many had problems finding the option to do this, as it was hidden in the context menu of the suggestion.

To fix these problems, both actions (executing as well as judging) will be made available to the users by easily accessible single-click buttons for each suggestion.

5   Summary and Conclusion

In this paper we have presented a new tool for giving automated advice during the search process. This tool has been successfully integrated into the Digital Library toolbox DAFFODIL, and uses case-base reasoning techniques to find and rank the most appropriate suggestions for the user's current search situation.

A small scale pilot evaluation was conducted, during which 12 users worked with the Suggestion Tool for a total of 560 minutes. The evaluation pointed out a number of possible improvements, but overall user reception of the suggestions was positive. The users found the automated, non-intrusive advice to be helpful, and implemented suggested tactics with success to further their search task. These results confirmed similar findings from [10] and [15].

Several problems that were found during the evaluation were fixed and the list of available suggestions has been extended since and continues to be extended. Further evaluations need to be conducted to improve the ranking of suggestions and determine a cut-off point were the similarity of cases no longer merits their inclusion in the list of possible advices. In addition, an extension of the tool from suggestions of single, isolated tactics and stratagems towards sequences or paths of actions, as originally proposed in [17], is being considered.

References

1. Aamodt, A., Plaza, E.: Case-based reasoning: Foundational issues, methodological variations, and system approaches. AI Communications 7(1), 39–59 (1994)
2. Bates, M.J.: Information search tactics. Journal of the American Society for Information Science 30(4), 205–214 (1979)
3. Bates, M.J.: The design of browsing and berrypicking techniques for the online search interface. Online Review 13(5), 407–424 (1989)
4. Bates, M.J.: Where should the person stop and the information search interface start? Information Processing and Management 26(5), 575–591 (1990)
5. Belkin, N.J.: Interaction with texts: Information retrieval as information seeking behavior. In: Knorz, G., Krause, J., Womser-Hacker, C. (eds.) Information Retrieval 1993. Von der Modellierung zur Anwendung. Konstanz, pp. 55–66 (1993)
6. Belkin, N.J., Cool, C., Stein, A., Thiel, U.: Cases, scripts, and information-seeking strategies: On the design of interactive information retrieval systems. Expert Systems with Applications 9, 1–30 (1995)
7. Belkin, N.J., Marchetti, P.G., Cool, C.: BRAQUE: Design of an interface to support user interaction in information retrieval. Information Processing and Management 29(3), 325–344 (1993)

8. Bhavnani, S.K., Drabenstott, K., Radev, D.: Towards a unified framework of IR tasks and strategies. In: Proceedings of ASIST 2001: 64th Annual Meeting, pp. 340–354 (2001)

9. Brajnik, G., Mizzaro, S., Tasso, C.: Evaluating user interfaces to information retrieval systems: A case study on user support. In: Proceedings of the SIGIR 1996, pp. 128–136 (1996)

10. Brajnik, G., Mizzaro, S., Tasso, C., Venuti, F.: Strategic help in user interfaces for information retrieval. Journal of the American Society for Information Science and Technology 53(5), 343–358 (2002)

11. Drabenstott, K.M.: Do nondomain experts enlist the strategies of domain experts. Journal of the American Society for Information Science and Technology 54(9), 836–854 (2003)

12. Fuhr, N., Klas, C.-P., Schaefer, A., Mutschke, P.: Daffodil: An integrated desktop for supporting high-level search activities in federated digital libraries. In: Agosti, M., Thanos, C. (eds.) ECDL 2002. LNCS, vol. 2458, pp. 597–612. Springer, Heidelberg (2002)

13. Gövert, N., Fuhr, N., Klas, C.-P.: Daffodil: Distributed agents for user-friendly access of digital libraries. In: Borbinha, J.L., Baker, T. (eds.) ECDL 2000. LNCS, vol. 1923, pp. 352–355. Springer, Heidelberg (2000)

14. Hsieh-Yee, I.: Effects of search experience and subject knowledge on online search behavior: Measuring the search tactics of novice and experienced searchers. Journal of the American Society for Information Science 44(3), 161–174 (1993)

15. Jansen, B.J.: Seeking and implementing automated assistance during the search process. Information Processing and Management 41(4), 909–928 (2005)

16. Klas, C.-P., Albrechtsen, H., Fuhr, N., Hansen, P., Kapidakis, S., ó Kovács, L., Kriewel, S., Micsik, A., Papatheodorou, C., Tsakonas, G., Jacob, E.: A logging scheme for comparative digital library evaluation. In: Gonzalo, J., Thanos, C., Verdejo, M.F., Carrasco, R.C. (eds.) ECDL 2006. LNCS, vol. 4172, pp. 267–278. Springer, Heidelberg (2006)

17. Kriewel, S.: Finding and using strategies for search situations in digital libraries. Bulletin of the IEEE Technical Committee on Digital Libraries 2(2) (2006), http://www.ieee-tcdl.org/Bulletin/v2n2/kriewel/kriewel.html

18. Kriewel, S., Klas, C.-P., Schaefer, A., Fuhr, N.: Daffodil - strategic support for user-oriented access to heterogeneous digital libraries. D-Lib Magazine 10(6) (June 2004), http://www.dlib.org/dlib/june04/kriewel/06kriewel.html

19. Wildemuth, B.M.: The effects of domain knowledge on search tactic formulation. Journal of the American Society for Information Science and Technology 55(3), 246–258 (2004)

A Ranking Scheme for XML Information Retrieval Based on Benefit and Reading Effort

Toshiyuki Shimizu and Masatoshi Yoshikawa

Graduate School of Informatics, Kyoto University
shimizu@soc.i.kyoto-u.ac.jp, yoshikawa@i.kyoto-u.ac.jp

Abstract. XML information retrieval (XML-IR) systems search for relevant document fragments in XML documents for given queries. In top-k search, users control the size of output by an integer k. In XML-IR, however, each output element varies widely in size. Consequently, total output size of top-k elements is uncontrollable by simply giving an integer k. In addition, search results may have nesting elements. If a system orders result elements simply by their relevance, we may browse the same content more than once due to the nestings. To handle these problems, we propose a new ranking method that enables us to browse search results of XML-IR systems efficiently by introducing the concepts of *benefit* and *reading effort*. We also propose an evaluation metrics based on *benefit* and *reading effort*, and compared the metrics with existing XML-IR metrics by experiments.

1   Introduction

When we want to retrieve information about a topic from large amount of XML documents, keyword search is one solution to retrieve document fragments relevant to the topic. XML information retrieval (XML-IR) systems generally use elements as search units, and output ranked elements relevant to given queries. For example, in the case of scholarly articles marked up in XML, XML-IR systems retrieve and rank such elements corresponding to sections, subsections, and paragraphs.

INEX 2005 [1] defines three element retrieval strategies for the purpose of evaluating the effectiveness of XML-IR systems. A system with the *Thorough* strategy simply retrieves relevant elements from all elements and ranks them in order of relevance. The retrieved elements using the *Thorough* strategy may overlap due to nestings. By selecting the element with the highest score in a path and removing the overlapping elements, the system with the *Focussed* strategy retrieves only focused elements. Though the systems can exclude redundancy by using the *Focussed* strategy, we can not find the non-focused elements in the result and may lose some possible benefits of XML-IR [2]. A system with the *FetchBrowse* strategy first identifies relevant documents (the fetching phase) and then identifies relevant elements within a fetched document (the browsing phase).

We considered that following problems exist in the element retrieval of XML-IR.

D.H.-L. Goh et al. (Eds.): ICADL 2007, LNCS 4822, pp. 230–240, 2007.

- *Variety of element size*
  Each element retrieved by XML-IR systems varies widely in size. An element may be large one such as root element, which is corresponding to whole document, or small one. Therefore, the cost for reading the content of a retrieved element is unknown beforehand.
- *Handling nesting elements*
  When users browse the content of nesting elements, by browsing the content of an ancestor element, users can browse the content of all its descendant elements. Though it is important to take the nestings into consideration, the *Focussed* strategy, which is the only strategy in INEX 2005 that considers about nestings, lacks flexibility because it does not retrieve elements except focused elements.

In general, users of XML-IR systems browse search results from top ranked element to lower ranked elements. Therefore, fast retrieval of high ranked elements is important and there are some researches on top-k search of XML-IR [3,4]. However, total output size of top-k elements is uncontrollable by simply giving an integer k. We considered that top-k search is not suited for XML-IR, and using the cost for reading the content of a retrieved element, which we call *reading effort*, instead of an integer k is better alternative to control the total output size. In addition, the search results of top-k search using the *Thorough* strategy may contain nesting elements; therefore top-k search is not suited for XML-IR also in this point.

For example, when we retrieve top-100 result elements using our system proposed in [5] with the *Thorough* strategy, The total result size of top-100 elements are varied about 30 times between the largest and the smallest for 40 queries of INEX 2005. We also observed that about 15% of top-100 total result size was overlapping content on average.

To overcome these problems, we introduced *benefit* which is the amount of gain by reading the element and *reading effort* which is the cost of reading elements, and used them in result retrieval and ranking algorithm. By considering that the *benefit* does not increase even if we read the same content repeatedly, we handle the problem of content overlapping by nestings. We supposed that users specify the threshold amount of *reading effort* they can spend in reading the result elements. The system retrieves elements that have larger *benefit* within the specified *reading effort*. The scores for result elements can be viewed as efficiency of obtaining *benefit* from the element, and we supposed the score of an element is calculated by dividing *benefit* by *reading effort*. We also propose an evaluation metrics based on *benefit* and *reading effort*.

To handle nestings in search results of XML-IR, Clarke [2] proposed to control overlapping by re-ranking the descendant and ancestor elements of the reported element. Clarke supposes each result element can be browsed with the same cost, whereas we introduced *reading effort* for the ranking algorithm and the evaluation metrics.

For evaluation metrics, we used assessments for XML test collection of INEX 2005. We compared the metrics with existing XML-IR metrics by experiments, and found there is not strong correlation between them.

2   Benefit and Reading Effort

To handle nesting elements and variety of element size, we introduced *benefit* and *reading effort*. In this section, we discuss the properties of *benefit* and *reading effort*.

2.1   Benefit

For a given query, the *benefit* of an element is the amount of gain about the query by reading the element. We describe the *benefit* of element e as $e.benefit$. Basically, it is natural to consider that the *benefit* of an element is sum of the *benefit* of the child elements. However, by reading all child elements together, the content of the child elements may complement each other; hence the *benefit* of the parent element may larger than sum of the *benefit* of the child elements. We considered that the following assumption holds on *benefit*.

Assumption 1. Property of Benefit
The *benefit* of an element is greater than or equal to the sum of the *benefit* of the child elements.

2.2   Reading Effort

The *reading effort* of an element is the amount of cost by reading the content of the element. We describe the *reading effort* of element e as $e.reading\_effort$. Note that *reading effort* does not depend on queries and can be calculated based on the element itself. Basically, it is natural to consider that the *reading effort* of an element is sum of the *reading effort* of the child elements. However, by reading all together, we can continuously read with the same context; hence the *reading effort* of the parent element may smaller than sum of the *reading effort* of the child elements. We considered that the following assumption holds on *reading effort*.

Assumption 2. Property of Reading Effort
The *reading effort* of an element is less than or equal to the sum of the *reading effort* of the child elements.

3   Ranking Method Based on Benefit and Reading Effort

XML-IR systems we propose calculate the *benefit* of elements for the given query, and rank the result elements in order of efficiency to obtain *benefit*. We supposed the score is calculated by dividing *benefit* by *reading effort*, and it corresponds to the efficiency to obtain *benefit*. We describe the score of element e as $e.score$. We also supposed that users specify the threshold amount of *reading effort*, and the system retrieves elements that have larger *benefit* within the specified *reading effort*.

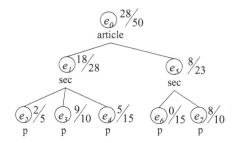

Fig. 1. An example of calculated *benefit* and *reading effort*

Figure 1 shows an example of *benefit* and *reading effort* calculated by a system for a query. In Figure 1, the tree structure of an XML document is represented, and *benefit* and *reading effort* are shown in the form of *benefit/reading effort* adjacent to the element. For the sake of simplicity, we do not assume any concrete calculation formula for *benefit* and *reading effort* in Figure 1, however the values meet Assumption 1 and Assumption 2.

When a system calculate *benefit* and *reading effort* for a query as Figure 1, if a user specifies the threshold of *reading effort* to 15, the element set that maximize *benefit* is $\{e_3, e_2\}$, whereas if 20 is specified, the element set that maximize *benefit* is $\{e_3, e_7\}$. The problem of maximizing *benefit* is a variant of knapsack problems that has restriction of nestings. However, the system in the running example that maximize *benefit* does not output the content of e_2 when 20 is specified as the threshold of *reading effort*, though the content of e_2 is output when 15 is specified. Therefore, a user who specifies the threshold of *reading effort* to 20 can not obtain information from e_2 though she/he pays more *reading effort* than a user who can obtain information from e_2 by specifying 15. To avoid such situation, we considered it is important that systems have the property of the search result continuity. The search result continuity is the property defined as following.

Definition 1. *Search result continuity*
When we describe the result element set for the threshold r of reading effort as $E^r = \{e_1^r, e_2^r, ..., e_n^r\}$, and the result element set for the threshold r' as $E^{r'} = \{e_1^{r'}, e_2^{r'}, ..., e_m^{r'}\}$, the system has the property of the search result continuity if the following holds for any r and r'. The function ancestor-or-self (e) returns element set that consist of ancestor elements of e and e itself.

$$\text{if } r \leq r' \text{ then } \forall e \in E^r, \exists e' \in E^{r'} \text{ s.t. } e' \in \text{ancestor-or-self} (e) \qquad \Box$$

In other words, the content of element set for *reading effort* r must be contained in the content of element set for *reading effort* r' if we increase the threshold value of *reading effort* from r to r'.

In addition, when we consider about ranking, we need to take the overlapping of content by nestings into account. Two patterns are possible for overlapping.

1. **Contain:** The content of a result element contains the content of upper ranked result element.
2. **Contained:** The content of a result element is contained by the content of upper ranked result element.

We think practical systems must hold search result continuity, and the system greedy retrieve result elements form higher scored elements considering nestings. The system first retrieves result elements by the *Thorough* strategy using the score based on *benefit* and *reading effort*. Then, while removing the overlapping of content by nestings, the system retrieves result elements from higher scored elements up to the threshold value of *reading effort* user input.

For removing the overlapping of content by nestings, if the pattern of **Contain** occurs the system removes the descendant elements that are already reported from the result list, and add the result to result list. If the pattern of **Contained** occurs the system simply skips the result element because the content of the element is already obtained.

When a result element e is retrieved, the *benefit* and *reading effort* of the ancestor element e_a is affected by e. In the case that the system retrieves e_a after e, the increment of *benefit* by e_a is $e_a.benefit - e.benefit$, and the increment of *reading effort* to obtain the increment of *benefit* by e_a is $e_a.reading\_effort - e.reading\_effort$.

When we suppose that users specify the threshold value of *reading effort*, the ranking algorithm based on *benefit* and *reading effort* considering nestings is shown in Figure 2.

As an example of ranking scheme, we show the case when a user specifies the threshold of *reading effort* to 40, and *benefit* and *reading effort* are calculated like Figure 1. The system retrieves ranked result element list using the *Thorough* strategy. In this case, the list is $\{e_3(e_3.score = 9/10 = 0.9), e_7(0.8), e_1(0.64), e_0(0.56), e_2(0.4), e_5(0.35), e_4(0.33)\}$. This list and threshold of *reading effort* are input of the ranking algorithm, and the list is processed from the top ranked result to lower ranked results. First, e_3 is processed and added to the output list $list_{out}$. At the same time, the *benefit* and *reading effort* of the ancestor elements e_1 and e_0 are adjusted. For e_1, $e_1.benefit$ is decreased to 9 and $e_1.reading\_effort$ is decreased to 18, and thereby $e_1.score$ is set to 0.5 (9/18). For e_0, $e_0.benefit$ is decreased to 19 and $e_0.reading\_effort$ is decreased to 40, and thereby $e_0.score$ is set to 0.48 (19/40). The system reflects these adjustments, and re-rank the elements in $list_{in}$. In this case, $list_{in}$ becomes $\{e_7(0.8), e_1(0.5), e_0(0.48), e_2(0.4), e_5(0.35), e_4(0.33)\}$. Then, e_7 is processed and $list_{out}$ becomes $\{e_3, e_7\}$, and $list_{in}$ becomes $\{e_1(0.5), e_2(0.4), e_0(0.37), e_4(0.33), e_5(0)\}$. Next, e_1 is processed and the system removes e_3 from $list_{out}$ because e_3 is the descendant of e_1. $list_{out}$ becomes $\{e_7, e_1\}$, and $list_{in}$ becomes $\{e_2(0.4), e_4(0.33), e_0(0.17), e_5(0)\}$. Subsequently, e_2 and e_4 are processed, however they are skipped because the ancestor element e_1 is already in $list_{out}$. Next, e_0 is processed, however if we retrieve e_0, the cumulated reading effort exceeds the specified *reading effort*, so the processing terminates and the system outputs $list_{out}$ $\{e_7, e_1\}$ as the final result.

Input: $list_{in}$, // result list of *Thorough*
                 $reading\_effort_t$ // threshold of *reading effort*
Output: $list_{out}$

$reading\_effort_c = 0$ // cumulated reading effort
while $((e = top(list_{in}))! = null)$ **do**
  remove e from $list_{in}$
  $skip = false$
  $remove = empty$
  for $(e_o$ in $list_{out})$ **do**
    if $(e_o$ is ancestor of $e)$ **then**
      $skip = true$
      break
    end if
    if $(e_o$ is descendant of $e)$ **then**
      add e_o to *remove*
    end if
  end for
  if $(skip)$ **then**
    continue
  end if
  $reading\_effort_c+ = e.reading\_effort$
  if $(reading\_effort_c > reading\_effort_t)$ **then**
    break
  end if
  for $(e_d \in remove)$ **do**
    remove e_d from $list_{out}$
  end for
  add e to $list_{out}$
  for $(e_a \in e.ancestors)$ **do**
    $e_a.benefit- = e.benefit$
    $e_a.reading\_effort- = e.reading\_effort$
    rerank e_a in $list_{in}$
  end for
end while
return $list_{out}$

Fig. 2. Ranking algorithm based on *benefit* and *reading effort*

4   Evaluation Metrics

For evaluating systems based on *benefit* and *reading effort*, we can compare cumulated *benefit* by the system with cumulated *benefit* by the system which knows actual *benefit* for each element for a certain threshold of *reading effort*. The system which knows actual *benefit* for each element can use the best list of *Thorough* as input, and we call this system BTIL (Best Thorough Input List) system. Implementers of XML-IR systems develop better system by guessing the *benefit* of each element close to the actual *benefit*. We supposed that we can use

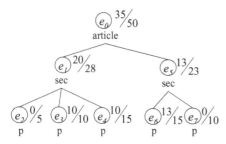
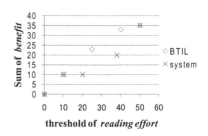

Fig. 3. Actual *benefit* and *reading effort*   **Fig. 4.** b/e graph

common *reading effort* value between the BTIL system and the system to be evaluated, because *reading effort* is the value that is not depend on queries.

As an example, we explain about the case the system calculates *benefit* and *reading effort* like Figure 1, however the actual *benefit* and *reading effort* are those shown in Figure 3. In this case, for the threshold value of *reading effort* 40, the system can obtain 20 *benefit* by retrieving $\{e_7, e_1\}$, however the BTIL system can obtain 33 *benefit* by retrieving $\{e_3, e_6, e_4\}$.

We can draw a graph by plotting the cumulated *benefit* by the system and the cumulated *benefit* by the BTIL system changing the threshold of *reading effort*. We call this graph benefit/effort graph (b/e graph, for short), and use for evaluation. Figure 4 shows the b/e graph for the running example. In Figure 4, 'BTIL' is for BTIL system and 'system' is for the system to be evaluated.

In this b/e graph, for a given *reading effort* r, the obtained *benefit* is the *benefit* of the point having maximum *reading effort* under r. For example, if the threshold value of *reading effort* was 30, the obtained *benefit* for 'BTIL' is 23 and 10 for 'system'. The b/e graph enables us to intuitively understand the performance of the system compared to the BTIL system.

5   Experiments

We used test collection for XML-IR provided by INEX 2005 project [1]. The test collection consists of XML documents, queries called topics, and relevance assessments. We implemented a system using *benefit* and *reading effort*, and obtained b/e graphs for some topics of INEX 2005. In addition, we examined the correlation between the metrics based on b/e graph and the existing XML-IR metrics.

5.1   Assessments of INEX 2005

The relevance assessments of INEX 2005 consists of two parts, Exhaustivity (ex) and Specificity (sp)[1]. Exhaustivity is the extent to which the element discusses the topic of request, and it has three levels; Highly exhaustive (HE), Partially

---

[1] The assessments of INEX 2006 only use Specificity.

exhaustive (PE), and Not exhaustive (NE) [2]. We converted HE, PE, and NE to numeric as 1, 0.5, 0, respectively. Specificity is the extent to which the element focuses on the topic of request, and it is calculated by dividing $rsize$, which is the length of the content relevant to the topic, by $size$, which is the whole length of the element.

We describe ex, sp, $rsize$, and $size$ of element e as $ex(e)$, $sp(e)$, $rsize(e)$, $size(e)$. The following formulas hold from the properties of ex, sp, $rsize$, and $size$. $e.parent$ is the parent element of e and $e.children$ is the child element set of e.

$$sp(e) = rsize(e)/size(e) \tag{1}$$

$$ex(e) \leq ex(e.parent) \tag{2}$$

$$rsize(e) = \sum_{e_i \in e.children} rsize(e_i) \tag{3}$$

$$size(e) = \sum_{e_i \in e.children} size(e_i) \tag{4}$$

5.2 Calculation of Actual Benefit and Reading Effort

We considered calculating actual *benefit* and *reading effort* from assessments of INEX 2005. We used following equations.

$$e.benefit = ex(e)^{\alpha} * rsize(e)^{\beta} \quad (\alpha \geq 0,\ \beta \geq 1) \tag{5}$$

$$e.reading\_effort = size(e)^{\gamma} \quad (0 \leq \gamma \leq 1) \tag{6}$$

Equation 5 satisfies Assumption 1 from Equation 2 and 3, and Equation 6 satisfies Assumption 2 from Equation 4. Though the assessments of INEX 2006 only use Specificity, the above equation is compatible with them by setting $\alpha = 0$.

5.3 System Implementation

The focus of system implementation is how to calculate *benefit* because we supposed that we can use common *reading effort* value, which is not depend on queries, with the BTIL system. We need the calculation formula for *benefit* that satisfies Assumption 1.

For this experiment, we used a formula for *benefit* based on $tf - ief$ [3]. When the $tf - ief$ is large, it is considered that we can obtain much information about the topic in the input query, so we can guess *benefit* is large. In addition, when multiple terms are used in the input query, we considered that *benefit* becomes larger if the element contains more terms in the query.

$$e.benefit = \frac{n}{|q|} * \sum_{t \in q}(tf * ief) \tag{7}$$

---

[2] Too Small (TS) is introduced for small elements, however we regard TS is equal to NE.

[3] ief stands for inverse element frequency.

$$ief = ln\frac{N+1}{ef} \tag{8}$$

where tf is the term frequency, ief is the inverse element frequency. Here, n is the number of terms occurring in both q and e, q is the input query, $|q|$ is the number of terms in q, N is the number of all elements, and ef is the number of elements that the term occurs. Though some term weighting schemes for XML documents are proposed [5,6,7], we used simple formula which satisfies Assumption 1, as we considered it is important to satisfy Assumption 1.

5.4   b/e Graph

We obtained b/e graphs of the system in Section 5.3, considering that the actual *benefit* and *reading effort* are given by the scheme in Section 5.2. As examples, we show the b/e graphs of the four topics; Topic 203, Topic 206, Topic 207, and Topic 210 of INEX 2005 in Figure 5, Figure 6, Figure 7, and Figure 8, respectively. In this experiment, we set $\alpha = 0.5$, $\beta = 1$, and $\gamma = 1$. Although the sum of *benefit* by the system we implemented is rather low compared to BTIL system especially in Topic 206, note that the performance of the system is not related to the usefulness of the proposed scheme.

We examined the comparison of the metrics based on b/e graph with existing XML-IR metrics. Using 29 Topics of INEX 2005, we calculated for each topic iMAep (interpolated Mean Average effort precision) [8], which is one of the existing XML-IR metrics using result list of *Thorough*, and iMArep (interpolated Mean Average reading effort precision), which is calculated based on b/e graph

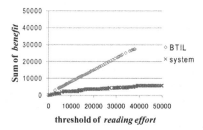

Fig. 5. b/e graph of Topic 203

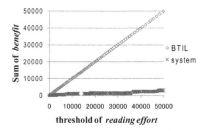

Fig. 6. b/e graph of Topic 206

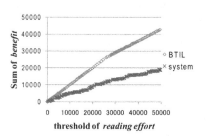

Fig. 7. b/e graph of Topic 207

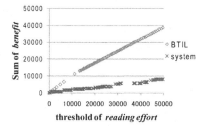

Fig. 8. b/e graph of Topic 210

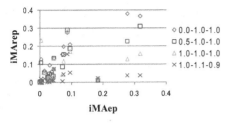

Fig. 9. Relationship between iMAep and iMArep

Table 1. Correlation coefficient between iMAep and iMArep

α-β-γ	correlation coefficient
0.0-1.0-1.0	0.81
0.5-1.0-1.0	0.75
1.0-1.0-1.0	0.48
1.0-1.1-0.9	0.39

with similar concept of iMAep. We used linear interpolation of BTIL plots on b/e graph as ideal for iMArep. iMAep is calculated based on the rank, while iMArep is calculated based on the *reading effort*. The relationship between iMAep and iMArep is shown in Figure 9. In Figure 9, we examined with four patterns of parameters α, β, and γ, and they are shown in the form of α-β-γ. Then, we examined correlation between iMAep and iMArep for one measure of effectiveness that we evaluate systems based on b/e graph. Table 1 shows the correlation coefficients of each pattern. In the case of 0.0-1.0-1.0 or 0.5-1.0-1.0, the correlation is relatively strong, however in the case of 1.0-1.0-1.0 or 1.0-1.1-0.9, we can say the correlation is weak. As the correlation of the pattern 1.0-1.1-0.9, which is considered to be close to the actual situation, is weak, we think the system can be evaluated with different measure to existing metrics by using the b/e graph based evaluation.

6   Conclusions

We introduced the concept of *benefit* and *reading effort* for XML-IR systems, and proposed the ranking algorithm and evaluation metrics based on them. The system retrieves the result elements considering efficiency and removing the overlapping of content by nestings.

Future works include introducing the concept of *switching effort*, which is the cost of switch the result item in the result list, as many results will increase the cost of browsing. Furthermore, for the XML documents created by marking up original PDF files, it is natural to show search result elements mapped on a physical page image [9], and integration with such user interface is also one of our future works. A major drawback of our current scheme is that users must specify the threshold of *reading effort*. We believe that developing user interfaces that can smoothly retrieve result elements when users change the threshold value of *reading effort* is a promising solution.

References

1. Malik, S., Kazai, G., Lalmas, M., Fuhr, N.: Overview of INEX 2005. In: Fuhr, N., Lalmas, M., Malik, S., Kazai, G. (eds.) INEX 2005. LNCS, vol. 3977, pp. 1–15. Springer, Heidelberg (2006)

2. Clarke, C.L.A.: Controlling overlap in content-oriented XML retrieval. In: SIGIR, pp. 314–321 (2005)
3. Theobald, M., Schenkel, R., Weikum, G.: An efficient and versatile query engine for TopX search. In: VLDB, pp. 625–636 (2005)
4. Kaushik, R., Krishnamurthy, R., Naughton, J.F., Ramakrishnan, R.: On the integration of structure indexes and inverted lists. In: ACM SIGMOD, pp. 779–790 (2004)
5. Shimizu, T., Terada, N., Yoshikawa, M.: Kikori-KS: An effective and efficient keyword search system for digital libraries in XML. In: Sugimoto, S., Hunter, J., Rauber, A., Morishima, A. (eds.) ICADL 2006. LNCS, vol. 4312, pp. 390–399. Springer, Heidelberg (2006)
6. Grabs, T., Schek, H.-J.: ETH Zürich at INEX: Flexible information retrieval from XML with PowerDB-XML. In: INEX, pp. 141–148 (2002)
7. Amer-Yahia, S., Curtmola, E., Deutsch, A.: Flexible and efficient XML search with complex full-text predicates. In: ACM SIGMOD, pp. 575–586 (2006)
8. Kazai, G., Lalmas, M.: INEX 2005 evaluation measures. In: Fuhr, N., Lalmas, M., Malik, S., Kazai, G. (eds.) INEX 2005. LNCS, vol. 3977, pp. 16–29. Springer, Heidelberg (2006)
9. Shimizu, T., Yoshikawa, M.: XML information retrieval considering physical page layout of logical elements. In: WebDB (2007)

Improving MEDLINE Document Retrieval Using Automatic Query Expansion

Sooyoung Yoo and Jinwook Choi

Dept of Biomedical Engineering, College of Medicine,
Seoul National University, 28 Yongon-Dong Chongro-Gu,
Seoul, Korea
{yoosoo0,jinchoi}@snu.ac.kr

Abstract. In this study, we performed a comprehensive evaluation of pseudo-relevance feedback technique for automatic query expansion using OHSUMED test collection. The well-known term sorting methods for the selection of expansion terms were tested in our experiments. We also proposed a new term reweighting method for further performance improvements. Through the multiple sets of test, we suggested that local context analysis was probably the most effective method of selecting good expansion terms from a set of MEDLINE documents given enough feedback documents. Both term sorting and term reweighting method might need to be carefully considered to achieve maximum performance improvements.

Keywords: Pseudo-relevance feedback, MEDLINE, Term sorting method.

1 Introduction

Automatic query expansion and relevance feedback techniques have been proposed to address the query-document mismatch problem. Relevance feedback (RF) expands terms from the user-identified relevant documents. Pseudo-relevance feedback (PRF) expands terms from the top documents initially retrieved. Although RF is useful for searchers, the overall performance of PRF is better in terms of search performance and searcher satisfaction [1]. In this paper, we focus on PRF technique for improving MEDLINE document retrieval.

Typically, PRF assumes that the initially retrieved top R documents are relevant. It extracts candidate expansion terms from the top R documents, sorts them using a term sorting (scoring) technique, and appends the top-ranked E terms to the initial query with modified weights. However, the performance of PRF can be affected by the quality of the initial retrieval result, such as the number of pseudo-relevant documents (R), the number of expansion terms (E), the term sorting method, and the term reweighting method applied [2-5]. The R and E parameters are usually chosen by experiments on a particular test collection. For the domain-specific test collection called OHSUMED where the documents are short references to medical literature, the performance of PRF therefore needs to be evaluated against various factors affecting the retrieval accuracy.

D.H.-L. Goh et al. (Eds.): ICADL 2007, LNCS 4822, pp. 241–249, 2007.
© Springer-Verlag Berlin Heidelberg 2007

In this study, using the OHSUMED test collection, we perform a comprehensive experimental evaluation for various well-known term sorting methods and different term reweighing methods. For each term reweighting method, the characteristics among different term sorting algorithms will be discussed.

2 Methods

2.1 Test Collection

We used OHSUMED [6] as a test collection. The test collection is a subset of the MEDLINE database, which is a bibliographic database of important, peer-reviewed medical literature maintained by the National Library of Medicine (NLM). It contains 348,566 MEDLINE references from 1987 to 1991, and 106 topics (queries) generated by actual physicians in the course of patient care. About 75% of the references contain title and abstracts, while the remainder has only titles. Each reference also contains human-assigned subject headings from the Medical Subject Headings (MeSH). Each query contains a brief statement about a patient, followed by the information need. The queries are generally terse. The relevance is judged to be "definitely relevant", "possibly relevant", or "non-relevant". For our experiments we assume only "definitely relevant" are relevant. Therefore, only 101 queries which have definitely relevant documents are used for our evaluation. We use title, abstract and MeSH fields to represent each document and the information need field to represent each query.

2.2 Baseline Retrieval System

The baseline retrieval system was developed using SMART stopwords and Lovins' stemmer [7]. We simply used single terms as index terms. It had been shown that the best document-query weighting scheme was ann.atn for OHSUMED collection [6]. However, in our preliminary experiments, we found out that Okapi BM25 similarity measure [8] worked 9.4% significantly better than ann.atn in terms of precision at 10 documents (absolute precision at 10 documents was 0.2861 for Okapi BM25 and 0.2614 for ann.atn) although there was no significant difference for other evaluation measures (pared t-test, p=0.05).

Therefore, we chose Okapi BM25 weighting scheme as our unexpanded baseline retrieval model. In Okapi BM25 formula, the initial top-ranked documents are retrieved by computing a similarity measure between a query q and a document d as follows:

$$sim(q,d) = \sum_{t \in q \wedge d} w_{d,t} \cdot w_{q,t} \,. \tag{1}$$

with $w_{d,t} = \dfrac{(k_1 + 1) \cdot f_{d,t}}{K + f_{d,t}}$ and $w_{q,t} = \dfrac{(k_3 + 1) \cdot f_{q,t}}{k_3 + f_{q,t}} \cdot \log \dfrac{N - f_t + 0.5}{f_t + 0.5}$

where t is a term of query q, f_t is the number of documents containing the term t across the document collection that contains N documents and $f_{d,t}$ is the frequency of the term t in document d. K is $k_1((1-b) + b \times dl/avdl)$. k_1, b, and k_3 are parameters set to 1.2, 0.75, and 1,000 respectively. dl and $avdl$ are respectively the document length and average document length measured in some suitable unit.

2.3 Selection of Expansion Terms

After we extracted all candidate expansion terms from the top R documents initially retrieved, we selected high-ranked E expansion terms to be added to the original query. In order to rank all candidate terms, we evaluated various term sorting methods in our preliminary experiments. From the experiments, we chose six competing methods with different properties (i.e. low term overlapping) to be evaluated further in this paper. Following term sorting algorithms were not considered in this paper: frequency [4], modified F4point-5 (F4MODIFIED) [9], the new term selection value based on significance measure [8], Doszkocs' variant of CHI-squared (CHI1) [5], r_lohi [10], and idf.

The six term sorting methods to be compared were Rocchio weight based on the Vector Space Model [5], Kullback-Leibler Divergence (KLD) based on the information theory [5], Robertson Selection Value (RSV) [5], CHI-squared (CHI2) [5], Expected Mutual Information Measure (EMIM) based on probabilistic distribution analysis [10], and Local Context Analysis (LCA) utilizing co-occurrence with all query terms [11]. In RSV, we did not ignore the probability that a nonrelevant document contain a candidate term t since the performance was better than the performance of ignoring it. We replaced the non-relevant documents statistics with the collection level statistics because we did not have any information about non-relevant documents.

After sorting all candidate terms including original query terms using one of the above methods, top-ranked E new terms (threshold score > 0) were finally selected for query expansion.

2.4 Traditional Term Reweighting Techniques

We evaluate two popular traditional term reweighting methods and our variants described in the next section.

For probabilistic feedback, we use the modified Robertson/Sparck-Jones weight [8]. It reweights expansion terms as follows:

$$\frac{1}{3} \times \log\left(\frac{(r_t + 0.5)/(R - r_t + 0.5)}{(f_t - r_t + 0.5)/(N - f_t - R + r_t + 0.5)}\right). \tag{2}$$

where r_t is the number of pseudo-relevant documents containing term t and the same definitions are used as in the above Okapi BM25 formula. The original query terms are reweighted by the original Okapi weight. Our preliminary experiments showed that 1/3 downgrading of its original Okapi weight for expansion terms was significantly better than using itself on the OHSUMED test collection.

For vector space feedback, we use standard Rocchio's formula. In original Rocchio formula, the new weight w'_{qt} of term t after query expansion is assigned as: (we assume a positive feedback)

$$w'_{q,t} = \alpha \cdot w_{q,t} + \frac{\beta}{R} \cdot \sum_{k=1}^{R} w_{k,t} \cdot$$ (3)

where $w_{q,t}$ is the weight of term t in the unexpanded query and $w_{k,t}$ is the weight of term t in a pseudo-relevant document k (in our retrieval system, that is, $w_{d,t}$ component of the Okapi BM25 formula). The α and β tuning constants are set to 1.

2.5 New Term Reweighting Techniques

Within Rocchio feedback formula, two variants of term reweighting were devised by extending the ideas of [12] for comparison. The main idea is to reflect the result of a term sorting algorithm on term reweighting process.

First, instead of using original Rocchio weight reflecting term importance within the pseudo-relevant documents, we utilized rank position of a term in the sorted term list for assigning the relevance weight as follows.

$$w'_{q,t} = \alpha \cdot w_{q,t} + \beta \cdot rank\_norm\_score_t \cdot$$ (4)

The $rank\_norm\_socre_t$ is evenly decreasing score according to the rank position of term t in the sorted term list. The $rank\_norm\_score_t$ of term t is calculated as $1 - (rank_t - 1) / |term\_list|$ where $rank_t$ is the rank position of term t in the sorted term list and $|term\_list|$ is the number of terms in the term list expanded. We call this approach "$rank\_norm$".

Second, we intended to reflect the phenomenon that "ordinary" is shared among many, while "outstanding" is less frequent [13] on deciding the relevance weight of a term. Through multiple sets of preliminary test, we hypothesized that only the small number of high ranked terms would be enough more important in terms of relevance. Based on the hypothesis, the following formula was devised.

$$w'_{q,t} = \alpha \cdot w_{q,t} + \beta \cdot rank\_group\_score_t \cdot$$ (5)

For calculating $rank\_group\_score_t$ of term t, the sorted terms are firstly divided into k groups. A group of terms is then simply given $rank\_group\_score_t$ from k to 1. We assign a relatively small number of high scores, and a relatively large number of small scores using the term partitioning method used in the referenced paper [13]. We call this approach "$rank\_group\_kX$" where X is the number of groups of terms. In our preliminary experiments, small values of k performed better for OHSUMED test collection. We therefore used $k = 2$ in this paper, i.e. giving terms of the first group twice $rank\_group\_score$ score than terms of the second group.

In this study, we also fixed $\alpha = \beta = 1$ for both $rank\_norm$ and $rank\_group\_kX$ term reweighting methods.

3 Results

We retrieved the top-ranked 100 documents for 101 queries, and evaluated the performance using mean average precision (MAP). The unexpanded baseline MAP was 0.2163. We measured the performance of PRF for a wide range of R (1,2,3,4, and

5 to 50 by 5) and E (5 to 80 by 5) parameters. From our experiments, the performance was generally the best when E was between 10 and 15 in OHSUMED test collection. However, R parameter could not be fixed easily. Given a fixed number of expansion terms (E = 15), we therefore showed the performance improvements over the unexpanded baseline against different number of pseudo-relevant documents.

Fig. 1 to 4 display the MAP percentage change over the unexpanded baseline on various number of pseudo-relevance documents for different term sorting algorithms where probabilistic feedback, standard Rocchio's feedback, *rank_norm*, and *rank_group_k2* term reweighting were applied respectively.

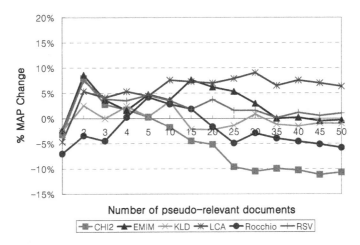

Fig. 1. Percent improvement in mean average precision with fixed E parameter (15 terms) for probabilistic term reweighting

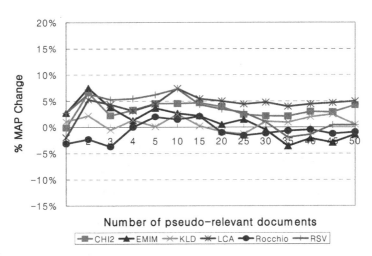

Fig. 2. Percent improvement in mean average precision with fixed E parameter (15 terms) for original Rocchio term reweighting

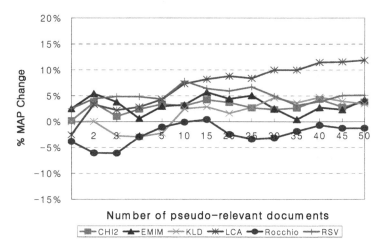

Fig. 3. Percent improvement in mean average precision with fixed E parameter (15 terms) for *rank_norm* term reweighting within Rocchio framework

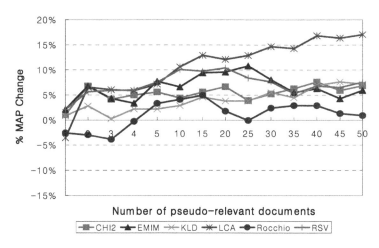

Fig. 4. Percent improvement in mean average precision with fixed E parameter (15 terms) for *rank_group_k2* term reweighting within Rocchio framework

In probabilistic term reweighting, the performance of competing term sorting algorithms was greatly affected by the R parameter settings as can be seen in Fig. 1. There was a noticeable decrease in the performance for CHI2 term sorting method when more than 25 documents were used for feedback on OHSUMED test collection. Overall, LCA term sorting method was less sensitive to R parameter settings with comparable or better performance than other term sorting strategies.

In standard Rocchio's term reweighting, well-known term sorting algorithms did not produce different performance patterns on a wide rage of R parameter as shown in Fig. 2. Although remarkable performance improvement could not be achieved for all term sorting algorithms, better performance improvement was expected for LCA and RSV.

In *rank_norm* and *rank_group_k2* term reweighting, the performance differences among different term sorting algorithms were distinguishable. As can be seen in Fig. 3 and 4, the performance of LCA was much better than the other methods on large R settings. On the other hand, Rocchio term sorting method showed worst performance. It seems that Rocchio method as a term sorting might not select good expansion terms from a set of MEDLINE documents.

It also can be seen that *Rank_group_k2* term reweighing method is better for the same term sorting method compared to *rank_norm* method. It supports our hypothesis that only top few terms of the sorted term list can be considered to be most important in determining their relevance weight. Therefore, it may be reasonable to divide terms into groups of more "good" terms and "less meaning" terms, rather than to differentiate their weight. It also seems that the difference of the relevance weights is less important.

Consequently, our experimental results suggest that LCA is probably the most effective method of selecting good expansion terms from a set of MEDLINE documents when feedback documents are given enough large. In addition, maximum performance improvements may be obtained by employing our *rank_gorup_k2* term reweighting rather than traditional feedback methods.

For the further analysis of individual queries, per-query improvements in MAP are given in Fig. 5. The differences in MAP between expanded query using LCA term sorting method and queries without expansion (baseline) are shown. Given fixed R = 50 and E = 15 parameters, each line is the performance differences for different term reweighting method. Our term rank-based reweighting scheme shows better performance than traditional probabilistic or Rocchio reweighting formula for more individual queries. It is proven that the reweighting methods affect the performance of individual queries and our reweighting methods are effective for more individual queries.

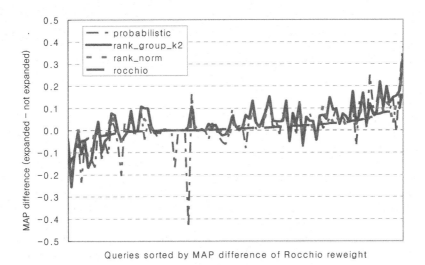

Fig. 5. Queries sorted by difference in mean average precision of original Rocchio term reweight for LCA term sorting method (R=50, E=15)

4 Discussion

For comparing well-known term sorting methods, LCA showed better performance than the other methods. It may be mainly due to the characteristics of OHSUMED queries itself. The queries frequently contain terms which represent a special medical task (e.g., "diagnosis", "treatment", "etiology", etc). These terms are typically general. However, they can be effectively used for restricting query context. Since LCA considers co-occurrence with all query terms, it seems to implicitly restrict expansion terms to a specific medical task. Therefore, LCA will be suitable method for selecting expansion terms from a set of MEDLINE documents.

We tried to combine all pair-wise term sorting methods using standard combination methods [14] for further performance improvements. However, combing term sorting algorithms did not give any significant improvements over single best method in our experiments. More careful considerations may be needed when combining different methods in OHSUMED test collection.

5 Conclusion

In this paper, we performed a comprehensive experiment on PRF technique for a wide range of parameter choices using OHSUMED test collection. For the selection of expansion terms, LCA method utilizing co-occurrence with all query terms showed best performance when the pseudo-relevant documents were given large enough. Further performance improvements were achieved by applying our term rank-based reweighing variants within Rocchio framework rather than traditional probabilistic or original Rocchio formula. Therefore, both term sorting and term reweighting method might need to be carefully considered to achieve maximum performance improvements.

Acknowledgments. This work was supported in part by the Advanced Biomedical Research Center (ABRC) funded by KOSEF, and in part by the MIC (Ministry of Information and Communication), Korea, under the ITRC (Information Technology Research Center) support program supervised by the IITA (Institute of Information Technology Advancement) (IITA-2006-(C1090-0602-0002)).

References

1. White, R.W.: Implicit feedback for interactive information retrieval. In: SIGIR Forum 2005, p. 70 (2005)
2. Fan, W., Luo, M., Wang, L., Xi, W., Fox, E.A.: Tuning before feedback: combining ranking discovery and blind feedback for robust retrieval. In: 27th annual international ACM SIGIR conference on Research and development in information retrieval, pp. 138–145. ACM Press, New York (2004)
3. Jimmny, L., Murray, G.C.: Assessing the term independence assumption in blind relevance feedback. In: 28th annual international ACM SIGIR conference on Research and development in information retrieval, pp. 635–636. ACM Press, New York (2005)

4. Harman, D.: Relevance feedback revisited. In: 15th annual international ACM SIGIR conference on Research and development in information retrieval, pp. 1–10. ACM Press, New York (1992)
5. Carpineto, C., Mori, R., Romano, G., Bigi, B.: An Information-Theoretic Approach to Automatic Query Expansion. ACM Trans. Inf. Syst. 19, 1–27 (2001)
6. Hersh, W., Buckley, C., Leone, T.J., Hickam, D.: OHSUMED: an interactive retrieval evaluation and new large test collection for research. In: 17th annual international ACM SIGIR conference on Research and development in information retrieval, pp. 192–201. Springer, Heidelberg (1994)
7. Lovins, J.B.: Development of a stemming algorithm. Mechanical Translation and Computational Linguistics. 11, 22–31 (1968)
8. Robertson, S.E., Walker, S.: Okapi/Keenbow at TREC-8. In: 8th Text REtrieval Conference (TREC-8), pp. 151–161 (1999)
9. Robertson, S.E.: On relevance weight estimation and query expansion. Journal of Documentation 42, 182–188 (1986)
10. Efthimiadis, E.N., Brion, P.V.: UCLA-Okapi at TREC-2: Query Expansion Experiments. In: 2nd Text REtrieval Conference (TREC.2), pp. 200–215, NIST Special Publication (1994)
11. Xu, J., Croft, W.B.: Improving the effectiveness of information retrieval with local context analysis. ACM Trans. Inf. Syst. 18, 79–112 (2000)
12. Carpineto, C., Romano, G.: Improving retrieval feedback with multiple term-ranking function combination. ACM Trans. Inf. Syst. 20, 259–290 (2002)
13. Anh, V.N., Moffat, A.: Simplified similarity scoring using term ranks. In: 28th annual international ACM SIGIR conference on Research and development in information retrieval, pp. 226–233. ACM Press, New York (2005)
14. JH, L.: Analyses of multiple evidence combination. In: 20th annual international ACM SIGIR conference on Research and development in information retrieval, pp. 267–276. ACM Press, New York (1997)

Personal Name Disambiguation in Web Search Results Based on a Semi-supervised Clustering Approach

Kazunari Sugiyama and Manabu Okumura

Precision and Intelligence Laboratory, Tokyo Institute of Technology,
4259 Nagatsuta, Midori, Yokohama, Kanagawa 226-8503, Japan
sugiyama@lr.pi.titech.ac.jp, oku@pi.titech.ac.jp

Abstract. Most of the previous works that disambiguate personal names in Web search results often employ agglomerative clustering approaches. In contrast, we have adopted a semi-supervised clustering approach in order to guide the clustering more appropriately. Our proposed semi-supervised clustering approach is novel in that it controls the fluctuation of the centroid of a cluster, and achieved a purity of 0.72 and inverse purity of 0.81, and their harmonic mean F was 0.76.

Keywords: Information retrieval, Semi-supervised clustering, Personal name disambiguation.

1 Introduction

Personal names are often submitted to search engines as query keywords. However, in response to a personal name query, search engines return a long list of search results containing Web pages about several namesakes. For example, when a user submits a personal name such as "William Cohen" to the search engine Google[1], the returned results contain more than one person named "William Cohen." The results include a computer science professor, an U.S. politician, a surgeon, and others; these results are not classified into separate clusters but are mixed together.

Most of the previous works on disambiguating personal names in Web search results employ several types of unsupervised agglomerative clustering approaches [1], [2], [3], [4], [5]. However, it is hard for these approaches to guide the clustering process appropriately. Therefore, if some Web pages that describe the entity of a person are introduced in a semi-supervised manner, the clustering for personal name disambiguation would be much more accurate. Hereafter, we refer to such a Web page as the "*seed page.*" Then, in order to disambiguate personal names in Web search results, we introduce semi-supervised clustering that uses the seed page to improve the clustering accuracy. Existing methods for semi-supervised clustering can be classified into the following two categories: (1) *constraint-based* [6], [7], [8] and (2) *distance-based* [9], [10]. These approaches aim at refining pure K-means algorithm [11] that needs to set the number of clusters K in advance. However, in our study, the number of namesakes in the Web search results is not known previously. Moreover, they do not consider controlling the fluctuation of the centroid of a cluster although these algorithms focus on introducing

---

[1] http://www.google.com/

D.H.-L. Goh et al. (Eds.): ICADL 2007, LNCS 4822, pp. 250–256, 2007.

constraints and learning distances. We believe that in semi-supervised clustering, it is important to control the fluctuation of the centroid of a cluster that contains a seed page as well as to introduce constraints in order to obtain highly accurate clustering results. Focusing on this point, we propose a novel semi-supervised clustering approach that controls the fluctuation of the centroid of a cluster that contains a seed page.

2 Our Proposed Semi-supervised Clustering

In the following discussion, we denote the feature vector $\boldsymbol{w}^p$ of a Web page p in a set of search results as follows:

$$\boldsymbol{w}^p = (w_{t_1}^p, w_{t_2}^p, \cdots, w_{t_m}^p), \tag{1}$$

where m is the number of distinct terms in the Web page p and t_k $(k = 1, 2, \cdots, m)$ denotes each term. In our preliminary experiments for generating feature vectors for clustering in our task, we found that gain [12] is the most effective term weighting scheme. Using the gain scheme, we also define each element $w_{t_k}^p$ of $\boldsymbol{w}^p$ as follows:

$$w_{t_k}^p = \frac{df(t_k)}{N} \left(\frac{df(t_k)}{N} - 1 - \log \frac{df(t_k)}{N} \right),$$

where $df(t_k)$ is the number of search-result Web pages in which term t_k appears and N is the total number of search-result Web pages. In addition, we also define the centroid vector of a cluster G as follows:

$$G = (g_{t_1}, g_{t_2}, \cdots, g_{t_m}), \tag{2}$$

where g_{t_k} is the weight of each term in the centroid vector of a cluster and t_k $(k = 1, 2, \cdots, m)$ denotes each term.

Our proposed approach controls the fluctuation of the centroid of a cluster that contains a seed page when a new cluster is merged into it. In this process, when we merge the feature vector $\boldsymbol{w}^p$ of a search-result Web page into the most similar cluster that contains a seed page, we weight each element of $\boldsymbol{w}^p$ by the distance $D(G, \boldsymbol{w}^p)$ between G and $\boldsymbol{w}^p$. We employ the following as a measure of the distance: (i) Euclidean distance, (ii) Mahalanobis distance, and (iii) adaptive Mahalanobis distance. The adaptive Mahalanobis distance is a measure that overcomes the drawback of Mahalanobis distance in that the value of covariance tends to be large when the number of members of a cluster is small. Using Equations (1) and (2), we define the new centroid vector of cluster G^{new} after merging a certain cluster into its most similar cluster as follows:

$$G^{new} = \frac{\left(\sum_{\boldsymbol{w}^{p^{(G)}} \in G}^q \boldsymbol{w}^{p^{(G)}} + \frac{\boldsymbol{w}^p}{D(G, \boldsymbol{w}^p)} \right)}{q + 1}, \tag{3}$$

where $\boldsymbol{w}^{p^{(G)}}$ and q are the feature vector $\boldsymbol{w}^p$ of a search-result Web page and the number of search-result Web pages $(q < n)$ in the cluster, respectively. When we merge clusters

---

Algorithm: Semi-supervised clustering

Input: Set of search-result Web page $p_i\,(i = 1, 2, \cdots, n)$, and seed pages $p_{s_j}\,(j = 1, 2, \cdots, u)$,
$\quad Wp = \{p_1, p_2, \cdots, p_n, p_{s_1}, p_{s_2}, \cdots, p_{s_u}\}$.

Output: Clusters that contain the Web pages that refer to the same person.

Method:

1. Set each element in Wp as an initial cluster.

2. Repeat the following steps for all $p_i\,(i = 1, 2, \cdots, n)$ in Wp

    2.1 Compute the similarity between p_i and p_{s_j}.

        if the maximum similarity is obtained between p_i and p_{s_j},

            then merge p_i into p_{s_j} and recompute the centroid of the cluster using Equation (3),

        else p_i is stored as other clusters Oth, namely, $Oth = \{p_i\}$.

3. Repeat the following steps for all $p_h\,(h = 1, 2, \cdots, m, (m < n))$ in Oth

        until all of the similarities between two clusters are less than the predefined threshold.

    3.1 Compute the similarity between p_h and $p_r\,(r = h + 1, \cdots, m)$

        if the maximum similarity is obtained between p_h and p_r,

            then merge p_h and p_r and recompute the centroid of the cluster using Equation (4),

        else p_h is an independent cluster.

    3.2 Compute all of the similarities between two clusters.

Fig. 1. Our proposed semi-supervised clustering algorithm

that do not contain seed pages, we do not control the centroid of a cluster, and define the centroid vector of the cluster as follows:

$$G^{new} = \frac{\left(\sum_{w^{p^{(G)}} \in G}^{q} w^{p^{(G)}} + w^p\right)}{q + 1}, \tag{4}$$

Figure 1 shows the detailed algorithm of our proposed semi-supervised clustering approach.

3 Experiments

3.1 Experimental Data

In our experiments, we used the WePS corpus established for Web People Search Task [13]. The WePS corpus comprises 79 person sets, each of which corresponds to the top 100 search results of Yahoo!2 via its search API for a person name query. In other words, it contains approximately 7900 Web pages, and 49 and 30 personal names in the training and test sets, respectively.

3.2 Evaluation Measure

We evaluate clustering accuracy based on the *purity*, *inverse purity* and their harmonic mean F adopted in the Web People Search Task. Given a manual classification of the documents into a set of labels, the precision of each cluster P with respect to a label

---

2 http://www.yahoo.com/

Table 1. Clustering accuracy obtained using agglomerative and our proposed semi-supervised clustering with one seed page

Clustering approach	Type of seed page	Purity	Inverse purity	F
Agglomerative clustering	no seed page	0.66	0.49	0.51
Semi-supervised clustering				
(i) Euclidean distance	(a) Wikipedia article	0.39	0.90	0.54
	(b) Top-ranked Web page	0.40	0.82	0.54
(ii) Mahalanobis distance	(a) Wikipedia article	0.44	0.96	0.55
	(b) Top-ranked Web page	0.47	0.81	0.60
(iii) Adaptive Mahalanobis distance	(a) Wikipedia article	0.48	0.88	**0.62**
	(b) Top-ranked Web page	0.50	0.78	*0.61*

partition L containing all documents assigned to the label, is the fraction of documents in P which belong to L. The purity is then defined as the weighted average of the maximum precision values of each cluster P, and the inverse purity is defined as the weighted average of the maximum precision values of each partition L over the clusters. Purity and inverse purity achieves maximum value of 1 when every cluster has one single member and when there is only one single cluster, respectively.

3.3   Experimental Results

3.3.1   Experimental Results Using Full Text in the Documents

We compare clustering accuracy obtained using agglomerative and our proposed semi-supervised clustering using full text in seed pages and search-result Web pages. In both approaches, we first determine the optimal similarity for merging similar clusters using the training set in the WePS corpus and then apply it to the test set in the corpus. This similarity is set to 0.0065. Moreover, in our semi-supervised clustering approach, we use the following two types of seed pages: (a) an article on each person in Wikipedia [14] and (b) the top-ranked Web page in the Web search results. We first conducted experiments using one seed page. However, every personal name in the test set of the WePS corpus does not have a corresponding article in Wikipedia. Therefore, if a personal name has an article in Wikipedia, we used it as the seed page. Otherwise, we used the top-ranked Web page in the Web search results as the seed page. We used Wikipedia article as a seed page for 16 persons and the top-ranked Web page for 14 persons in the test set of the WePS corpus. In a recent work that applies Wikipedia to personal name disambiguation, Bunescu and Paşca [15] identify and disambiguate named entities by using the structures of Wikipedia. Table 1 lists the clustering accuracies obtained using agglomerative and our semi-supervised clustering approach with one seed page.

Moreover, with regard to the adaptive Mahalanobis distance where the best F is obtained in the experiments using one seed page, we conduct further experiments by varying the number of seed pages. Figures 2 and 3 show the clustering accuracies obtained using multiple Wikipedia articles, Web pages ranked up to the top 5, respectively.

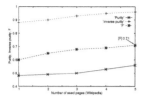 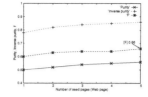 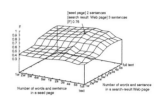

Fig. 2. Clustering accuracy obtained using multiple seed pages (5 Wikipedia articles)

Fig. 3. Clustering accuracy obtained using multiple seed pages (Web pages ranked up to the top 5)

Fig. 4. Clustering accuracy obtained varying the number of words and sentences backward and forward from a personal name in a seed page and a search-result Web page in the case of the 5 seed pages (Wikipedia articles) shown in Fig. 2 ("w" and "s" denote "word" and "sentence," respectively)

3.3.2 Experimental Results Using Fragments in the Documents

We observed that the words that characterize the person often appear around a personal name. Therefore, we vary the numbers of words and sentences backward and forward from a personal name in the case where we used 5 Wikipedia articles as seed pages; in other words, the best value of F (0.71) is obtained in our experiment. In this experiment, using training set in the WePS corpus, we first search for the number of words or sentences around a personal name in a seed page and a search-result Web page that gives the best F. Figure 4 shows that the best F (0.76) is obtained when we use 2 and 3 sentences around a personal name in a seed page and a search-result Web page, respectively. After applying these number of sentences around a personal name to the test set of WePS corpus, we finally obtained the clustering accuracy, (purity:0.72, inverse purity:0.81, F:0.76).

3.4 Discussion

In the agglomerative clustering approach, in Table 1, the high purity (0.66) with low inverse purity (0.49) indicates that the agglomerative clustering tends to generate clusters that contain only one search-result Web pages.

In our proposed semi-supervised clustering approach, Table 1 shows that all the approaches outperform agglomerative clustering with regard to the values of inverse purity and F, although most of the purity values cannot outperform those obtained using agglomerative clustering. We consider that this is due to the effect of controlling the fluctuation of the centroid of a cluster that contains a seed page. In our proposed semi-supervised clustering approach, the best value of F (0.62) is obtained in the case where we employ the adaptive Mahalanobis distance with an Wikipedia article as a seed page. Moreover, in the semi-supervised clustering approach using multiple seed pages, Figures 2 and 3 indicate that the values of both purity and inverse purity improve as the number of seed pages increases. This shows that introducing seed pages can guide the clustering process more appropriately.

In the experiments using fragments in the documents, we found that we can disambiguate a personal name more effectively by using several sentences than words around a personal name in a seed page and a search-result Web page in the training set of the WePS corpus. This is because we could acquire useful information from sentences that characterize an entity of a person. Moreover, the obtained clustering accuracy (purity:0.72, inverse purity:0.81, F:0.76) is comparable to the top result (purity:0.72, inverse purity:0.88, F:0.78) among the participant systems in Web People Search Task [13].

4 Conclusion

In this paper, we have proposed a semi-supervised clustering approach for disambiguating personal names in Web search results. Our approach is novel in that it realizes highly accurate semi-supervised clustering by controlling the fluctuation of the centroid of a cluster that contains a seed page. In our proposed semi-supervised clustering approach, we introduced some distance measures to control the centroid fluctuation. Experimental results show that our proposed approach achieved the best value of F (0.76) when we simultaneously used 2 sentences backward and forward from an ambiguous name in a seed page and 3 sentences backward and forward from an ambiguous name in a search-result Web page. In future work, we plan to use Web pages hyperlinked from a target page to disambiguate personal names in Web search results and extend our approach to disambiguate place names.

References

1. Mann, G.S., Yarowsky, D.: Unsupervised Personal Name Disambiguation. In: Proc. of the 7th Conference on Natural Language Learning (CoNLL-2003), pp. 33–40 (2003)
2. Pedersen, T., Purandare, A., Kulkarni, A.: Name Discrimination by Clustering Similar Contexts. In: Gelbukh, A. (ed.) CICLing 2005. LNCS, vol. 3406, pp. 226–237. Springer, Heidelberg (2005)
3. Wan, X., Gao, J., Li, M., Ding, B.: Person Resolution in Person Search Results: WebHawk. In: Proc. of the 14th International Conference on Information and Knowledge Management (CIKM 2005), pp. 163–170 (2005)
4. Bekkerman, R., McCallum, A.: Disambiguating Web Appearances of People in a Social Network. In: Proc. of the 14th International World Wide Web Conference (WWW2005), pp. 463–470 (2005)
5. Bollegala, D., Matsuo, Y., Ishizuka, M.: Extracting Key Phrases to Disambiguate Personal Names on the Web. In: Gelbukh, A. (ed.) CICLing 2006. LNCS, vol. 3878, pp. 223–234. Springer, Heidelberg (2006)
6. Wagstaff, K., Cardie, C.: Clustering with Instance-level Constraints. In: Proc. of the 17th International Conference on Machine Learning (ICML 2000), pp. 1103–1110 (2000)
7. Wagstaff, K., Rogers, S., Schroedl, S.: Constrained K-means Clustering with Background Knowledge. In: Proc. of the 17th International Conference on Machine Learning (ICML 2001), pp. 577–584 (2001)

8. Basu, S., Banerjee, A., Mooney, R.: Semi-supervised Clustering by Seeding. In: Proc. of the 19th International Conference on Machine Learning (ICML 2002), pp. 27–34 (2002)
9. Klein, D., Kamvar, S.D., Manning, C.D.: From Instance-level Constraints to Space-level Constraints: Making the Most of Prior Knowledge in Data Clustering. In: Proc. of the 19th International Conference on Machine Learning (ICML 2002), pp. 307–314 (2002)
10. Xing, E.P., Ng, A.Y., Jordan, M.I., Russell, S.J.: Distance Metric Learning with Application to Clustering with Side-Information. Advances in Neural Information Processing Systems 15, 521–528 (2003)
11. MacQueen, J.: Some Methods for Classification and Analysis of Multivariate Observations. In: Proc. of the 5th Berkeley Symposium on Mathmatical Statistics and Probability, pp. 281–297 (1967)
12. Papineni, K.: Why Inverse Document Frequency? In: Proc. of the 2nd Meeting of the North American Chapter of the Association for Computational Linguistics (NAACL 2001), pp. 25–32 (2001)
13. Artiles, J., Gonzalo, J., Sekine, S.: The SemEval-2007 WePS Evaluation: Establishing a Benchmark for the Web People Search Task. In: Proc. of the 4th International Workshop on Semantic Evaluations (SemEval-2007), pp. 64–69 (2007)
14. Remy, M.: Wikipedia: The Free Encyclopedia. Online Information Review 26(6), 434 (2002)
15. Bunescu, R., Paşca, M.: Using Encyclopedic Knowledge for Named Entity Disambiguation. In: Proc. of the 11th Conference of the European Chapter of the Association for Computational Linguistics (EACL 2006), pp. 9–16 (2006)

Preserving Interactive Multimedia Art:
A Case Study in Preservation Planning

Christoph Becker[1], Günther Kolar[2], Josef Küng[3], and Andreas Rauber[1]

[1] Vienna University of Technology, Vienna, Austria
http://www.ifs.tuwien.ac.at/dp
[2] Ludwig Boltzmann Institute Media.Art.Research., Linz, Austria
http://media.lbg.ac.at/en
[3] Johannes Kepler University of Linz, Austria
http://www.faw.uni-linz.ac.at/faw/fawInternet/en/_/index.html

Abstract. Over the last years, digital preservation has become a particularly active research area. While several initiatives are dealing with the preservation of standard document formats, the challenges of preserving multimedia objects and pieces of electronic art are still to be tackled. This paper presents the findings of a pilot project for preserving born-digital interactive multimedia art. We describe the specific challenges the collection poses to digital preservation and the results of a case study identifying requirements on the preservation of interactive artworks.

1 Introduction

The last decades have made digital objects the primary medium to create, shape, and exchange information. An increasing part of our cultural and scientific heritage is being created and maintained in digital form; digital content is at the heart of today's economy, and its ubiquity is increasingly shaping private lives.

Furthermore, the field of arts has more and more adopted the new media. Digital photography has long exceeded the analog pendant in popularity, and more and more artists focus on digital media in their work. As opposed to initial expectations, however, digital content has a short live span.

The ever-growing complexity and heterogeneity of digital file formats together with rapid changes in underlying technologies have posed extreme challenges to the longevity of information. So far, digital objects are inherently ephemeral. Memory institutions such as national libraries and archives were amongst the first to approach the problem of ensuring long-term access to digital objects when the original software or hardware to interpret them correctly becomes unavailable [22].

Traditional memory institutions primarily own collections of digitised material and large homogeneous collections of electronic documents in widely adopted and well-understood file formats. In contrast, collections of born-digital art pose a whole new problem field. Electronic art is extremely complex to preserve due

D.H.-L. Goh et al. (Eds.): ICADL 2007, LNCS 4822, pp. 257–266, 2007.

to the heterogeneity of employed media as well as the complexity of file formats. Moreover, artists cannot be obliged to conform to submission policies that prescribe formats and standards, yielding to highly heterogeneous collections of proprietary file formats.

This paper presents findings of a pilot project dealing with the preservation of born-digital multimedia art. Specifically, we focus on a collection of interactive artworks held by the Ars Electronica[1]. We describe the context of the collection and the specific challenges that interactive multimedia art poses to digital preservation. We then focus on the requirements that potential preservation strategies have to fulfil in order to be fit for purpose in the given setting.

The remainder of this paper is organised as follows. Section 2 provides an overview of previous work in the area of digital preservation and introduces the challenges born-digital artworks pose to digital preservation, while Section 3 introduces the PLANETS approach to preservation planning. We then describe the case study and the results we obtained in Section 4. In Section 5 we draw conclusions and give a short outlook on future work.

2   Related Work

Digital preservation is a pressing matter – large parts of our cultural, scientific, and artistic heritage are exposed to the risks of obsolescence. The rising awareness of the urgency to deal with the obsolescence that digital material is facing has led to a number of research initiatives over the last decade. Research has mainly focussed on two predominant strategies – migration[21,14] and emulation[17,23]. Migration, the conversion of a digital object to another representation, is the most widely applied solution for standard object types such as electronic documents or images. The critical problem generally is how to ensure consistency and authenticity and preserve all the essential features and the conceptual characteristics of the original object whilst transforming its logical representation. Lawrence et. al. presented different kinds of risks for a migration project [13].

In contrast to migration, emulation operates on environments for objects rather than the objects themselves. Emulation aims at mimicking a certain environment that a digital object needs, e.g. a certain processor or a certain operating system. Rothenberg [17] envisions a framework of an ideal preservation surrounding for emulation. Recently, Van der Hoeven presented an emerging approach to emulation called *Modular emulation* in [23].

The challenge of preserving born-digital multimedia art, which is inherently interactive, virtual, and temporary, has been an actively discussed topic in the last years. In 2004, the ERPANET project organised a workshop [7] on archiving and preservation of born-digital art. Besser reports on the longevity of electronic art in [4]. The Variable Media Network, a joint effort founded by institutions such as the Guggenheim Museum New York and the Berkeley Art Museum/Pacific Film Archives, investigated properties of an artwork that are

---

[1] http://www.aec.at

subject to change and develops tools, methods and standards to implement new preservation strategies for unstable and mixed media[6]. The most prominent results of this initiative is the Variable Media Questionnaire [15], developed at the Guggenheim Musem New York, which assists artists and curators in understanding which properties of an artwork are subject to change and how these should be handled in the best possible way. Guggenheim is also participating in the project "Archiving the Avantgarde"[1] together with the Pacific Film Archive, which develops ways to catalog and preserve collections of variable media art.

In the field of computer science, the most notable work has been carried out in the PANIC project [11,10] which developed preservation strategies for multimedia objects [9]. However, they focus on dealing with composite objects that contain different content; interaction is not covered.

Preserving the inherent complexities of interactive multimedia is a very difficult task, particularly because formats used in multimedia art are ephemeral and unstable. It also poses a conflict between the transformation necessary to keep the work accessible, and desired authenticity of each piece of art [5]. Jones [12] reports on a case study which used hardware emulation to recreate one of the first interactive video artworks. Emulation is often able to retain the original appearance of the digital object, and its proponents claim it is the ideal preservation solution [17]. But there are also a lot of critical voices. The main points of criticism are its complexity and the fact that intellectual property rights might prevent the creation of emulators [3,8].

The main obstacle to the second prominent approach, migration, in this context is the diversity and complexity of obsolete file formats that are used in the field of digital art. Depocas [5] argues that efforts to preserve born-digital media art always have to be based on structured documentation and adds that often the documentation is the only thing that remains.

3   Preservation Planning

A range of tools exist today to support the variety of preservation strategies such as migration or emulation. The selection of the preservation strategy and tools is often the most difficult part in digital preservation endeavours; technical as well as process and financial aspects of a preservation strategy form the basis for the decision on which preservation strategy to adopt. Both object characteristics and different preservation requirements across institutions and settings influence the selection of the most appropriate strategy for digital preservation and make it a crucial decision process. This process is called Preservation Planning.

The preservation planning approach of PLANETS[2] allows the assessment of all kinds of preservation actions against individual requirements and the selection of the most suitable solution. It enforces the explicit definition of preservation requirements and supports the appropriate documentation and evaluation by assisting in the process of running preservation experiments. The methodology is based on work described in [16,20] and was recently revised and described in

---

[2] http://www.planets-project.eu

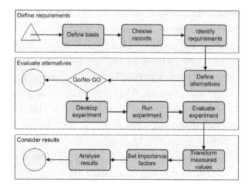

Fig. 1. Overview of the PLANETS preservation planning workflow

detail in [19]. It combines utility analysis [24] with a structured workflow and a controlled environment to enable the objective comparison and evaluation of different preservation strategies. The workflow consists of three phases, which are depicted in Figure 1 and described in more detail below.

1. **Define requirements** describes the scenario, the collection considered as well as institutional policies and obligations. Then the requirements and goals for a preservation solution in a given application domain are defined. In the so-called objective tree, high-level goals and detailed requirements are collected and organised in a tree structure.

    While the resulting trees usually differ through changing preservation settings, some general principles can be observed. At the top level, the objectives can usually be organised into four main categories:
    - *File characteristics* describe the visual and contextual experience a user has by dealing with a digital record. Subdivisions may be "Content", "Context", "Structure", "Appearance", and "Behaviour" [18], with lowest level objectives being e.g. color depth, image resolution, forms of interactivity, macro support, or embedded metadata.
    - *Record characteristics* describe the technical foundations of a digital record, the context, interrelationships and metadata.
    - *Process characteristics* describe the preservation process. These include usability, complexity or scalability.
    - *Costs* have a significant influence on the choice of a preservation solution.
    The objective tree is usually created in a workshop setting with experts from different domains contributing to the requirements gathering process. The tree documents the individual preservation requirements of an institution for a given partially homogeneous collection of objects. Strodl et. al. [19] report on a series of case studies and describe objective trees created in these.

    An essential step is the assignment of measurable effects to the objectives. Wherever possible, these effects should be objectively measurable (e.g. € per year, frames per second). In some cases, such as degrees of openness and stability or support of a standard, (semi-) subjective scales will need to be employed.

2. **Evaluate alternatives** identifies and evaluates potential alternatives. The alternatives' characteristics and technical details are specified; then the resources for the experiments are selected and the required tools are set up, and a set of experiments is performed. Based on the requirements defined in the beginning, the results of the experiments are evaluated to determine the degree to which the requirements defined in the objective tree were met.

3. **Consider results** aggregates the results of the experiments to make them comparable. The measurements taken in the experiments might all have different scales. In order to make these comparable, they are transformed to a uniform scale using transformation tables. Then the importance factors are set, as not all of the objectives of the tree are equally important, and the alternatives are ranked. The stability of the final ranking is analysed with respect to minor changes in the weighting and performance of the individual objectives using Sensitivity Analysis. After this a clear and well argued accountable, recommendation for one of the alternatives can be made.

PLANETS is developing a decision support tool for preservation planning, which supports the process and integrates distributed software services for preservation action and object characterisation. In the future this system will also allow for semi-automatic evaluation of alternative preservation strategies.

While the digital preservation community is increasingly developing solid methods of dealing with common digital objects such as electronic documents and images, dealing with complex interactive content is still an open issue. The next section will outline the challenges of interactive multimedia art in the context of digital preservation, referring to a real-word case. We will then apply the preservation planning methodology outlined above and analyse the requirements for preserving interactive multimedia art.

4   A Real-World Case: Ars Electronica

More and more modern museums hold pieces of born-digital art. The Ars Electronica in Linz, Austria, is one of the most prominent institutions in the field of electronic art. It owns one of the world's most extensive archives of digital media art collected over the last 25 years.

The collection of the Ars Electronica contains more than 30.000 works and video documentation and is growing at a rate of over 3000 pieces per year. Of these works, about 6200 pieces have been deposited as CDs containing multimedia and interactive art in different formats like long-obsolete presentation file formats with interactive visuals, audio and video content. The CDs are divided into the categories Digital Music (4000), Computer Animation (1000), and Interactive Art (1200). These collections pose extreme problems to digital preservation due to their specific and complex characteristics. The main issues arising in this context are the following.

1. The collections are highly heterogenous, there is no common file format. Instead, digital art ranges from standard image and video files to specifically designed, proprietary software pieces which are sometimes highly dependent on a specific environment. Some pieces of art even deal with the issues of digital deterioration, damaged content and the like.
2. Often artists object to the idea of preserving their artwork, because they feel its value lies in the instantaneous situation, it should be volatile or they want to retain control about the original object.
3. Many of the artworks are integrated applications, for which the underlying file format can not easily be identified. For example, some interactive artworks combine multimedia content with viewer applications specifically designed to render the contained objects.

Contrary to traditional digital preservation endeavours, these specific issues also bring about the need for particular actions. For specific objects, it might be necessary to involve creators in the definition of requirements to ensure that their intentions are communicated through a probably transformed representation of their artworks. Other pieces might only be preservable by developing custom software particularly for this purpose.

In a joint effort, the Vienna University of Technology and the Ludwig Boltzmann Institute Media.Art.Research are investigating possible approaches to deal with the preservation of born-digital interactive art. The aim is to not only preserve these pieces of art over the long term, but also make them accessible in a satisfying form on the web. In a pilot study, we concentrated on a sub-collection of the large collection the Ars Electronica owns. This collection contains about 90 interactive presentations in the formats Asymetrix Compel, Asymetrix Toolbook, and Macromedia Director. The companies that created these formats have ceased to exist; only the latter format is supported by current software[2].

The aim of the pilot project is to find means of preserving the original intention of the artists as well as the user experience and thus truly preserve the original artwork. To achieve this, we applied the preservation planning approach outlined in Section 3 and analysed the requirements on preserving interactive

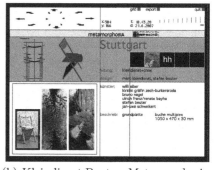

(a) Kolb: Cycosmos   (b) Kleindienst,Beuter: Metamorphosia

Fig. 2. Sample interactive artworks from 1997

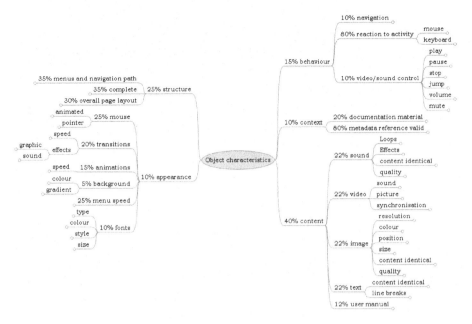

Fig. 3. High-level view of the essential object characteristics showing weights

art. In a series of workshops with curators, art historians, computer scientists, preservation specialists, and management, the first phase of the planning process was completed. Figure 2 shows screenshots of two exemplary sample records that were chosen as part of this process. These sample objects are used for identifying requirements and evaluating the performance of different preservation strategies.

Figure 3 provides an overview of the essential object characteristics that were identified, and also documents the weights that have been assigned to the upper levels of the tree hierarchy. Naturally, the primary focus lies on the content of the artworks, such as the contained text, images, and sounds. The second most important criterion is the completeness of the navigational structure that constitutes each interactive artwork. A purely linear recording of an interactive piece of art will most probably not prevail the true spirit and the spectator's experience. This interactivity is also by far the most important criterion when it comes to behavioural characteristics.

Figure 4 details some aspects of the object characteristics as they are displayed in the software and provides some measurement units to illustrate the quantitative nature of the evaluation process. A particularly relevant aspect is the measurement of interaction features and the degree they have been preserved by a preservation action. In principle, interactive presentations exhibit two facets: They have a graph-like navigation structure, and they allow the user to navigate along the paths.

Different strategies of preserving an interactive presentation will show different strengths and weaknesses in preserving these characteristics. For example, migrating an interactive presentation to a collection of images and videos and

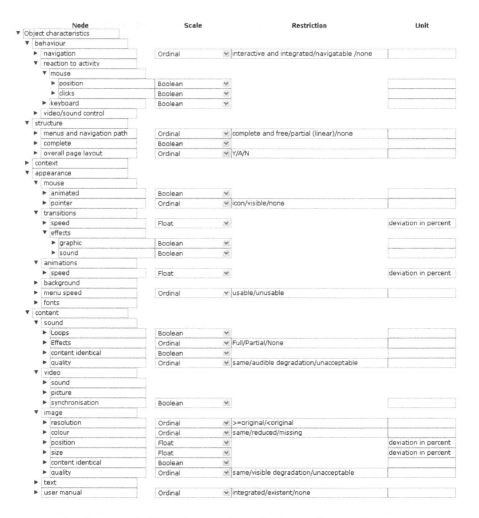

Fig. 4. Selected object characteristics for interactive multimedia art

documenting the navigational structure externally will preserve the complete structure and the possibility to navigate along the paths, but miss the interactivity. The structural aspect of this is measured in the criterion *menus and navigation path*, while the interaction is covered by the behavioural criterion 'navigation', which can take one of the values *interactive and integrated*, *navigatable*, or *none*.

Measuring the degree to which the content of pages such as images and sound is preserved is more straightforward. The lower part of Figure 4 provides some examples on the properties that describe images, sound, and video. While currently properties such as image quality have to be judged manually, PLANETS will integrate object characterisation services that are able to evaluate and quantify many of the described characteristics automatically.

5   Summary and Outlook

This paper presented a case study on preserving interactive multimedia art. We focussed on the specific problems that arise in the context of interactive digital art. In the main part of this paper, we concentrated on a collection of interactive artworks in mostly obsolete presentation file formats. We described the essential object characteristics that have been identified by the responsible curators and outlined how they can be quantified.

The next step in the pilot project is the implementation of potential preservation strategies on the sample records and the evaluation of outcomes. The results of this evaluation will then lead to a well-founded recommendation for the preservation of the complete collection of interactive art that we discussed.

Acknowledgements

Part of this work was supported by the European Union in the 6th Framework Program, IST, through the Digital Preservation Cluster (WP6) of the DELOS Network of Excellence on Digital Libraries, contract 507618, and the PLANETS project, contract 033789.

References

1. Archiving the Avant Garde: Documenting and Preserving Digital / Variable Media Art (February 2007), http://www.bampfa.berkeley.edu/about_bampfa/avantgarde.html
2. Adobe.: Macromedia Director and Adobe Shockwave Player: FAQ (June 2007), http://www.adobe.com/products/director/special/crossproduct/faq.html
3. Bearman, D.: Reality and chimeras in the preservation of electronic records. D-Lib Magazine 5(4) (April 1999)
4. Besser, H.: Longevity of electronic art. submitted to International Cultural Heritage Informatics Meeting (February 2001)
5. Depocas, A.: Digital preservation: Recording the Recoding. The Documentary Strategy. In: Ars Electronica 2001: Takeover. Who's doing the Art of Tomorrow? (2001), http://www.aec.at/festival2001/texte/depocas_e.html
6. Depocas, A., Ippolito, J., Jones, C.(eds.): Permanence Through Change: The Variable Media Approach. Guggenheim Museum Publications (February 2004)
7. ERPANET.: The archiving and preservation of born-digital art workshop. Briefing Paper for the ERPANET workshop on Preservation of Digital Art (2004)
8. Granger, S.: Emulation as a digital preservation strategy. D-Lib Magazine 6(10) (October 2000)
9. Hunter, J., Choudhury, S.: Implementing preservation strategies for complex multimedia objects. In: Koch, T., Sølvberg, I.T. (eds.) ECDL 2003. LNCS, vol. 2769, pp. 473–486. Springer, Heidelberg (2003)
10. Hunter, J., Choudhury, S.: A semi-automated digital preservation system based on semantic web services. In: JCDL 2004. Proceedings of the Joint Conference on Digital Libraries, Tucson, Arizona, pp. 269–278. ACM Press, New York (2004)
11. Hunter, J., Choudhury, S.: PANIC - an integrated approach to the preservation of complex digital objects using semantic web services. International Journal on Digital Libraries: Special Issue on Complex Digital Objects 6(2), 174–183 (2006)

12. Jones, C.: Seeing double: Emulation in theory and practice. the erl king case study. In: Electronic Media Group, Annual Meeting of the American Institute for Conservation of Historic and Artistic Works. Variable Media Network, Solomon R. Guggenheim Museum (2004)

13. Lawrence, G.W., Kehoe, W.R., Rieger, O.Y., Walters, W.H., Kenney, A.R.: Risk management of digital information: A file format investigation. CLIR Report 93, Council on Library and Information Resources (June 2000)

14. Mellor, P., Wheatley, P., Sergeant, D.M.: Migration on request, a practical technique for preservation. In: Agosti, M., Thanos, C. (eds.) ECDL 2002. LNCS, vol. 2458, pp. 516–526. Springer, Heidelberg (2002)

15. Variable Media Network.: Variable media questionnaire. (June 2007), http://www.variablemedia.net/e/welcome.html

16. Rauch, C., Rauber, A.: Preserving digital media: Towards a preservation solution evaluation metric. In: Chen, Z., Chen, H., Miao, Q., Fu, Y., Fox, E., Lim, E.-p. (eds.) ICADL 2004. LNCS, vol. 3334, pp. 203–212. Springer, Heidelberg (2004)

17. Rothenberg, J.: Avoiding Technological Quicksand: Finding a Viable Technical Foundation for Digital Preservation. Council on Library and Information Resources (January 1999), http://www.clir.org/pubs/reports/rothenberg/contents.html

18. Rothenberg, J., Bikson, T.: Carrying authentic, understandable and usable digital records through time. Technical report, Report to the Dutch National Archives and Ministry of the Interior (1999)

19. Strodl, S., Becker, C., Neumayer, R., Rauber, A.: How to choose a digital preservation strategy: Evaluating a preservation planning procedure. In: Proceedings of the 7th ACM IEEE Joint Conference on Digital Libraries (JCDL 2007), pp. 29–38 (June 2007)

20. Strodl, S., Rauber, A., Rauch, C., Hofman, H., Debole, F., Amato, G.: The DELOS testbed for choosing a digital preservation strategy. In: Sugimoto, S., Hunter, J., Rauber, A., Morishima, A. (eds.) ICADL 2006. LNCS, vol. 4312, pp. 323–332. Springer, Heidelberg (2006)

21. Digital Preservation Testbed.: Migration: Context and current status. White paper, National Archives and Ministry of the Interior and Kingdom Relations (2001)

22. UNESCO charter on the preservation of digital heritage. Adopted at the 32nd session of the General Conference of UNESCO (October 17, 2003), http://portal.unesco.org/ci/en/files/13367/10700115911Charter_en.pdf/Charter_en.pdf

23. van der Hoeven, J., van Wijngaarden, H.: Modular emulation as a long-term preservation strategy for digital objects. In: 5th International Web Archiving Workshop (IWAW05) (2005)

24. Weirich, P., Skyrms, B., Adams, E.W., Binmore, K., Butterfield, J., Diaconis, P., Harper, W.L.: Decision Space: Multidimensional Utility Analysis. Cambridge University Press, Cambridge (2001)

Identification of FRBR Works Within Bibliographic Databases: An Experiment with UNIMARC and Duplicate Detection Techniques

Nuno Freire, José Borbinha, and Pável Calado

INESC-ID, Rua Alves Redol 9, Apartado 13069
1000-029 Lisboa, Portugal
nuno.freire@ist.utl.pt, jlb@ist.utl.pt,
pavel.calado@tagus.ist.utl.pt

Abstract. Many experiments and studies have been conducted on the application of FRBR as an implementation model for bibliographic databases, in order to improve the services of resource discovery and transmit better perception of the information spaces represented in catalogues. One of these applications is the attempt to identify the FRBR work instances shared by several bibliographic records. In our work we evaluate the applicability to this problem of techniques based on string similarity, used in duplicate detection procedures mainly by the database research community. We describe the particularities of the application of these techniques to bibliographic data, and empirically compare the results obtained with these techniques to those obtained by current techniques, which are based on exact matching. Experiments performed on the Portuguese national union catalogue show a significant improvement over currently used approaches.

Keywords: Functional Requirements for Bibliographic Records, FRBR, Bibliographic databases, string similarity, duplicate detection.

1 Introduction

FRBR - Functional Requirements for Bibliographic Records [1], is a conceptual model developed by the IFLA - International Federation of Library Associations and Institutions, proposing how the bibliographic information should be represented. A purpose of the FRBR model is to bring the bibliographic world closer to the actual multimedia, digital and more heterogeneous world. Information systems supporting information schemas compatible with the FRBR model are supposed to assure e richer representation of the information, and therefore, to provide better services of resource discovery and transmit better perception of the information spaces represented in catalogues.

FRBR is widely recognized as a valuable model, but with an important constraining: as it represents a richer semantic model, it is not easy to "upgrade" to its level the existing bibliographic catalogues (it is not trivial to extract new semantics concepts from information structures that were not designed to hold them...). Considering the millions of MARC related records existing nowadays all over the works, created at a

D.H.-L. Goh et al. (Eds.): ICADL 2007, LNCS 4822, pp. 267–276, 2007.

very high cost, made the problem of converting traditional catalogues to FRBR structures a very relevant one (while on the same time very challenging).

This is the main motivation for the work reported in this paper, where we try to contribute to the solution of the problem of building FRBR structures from inherited UNIMARC bibliographic records by applying techniques of detection of duplicate information.

The detection of duplicate records is an area of great interest in the traditional database research community. It is a problem faced many times when implementing a data warehouses or any other system that aggregates data from heterogeneous data sources. Often, the same entities in the real world have two or more representations in such databases. These duplicate records do not share a common key, they may differ in structure and in lexicon, and they may even contain errors, making their detection a very difficult task.

In fact, this problem is common to many research communities, although the term used is not always the same [1]: record linkage, record matching, merge-purge, data deduplication, instance identification, database hardening, and name matching. In our work we evaluated the applicability of some of the techniques developed in these communities to the purpose of detecting common expressions of works within bibliographic databases, according to the FRBR definition.

This paper follows with a description of the main techniques for the detection of duplicate records. After that we analyse the problem of the detection of FRBR works in groups of UNIMARC records, and we formulate our hypothesis to address it. In the following section we describe the experiment designed for this purpose, as also the results achieved. The paper continues with a discussion of the results, a description of related work, and finished with the conclusions and references to future work.

2   Techniques for the Detection of Duplicate Records

When setting up a duplicate detection process, several issues have to be addressed. Which data will be used for comparison, how should it be coded, how fields are to be matched individually, and on what conditions the comparison results of the individual fields identify two records as duplicates. In the rest of this section we describe these issues and the general approaches widely used to solve them.

The process starts with the preparation of data for further processing. This step comprises tasks for selecting the relevant data from the data sources, and parsing and transforming it so that in conforms to a standardized data schema. Data preparation greatly reduces the structural heterogeneity of the source data, but misspellings and different conventions for recording the same information continue to result in different, multiple representations of a unique object in the database. For this reason, records have to be compared for their similarity, by measuring similarities of the fields, one by one. These similarity results can then be further processed to decide if the records match.

One of the main obstacles to the detection of duplicates is the typographical variations of string data. Therefore string comparison techniques have bean a very active topic in research. Among the many techniques for matching string fields, three main types of similarity metrics can be considered: character based, token based and phonetic. How well each of these techniques works on evaluating the similarity between strings depends on the characteristics of the data they are applied to.

Character based similarity metrics are most suitable to handle typographical errors. They measure the amount of edit operations (insert, delete, replace) that are necessary to transform one string into the other being compared. These techniques don't work well in cases where typographical conventions lead to rearrangement of words (people's names, for example, may be entered by their surname or by their first name). Token-based metrics, on the other hand, try to compensate for this problem by matching tokens in the strings (typically words) independently of their location within the strings. These techniques usually make use of token-weighting schemes, such as the "term frequency–inverse document frequency" (TFIDF) [2]. Finally, phonetic techniques address those cases where strings may be phonetically similar even if they are not similar at character or token level. These are widely used to match fields containing person surnames.

Several algorithms exist for all these kinds of metrics and the decision of which field comparison techniques to use is not an easy one. Analysis of the few existing studies seams to indicate that no single metric is suitable for all data sets [1, 4]. In many cases, using flexible metrics that can accommodate multiple similarity comparisons may lead to the best results.

The final decision to match two records is made by reasoning on the similarity scores obtained by comparing the individual fields. Depending on the complexity of the record structure and on the possibility of creating a training set of data, two types of techniques may be implemented: declarative techniques or machine learning techniques. In general, better results can be obtained by using machine learning techniques [1, 6].

3   Detection of FRBR Works in UNIMARC Records

Previous experiments on the identification of works within bibliographic databases have not fully explored the applicability of duplicate detection techniques. The methods deployed usually consist of algorithms that have a strong emphasis on the data preparation phase, to create keys that identify the work from data in the bibliographic record. These keys are then used to match the records using declarative techniques, with decision tables or sets of rules [7, 8]. However, the fields that make up the keys are compared using exact comparison without, resorting to similarity metrics. Although some of the heterogeneity of data is handled quite well in the data preparation phase, a simple exact comparison may be insufficient. This causes record matches to be missed.

The two most important fields for matching works in UNIMARC records are the title (including subtitles) and the authors. Both fields are prone to typing errors, and the use of abbreviations is frequent, which is enough to disable exact matching of records (two examples are shown in Table 1). Additionally, in PORBASE, we observed that subtitles may be recorded in different forms: some are omitted, others are recorded separately from the title, others together with the title, and, sometimes, in records with more than one subtitle, they may be recorded in different orders.

The authors' field may not be as problematic as titles, since authority control practices used in libraries share the same cataloguing rules and procedures. However, when trying to identify works across different data sets from unrelated libraries (especially if these are from different countries) exact matches will be much harder to find for authors, mainly because of different spelling of names across different languages.

From the previous analysis, we formulate the hypotheses that the use of similarity metrics applied to titles and authors could improve the identification of works across existing bibliographic records, by increasing the number of relevant record matches without a significant increase in the number of false matches. To test our hypotheses we designed a set of experiments to compare the use of exact matching techniques to the use of matching based on similarity metrics.

Table 1. Example of two works described in bibliographic records extracted from PORBASE. We can observe the occurrence of typing errors in both title and author fields and different structures for the subtitle field.

Rec.	Title	Subtitles	Authors
A	Grammar's great!	exercícios com soluções, 5°, 6°, 7° anos	Sottomayor, Maria Manuela
B	Grammar's great!	5°, 6° e 7° anos	Sotomayor, Maria Manuela
		exercícios com soluções	
C	Anti-gadouel	français, niveau 6-8, 12ème année	Gueidão, Ana
			Crespo, Idalina
D	Anti gadoue	francês	Gueidão, Ana
		niveau 6-8	Crespo, Idalina
		12ème année	

4   Related Work

Few works have explored the computer aided identification of FRBR work entities in bibliographic databases. The major reference works in this area are the several experiments that have been conducted by OCLC [7] and that have been applied by several other projects.

Of particular interest is the approach from the Melvyl Recommender Project [8] that tries to match records, also when titles and authors don't match exactly, by using other data in the bibliographic records, such as dates, identifiers, and by taking in consideration partial matches in author names and subtitles. However we don't know of any other experiment or study that tried to use similarity metrics in this specific task.

Our work has some overlap with the work carried out in citation indexing systems, that autonomously index the citations found in research papers [12, 13, 14]. These works also use similarity metrics to match author names and titles. However, similarity metrics are very data sensitive, and the FRBR work instance identification has patterns of data heterogeneity different from citation matching.

5   The Experiment

The experiments were carried out in two data sets of UNIMARC bibliographic records: (1) PORBASE, the full Portuguese National Bibliographic Database; and (2) the record set for Porto Editora, a major Portuguese book publisher, which is a subset of records also taken from PORBASE.

The data set from Porto Editora consists of 6.492 records. This publisher is focused on educational works (school manuals, classic works, dictionaries, etc.) which typically have multiple editions, making this data set very appropriate to validate similarity techniques, as we can measure precision and recall. To measure them, we used a similarity metric with a low similarity threshold and manually classified the record matches. This classification was based on a summarized version of the bibliographic records containing only the titles, authors, ISBNs, editions and publication dates.

The PORBASE data set, containing 1.360.686 records, was our real target. It is the largest bibliographic database in Portugal, with collections from nearly 200 different libraries. Our assumption was that once we had our techniques tuned with the Porto Editora set, we could accept the results with a higher level of confidence in PORBASE.

The data preparation phase followed a similar process to the one defined in the OCLC FRBR work-set algorithm [7], now adapted to the UNIMARC format.

5.1 Similarity of the Titles

Some data heterogeneity was still evident after data preparation. Problems such as misspelling and typing errors, lack of spaces between words, abbreviations, various ways of recording subtitles, and missing words would still occur.

A survey on the comparison studies that where conducted on data with similar characteristics indicated that a token based metric should be used [9]. Preliminary experiments were performed to determine the most appropriate similarity metric for titles. A combination of the Jaro-Winkler metric [10] with a TFIDF weighting scheme[1] gave the best results on our preliminary tests. In fact, the results obtained with any single metric were not very satisfactory, which lead to the adoption of combined metrics. Although several metrics resulted in high recall in the matching records, their precision was lower than expected. This came from the fact that small variations in the title of different works from the same authors are very common.

We created a similarity metric that adjusts the similarity score given by Jaro-Winkler-TFIDF to better distinguish between similar titles that refer to the same work from those that do not. We will refer to this metric as BRT metric for the remainder of this paper.

When processing titles, the main problems found were the following. Similarity is significantly lowered when a difference in numeration existed between titles (number, year, roman numerals, and single letter like 'A' 'B' 'C'). Examples of these cases where found in school manuals (as, for example, a mathematics manual of different levels by the same authors: "Mathematics 7" and "Mathematics 8").

Title length greatly influences the similarity values. If two long titles differ in just one or two words, the similarity score will still be high. Therefore, these non-matching words should be compared one by one and, if a word was significantly

---

[1] An implementation of Jaro-Winkler with TFIDF from the SecondString project (http://secondstring.sourceforge.net/) was used for this purpose.

different from the other (as given by the Smith-Waterman metric), the similarity score was lowered.

Finally, we observed that sometimes some words were omitted in the titles. However, these were typically stop words, i.e., very frequent words that carry very little information, thus the penalty to the similarity score was only marginal.

5.2 Similarity of Author Names

The similarity between author names was measured using the Jaro-Winkler metric. In fact, the only discrepancies found where caused by typing errors or missing or abbreviated middle names, which are easily matched by Jaro-Winkler. For this reason, no further algorithms where tested.

5.3 Final Matching

The final decision to match the records was done by a declarative rule that used, as input, the similarity scores for title and author names. The similarity metrics gave results between 0 (no similarity exists) and 1 (identical). The matching rule defined a minimum threshold of 0,7 for authors, of 0,65 for titles, and of 0,6 for the product of both the similarity scores. The choice of these similarity thresholds was based on the results obtained in our experiments at different thresholds (shown in section 4.5).

Because comparing the similarity of fields is a time consuming process, it is imperative to avoid comparing every record to every other record in the database. To solve this problem, we took a clustering approach. Clusters of the titles were created based on cosine similarity of the titles, and only the records within the same cluster were compared for similarity. This technique improved the performance because the creation of the clusters is much faster than measuring the similarity for all records, reducing the number of record similarity comparisons to a great extent.

5.4 Exact Matching Process

A second independent process to detect duplicate works was implemented without resorting to similarity metrics. The purpose was to compare the results from exact matching to those of similarity matching.

The titles and authors were stored in a relational database after the data preparation phase, which was the same as for both processes. Matching of titles and authors was done using SQL queries and fields would only match if they had exactly the same data.

5.5 The Results

When using the similarity metrics, we obtained the following results. The Jaro-Winkler method identified all cases of similar author names in the test data sample. Matching of authors by similarity accounted for less than 1% of the total matched authors. Further analysis or comparison with other similarity metrics was not performed on authors, since the data set size was too small to draw any meaningful conclusions.

For the titles the sample proved to be an excellent test case, with a very high number of cases that exact comparison missed and were matched by the similarity metrics.

The measured recall for the exact matching process was 63.84%. Figure 1 shows the recall and precision results of three experiments with similarity metrics on the Porto Editora data set. In all three experiments, authors were always compared using the Jaro-Winkler metric, while the similarity metric used for titles varied.

The first test was performed by comparing the titles using the Jaro-Winkler with TFIDF. This metric greatly improved recall, when compared to exact matching, yielding recall values above 98%. Precision was lower than exact matching, as was expected. To obtain gains in recall above 98%, precision started to deteriorate to unsatisfactory levels.

The analysis of the precision failures of the previous method led to the development of the BRT metric that is adapted to the comparison of titles in bibliographic records (as described in section 4.1). This similarity metric obtained the best results of our tests, with both precision and recall levels very close to 100%.

A third test was performed in an attempt to further improve the recall of the previous test. During the experiments, we observed that most of the missed matches where caused by a missing subtitle on one of the records. We therefore adapted the BRT metric by slightly increasing the similarity score in these cases. This change yielded little gain in recall at some similarity thresholds (data not shown) but the best result was still obtained by the BRT metric.

Figure 1 shows the results of the two metrics tested, at 5 levels of similarity thresholds for each metric. Table 2 shows the best results obtained for each method and the similarity threshold at which they were obtained.

We also tested the clustering method used to reduce the number of record similarity calculations necessary. We checked for any missed matches and to what extent it reduced the number of comparisons on the Porto Editora data set. We observed that the number of records comparisons was reduced from 21.063.295 to 782.853 and no records matches where missed.

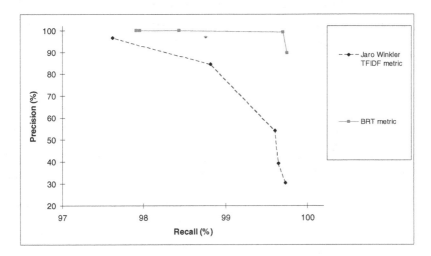

Fig. 1. Recall and precision results of the two methods used on titles, at the most relevant similarity thresholds, in the Porto Editora data set

Table 2. Best recall/precision relation obtained with the three matching methods used for matching titles on the Porto Editora data set

Matching method	Recall (%)	Precision (%)	Similarity threshold
Exact matching	63,84	100,00	-
Jaro Winkler TFIDF metric	98,82	84,63	0,90
BRT metric	98,43	99,85	0,65

A second experiment, with the total number of records from PORBASE, was then used to test our similarity metric in a more realistic environment. In this case, we applied the exact matching and the BRT metric. For both cases, we measured the number of record matches and the corresponding number of FRBR works detected. The results are shown in Table 3.

Exact matching matched a total of 290.955 records, with 104.648 distinct works with an average of 2,78 records/work. Similarity matched a total of 355.840 records, forming 126.458 distinct works. It resulted in an increase of 22,3% in the number records matched and an increase in the number of groups of 20,8%, with an average of 2,82 records per work. The distribution of works by number of matched records on both methods can be seen in Table 4.

Table 3. Number of records matched and total sets of records created with exact matching and with the defined similarity metric in the PORBASE data set

	Exact matching	Similarity matching
Records matched	290.955	354.773
Works	104.648	126.457
Records per work (standard deviation)	2,78(2,86)	2,82(3.98)

Table 4. Distribution of works by number of records

Record per work	1	2	3	4	5	6	7-9	10+
Exact matching	1.069.731	73.980	16.276	6.059	2.899	1.668	2.077	1.689
Similarity matching	1.005.913	87.613	20.303	7.674	3.637	2.049	2.587	2.078

6 Discussion

Measuring the similarity between titles with a simple application of a generic similarity metric will result in good recall but low precision. On the other hand, exact matching results in good precision but low recall. The low recall obtained using exact matching on the Porto Editora data set was probably due to high number of records of school manual records that contained several subtitles, because the results obtained from the experiment with PORBASE revealed a smaller difference in the number of record matches between the exact and similarity methods.

Due to the lack of heterogeneity in author names in the Porto Editora dataset, it was not possible to evaluate the performance of similarity techniques for matching author names, and it was not possible to try to find a suitable data set due to lack of resources to manually check the record matches. However we don't believe that this leads to the conclusion that similarity analysis of authors is of little use in the identification of work entities. It may still be relevant for records from unrelated sources. An example of such case is the LEAF project (Linking and Exploring Authority Files), which attempted to link authority records from libraries and archives from various European countries. This project, however, has only used exact matching for comparing person names [11].

For the above reasons, our experiment had a more focused approach on title similarity. The close observation of the mismatches that caused low recall in exact comparison and low precision in similarity metrics lead us to the development of a similarity metric specific for titles. The results obtained for both recall and precision were very close to 100%, leading us to conclude that it can be used in real world applications with significant increases in the usability of the library systems, with an insignificant introduction of errors by wrong record matching.

An interesting result was the high number of matches found by both matching methods. Matches by similarity were found in 354.773 records, representing 26% of the records in PORBASE.

7 Conclusions and Future Work

Our work has shown that similarity metrics can be used in the task of identifying FRBR works within bibliographic databases with a low error margin, while managing to identify most matches. When compared with exact matching the number of matches increases by a significant proportion. This proportion is very likely to be higher when trying to identify the works in more heterogenic environments, such as within libraries from different countries or in organizations of different types, archives and entertainment (theatre, cinema, etc.). We plan to further test our method in such environments. Likely points for improvement are in the matching of author names and on adding machine learning techniques to fine tune the final reasoning that matches the records.

The same techniques used in our work are likely to have applicability in other tasks related to FRBRization of bibliographic databases. These tasks include, for instance, identifying different expressions of the same work. It can also complement the work in [15] by identifying the duplicate entity instances extracted from the bibliographic records individually.

Our experience with PORBASE will also evolve to be integrated in a new prototype of an FRBR aware OPAC that is now under development.

References

1. IFLA Study Group on the Functional Requirements for Bibliographic Records: Functional requirements for bibliographic records: final report. München: K.G. Saur, UBCIM publications, new series, vol. 19 (1998), www.ifla.org/VII/s13/frbr/frbr.pdf ISBN 3-598-11382-X
2. Salton, G., Buckley, C.: Term-weighting approaches in automatic text retrieval. Information Processing & Management 24(5), 513–523 (1988)
3. Elmagarmid, A.K., Ipeirotis, P.G., Verykios, V.S.: Duplicate Record Detection: A Survey. IEEE Transactions on knowledge and data engineering 19(1), 1–16 (2007)

4. Bilenko, M., Mooney, R.J., Cohen, W.W., Ravikumar, P., Fienberg, S.E.: Adaptive name matching in information integration. IEEE Intelligent Systems 18(5), 16–23 (2003)
5. Zhao, M.: Semantic matching across heterogeneous data sources. Communications of the ACM 50(1), 45–50 (2007)
6. Zhao, H., Ram, S.: Entity identification for heterogeneous database integration: A multiple classifier system approach and empirical evaluation. Information Systems 30(2), 119–132 (2005)
7. Hickey, T.B., O'Neill, E.T., Toves, J.: Experiments with the IFLA Functional Requirements for Bibliographic Records (FRBR). D-Lib Magazine 8, 9 (2002), http://www.dlib.org/dlib/september02/hickey/09hickey.html
8. California Digital Library.: The Melvyl Recommender Project. Full Text Extension. Supplementary Report (2006), http://www.cdlib.org/inside/projects/melvyl_recommender/report_docs/mellon_extension.pdf
9. Cohen, W.W., Ravikumar, P., Fienberg, S.E.: A Comparison of String Distance Metrics for Name-Matching Tasks. American Association for Artificial Intelligence (2003), http://www.isi.edu/info-agents/workshops/ijcai03/papers/Cohen-p.pdf
10. Jaro, M.A.: Advances in record linking methodology as applied to the 1985 census of Tampa Florida. Journal of the American Statistical Society 64, 1183–1210 (1989)
11. Kaiser, M., Lieder, H.J., Majcen, K., Vallant, H.: New Ways of Sharing and Using Authority Information. D-Lib Magazine 9, 11 (2003), http://www.dlib.org/dlib/november03/lieder/11lieder.html
12. Lawrence, S., Giles, C.L., Bollacker, K.D.: Autonomous Citation Matching. In: Proceedings of the Third International Conference on Autonomous Agents, ACM press, New York (1999)
13. Pasula, H., Marthi, B., Milch, B., Russell, S., Shpitser, I.: Identity Uncertainty and Citation Matching. In: Advances in Neural Information Processing (2002), http://people.csail.mit.edu/milch/papers/nipsnewer.pdf
14. Lee, D., On, B.W., Kang, J., Park, S.: Effective and Scalable Solutions for Mixed and Split Citation Problems in Digital Libraries. In: Proceedings of the 2nd international workshop on Information quality in information systems, pp. 69–76 (2005)
15. Aalberg, T.: A process and tool for the conversion of MARC records to a normalized FRBR implementation. Digital Libraries: Achievements, Challenges and Opportunities. In: 9th International Conference on Asian Digital Libraries, pp. 283–292 (2006)

Predicting Social Annotation by Spreading Activation

Abon Chen[1], Hsin-Hsi Chen[2,*], and Polly Huang[1]

[1] Department of Electrical Engineering
[2] Department of Computer Science and Information Engineering
National Taiwan University
Taipei, Taiwan
{r94921033,hhchen}@ntu.edu.tw, phuang@cc.ee.ntu.edu.tw

Abstract. Social bookmark services like *del.icio.us* enable easy annotation for users to organize their resources. Collaborative tagging provides useful index for information retrieval. However, lack of sufficient tags for the developing documents, in particular for new arrivals, hides important documents from being retrieved at the earlier stages. This paper proposes a spreading activation approach to predict social annotation based on document contents and users' tagging records. Total 28,792 mature documents selected from *del.icio.us* are taken as answer keys. The experimental results show that this approach predicts 71.28% of a 100 users' tag set with only 5 users' tagging records, and 84.76% of a 13-month tag set with only 1-month tagging record under the precision rates of 82.43% and 89.67%, respectively.

Keywords: Collaborative Tagging, Social Annotation, Spreading Activation.

1 Introduction

Collaborative tagging is a very common application on the web. When reading interesting documents, web users are often willing to share their understanding of the documents with others by social annotation. Social bookmark tools [1] facilitate flexible information organization. Consider a social bookmark service *del.icio.us* as an example. It provides online resource organization tools for users, and works as an online collection, just like "my favorites" in a local browser on our own device. Rather than keeping "my favorite" on a local device, it also makes "my favorite" reachable when we surf on Internet. At the same time, we can input our own short description and tags for the resources collected as shown in Figure 1. Initially, tags are only for one's convenience. With the public sharing, users are able to discover and tag their own collection by browsing others.

A resource named an URL receives more and more tags when it is bookmarked by multiple users. Surprisingly, the freedom of annotation does not drive to chaos. Instead, the tag distribution shows a sense of consensus over time, and the stability reveals the collaborative behavior of users [2]. The visualization web service [3] demonstrates the tagging activity in time. The results of social annotation can be

---

* Corresponding author.

D.H.-L. Goh et al. (Eds.): ICADL 2007, LNCS 4822, pp. 277–286, 2007.

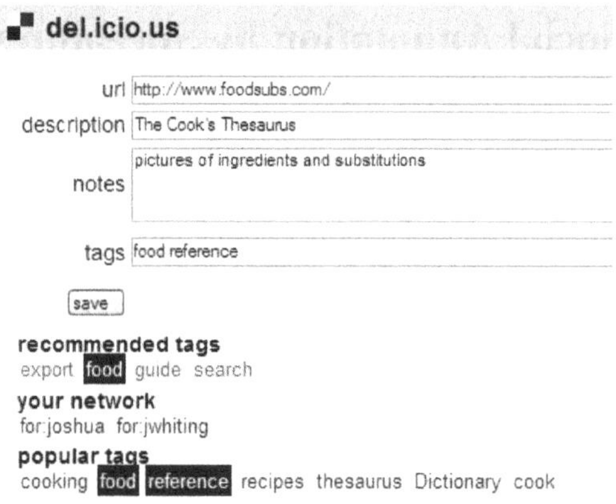

Fig. 1. Scenario of social annotation

employed to social network analysis [4], semantic web construction [5], enterprise search [6][7], and so on.

Annotation may facilitate recommendation and effective retrieval. However, not all resources can gain the benefits from that. Ill-tagged period of URLs prevents them from being retrieved. The retrieval performance for new-coming URLs degrades inevitably. Thus, how to predict a quality tagging set for a resource is an important issue. Indexing in traditional information retrieval [8] captures content of documents for effective retrieval. It focuses on document contents only. This paper will consider tagging records of users as additional cues. In information retrieval, spreading activation methods have been used to expand search vocabulary and complement the retrieved document [9]. These papers [10][11] adopt this methodology to select useful concepts from outside resources like WordNet and ConceptNet for query expansion. Here, we will employ it to model tag recommendation from users' tagging records.

This paper is organized as follows. Section 2 proposes our methods. Baseline and two alternatives of spreading activation are specified. Section 3 introduces the test material and discusses the experimental results. Section 4 concludes the remarks.

2   Tag Prediction

Given an URL denoting a document and the tagging records T_1, T_2, ..., T_h posted by h users, tag prediction aims to recommend suitable tags for this URL. Two possible tasks may be done depending on whether the tagging records are available or not.

(1)   Initial tagging: traditional indexing.
(2)   History-based tagging: spreading activation.

In the initial tag assignment, terms of larger weights are selected from URL address, document content, outgoing link address and content of the outgoing link, and are regarded as *recommended tags*. Traditional *tf-idf* scheme may be adopted to compute the weight of each term.

In the history-based tag assignments, spreading activation triggered by tagging records is employed. The concept of spreading activation can be explained by a natural phenomenon. When we drop a stone in a pond, oscillation on surface transfers energy to neighborhood, and becomes smaller and smaller in amplitude due to water resistance. In this model, we can imagine a *posted tag* by a user as a stone. Its energy propagates from the most related tags to less relevant ones. A tag has an energy level indicating its relatedness to the posted tag. In general, a user may post more than one tag in a tagging record. In this way, a tag may receive energy contributed from posted tags through different paths. Tags of higher energy are selected and recommended.

In spreading activation, tags are linked as a network. Two tags are linked when they have an association. The degree of the tag association is measured by a weight. A tag t_i may have n outgoing links to tags $t_{i1}, t_{i2}, ..., t_{in}$ with weights $w_{i1}, w_{i2}, ..., w_{in}$. Assume e_i is an energy level of tag t_i. During spreading activation, t_i will keep portion of energy, say, $\alpha \cdot e_i$ (where α is a decay factor, e.g., 0.8 in our experiments). The fraction of energy, $(1-\alpha) \cdot e_i$, is distributed to the neighbor tags based on their weights. For example, t_{ij} will receive the amount of energy, $(1-\alpha) \cdot e_i \cdot w_{ij}/(w_{i1}+w_{i2}+...+w_{in})$. A tag may have more than one incoming links, so that it may receive energy from different neighbors. We can use mutual information to compute the association of two tags.

An energy spreading matrix $[M]_{n \times n}$ shown as follows defines how much energy is distributed in a spreading activation cycle.

$m_{ij} = \alpha$ when $i=j$;
$m_{ij} = (1-\alpha) \cdot w_{ij}/(w_{i1}+w_{i2}+...+w_{in})$ when $i \neq j$.

Assume $E^i=[e^i_1, e^i_2, ..., e^i_n]$ denotes a vector of energy levels $e^i_1, e^i_2, ..., e^i_n$ for tags $t_1, t_2, ..., t_n$ after i-th user's annotation, but before spreading activation. After one cycle of spreading activation, the new energy vector $E^{i*}=[e^{i*}_1, e^{i*}_2, ..., e^{i*}_n]$ is computed by using the formula $E^{i*}= E^i \times M$.

During initial tagging, tags are assigned initial energy E^0. Assume a tagging record $T_i=(b_1, b_2, ..., b_n)$ is an n-tuple binary vector. Here b_j is set to 1 when tag t_j is in the i-th user's annotation. In this way, $E^i= E^{(i-1)'}+ T_i$, where $E^{(i-1)'}$ denotes the energy vector after $(i-1)$-th automatic tagging.

Spreading activation is triggered by the posted tags, and propagates energy to the relevant tags. Two strategies, called *SA* and *ET*, are shown as follows to control the propagation.

(1) *SA*: Propagation cannot out of a specific number of cycles, e.g., 3.
(2) *ET*: When the energy through a link is below a given threshold, e.g., 0.1, it is too low to be propagated.

After spreading activation, the final energy is stored in each tag. Tags are sorted by the energy, and top-p tags of higher energy are recommended.

Figure 2 illustrates the process of spreading activation by an example. Figure 2(a) shows part of the initial tagging. The arrival of the first user's tagging is depicted by the dark dots in Figure 2(b). The spreading activation goes on by propagating the energy with probability shown in Figure 2(c). The 1st propagation result is specified by Figure 2 (d).

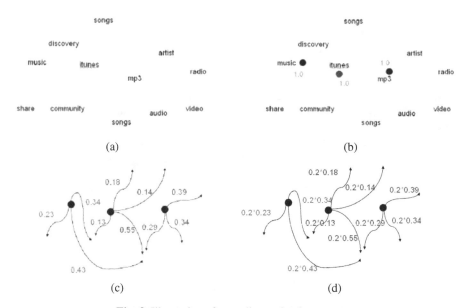

Fig. 2. Illustration of spreading activation process

Algorithms 1 and 2 show two possible implementation of the tag prediction based on these two strategies, respectively.

Algorithm 1. Spreading Activation Method with Limited Cycles
h {total tagging records}
c {maximum number of propagation cycles}
p {total recommendation tags}
n {total number of tags}
E^i {a vector of energy levels for tags}
M {an $n{\times}n$ energy spreading matrix}
$i = 1$
$E^0 = InitialTagging()$
while $i <= h$ **do**
　　$E^i = E^{(i-1)} + T_i$
　　$E^i = E^i \times M$
　　$j = 1$
　　while $j < c$ **do**
　　　　$E^i = E^i \times M$
　　　　$j = j + 1$
　　end while
　$i = i + 1$
end while
sort the n tags in the descending order of their energy in E^h
return top-p tags

Algorithm 2. Spreading Activation Method with Energy Threshold

h {total tagging records}
t {energy threshold}
n {total number of tags}
E^i {a vector of energy levels for tags}
M {an $n \times n$ energy spreading matrix}
$i = 1$
$E^0 = InitialTagging()$
while $i <= h$ **do**
    $E^i = E^{(i-1)} + T_i$
    $E^i = E^i \times M$
    $block = 0$
    repeat
    $j = 1$
    while $j <= n$ **do**
        if $E^i \times j$-th column of $M > t$
            then $e^i_j = E^i \times j$-th column of M
            else $block = 1$
        $j = j + 1$
    end while
    until $block = 1$
  $i = i + 1$
end while
sort the n tags in the descending order of their energy in E^h
return top-p tags

3 Results and Discussion

3.1 Experimental Material

We collected a sample of *del.icio.us* data by crawling its popular feed every 30 minutes during March 27 and April 19, 2007. The data set consists of 2,475,999 taggings made by 10,109 different users on 31,025 different URLs with 125,092 different tags. For evaluation, we extract the mature URLs from the gathered data set by the criteria [5][12], i.e., (1) a mature URL should have its tag distribution remaining stable, and (2) a mature URL should have enough amount of tags applied by users. In this way, we have 28,792 mature URLs for experiments.

3.2 Performance Evaluation

For each URL in the test set, we have its i-th user's tagging record as input at the corresponding suggestion stage. The mature tag set is considered as answer keys in the evaluation. The tags are said to be correctly recommended when they are also listed in the mature tag set. Conventional recall rate and precision rate are adopted to measure the coverage and the quality of recommended tag set. Recall rate is the number of tags correctly recommended divided by total mature tags. Precision rate is the number of tags correctly recommended divided by total recommended tags.

Figure 3 and Figure 4 show the recall rate and the precision rate of the proposed methods after the 1st, 2nd, 3rd, 4th, and 5th user tagging records have been read. SA_i denotes energy spreads at most i cycles. ET means spreading activation controlled by energy threshold. That is, it stops spreading when energy being propagated below a threshold. In the experiments, 0.1 is adopted.

The recall rate of the baseline system stays around 10% even tagging record grows. This is because the tag set proposed by the baseline system is just the union of the initial recommendation and the tagging records collected up to now. Comparatively, the recall rate of all the spreading activation methods improves steadily. ET strategy is better than SA strategy. With SA strategy, even though there is still enough energy for propagation, the spreading is stopped due to the restriction of maximum cycles. Spreading at most 3 cycles outperforms the other four SA methods.

Because the baseline system only conservatively includes the tags provided by the users, i.e., it performs without any expansion, its precision rate is higher than its recall rate trivially. The precision rates of ET are the best of all the methods. Both the precision rate and the recall rate increase when more user tagging records are posted.

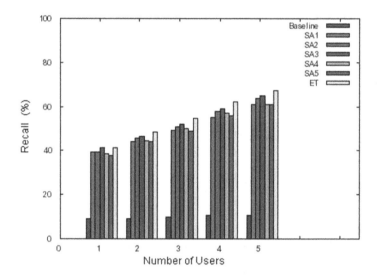

Fig. 3. Comparison of baseline and spreading activation methods from recall perspective

3.3 Coverage over Users and Time

This section discusses how many efforts our system saves from two aspects, i.e., users and time. The spreading activation with energy threshold strategy is the best, so that it is adopted in the latter experiments. In Figures 5 and 6, we assume the mature tag set is achieved when 100 users are involved in social annotation. The upper line and the lower line in Figure 5 show the recall rates of the tag prediction and the social

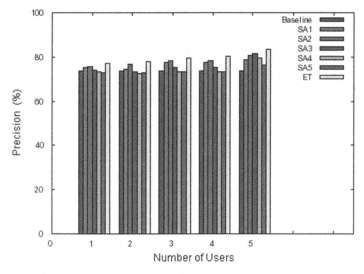

Fig. 4. Comparison of baseline and spreading activation methods from precision perspective

annotation, respectively. The automatic annotation method catches up to the mature tag set much faster than the manual annotation only. For example, the tag records of the first 5 users occupy 20.48% of the mature tag set. In contrast, the spreading activation method can achieve 71.28% of the mature tag set.

The precision rates of manual tagging in Figure 6 are 100%. The precision rates of automatic tagging are also very high, i.e., from 82.43%, 89.67%, ..., to 93.42%, under different number of user involvements. That confirms the quality of the recommended tags. From user perspective, the tag prediction method saves 75% of human cost under the recall rate of 71.28% and precision rate of 82.43%.

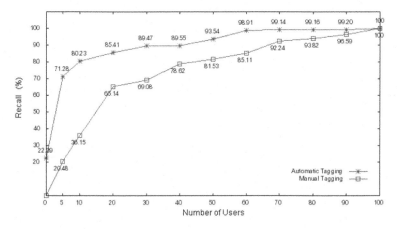

Fig. 5. Recall rate of tagging from user perspective

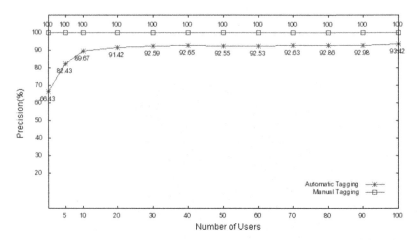

Fig. 6. Precision rate of tagging from user perspective

Figure 7 and Figure 8 show the coverage and the quality of the tag set from the time aspect. Here the developed tag set after 13 months is regarded as mature. The upper line and the lower line of Figure 7 denote the recall rates of the automatic annotation and the manual annotation, respectively. In the first 0.2 month, the corresponding coverage is 70.28% and 20.48%, respectively. After 1 month, the coverage of the spreading activation method increases to 84.76%. It means 12 months can be saved under the coverage of 84.76%. The precision rates shown in Figure 8 ensure the quality of the recommended tags. They are more than 90% after 1-month social annotation.

In summary, the spreading activation method recommends high quality tags without much delay. That makes resources searchable at the earlier stage.

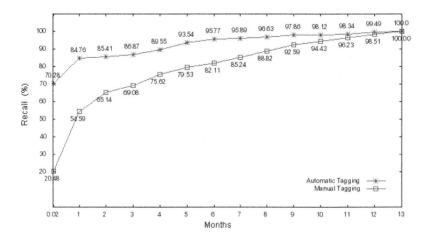

Fig. 7. Recall rate of tagging from time perspective

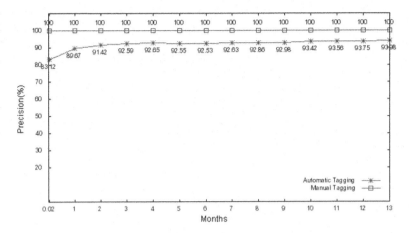

Fig. 8. Precision rate of tagging from time perspective

4 Concluding Remarks

In this paper, a spreading activation method is proposed to predict the tag set of a mature URL based on document content and users' tagging records. The strategies of limited cycles and energy thresholds are explored. The experimental results show that this approach with energy threshold predicts 71.28% of a 100 users' tag set with only 5 users' tagging records, and 84.76% of a 13-month tag set with 1-month tagging record under the precision rates of 82.43% and 89.67%, respectively. Users will benefit from the retrieval performance enhanced by sufficient tags a lot earlier. Currently, only contents of resources and annotation histories are considered. We will investigate more cues like the categorization of resources and the link relationships among resources to predict the social annotation in the future.

Acknowledgments. Research of this paper was partially supported by National Science Council, Taiwan, under the contract NSC 96-2752-E-001-001-PAE.

References

1. Hammond, T., Hannay, T., Lund, B., Scott, J.: Social Bookmarking Tools (I): A General Review. D-Lib Magazine 11(4) (2005)
2. Golder, S.A., Huberman, B.A.: Usage Patterns of Collaborative Tagging Systems. Journal of Information Science. 32(2), 198–208 (2006)
3. Russell, T.: Cloudalicious: Folksonomy over Time. In: Proceedings of the 6th ACM/IEEE-CS Joint Conference on Digital Libraries, pp. 364–364. ACM Press, New York (2006)
4. Mika, P.: Ontologies are Us: A Unified Model of Social Networks and Semantics. In: Gil, Y., Motta, E., Benjamins, V.R., Musen, M.A. (eds.) ISWC 2005. LNCS, vol. 3729, pp. 522–536. Springer, Heidelberg (2005)
5. Wu, X., Zhang, L., Yu, Y.: Exploring Social Annotations for the Semantic Web. In: Proceedings of the 15th International Conference on World Wide Web, pp. 417–426. ACM Press, New York (2006)

6. Dmitriev, P.A., Eiron, N., Fontoura, M., Shekita, E.: Using Annotations in Enterprise Search. In: Proceedings of the 15th International Conference on World Wide Web, pp. 811–817. ACM Press, New York (2006)

7. Hotho, A., Jaschke, R., Schmitz, C., Stumme, G.: Information Retrieval in Folksonomies: Search and Ranking. In: Sure, Y., Domingue, J. (eds.) ESWC 2006. LNCS, vol. 4011, pp. 411–426. Springer, Heidelberg (2006)

8. Baeza-Yates, R., Riberiro-Neto, B.: Modern Information Retrieval. Addison-Wesley, Reading (1999)

9. Salton, G., Buckley, C.: On the Use of Spreading Activation Methods in Automatic Information Retrieval. In: Proceedings of the 11th Annual International ACM SIGIR Conference on Research and Development in Information Retrieval, pp. 147–160. ACM Press, New York (1988)

10. Hsu, M.H., Chen, H.H.: Information Retrieval with Commonsense Knowledge. In: Proceedings of 29th Annual International ACM SIGIR Conference on Research and Development on Information Retrieval, pp. 651–652. ACM Press, New York (2006)

11. Hsu, M.H., Tsai, M.F, Chen, H.H.: Query Expansion with ConceptNet and WordNet: An Intrinsic Comparison. In: Ng, H.T., Leong, M.-K., Kan, M.-Y., Ji, D. (eds.) AIRS 2006. LNCS, vol. 4182, pp. 1–13. Springer, Heidelberg (2006)

12. Halpin, H., Robu, V., Shepherd, H.: The Complex Dynamics of Collaborative Tagging. In: Proceedings of the 16th International Conference on World Wide Web, pp. 211–220. ACM Press, New York (2007)

Mobile Tagging and Accessibility Information Sharing Using a Geospatial Digital Library

Dion Hoe-Lian Goh[1], Louisiana Liman Sepoetro[1],
Ma Qi[1], Ramaravikumar Ramakhrisnan[1], Yin-Leng Theng[1],
Fiftarina Puspitasari[2], and Ee-Peng Lim[3]

[1] Wee Kim Wee School of Communication and Information,
Nanyang Technological University
{ashlgoh,loui0003,maqi0002,rama0017,tyltheng}@ntu.edu.sg
[2] Center for Research in Pedagogy and Practice, National Institute of Education
puspitf@nie.edu.sg
[3] School of Computer Engineering, Nanyang Technological University
aseplim@ntu.edu.sg

Abstract. Mobile tagging is an extension of social tagging that allows users to associate location-sensitive information with physical objects in the real world. This paper presents MoTag, a mobile tagging application that is used to help people with disabilities share up-to-date accessibility information about buildings and other physical structures to help them navigate their environment. MoTag integrates with G-Portal, a geospatial digital library for storing, managing and retrieving tags.

Keywords: Mobile tagging, Metadata, Accessibility, Information sharing, Geospatial digital library.

1 Introduction

People with disabilities face many obstacles as they navigate their environment and would welcome information that could help make this task easier. One category of information is the facilities that buildings and other physical structures provide for people with disabilities. Examples include wheelchair ramps, Braille numbers on elevators, and parking lots for the disabled. In this work, we term such information as "accessibility information". It is important to note that accessibility information helps not only people with disabilities but a wider cross section of people which includes the elderly and children as well.

Accessibility information may be obtained within a building or structure, through Web sites or printed guides. However, these sources of information may not be the most updated due to facility break downs, remodeling work or repairs being carried out. The lack of real-time information may lead to navigation problems if people with disabilities rely solely on such sources. In this paper, we discuss a possible solution to this problem through a mobile tagging application (MoTag) that interfaces with a geospatial digital library to allow users to share and retrieve accessibility information in real-time. In MoTag, users create, update and receive tags, which are metadata

D.H.-L. Goh et al. (Eds.): ICADL 2007, LNCS 4822, pp. 287–296, 2007.

describing accessibility information that are associated with buildings and physical structures. The geospatial digital library used is G-Portal [8, 9] and serves as a backend service to support tag management, processing and retrieval.

This paper follows with a review of the ideas behind mobile tagging and the G-Portal system. A description of the design and implementation of MoTag is then presented and findings of an initial user study of the system are provided. The paper concludes with a summary of our work and opportunities for future research.

2 Background

This section discusses mobile tagging and the G-Portal digital library which serve as the foundation for the MoTag project.

2.1 Mobile Tagging

The Web has evolved from a unidirectional information repository where access to information by users is the main focus, to a platform for collaboration in which content is generated and shared among users. Also known as Web 2.0, examples of such applications include blogs, wikis, social networking, media sharing and social tagging, among many others. As this new avenue for content-generation becomes increasingly popular, the resulting information explosion requires new techniques to manage, search and access such content.

Social tagging is one such approach for managing and discovering content on the Web and refers to the assignment of uncontrolled keywords to resources by users [10]. Such keywords are known as tags and are a simplified form of metadata. Tags are used to organize information, and because they are shareable, users have an alternative way to access content apart from search engines and Web taxonomies. Tags are a form of user-generated content, and popular applications include del.icio.us for tagging Web sites, and Connotea for research content. Besides these purpose-built applications, social tagging has also been used in blogs, wikis, social networking, media sharing and other sites because they have become an accepted way of managing and discovering content. Examples include Flickr and YouTube.

The use of Web 2.0 applications have thus far been mainly confined to desktop computers. However, the popularity of mobile devices and increasing availability of wireless networking access on these devices (e.g. GPRS, 3G, WIFI) suggests new opportunities for deploying similar Web 2.0 collaborative applications on these devices. One important characteristic of the mobile device that distinguishes it from a desktop computer is its mobility and this changing location creates a new dimension in terms of user-generated content. In particular, mobile tagging is one such application that extends Web-based social tagging by taking advantage of mobility. Here, tags (keywords, media elements and other metadata attributes) are applied to physical objects in the real-world as opposed to content (such as Web pages) in the virtual world. Mobile tagging is a promising area and in the research literature, has been applied to education [13], entertainment [4], tourism [3] and many others.

In this project, we apply the concept of mobile tagging to the provision of accessibility information for people with disabilities. While mobile tagging has been employed in many domains, little known work has been done in this area even though there are many benefits that can be reaped. Our work introduces a mobile tagging application, MoTag (Mobile Tagger), that allows users with personal digital assistants (PDAs) to tag buildings and other structures with accessibility information, and to also receive such information. A crucial component of the application is its backend for managing, processing and retrieving tags and associated information. These operations are achieved using a geospatial digital library system known as G-Portal.

2.2 G-Portal

G-Portal is a digital library of geospatial and georeferenced resources, providing a variety of services to access and manage them [8, 9]. The resources maintained comprise mainly metadata records that describe actual resources, such as Web pages, images and other objects that are accessible on the Web. Other types of information managed by G-Portal include semi-structured data records and annotations.

Each resource contains among other attributes, a location attribute (if available) storing its geospatial shape and position, and a link to the corresponding actual resource. G-Portal provides a map-based interface that visualizes resources with location attributes on a map. This interface makes resources with known geographical locations easily and intuitively accessible and helps users discover the spatial relationships between resources. For resources without a location attribute, G-Portal provides a classification-based interface that organizes resources based on a customizable taxonomy. A query interface that supports searches for resources based on keywords and spatial operators is also available.

G-Portal organizes resources into projects in which each project contains a collection of resources that are relevant to a specific topic or learning activity. Within each project, resources are further grouped into layers for finer grained organization. Each layer serves as a category to store logically related resources. For example, a project studying flora and fauna in nature trails may include rivers, lakes and hills in a map layer, flora and fauna information in another, and annotations in a separate layer.

The G-Portal client is developed as a Java applet with all projects, layers and resources stored within a database server that supports XML and spatial operations. G-Portal can therefore be accessed from any Java-enabled Web browser, making it possible for users to easily access and manipulate personalized project space anywhere, anytime.

3 MoTag: Design and Implementation

MoTag is a mobile tagging application in which user-generated tags describing accessibility information are applied to buildings and other physical structures. Each building can have one or more tags. In MoTag, we extend the concept of keyword-based tags to include a richer form of metadata encompassing: the keyword that describes the physical structure, similar to Web-based tags; location of the object (GPS coordinates); one or more media elements that describe the object (e.g. image or video of an unusable wheelchair ramp); comments associated with the tag; and other implicit attributes captured at tag creation time including creator and creation time.

Since tags are associated with locations in the form of GPS coordinates, we require a system to efficiently store, manage and retrieve them. Here, we employ G-Portal for these tasks. The G-Portal digital library is ideally suited for our work because it is also designed to allow users to contribute resources, making the system a common platform for sharing mobile tags of accessibility information. MoTag is implemented as a Pocket PC-based front-end that communicates with the G-Portal server via sockets. Figure 1 shows the architecture of system. The client is the MoTag application running on a personal digital assistant (PDA) or other mobile device. The MoTag client interacts with the G-Portal digital library server for tag management, processing and retrieval functionality.

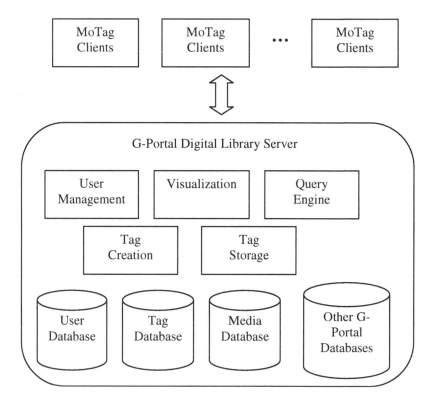

Fig. 1. MoTag system architecture

In a typical usage scenario, a user with a physical disability visits a shopping mall and discovers that the advertised wheelchair ramp on the building's front entrance is not usable and in need of repair. The user launches the MoTag application on his PDA and tags the offending facility with a keyword and some comments describing the state of the restroom. Using the camera mounted on his PDA, a picture of the ramp is also taken and attached with the tag. MoTag uses the PDA's GPS unit to capture the shopping mall's location as well. This information, constituting the tag is packaged by MoTag and uploaded to G-Portal. Some time later, another user planning to visit the same mall browses for accessibility information on it. On launching MoTag, this user discovers the wheelchair ramp problem with this particular mall and considers going elsewhere.

Fig. 2. MoTag's map-based view (left) and tag creation interface (right)

Fig. 3. Viewing official accessibility information using MoTag

Figure 2 shows two of MoTag's screens running on a PocketPC emulator. The map-based view (on the left) gives the user an overview of the coverage area including the available tags (represented as circles). Our coverage area currently encompasses Orchard Road, which is Singapore's main shopping belt with many large malls and other buildings frequented by locals and tourists. From here, users are able to navigate the map by panning and zooming, view tags and create new tags. The tag creation screen (right of Figure 2) requires users to enter the tag, comments and optional attachments which are media elements associated with the tag. The physical location of the tag (coordinates) and building name are also shown.

For viewing tags, a screen similar to the tag creation interface is used except that content is read-only. In addition to the tag information, MoTag provides users with official accessibility information obtained from government agencies. Such information includes entrances to the buildings, lift access, restrooms and so on (see Figure 3). Since the number of tags that users create could be large, several methods are implemented in MoTag to support the retrieval of tags:

- Manual selection from the map. Here, users select a building on MoTag's map-based interface. As shown in Figure 2, buildings that have been tagged are indicated with circular icons. Selecting an icon causes MoTag to present a list of tags for browsing (see Figure 4).
- Browse tag list. MoTag displays an alphabetical list of tags (see Figure 4). Selecting a tag results in a list of associated buildings being displayed. Users may also filter the list by specifying the tag's starting alphabet.
- Search for tags. Users enter terms and receive a list of matching tags. From here, users may further browse and filter the retrieved list.

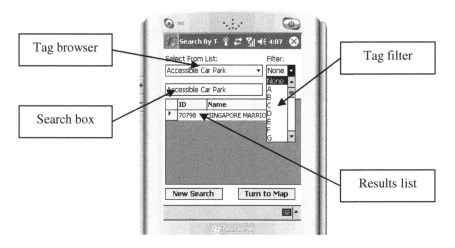

Fig. 4. MoTag's browse and search interface

In all cases, MoTag interacts with the G-Portal server to retrieve the relevant tags. To reduce latency and improve response times, tags that are created by the current user or previously retrieved are cached in the mobile device. At any point in time, users can then choose to synchronize with the G-Portal server to obtain the latest updates. During synchronization, only textual tag information is retrieved. Multimedia elements associated with each tag are only retrieved if users elect to do so. This design reduces the communication time with G-Portal and also reduces network charges (if applicable, for example when using GPRS). Communication is accomplished via XML for maximum portability. Doing so allows for different client-side access mechanisms, thus extending the reach and applicability of the digital library system. For example, G-Portal has been also been used in the desktop and mobile environments for educational applications [5, 12].

4 Evaluation

A pilot study of MoTag was conducted to evaluate the usability of the system using the heuristic evaluation approach. Heuristic evaluation [11] is a usability engineering method for discovering usability problems in a user interface design so that they can be attended to as part of an iterative design process. Heuristic evaluation involves examining the interface and judging its compliance with recognized usability principles (known as the "heuristics"). In the evaluation, Nielsen's [11] 10 usability heuristics were adopted (see Table 1).

4.1 Participants and Tasks

Twelve volunteers (four females and eight males) were recruited for the evaluation. Participants' ages ranged between 16-45 and all owned a mobile device or at least had experience using one.

Four tasks were performed by the participants. The first three tasks involved searching and viewing specific tags while the fourth task required participants to create a new tag. The Orchard Road area was selected as the site of the evaluation because it was a popular shopping belt and provided a realistic setting for the application. Specific buildings (not named in this paper) were also selected based on popularity with shoppers. Participants were first briefed about the ideas behind mobile tagging and the MoTag system. They were then issued with PDAs and after giving them an opportunity to experiment with the system, they were asked to visit the Orchard Road area to complete the tasks. The time taken to complete each task was also noted. The four tasks were:

A. Determine if Building A has accessible toilets.
B. Determine if Building B has reserved parking facilities for people with disabilities.
C. Determine if Building C is accessible to people with disabilities.
D. Select a building and create a tag for accessibility information.

Upon completion of the four tasks, participants completed a questionnaire to rate the conformance of MoTag to each of the 10 usability heuristics on a scale of 1 (low conformance) to 5 (high conformance). They were also asked to provide qualitative feedback on the system.

4.2 Results and Analyses

Table 1 shows the results of the evaluation. The values in the rating column (maximum of 5, minimum of 1) were obtained by averaging the responses of the 12 participants. Values closer to 5 suggest strong conformance for a heuristic while values closer to 1 indicate weak conformance.

As shown in the table, participants rated most of the heuristics relatively highly with scores of around 4. This suggests that participants found MoTag to be a usable system as no major usability issues were noted. However, one heuristic, "Help and documentation", did not score well relatively. This was understandable for two reasons: (1) MoTag is a prototype application and as such did not have documentation; (2) the idea of mobile tagging is relatively new and thus users might need more assistance to accomplish their tasks.

The time taken to complete each of the four tasks was also recorded. This was approximately 30 seconds for Tasks A, B and C which involved similar operations. Task D took about 65 seconds. Although there are no established benchmarks for comparison, these times appear to be reasonable because apart from MoTag, the only way to obtain accessibility information currently along Orchard Road is either by actually visiting the building or by consulting printed sources. These two alternatives are however rather cumbersome especially for people with disabilities.

In addition, many participants commented that the limited screen sizes and keypads made navigation, searching and data entry difficult. For example, many found that map-based view provided only a small coverage area, thus requiring some amount of panning and zooming. Other participants commented on the need to switch between multiple screens in order to search and view tags. A few remarked on the slow response times of the PDA, while others mentioned the difficulty of entering tag data or search terms due to the limited input facilities. We note however that many of these issues are inherent in applications running on mobile devices and are not unique to MoTag. In designing and implementing mobile applications, there is a recognized trade-off between mobility and the device capability [1]. Nevertheless, because these are identified problems raised during the evaluation, various input/output alternatives could be experimented with in future work.

Table 1. Evaluation results using Nielsen's 10 usability heuristics

Heuristic	Rating
Visibility of system status	4.0
Match between system and the real world	4.1
User control and freedom	3.9
Consistency and standards	4.0
Error prevention	4.0
Recognition rather than recall	4.0
Flexibility and efficiency of use	3.5
Aesthetic and minimalist design	4.0
Help users recognize, diagnose, and recover from errors	4.0
Help and documentation	2.9

5   Discussion and Conclusion

In this paper, we present MoTag, a mobile tagging application targeted at helping users share accessibility information by associating buildings and other physical structures with tags or metadata, consisting of both textual attributes and media elements. The location-based nature of the application requires efficient geospatial data management functionality and in our design, we integrate MoTag with G-Portal, a geospatial digital library. G-Portal lends itself well to the task because the system is designed for managing geospatial and georeferenced resources. Further, because G-Portal is also designed for information sharing among digital library users, it is able to support the creation and sharing of tags among MoTag users. As societies begin to recognize the need for helping people with disabilities using assistive technologies, MoTag has the potential to benefit this segment of users by providing timely information that could help them navigate their environment.

MoTag shares similar objectives with existing mobile tagging systems in the provision of services for creating and sharing tags describing physical objects. For example, AURA (Advanced User Resource Annotation) [2] links physical objects and the virtual world using a barcode scanner attached to a PDA. By scanning the barcode, information (if available) about the item is displayed. Users may also add comments which are then uploaded to a server. Urban Tapestries [7] allows users to author location specific multimedia information, similar to the concept of tagging. Using a PDA equipped with a GPS unit, users can tag a location with text, sound, images and video. These tags can be shared with other users. In contrast to these systems, the advantage of MoTag is that it integrates with a geospatial digital library backend (G-Portal) to provide a richer range of services for tag management, processing and retrieval such as strong querying facilities including the ability to perform spatial queries, collection building, and the ability to share tags across both the mobile and Web platforms. The integration of MoTag with a digital library is similar to the work of the TIP/Greenstone bridge project [6] that combines a mobile tourist guide with the Greenstone digital library. Like G-Portal, Greenstone provides an array of digital library services to manage and deliver information, but it relies on the TIP (Tourist Information Provider) system for geospatial data operations. G-Portal, on the other hand, was built with geospatial data management from the start and hence is ideally suited for mobile tagging tasks.

Work on MoTag is ongoing. As revealed in the pilot study, improving the user interface to overcome the limited I/O capabilities of PDAs and other mobile devices is one area of research. We are investigating techniques for automating tag recommendation to reduce the burden of manual searching and browsing. In addition, the results of the pilot study may not be generalizable due to the small sample size. Further work could involve a greater variety of tasks and different types of users such as novices and experts, people with different disabilities, and varying age groups. As part of this larger evaluation, a comparison of MoTag against the use of existing resources for accessibility information would also be conducted to determine the effectiveness of the system. Next, as designed, MoTag is not suitable for people with visual impairments. One possible area of future work could investigate alternative interfaces for such people. Finally, because MoTag uses GPS, participants found that the system could only capture coordinates outdoors making it difficult to use at times. Relying on GPS alone also affords only a coarse grained form of information organization since tags are applied to the building level and not within buildings. This is a recognized limitation of GPS and in future work, other positioning technologies such as WIFI triangulation, radio beacons, and RFID tags could be employed.

Acknowledgments. This work is partly funded by A*STAR grant 062 130 0057.

References

[1] B'Far, R.: Mobile Computing Principles. Cambridge University Press, Cambridge (2004)

[2] Brush, A.J.B., Turner, T.C., Smith, M.A., Gupta, N.: Scanning objects in the wild: Assessing an object triggered information system. In: Beigl, M., Intille, S.S., Rekimoto, J., Tokuda, H. (eds.) UbiComp 2005. LNCS, vol. 3660, pp. 305–322. Springer, Heidelberg (2005)

[3] Carboni, D., Sanna, S., Zanarini, P.: GeoPix: Image retrieval on the Geo Web, from camera click to mouse click. In: Proceedings of the 8th Conference on Human-Computer Interaction with Mobile Devices and Services, pp. 169–172 (2006)

[4] Garner, P., Rashid, O., Coulton, P., Edwards, R.: The mobile phone as a Digital SprayCan. In: Proceedings of the 2006 ACM SIGCHI International Conference on Advances in Computer Entertainment Technology (2006)

[5] Goh, D.H., Sun, A., Zong, W., Lim, E.P., Theng, Y.L., Hedberg, J.G., Chang, C.H.: Managing geography learning objects using personalized project spaces in g-portal. In: Rauber, A., Christodoulakis, S., Tjoa, A.M. (eds.) ECDL 2005. LNCS, vol. 3652, pp. 336–343. Springer, Heidelberg (2005)

[6] Hinze, A., Gao, X., Bainbridge, D.: The TIP/Greenstone bridge: A service for mobile location-based access to digital libraries. In: Gonzalo, J., Thanos, C., Verdejo, M.F., Carrasco, R.C. (eds.) ECDL 2006. LNCS, vol. 4172, pp. 99–110. Springer, Heidelberg (2006)

[7] Lane, G.: Urban Tapestries: Wireless networking, public authoring and social knowledge. Personal and Ubiquitous Computing 7(3-4), 169–175 (2003)

[8] Lim, E.P., Liu, Z.H., Goh, D.H., Theng, Y.L., Ng, W.K.: On organizing and accessing geospatial and georeferenced Web resources using the G-Portal system. Information Processing and Management 41(5), 1277–1297 (2005)

[9] Liu, Z.H., Yu, H., Lim, E.P., Yin, M., Goh, D., Theng, Y.L., Ng, W.K.: A Java-based digital library portal for geography education. Science of Computer Programming 53(1), 87–105 (2004)

[10] Macgregor, G., McCulloch, E.: Collaborative tagging as a knowledge organization and resource discovery tool. Library Review 55(5), 291–300 (2006)

[11] Nielsen, J.: Finding usability problems through heuristic evaluation. In: Proceedings of the ACM CHI 1992 Conference, pp. 373–380 (1992)

[12] Theng, Y.L., Tan, K.L., Lim, E.P., Zhang, J., Goh, D.H., Chatterjea, K., Chang C.H., Sun, A., Han, Y., Dang, N.H., Li, Y.Y., Vo, M.C.: Mobile G-Portal supporting collaborative sharing and learning in geography fieldwork: An empirical study. Paper accepted to the ACM/IEEE 2007 Joint Conference on Digital Libraries (JCDL 2007) (2007)

[13] Yeh, R.B., Liao, C., Klemmer, S.R., Guimbretière, F., Lee, B., Kakaradov, B., Stamberger, J., Paepcke, A.: ButterflyNet: A mobile capture and access system for field biology research. In: Proceedings of the SIGCHI Conference on Human Factors in Computing Systems 2006, pp. 571–580 (2006)

Social Navigation in Digital Libraries by Bookmarking*

Fiftarina Puspitasari, Ee-Peng Lim, Dion Hoe-Lian Goh, Chew-Hung Chang, Jun Zhang, Aixin Sun, Yin-Leng Theng, Kalyani Chatterjea, and Yuanyuan Li

Nanyang Technological University, Singapore 639798

Abstract. In the age of Web 2.0, users are increasingly familar with social tagging or bookmarking where comments and ratings are added by users to objects on the web for public consumption. Such comments and ratings are represented in bookmarks which can be used for information or opinion sharing, user interest discovery, and content recommendation. In this paper, we investigate social bookmarking in digital libraries and derive the design requirements for digital library incorporating social bookmarking. Instead of implementing social bookmarking functions in digital library systems from ground zero, we have chosen to explore the possibilities of integrating pre-existing digital library systems with pre-existing social bookmarking systems, and to derive a feasible system architectural design. We also present a case study where G-PORTAL, a geography digital library system, is integrated with SCUTTLE, an open source social bookmarking system.

1 Introduction

Social bookmarking[5] is no stranger to web users. Many users employ social bookmarking (e.g., del.icio.us[1], CONNOTEA[2], SCUTTLE[3]) to tag web objects for future references, for sharing, as well as for expressing opinions on the bookmarked objects. With social bookmarking, users can more easily collaborate by tracing each other's activities recorded by the social bookmarks, and this is also known as *social navigation*[4]. Moreover, ratings given in bookmarks can be used to evaluate the quality of bookmarked objects[7].

Most existing digital library systems consist of some content and metadata repositories created by professional librarians. They address mainly the information search and browsing issues but not issues related to user collaboration. Hence, social bookmarking is so far not a core feature in digital libraries despite its popularity among Web 2.0 applications. There have also been very little study on system issues related to extending digital library systems with social bookmarking.

---

* This work is funded by the Centre for Research in Pedagogy and Practice, National Institute of Education, Singapore.
[1] http://del.icio.us
[2] http://www.connotea.org
[3] http://scuttle.org

D.H.-L. Goh et al. (Eds.): ICADL 2007, LNCS 4822, pp. 297–306, 2007.

There are several obvious benefits introducing social bookmarking to digital libraries. Apart from giving users alternative means to search and browse content collections, social bookmarking creates a space for users to interact with one another forming online communities that are essential in knowledge creation, sharing and dissemination[9,12].

In this paper, we study the provision of social bookmarking in a digital library system from an architectural standpoint. To the best of our knowledge, there have not been much technical study on how social bookmarking systems can be extended to bookmark digital library objects, and how a digital library system can be extended to provide social bookmarking feature. We therefore adopt a fresh approach to determine the design requirements for social bookmarking in digital libraries. This leads us to an integration framework with a set of strategies to incorporate social bookmarking into a digital library system. We also present a case study where G-PORTAL[8], a digital library system for geography learning, is integrated with SCUTTLE, an open source social bookmarking system[11].

The remaining sections of this paper are organized as follows. Section 2 gives an overview of related research. In Section 3, the requirements of social bookmarking in a digital library system are derived, followed by our proposed integration framework in Section 4. Section 5 presents the integration between G-PORTAL and SCUTTLE as a case study. Finally, Section 6 gives our conclusions and outlines the future research directions.

2 Related Work

2.1 Digital Libraries with Social Bookmarking Support

Very few existing digital libraries provide bookmarking or more often called annotation services for users to mark useful content objects. DLESE, for example, introduces an annotation metadata format and encapsulates annotations within its metadata records[2]. Based on the annotation metadata format, DLESE could incorporate a few annotated sub-collections. The bookmarking functions of DLESE however have not been reported.

In the DiLAS project, a decentralized framework that manages social bookmarks or annotations independently from a digital library is proposed[1]. The main idea here is to provide social bookmarking functions to any digital libraries with a very loose coupling architecture. Instead of using an existing social bookmarking systems, DiLAS has custom-built a social bookmarking system.

Our work is unique is that it introduces a framework for integrating social bookmarking and digital library systems. The framework includes both a reference architecture and the common database integration strategy. We also present a example case study to augment an existing digital library system with social bookmarking features.

2.2 Social Bookmarking Systems

Social bookmarking systems have become popular with their attractive features on storing and organizing online bookmarks. Most of these systems focus on sharing and visualizing bookmarks on web page URLs. Some of them are designed for certain target user groups. For example, CONNOTEA is designed to help scientists and researchers manage useful references to research literatures. Social bookmarking has also been incorporated into many different Web 2.0 applications, e.g., Flickr[4], and YouTube[5].

Despite their flexibility to bookmark any web content, social bookmarking systems has little control over the quality of bookmarked objects as well as the quality of bookmarks themselves. Digital library systems offer a solution to address the quality issues. With the digital library content carefully constructed and reviewed by library professionals, the quality of content and metadata objects and the consistency in controlled vocabulary classification in digital libraries are expected to be high. Hence, social bookmarking digital library content and metadata should be an attractive option when knowledge sharing is concerned.

3   Design Requirements for Digital Libraries with Social Bookmarking

In this section, we discuss digital library objects that can be good candidates for objects of bookmarking and how the bookmarks can be represented in the digital library context. Subsequently, we suggest some bookmarking features desired in a digital library. We divide the design requirement of a digital library with social bookmarking capabilities into several sub-areas as described below.

Objects for Bookmarking

The objects to be bookmarked can be divided into two aspects, namely *object type* and *granularity*. The object type refers to the kind of objects that can be bookmarked. Digital library systems usually maintain two main classes of objects, (i) content and (ii) metadata objects. For digital libraries that use content objects residing at public websites, the metadata objects can serve as surrogates and be used for bookmarking.

From the granularity standpoint, we examine three possible groupings of objects for bookmarking. One may want to bookmark: (a) a single object, (b) a collection of objects or (c) a fragment of an object. The granularity chosen will depend on the amount of information users want to bookmark. For example, a user may wish to bookmark a collection of news reports related to United States Presidential Election (type (b)), or a biography of some music composer appearing in a music history article (type (c)). Most existing social bookmarking systems support bookmarking of single objects only.

Bookmark Representation

Within a digital library, bookmarks can be treated as content or metadata objects and be represented in some record structure. There are mainly two possible record structures, namely:

---

[4] http://www.flickr.com

[5] http://www.youtube.com

- *Single fixed record structure:* A fixed set of attributes are defined to store a bookmark. This approach has been adopted by most existing social bookmarking systems. The attributes may include owner, reference to bookmarked object, date, keywords, ratings and comments.
- *Multiple record structures:* This bookmark representation allows a customized set of attributes to be used for representing a bookmark. This is necessary for bookmarks with special purposes. To allow bookmarks to be processed uniformly, these record structures will have to share a common set of mandatory attributes (e.g., owner, date, keywords, etc.).

Bookmark Access

This set of design requirements concerns the access to bookmarks. In the existing social bookmarking systems, bookmarks are either "owner accessible only" or "publicly accessible". In a digital library, there is however additional access control over the digital library objects that may affect the access to the bookmarks. For example, the bookmarks of a bookmarked object should not be accessible by all public users if the latter is only accessible to a small group of users.

Bookmark Query

Social bookmarks are essentially structured information that can be queried based on their attributes, including keywords, references to bookmarked objects, owners, and others. In the digital library context, querying bookmarks can be associated with querying digital library objects. For example, one can search content or metadata by querying their bookmarks and vice versa.

Bookmark Organization

Bookmark organization allows bookmarks to be navigated based on bookmark attributes as well as bookmark categories. We further divide the latter classification approach into the following schemes.

- *Folksonomy-based classification.* This scheme allows users to classify bookmarks by assigning one or more keywords to each bookmark. This keyword serves as an open category label. The set of category labels is open since any keyword can be used as opposed to a controlled set. Hence, a bookmark can be classified into multiple keyword categories. This kind of classification is popular among existing social bookmarking tools. When a user browses a category, s/he will not only view his/her bookmarks in the category, but also other users' bookmarks under the same category.
- *Controlled-vocabulary based classification.* This scheme uses hierarchical structure to organize bookmarks. In this case, the categories are predefined by some digital library administrator(s) or designated experts. This classification scheme is similar to the one implemented in existing digital library systems. The main advantage of such a system lies in the quality of classification.
- *Mixed classification.* This scheme allows the bookmarks being classified based on both folksonomy and hierarchical category structures. One example of mixed classification is *tag-bundle*, which was introduced by del.icio.us. Tag

bundle is used to group keywords into some user-defined categories. This allows users to define tag bundles of keywords for navigating bookmarks.

4   Integration Framework

Having identified the social bookmarking requirements of digital libraries, we now present a framework for integrating a digital library system with a social bookmarking system. The framework consists of a **reference architecture** to describe the generic architectural elements of a digital library system with social bookmarking functions, and some **integration strategy** to outline the possible integration approach. Due to space constraint, we will only present the common database integration strategy in this paper.

4.1   Reference Architecture

Figure 1 depicts the reference architecture with both digital library and social bookmarking system elements. Similar to digital library architectures adopted by National Science Digital Library (NSDL)[6] and DELOS[3], we divide the system into four levels, namely *user*, *user interface*, *service* and *storage*.

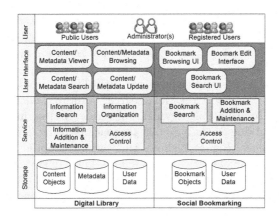

Fig. 1. Reference Architecture to Incorporate Social Bookmarking in DL

- **User Level.** A digital library system or a social bookmarking system serves a population of public users and registered users. In cases where the system is designed for internal use, only registered users may be supported. Administrators in a digital library or social bookmarking system are usually responsible for managing user accounts. For digital libraries, administrators may include librarians who maintain the digital collections and perform other library tasks.
- **Storage Level.** At the storage level, a digital library provides a repository for content objects, metadata and user data. While the object collections in

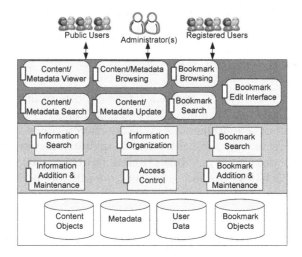

Fig. 2. Integration Strategy (Common Database)

most digital libraries are to be shared by all users, there are cases where digital library services are designed to maintain sub-collections for individuals or user groups[10,8]. For them, only the authorized users can access the sub-collections. A social bookmarking system, in contrast, maintains a collection of bookmarks. Usually, these bookmarks are open for public access except when they are specifically created for personal consumption by the owners.

– **Service level.** At the service level, the digital library and bookmarking systems provide similar services (i.e., search, addition and maintenance) over their objects. Information organization service is less important in bookmarking system due to the popular approach of using keywords to organize bookmarks. Hence, Figure 1 shows this service adopted only by digital libraries. Access control service exists in both kinds of systems to support user authentication and user access right maintenance (i.e., who owns what objects) and determines which users can perform what operations (e.g., view, update, delete) on which objects.

– **User interface level.** The digital library and social bookmarking systems adopt some user interface (UI) modules. Content/metadata viewer, one of these UI modules, is always present in a digital library system since each content/metadata object carries richer amount of information compared to a bookmark. The social bookmarking system, on the other hand, has more emphasis on social navigation and therefore has a more comprehensive bookmark browsing/search interface that allows users to find bookmark objects easily.

4.2   Common Database Integration Strategy

The common database strategy is suitable for an existing digital library system that needs to be extended with bookmarking features. The strategy involves the construction of a single system with both the digital library and bookmarking services adapted to the common database design as shown in Figure 2. The integrated system requires only one access control service. To minimize development efforts, the modules from the digital library and social bookmarking systems still have to provide APIs for them to interoperate with each other.

Having a single system clearly reduces confusions to the end users. The common database strategy however may also pose some inconsistencies in UI design. However, the use of common databases also makes it possible for some of the UI modules to be modified to suit the user needs. Such modifications will be further elaborated in our case study (see Section 5).

5 Case Study: G-PORTAL and SCUTTLE

5.1 Overview of G-PORTAL

As mentioned earlier, G-PORTAL is developed for learning purpose. It provides a repository of resources surrogated by metadata records and shared among G-PORTAL users. To support learning, G-PORTAL introduces a concept of *project*, which is a set of metadata records identified and assembled for a specific learning task. Users participate in the learning task by contributing/browsing metadata records within the same project space as well as organizing metadata based on user-defined classification schemes. A project can be configured to be accessible by all users, selected users, or only the project owner. When a project is assigned shared access to multiple users, these users can be grouped into one user group. A user group can be granted to access several different projects since it represents users sharing common interest and learning tasks.

5.2 Social Bookmarking to G-PORTAL

To support better learning, we have decided to augment G-PORTAL with bookmarking capability for metadata records in G-PORTAL using the common database strategy. This strategy reduces the amount of development efforts involved as we can keep most G-PORTAL modules intact while modifying some for the integrated system. The integration makes use of SCUTTLE, a open source social bookmarking system implemented using PHP[11]. The bookmarking services offered by SCUTTLE include bookmark navigation, adding bookmark, updating and deleting bookmarks.

After integration, the combined system contains modules from G-PORTAL and SCUTTLE share a common set of databases. The user and access control data of bookmark objects and digital library metadata objects are combined at the storage level. The service level of the combined system consists of a common access control module, metadata-related modules from the original G-PORTAL system and bookmarking modules from SCUTTLE. All these modules are extended with APIs to allow inter-module calls. At the UI level, similar merger of modules also takes place. In particular, the metadata viewer of G-PORTAL is combined with the bookmark edit interface of SCUTTLE so as to facilitate bookmarking when viewing metadata of G-PORTAL.

5.3   Bookmark Representation

We define G-PORTAL's bookmark representation to contain mandatory attributes and optional attributes, as shown below (with attribute type indicated in brackets):

- **Bookmarked object name and URL (mandatory)** together identify a bookmarked object. In G-Portal, a metadata object name refers to its resource name, and the metadata object's URL is the web reference to the metadata object. As the same metadata object may appear in different projects, it is important to keep the project information in the bookmark. Therefore, the URL contains the metadata object id and the id of project where the metadata object is bookmarked.
- **Bookmark owner(mandatory)** attribute stores the user id of the bookmark contributor.
- **Date of creation(mandatory)** records when the bookmark is created.
- **Description(optional)** stores users' comment or description on a bookmarked object.
- **Rating(optional)** is a value between 1 and 5 inclusive, where 1 represents 'least useful' and 5 represents 'most useful' respectively. This attribute can be used by users to rate the usefulness of metadata records.
- **Media object(optional)** attribute allows users to associate one or more media objects (e.g., images, videos, etc.) with the bookmark. These media objects have to be made available on the web and be identified by their URLs.
- **Bookmark keywords(optional)** is a set of keywords associated to the bookmark. Like other social bookmarking systems, the list of keywords is left open so that a folksonomy based classification of bookmarks can be supported.

5.4   Bookmark Access

The bookmark access control in the combined system is designed to be consistent with that of G-PORTAL. G-PORTAL enforces access control over its metadata objects by designating for each project a coordinator who specifies the group(s) of users allowed to access the project. Hence, the access control module only allows a user to add bookmarks to metadata records in a project when the user is authorized to access the project. Furthermore, an authorized user of a project is also allowed to view bookmarks on any metadata records in the project.

5.5   Bookmark Browsing and Search

Bookmark browsing and search interface is one single module in the integrated system. A user can browse and navigate bookmarks both in a bookmark list and a tag cloud as indicated in Figure 3. The bookmark list shows the result of a bookmark query, whereas a tag cloud shows a visual summary of keywords such that more frequently used keywords are shown in larger font sizes and rarely used ones in smaller font sizes. Hence, one can quickly identify recently used and popular keywords from the tag cloud.

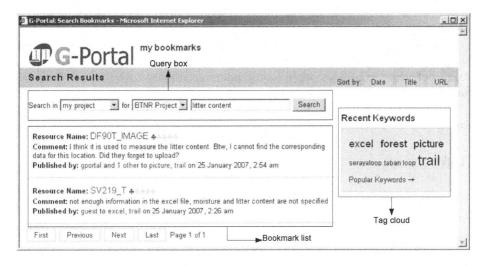

Fig. 3. Query Bookmarks by Project

Using the query box, a user can select the desired query option in a drop down list and enter query term(s). The query options supported include:

- *Query on all accessible bookmarks.* This option allows users to view all bookmarks belonging to projects accessible to the users.
- *Query on self created bookmarks.* Users can browse and query all bookmarks he or she has created.
- *Query by project.* This option caters to users who want to query bookmarks created for metadata records within a project. Figure 3 illustrates how a user searches bookmarks by project. In this case, the user first selects the *Search in my project* query option followed by selecting one of the accessible projects. In this example, the user selects the *BTNR Project*. The user finally provides the query term *"litter content"*. Upon submission, the query results are returned as a bookmark list.
- *Query by user group.* As mentioned in Section5.1, a user can belong to multiple user groups. To query bookmarks created by group members that may be relevant to users' need, users can choose *Search in my groups* option followed by specifying the group name and query term(s).

6   Conclusion

Social bookmarking is essential to digital libraries. This paper outlines the important design requirements for including social bookmarking in a digital library system. An integration framework consisting of a reference system architecture and a common database integration strategy has been developed. With this framework, a systematic approach to incorporate social bookmarking into digital library systems can be adopted. The paper also describes our experiences in integrating G-PORTAL digital library system with SCUTTLE, an open source social bookmarking tool.

Looking ahead, one will expect more digital libraries to incorporate social bookmarking to enhance collaboration among their users. There will be case studies of using strategies other than common database to be reported in the future. As part of our future work, we plan to conduct user evaluation on the usability of the G-PORTAL's social bookmarking functions and its impact on learning.

References

1. Agosti, M., Ferro, N., Panizzi, E., Trinchese, R.: Annotation as a support to user interaction for content enhancement in digital libraries. In: Working Conference on Advanced Visual Interfaces (2006)
2. Arko, R., Ginger, K., Kastens, K., Weatherley, J.: Using Annotations to Add Value to a Digital Library for Education. D-Lib Magazine 12(5) (2006)
3. Candela, L., Castelli, D., Pagano, P., Thanos, C., Ioannidis, Y., Koutrika, G., Ross, S., Schek, H.-J., Schuldt, H.: Setting the Foundation of Digital Libraries: The DELOS Manifesto. D-Lib Magazine 13(3/4) (2007)
4. Dieberger, A., Dourish, P., Hook, K., Resnick, P., Wexelblat, A.: Social Navigation: Techniques for Building More Usable Systems. Interactions 7(6), 36–45 (2000)
5. Hammond, T., Hannay, T., Lund, B., Scott, J.: Social bookmarking tools (i), a general review. D-Lib Magazine 2(4) (2005)
6. Lagoze, C., Arms, W., Gan, S., Hillmann, D., Ingram, C., Krafft, D., Marisa, R., Phipps, J., Saylor, J., Terrizzi, C., Hoehn, W., Millman, D., Allan, J., Guzman-Lara, S., Kalt, T.: Core services in the architecture of the national science digital library (NSDL). In: Proceedings of the 2nd ACM/IEEE-CS Joint Conference on Digital libraries, pp. 201–209 (2002)
7. Lauw, H.W., Lim, E.-P., Wang, K.: Summarizing review scores of unequal reviewers. In: SIAM International Conference on Data Mining (2007)
8. Lim, E.-P., Goh, D.H.-L., Liu, Z., Ng, W.-K., Khoo, C.S.-G., Higgins, S.E.: G-Portal: A Map-based Digital Library for Distributed Geospatial and Georeferenced Resources. In: ACM/IEEE-CS Joint Conference on Digital Libraries (2002)
9. Lomas, C.: Seven Things You Should Know About Social Bookmarking, Online (2005) (accessed January 3, 2007)
10. Shipman, F., Hsieh, H., Moore, J., Zacchi, A.: Supporting Personal Collections across Digital Libraries in Spatial Hypertext. In: ACM/IEEE-CS Joint Conference on Digital Libraries (2004)
11. SourceForge.net, http://sourceforge.net/scuttle
12. Yew, J., Gibson, F., Teasley, S.: Learning by tagging: group knowledge formation in a self-organizing learning community. In: International Conference on Learning Sciences (2006)

Blog Classification Using Tags: An Empirical Study*

Aixin Sun[1], Maggy Anastasia Suryanto[1], and Ying Liu[2]

[1] Nanyang Technological University, Singapore
axsun@ntu.edu.sg
[2] Hong Kong Polytechnic University, Hong Kong, China
mfyliu@polyu.edu.hk

Abstract. With an exponential growth of Weblogs (or blogs), many blog directories have appeared to help users to locate topical blogs. As tags are commonly used to describe blogs, we study the effectiveness of tags in blog classification. Compared with titles and descriptions, our experiments, using 24,247 blogs, showed that tags could lead to better classification accuracy. It is interesting to observe that more tags did not necessarily lead to better classification accuracy. To better describe blogs, we have also proposed a tag expansion algorithm that assigns a blog more tags that are often co-occur with those already associated with the blog. Our experiments showed that tag expansion helped to improve the recall of blog classification with the price of precision degradation.

1 Introduction

Blogs are online personal diaries managed by software packages that allow single-click publishing [5]. Each diary entry in a blog is also known as a blog post (or post). These blog posts are often displayed in reverse chronological order and their contents include personal views, observations, discussions, and other topics. The rapid growth of blogs has created new research opportunities in information retrieval, text mining, social studies, and many other areas.

Similar to Yahoo! Directory organizing Web sites/pages into topical categories, a number of blog directories are now available online. Examples are BlogFlux[1] to classify blogs into 161 flat topical categories; BlogCatalog[2] to organize blogs into hierarchical topical categories with 49 top-level categories; and BOTW[3] to list blogs in a hierarchy with 12 top-level categories. These blog directories provide an easy way of locating blogs of certain topic(s), in addition to blog searching. While many of these blog directories require manual assessment

---

* This research is supported by grant SUG7/06, Nanyang Technological University, Singapore.
[1] http://dir.blogflux.com/, accessed on Jun 24, 2007.
[2] http://www.blogcatalog.com/, accessed on Jun 24, 2007.
[3] http://blogs.botw.org/, accessed on Jun 24, 2007.

D.H.-L. Goh et al. (Eds.): ICADL 2007, LNCS 4822, pp. 307–316, 2007.

of blogs, which is labor intensive and time consuming, automatic blog classification methods offer an attractive option. *Blog classification* refers to the task of assigning blogs one or more pre-defined categories.

The problem of blog classification is different from most text/Web classification problems because of at least three reasons. Firstly, the object to be classified in blog classification are blogs where each blog consists of a set of blog posts and the number of posts may vary significantly from one blog to another. On the other hand, the objects to be classified in text/Web classification are individual text/Web documents, e.g., news articles [12]. Secondly, by its nature, a blog is frequently updated with newly published blog posts making blogs, the objects to be classified, rather dynamic compared with those documents involved in text/Web classification. Thirdly, as a type of user generated content, a blogger may write any topic of his/her interests and the topics of blog posts could be very diverse. The diversity of topics and dynamic updating nature make blog classification a much more challenging task compared with text/Web classification.

A blog could be described using features derived from its various properties, such as, title, description, tags, blog posts and so on. Among them, tags represent a new type of user-generated data that is not available for most text/Web documents. Although tags have been receiving much interests from researchers in various areas (see Section 2), the impact of using tags for blog classification has not been studied, to the best of our knowledge. On one hand, tags are often considered as "metadata" to describe the associated object (the blog in this case), which is believed to be indicative in blog classification. On the other hand, it is well known that tags are given by users without referencing to any controlled vocabulary. For this reason, different terms having similar semantics may be chosen by users to tag blogs of similar topics, and the same term maybe used to tag blogs of different topics as users may have different understanding on the scope of each topic.

In this paper, we study the effectiveness of tags in blog classification and try to answer the following three questions: (i) are tags more effective in blog classification than other type of data, e.g., title and description? (ii) is it true that more tags lead to more accurate classification? (iii) does tag expansion help in getting better classification accuracy? Tag expansion, similar to query expansion, refers to the process of expanding tags with terms having similar semantics. Tag expansion partially solves the problem of having different terms tagging blogs of the same topic.

To answer these questions, we conducted our experiments using 24,247 blogs collected from BlogFlux and classified them into 20 categories using Support Vector Machines (SVM) classifiers [8,15]. From our experimental results, we observed that tags were more effective in blog classification than features extracted from blog title and description although the latter usually contain more terms than tags. On the other hand, title and description are complementary to tags and the best classification accuracy was achieved when all these features were used together. Our experimental results surprisingly showed that more tags did not necessarily lead to better classification accuracy. To answer the third question, we proposed a tag expansion algorithm based on Personalized PageRank algorithm [7] using co-occurrence relationships among tags. Evaluated in our experiments, it is observed that the tag expansion algorithm could improve the classification accuracy only when the blogs have relatively more tags. Such an interesting observation calls for further study on the topic, which is also a part of our future work.

This paper serves as pilot study on automated blog classification using tags. The observations obtained from our experimental results could benefit future studies in this area. The rest of the paper is organized as follows. In Section 2, we survey the related studies on tagging and blog classification. We present our data corpus in Section 3 followed by the tag expansion algorithm in Section 4. Our experimental results are presented in Section 5. Finally, we conclude the paper in Section 6.

2   Related Work

Tagging has been receiving much attention from more and more researchers in various areas such as social studies and text/Web mining. Marlow *et al.* [10] summarized tagging systems used by various web sites and presented a taxonomy of tagging systems with 7 dimensions including the three (with their main categories) shown in Table 1. In the Table, we also present the characteristics of data corpus used in our experiments (see Section 3). Note that, although we state that the object type in our study is textual, the tags we are interested in are those attached to blogs (as a group of blog posts) rather than those attached to individual blog posts. This makes our research very different from many other works on tags associated with blog posts [1,2,6,13].

Berendt and Hanser in [1] compared the performance of blog post classification using features derived from tags, title, and body; it is argued that tags are not metadata but "more content" as (i) tags have a low similarity with post body and (ii) tags together with body yielded better classification accuracy than any of them alone. Brooks and Montanez studied the effectiveness of using tags to organize blog posts into clusters [2]. They found that posts sharing the same tag have a lower similarity than those sharing the same extracted term (for each blog post, the top 3 words having the highest $tf \cdot idf$ scores were extracted). Nevertheless, their results showed that tags are useful in grouping posts into broad categories. In [6], it is observed that frequently occurring tags are usually good meta-labels of a cluster produced using content clustering.

Table 1. Example taxonomy dimensions and characteristics of our corpus

Dimension	Main categories	Our corpus
Tagging right	self-tagging, permission-based, or free-for-all	self-tagging
Tagging support	blind, suggested, viewable	blind
Object type	textual, non-textual	textual

Our work is also related to those work in Web classification where the task is to assign Web pages one or more pre-defined categories. In Web classification, features derived from title, content, hyperlinks and anchor words of Web pages have been evaluated [3,14,15]. Among the classifiers evaluated, SVM classifiers have demonstrated good classification accuracy [12]. Web classification techniques have recently been applied to blogs to detect spam blogs [9] and to label blog posts to be informative or affective articles [11].

3   Data Corpus

Our corpus is collected from BlogFlux directory in June 2007. All blogs are organized in 161 flat categories arranged in alphabetical order with *Academic* and *Zookeeping* being the first and last categories (as of June 2007). For each blog, BlogFlux provides the following metadata: *title, description, blogger, geographic location, language*, one or more *categories* assigned to the blog, and *tags* associated with the blog. We collected the metadata of all 76,997 blogs listed in the directory and among them 52,709 (or 68%) are in English. As our main focus is the use of tags in classification, from the English blogs, we selected those having at least 2 tags to form our corpus. The corpus is known as BFE (BlogFlux English) dataset, consisting of 24,247 blogs. Note that, we did not obtain the posts of all blogs in BFE dataset as those blogs are hosted on various blog sites and extracting blog posts from a large number of blog sites is a challenging task.

3.1   Tag Normalization

When submitting a blog for possible listing in BlogFlux, one may give zero to five tags to describe the blog. These tags could be either word tags (consisting of exactly one atomic word) or phrase tags (consisting of more than one word). For easy processing of tags, we performed tag normalization as in [13]. Word tags, if not stopwords, are stemmed using Porter's stemming algorithm. Phrase tags are first tokenized; the non-stopword tokens (or words) are stemmed and indexed as single-word tags. All stemmed words originally from the same phrase are sorted in alphabetic order to form a normalized phrase tag. For example, phrase tag `real estate` becomes two word tags `real`, `estat`, and one phrase tag `estat real` after normalization. After normalization, 42,798 distinct tags were obtained from BFE dataset.

Figure 1 shows the distribution of the number of tags against the blog frequency of each tag (i.e., the number of blogs associated with the tag). It is clear that a power law distribution is demonstrated with majority of tags appears with very few blogs only, while a few tags having blog frequency greater than 1000. Such an observation shows that distribution of tags attached to blogs is similar to that attached with blog posts [6].

We define *tag degree* of a blog to be the number of normalized tags attached to a blog, and *category degree* to be the number of categories assigned to the blog. Figures 2(a) and 2(b) illustrate the distribution of tag degree and category degree among blogs in BFE dataset. On average, each blog has 6.3 normalized

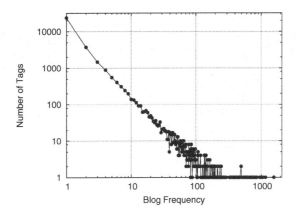

Fig. 1. Power law distribution of tags

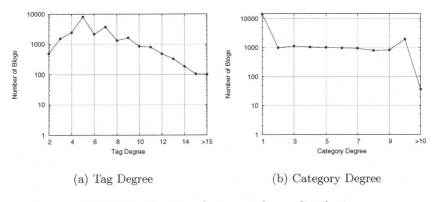

(a) Tag Degree   (b) Category Degree

Fig. 2. Tag degree and category degree distribution

tags with very few having more than 15 tags. Although most blogs are labeled to exactly one category, many of them are labeled to 2 to 10 categories with very few labeled to more than 15. On average, each blog is labeled to 3.1 categories. This suggests that topics demonstrated in blogs are often diverse and could be related to different categories.

3.2   Popular Tags and Category Names

As shown in Section 3.1, some tags are much more popular than others. We have therefore listed the 20 most popular word tags and phrase tags according to their blog frequencies, shown in Table 2. In the table, we also list the top 20 categories having most number of blogs belonging to them. It is interesting to observe that 11 of the top word tags, highlighted in bold, match the names among the top 20 categories. There are also another 12 tags, underlined, matching names among the rest 141 categories. This shows a strong relationship between the popular

Table 2. Popular word and phrase tags

Word Tag	BlogFreq	Phrase Tag	BlogFreq	Category	#Blogs
blog	1577	estat(e) real	249	personal	4276
new(s)	1166	internet market	180	internet	2297
polit(ics)	1067	design web	164	general	2262
music	1008	loss weight	152	humor	2077
marketing	956	make monei	110	entertainment	1977
art	858	current event	105	computers-tech	1726
travel	841	home work	95	business	1681
internet	800	engin optim search	88	technology	1607
life	784	cultur(e) pop	84	art	1554
busi(ness)	758	make monei onlin	80	politics	1516
humor	744	busi(ness) home	78	travel	1503
technolog(y)	740	game(s) video	78	music	1411
design	735	develop(ment) web	77	health	1320
person(al)	707	financ person	69	religion	1305
web	687	market onlin	67	sports	1196
photographi(y)	668	develop person	58	life	1168
review	623	busi small	55	photo-blog	1150
home	605	hip hop	52	food-drink	1130
video	602	affili market	52	commentary	1049
monei	595	creativ(e) write(ing)	48	opinion	995

tags and category names in blog directory, which suggests that tags could be effective features for blog classification.

4   Tag Expansion

As discussed in Section 1, different terms having similar semantics may be chosen by users to tag blogs of similar topics. To partially solve this problem, we propose a tag expansion algorithm using co-occurrence relationship among tags. The proposed tag expansion algorithm is based on the Personalized PageRank [4,7]. The basic idea is to expand the tags attached to a blog by bringing in those tags that are often used together with those former tags.

Let $\mathcal{T}$ be the set of tags. The directed graph with node set $\mathcal{T}$ and edges corresponding to co-occurrence relationships among tags is known as the *Tag Graph*. A directed edge from tag t_j to tag t_i exists if t_j and t_i are ever used together to tag one or more blogs, i.e., t_j co-occurs with t_i. Let T_b ($T_b \subseteq \mathcal{T}$) denote the set of tags attached to blog b. To expand T_b, each tag $t_i \in \mathcal{T}$ is scored using Equation 1 in an iterative manner.

$$s^{n+1}(t_i) = \alpha s^0(t_i) + (1 - \alpha) \sum s^n(t_j) \times w(t_j, t_i) \qquad (1)$$

In Equation 1, α is the teleportation probability and typically $\alpha = 0.15$; $s^n(t_j)$ is the score of tag t_j in the nth iteration; $s^0(t_i)$ is the initial score of t_i; and

$w(t_j, t_i)$ is the weight associated with the edge from t_j to t_i. In our experiment, $w(t_j, t_i)$ is defined by the ratio between the number of blogs tagged by both t_j and t_i and the number of blogs tagged by t_j, as shown in Equation 2 where $|C|$ refers to the number of elements in set C.

$$w(t_j, t_i) = \frac{|\{b | t_i \in T_b \wedge t_j \in T_b\}|}{|\{b | t_j \in T_b\}|} \tag{2}$$

$$s^0(t_i) = \begin{cases} \dfrac{tf(t_i)}{\sqrt{\sum_{t_\ell \in T_b} tf(t_\ell)^2}} & \text{if } t_i \in T_b \\ 0 & \text{otherwise} \end{cases} \tag{3}$$

The initial score of a tag t_i is given in Equation 3 where $tf(t_i)$ refers to the term frequency of tag t_i derived from all tags attached to blog b. Only those tags attached to blog b have non-zero scores; among them most of the tags have term frequency of 1. Very few tags may have term frequency more than one due to the tag normalization. For example tag web has term frequency of 2 if two tags are attached to a blog: web and web design.

5   Experiments

We conducted two sets of experiments. The first set is to evaluate the classification effectiveness of tags compared with other types of features derived from blogs. The second set is to evaluate the effectiveness of the proposed tag expansion algorithm in improving blog classification accuracy.

5.1   Experimental Setup

Recall that in our BFE dataset, each blog has its *title* and *description* besides its tags. We derived a text description, known as Des feature, for each blog using terms appearing in its title and description. After stopword removal and stemming, Des contains 14.8 terms on average for each blog. The number of terms in Des is about twice of the 6.3 terms contained in tags on average. In the first set of experiments, we report the classification accuracy of using features derived from tags, Des, and tags together with Des.

The BFE dataset was randomly partitioned into two sets: two-thirds blogs were used for training and the rest one-third for testing. The experiment was conducted on the top 20 categories with the largest number of blogs (see Table 2 last column) using SVM^{light} package[4]. Binary classification setting was applied. That is, one classifier was learned for each category and the positive (negative) examples were the blogs belonging (not belonging) to the category. Those blogs that do not belong to any of the top 20 categories always served as negative training/test examples. Binary weighting scheme was used in our experiment. The commonly used $tf \cdot idf$ weighting scheme yielded similar classification accuracy and we chose not to report the results due to space limitation.

---

[4] http://svmlight.joachims.org/, accessed 24 Jun 07.

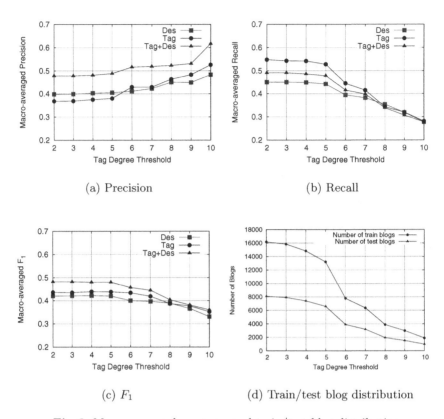

(a) Precision

(b) Recall

(c) F_1

(d) Train/test blog distribution

Fig. 3. Macro-averaged measures and train/test blog distribution

5.2   Experimental Results

We used *Precision, Recall* and F_1 to evaluate the classification accuracy. The results reported are macro-averaged measures [12].

To evaluate whether more tags lead to more accurate classification results, we obtained the classification results using different tag degree thresholds. For instance, if tag degree threshold is 5, then only those blogs having no less than 5 normalized tags will be involved in the classification. The number of train/test blogs against each tag degree threshold is given in Figure 3(d). It is clear that once the tag degree is above 5, the number of blogs in both training and test reduced sharply.

As shown in Figure 3(a), Tag together with Des, i.e., Tag+Des, achieved the best precision compared with either Tag or Des alone. With any of the three types of features, precision increased with the tag degree threshold. Tag achieved better precision than Des when tag degree was above 5. That is, more tags led to better precision. More tags, however, led to poorer recall as shown in Figure 3(b). Recall degraded sharply when tag degree is more than 5. Among all three types of features, Tag achieved the best recall and Des was the worst.

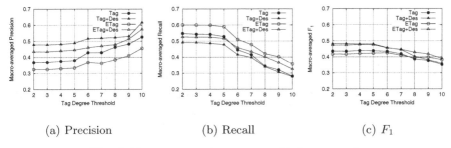

(a) Precision   (b) Recall   (c) F_1

Fig. 4. Macro-averaged measures with tag expansion

As a combined measure, F_1 show that Tag+Des achieved the best classification accuracy, followed by Tag. To summarize:

- Tag was more effective than Des in blog classification despite that average number of terms in Tag is half of that in Des.
- Tag combined with Des achieved the best classification accuracy.
- More tags led to better precision but poorer recall.

To evaluate the effectiveness of tag expansion, we applied the tag expansion algorithm to expand tags of blogs involved in both training and test. We used $\alpha = 0.15$, set number of iteration to be 2, and selected those tags having score $s^2(t_i) \geq 0.15$ as expanded tags. Study of the impact of number of iterations and score threshold is out of the scope of this paper due to space limitation. Figure 4 reports the performance of expanded tag (i.e., ETag) and ETag+Des. Results obtained using Tag and Tag+Des are plotted in the Figure for easy reference. From the results, it is clear that the expanded tag led to better recall but worse precision and slightly worse $F1$ when tag degree was less than 7. It is interesting to observe that when tag degree was high (e.g., ≥ 8) tag expansion achieved better F_1 compared with the non-expanded features. This may suggest that when too many tags are given to a blog, the tags are more specific to the blog and become less effective in determining the category of a blog. Nevertheless, the observation made from this experiment requires further study to better explain the reason behind, which is part of our future work.

6   Conclusion

We studied the problem of automatically classifying tagged objects (e.g., blogs) into pre-defined categories. Compared with title and description, which are often available for many objects, tags were more effective for accurate classification. Nevertheless, our experiments suggested that more tags did not necessarily lead to better classification, which calls for further study to better explain the reason behind. We have also evaluated a tag expansion algorithm which could improve the recall but hurt precision. As tags attached to blogs may not be strongly related to any particular posts, we believe our study could benefit the research on tags attached with other online objects including pictures, video and others.

References

1. Berendt, B., Hanser, C.: Tags are not metadata, but just more content - to some people. In: Proc. of Int'l Conf. on Weblogs and Social Media (ICWSM 2007), Colorado, USA (2007)
2. Brooks, C.H., Montanez, N.: Improved annotation of the blogosphere via autotagging and hierarchical clustering. In: Proc. of WWW 2006, Edinburgh, Scotland, pp. 625–632 (2006)
3. Dumais, S., Chen, H.: Hierarchical classification of web content. In: Proc. of SIGIR 2000, Athens, Greece, pp. 256–263 (2000)
4. Fogaras, D., Rácz, B., Csalogány, K., Sarlós, T.: Towards scaling fully personalized pagerank: Algorithms, lower bounds, and experiments. Internet Mathematics 2(3), 333–358 (2005)
5. Gruhl, D., Guha, R., Liben-Nowell, D., Tomkins, A.: Information diffusion through blogspace. In: Proc. of WWW 2004, New York, pp. 491–501 (2004)
6. Hayes, C., Avesani, P., Veeramachaneni, S.: An analysis of the use of tagging in a web blog recommender system. In: Proc. of IJCAI 2007, Hyderabad, India, pp. 2772–2777 (2007)
7. Jeh, G., Widom, J.: Scaling personalized web search. In: Proc. of WWW 2003, pp. 271–279. ACM Press, New York (2003)
8. Joachims, T.: Text categorization with support vector machines: learning with many relevant features. In: Proc. of 10th European Conf. on Machine Learning, Chemnitz, Germany, pp. 137–142 (1998)
9. Kolari, P., Finin, T., Joshi, A.: Svms for the blogosphere: Blog identification and splog detection. In: Proc. of AAAI 2006 Spring Symposium on Computational Approaches to Analysing Weblogs (2006)
10. Marlow, C., Naaman, M., Boyd, D., Davis, M.: Ht06, tagging paper, taxonomy, flickr, academic article, to read. In: Proc. of ACM HYPERTEXT 2006, Odense, Denmark, pp. 31–40 (2006)
11. Ni, X., Xue, G.-R., Ling, X., Yu, Y., Yang, Q.: Exploring in the weblog space by detecting informative and affective articles. In: Proc. of WWW 2007, Banff, Alberta, Canada, pp. 281–290 (2007)
12. Sebastiani, F.: Machine learning in automated text categorization. ACM Computing Surveys 34(1), 1–47 (2002)
13. Sood, S., Owsley, S., Hammond, K., Birnbaum, L.: Tagassist: Automatic tag suggestion for blog posts. In: Proc. of Int'l Conf. on Weblogs and Social Media (ICWSM 2007), Colorado, USA (March 2007)
14. Sun, A., Lim, E.-P.: Web unit mining – finding and classifying subgraphs of web pages. In: Proc. of ACM CIKM 2003, New Orleans, LA, USA, pp. 108–115 (2003)
15. Sun, A., Lim, E.-P., Ng, W.-K.: Web classification using support vector machine. In: Proc. of 4th WIDM held in conj. with CIKM 2002, Virginia, USA (2002)

Keyphrase Extraction in Scientific Publications

Thuy Dung Nguyen and Min-Yen Kan

Department of Computer Science, School of Computing,
National University of Singapore, Singapore, 117543
kanmy@comp.nus.edu.sg

Abstract. We present a keyphrase extraction algorithm for scientific publications. Different from previous work, we introduce features that capture the positions of phrases in document with respect to logical sections found in scientific discourse. We also introduce features that capture salient morphological phenomena found in scientific keyphrases, such as whether a candidate keyphrase is an acronyms or uses specific terminologically productive suffixes. We have implemented these features on top of a baseline feature set used by Kea [1]. In our evaluation using a corpus of 120 scientific publications multiply annotated for keyphrases, our system significantly outperformed Kea at the $p < .05$ level. As we know of no other existing multiply annotated keyphrase document collections, we have also made our evaluation corpus publicly available. We hope that this contribution will spur future comparative research.

1 Introduction

Keyphrases are defined as phrases that capture the main topics discussed in a document. As they offer a brief yet precise summary of a document content, they can be utilized for various applications. In an information retrieval (IR) environment, they serve as an indication of document relevance for users, as the list of keyphrases can quickly help determine whether a given document is relevant to their interest. As keyphrases reflect a document's main topics, they can be utilized to cluster documents into groups by measuring the overlap between the keyphrases assigned to them. Keyphrases also be used proactively in IR, in indexing. Good keyphrases supplement full-text indexing by assisting users in finding relevant documents.

Despite these known advantages of keyphrases, only a minority of documents have keyphrases assigned to them. This is because authors provide keyphrases only when they are instructed to do so [1], as manual assignment of keyphrases is expensive and time-consuming.

This need motivates research in finding automated approaches to keyphrase generation. Most existing automatic keyphrase generation programs view this task as a supervised machine learning classification task, where labeled keyphrases are used to learn a model of how true keyphrases differentiate themselves from other possible candidate phrases. The model is constructed using a set of features that capture the saliency of a phrase as a keyphrase.

In this work, we extend an existing state-of-the-art feature set with additional features that capture the logical position and additional morphological characteristics of

D.H.-L. Goh et al. (Eds.): ICADL 2007, LNCS 4822, pp. 317–326, 2007.
© Springer-Verlag Berlin Heidelberg 2007

keyphrases. Unlike earlier work that aim for a domain-independant algorithm, our work is tailored to scientific publications, where keyphrases manifest domain-specific characteristics. With our extended feature set, we demonstrate a statistically significant performance improvement over the well-known Kea algorithm [1] for scientific publications.

We first review previous approaches in automatic keyphrase generation next. We then describe the overall methodology for our system is described in Section 3, which details our new features used to enhance the baseline feature set. Evaluation, including our compilation of a suitable multiply-annotated corpus, is detailed in Section 4.

2 Related Work

Work on keyphrase generation can be categorized into two major approaches: *extraction* and *assignment*.

Keyphrase Extraction. Keyphrase extraction methods select phrases present in the source document itself. Such approaches usually consist of a candidate identification stage and a selection stage.

In the candidate identification stage, systems restrict the number of candidate phrases for later consideration in order to bound the computational complexity of the latter selection stage. Most systems we surveyed place either a length or phrase type restriction (e.g., noun phrases only). Kim and Wilbur [2] study this stage in more depth, proposing three statistical techniques for identifying content bearing terms, by examining the distributional properties of a candidate versus its context. Tomokiyo and Hurst [3] take a language modeling approach to keyphrase generation by calculating the phraseness of a candidate, which represents the extent to which a word sequence is considered to have a phrasal quality.

The bulk of the work comes in the selection stage, where the program judges whether a candidate is a keyphrase or not. In a supervised learning scenario, this stage critically hinges on the features used to describe a candidate. Barker and Cornacchia [4] used three features to build their model: candidate word length, occurrence frequency, and head noun frequency. Turney's GenEx [5] system computed a vector of nine features to represent candidates. These features captured candidate length and frequency like Barker and Cornacchia's system, but additionally modeled the candidate's position within the document. Frank et al. [1] introduced Kea keyphrasing system. Although they pursued numerous features, their final feature set only used three independent features for classification: 1) the TF×IDF score, 2) the position of the first occurrence, and 3) corpus keyphrase frequency, which measures how many times the candiate was assigned as a keyphrase in other training documents. Despite the reduced size of their feature set, Kea's performance is reported as comparable to GenEx.

Work by Turney [6] noted that candidate selection decisions are not independent. In other words, prior keyphrase selections should have an influence on the remaining selection decisions. He proposed to model the coherence of an entire set of candidate phrases using pointwise mutual information (PMI) between a candidate and k previously selected phrases. However, the PMI for these sets are difficult to obtain without sufficiently large datasets; Turney proposed using web search engine queries to obtain rough collocation estimates, although this has marked drawbacks in terms of network bandwidth and time inefficiency.

Supervised text classification is not the only method for keyword extraction. Probabilistic topic models [7] treat documents as a mixture of topics and topics as a probability distributions over words. Thus, topic models can be considered as generative models for documents, and dually, given a document one can infer the topic(s) responsible for generating that document. While quite potent, topic models also rely on large amounts of training data, and are ineffective for small corpora.

Keyphrase Assignment. In contrast to extraction, *keyphrase assignment* is typically used when the set of possible keyphrases is limited to a known, fixed set, usually derived from a controlled vocabulary or set of subject headings. Here, binary classifiers can be trained for each keyphrase k in the set, and the assignment of keyphrases for a document is given by running all k classifiers and assigning those which indicate a positive result. In essence, keyphrase assignment is the same as traditional multiclass text classification.

For such approaches, as the keyphrases are known *a priori*, mutual information between the keyphrase and other words in the document can be used to do feature selection [8]. If the keyphrases form a ontology with broader, narrower and related term linkages, these relations can also be harnessed to provide additional evidence for inference [9]. Medelyan and Witten [10] used thesaural relations as edges to calculate the connectivity degree of a candidate keyphrase, showing that this feature (in conjunction with others) also statistically improved assignment accuracy. A drawback of the keyphrase assignment method is that it requires a large annotated corpus, as suitable number of training examples need to be found for each possible keyphrase.

3   Methodology

Given the current state of keyphrase generation, we chose to use an extraction based approach, as no suitable compilation of subject headings or ontology exists that aim to facilitate retrieval effectiveness. Extraction-based methods also generate a more diverse set of keyphrases, which we believe would better support relevance assessment. We also chose to use a supervised approach, as other methods require large amounts of annotated corpora, which we did not have.

Among the surveyed related work, the Kea algorithm fits this specification quite well. Kea uses just a few domain-independent features that have been shown to yield robust yet state-of-the-art results. For these reasons, we chose it as the baseline system for comparison.

In developing a keyphrase method for scientific publications, we note that such documents distinguish themselves from others based on their use of technical language as well as their rich document structure. As such, we have tried to capitalize on these features in modeling as well. Key enhancements in our work is to compute such additional features that model keyphrases in terms of their 1) morphological status and 2) document-centric structural character.

Figure 1 shows the outline of our system and highlights our new contributions to keyphrase extraction in gray. Like the baseline system Kea, our system follows a

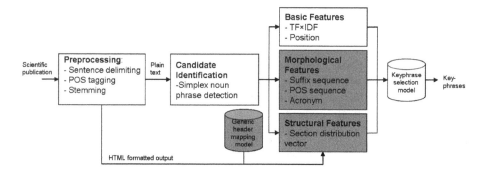

Fig. 1. System architecture. Contributions of this paper are highlighted in gray.

supervised machine learning approach. Training documents are used to generate linguistically motivated features and the extracted annotation from the training data serves as the class label $C = \{keyPhrase, \neg keyPhrase\}$.

Preprocessing is first done to convert the document from PDF to plain text and HTML formats, using the PDF995 utility suite. The plain text form is first processed to delimit sentences, then passed to a modern maximum entropy based part-of-speech (POS) tagger [11].

For candidate identification, all simplex noun phrases (i.e., ones without post modification, such as relative clauses and prepositional phrases) are deemed as keyphrase candidates. Case folding and stemming is also done to conflate statistics for variants, but only after the relevant morphological features for the individual candidate are calculated.

Candidate selection is the primary workhorse for keyphrase extraction. As stated, our key contribution is in introducing two additional sets of features that help to model the document structure of scientific publications as well as the characteristic terminological morphology. All extracted features (detailed in the next three subsections are used as evidence to create a keyphrase model using the standard Naïve Bayes learner implemented in the Weka machine learning toolkit [12].

3.1 Baseline Feature Set

We first review the two domain-independent features used by Kea and also in our enhanced system. Note that we did not use the **keyphrase frequency** feature of Kea, as this feature was reported only effective when sufficiently large training data is provided.

Term frequency $\times$ Inverse document frequency (TF$\times$IDF) - This is the standard salience metric used in information retrieval. Within a single document, frequently occurring terms are given high weight; over an entire corpus, terms that occur in few documents are given high weight. There are many specific formulations of tf$\times$idf; here we use a logarithm to dampen the inverse frequency term:

$$w_{ij} = \frac{f_{ij}}{max(f_{ij})} \times log_2 \frac{N}{df_i} \tag{1}$$

Position of first occurrence - This feature reflects the belief that keyphrases tend to appear at specific locations in the document (e.g., at the beginning). *Position* is calculated as the number of words that precede its first appearance, divided by the number of words in the document.

3.2 Extended Structural Features

Different logical sections of scientific publications contribute keyphrases at different rates. For example, few true keyphrases appear in experimental results but more occur in the *Abstract* or *Methods* sections. In a sense, the baseline position feature is a coarse-grained approximation of this, as academic publications tend to follow a consistent sequential structure: with an *Abstract*, followed by an *Introduction*, *Related Work*, *Methods*, *Evaluation*, *Conclusions* and *References*. We thus add an additional set of features to add this to our keyphrase model.

Section occurrence vector - We model the distribution of the keyphrase among different logical sections as a vector of frequency features for 14 generic section headers (as shown in Table 1. However, as headers in individual papers may deviate significantly from the norm (e.g., "Discussion" often should map to *Evaluation*, inferring how individual header instances map to generic headers is difficult. We created a maximum entropy (ME) based classifier that used four features – corresponding to 1) section number, 2) relative position, 3) previous section header and 4) current section header – to infer the generic section header (from our own list of 14 headers, as shown below in Table 1) for the input documents. The ME method was evaluated using ten fold cross validation on a corpus of 1020 annotated headers, garnering 938 correct assignments (92% accuracy). We also tried using the same features in a Hidden Markov Model (HMM) framework, but this only achieved 369 correct assignments, accruing a much lower accuracy (36%). We thus employ the ME version of the header mapper on an individual paper's headers (detected using orthography and numbering cues from the HTML converted format) to create the feature vector. Details of this header processing are omitted for space reasons; the interested reader is referred to the first author's thesis [13].

Table 1. The 14 generic headers used by our logical section detection module

Abstract	Categories and Subject Descriptors	General Terms
Introduction	Background	Methods
Conclusions	References	Evaluation
Related Work	Acknowledgments	Applications
Motivation	Implementation	

3.3 Extended Morphological Features

Jones and Paynter's study [14] has validated claims that authors often do choose good keyphrases for their own documents. We thus analyzed author-provided keyphrases of scientific publications to assess what characteristics a good keyphrase should possess. We focused on the linguistics characteristics of keyphrases assigned by authors.

POS sequence - We observed that almost all of the author assigned keyphrases are noun phrases, but whose part-of-speech tag sequence varies. For example, nominal modifiers to the headword feature occur more frequently than adjectival ones (e.g., "additive"/NN versus "additional"/JJ). This trend was observed for both bigram and trigram keyphrases. We use the POS tag sequence of the candidate as a single feature in our extended feature set.

Suffix sequence - In English, suffixes also hint at the terminological status of a candidate. Headwords of keyphrases manifest different suffix distributions than modifiers. We noticed that some suffixes such as $-ion, -ics, -ment$ often appear on headwords while others like $-ive, -al, -ic$ appear on modifiers. We use the sequence of morphological suffixes in a candidate as single feature. This feature partially overlaps with the POS sequence feature but is considerably more fine-grained.

Acronym status - Authors often introduce acronyms for phrases that are used many times in a document, saving space and making reference considerably easier. While there are considerably more sophisticated methods to detect acronyms, we found it sufficed to use use a simple approach. Our approach (Algorithm 1 scans for parenthetical expressions in the text and the preceding text can be considered a correspondance. We use a binary feature to indicate whether a candidate is an acronym.

Algorithm 1. Psuedocode for our simple acronym detection algorithm

Retrieve all the texts $T_1 \ldots T_N$ within parentheses () in document
for $i = 1$ to N **do**
  if length of $T_i < 2$ **then**
    Consider T_i as being neither acronym nor definition, continue
  end if
  if (T_i is in upper- or mixed-case) AND length of $T_i < MAX$ **then**
    Assume T_i is an acronym
    Move toward the left to get its definition def_i
    if def_i exists **then**
      Record the acronym T_i and its definition def_i
    end if
  else
    Assume T_i is the definition
    Move toward the left to get its acronym $acro_i$
    if $acro_i$ exists **then**
      Record the acronym $acro_i$ and its definition T_i
    end if
  end if
end for

---

4   Evaluation

Two main approaches to evaluation present themselves. The first approach involves the manual evaluation of generated keyphrases. Here, subjects are given the document and the generated keyphrase list and asked to rank the relevance of each phrase. A disadvantage of this approach is that it requires manual effort, but more significantly, such an approach does not aid any subsequent evaluation, as the relevant assessment needs to be done from scratch every time. The second approach adopts the standard IR metrics of precision and recall to measure how well the generated keyphrases match a gold-standard assigned keyphrases. We take this second approach, but a question of how to come up with a gold standard arises.

4.1 Data Collection

Evaluating keyphrases has shown to be subjective and difficult. Jones and Paynter (2001) proved that author keyphrases are good representations of the subject of a document. However, generate keyphrase extraction evaluation requires multiple judgments and cannot rely merely on the single set of author-provided keyphrases [10]. Although author assigned keyphrases are usually viewed as a good representation of the subject of a document, they may not be able to cover all the good keyphrases in a document as keyphrase assignment is inherently subjective: keyphrases assigned by one annotator are not the only correct ones.

Unfortunately, we could not find a publicly available scientific document dataset tagged by multiple reliable annotators with keyphrases[1]. We thus constructed our own data set that fits these qualities for the evaluation of our algorithm.

We first found suitable publications and then collected keyphrases from manual annotators. We first used the Google SOAP API to find documents using variants of the query "keywords general terms filetype:pdf". We downloaded over 250 of these PDF documents for futher processing. Documents were then manually restricted to scientific conference papers, with a length range of 4-12 pages. As our program only deals with textual input, we converted the PDF to plain text using the the PDF995 software suite as it handled two-columned text better than other programs tried. At the end of this process, we had 211 documents in plain text format which were converted successfully without problems.

We then recruited student volunteers from our department to participate in manual keyphrase assignment. Each volunteer was given three of PDF files (with author-assigned keyphrases hidden) to assign keyphrases to. To spur future research on automatic keyphrasing, we are making the full dataset and its details publicly available[2].

4.2 Results

For the experiments reported in this chapter, we used a subset of full dataset consisting of 120 documents, each of which has two keyphrase sets: one by the original author and the other by our volunteer. For each document, accuracy is the number of matches among keyphrases in the standard set and ten top-ranked extracted phrases.

To ensure clean separation between training and testing documents for our system and the trainable Kea baseline, all results reported here are obtained using ten-fold cross validation.

---

[1] We considered a corpus of socially "tagged" papers from citeulike.org, but rejected this as authors occasionally choose keyphrases for purposes other than document description.

[2] http://wing.comp.nus.edu.sg/downloads/keyphraseCorpus

Table 2. Evaluation results. Statistical signficance over the baseline shown in parentheses (2-tailed paired t-tests).

System	Average # of exact matches	Average score based on weight
Kea (baseline)	3.03	3.61
Our system	3.25 (0.024)	3.84 (0.033)

Table 2 shows the average number of exact matches of the two algorithms with respect to the gold standard in the second column. Aside from an exact match of keyphrase in the gold standard, we can calculate a weighted match score based on the number of keyphrase sets in which the keyphrase appears. Let n be the number of keyphrases set in which a phrase p appears. Its weight $w(p)$ is computed as $w(p) = 1 + ln(n)$. A corresponding average matching score based on this weight is shown also in Table 2 as the third column.

We perform two-tailed paired t-test to see whether the improvements are significant. The corresponding p-values are also shown in the table, which indicate that the results are significant at the $p < 0.05$ level.

4.3   Error Analysis

We performed some post-experimental analysis of the errors created by both systems that lead to the generation of poor keyphrases. Our analysis leads to two problematic areas for future improvement. One difficulty is in deciding whether a general term is a good keyphrase or not. This can be seen in Table 3 document. Phrases such as "data" and "cell" are too general to be useful keyphrases. These phrases appear many times in the document, having high TF×IDF scores, and also appear in important sections, such as the abstract and introduction, which results in their sectionrelated features are the same with those of correct keyphrases.

Table 3. Author and generated keyphrases for the sample document *Analysis of Soft Handover Measurements in 3G Network* (36.pdf) in our keyphrase corpus. Only the "soft handover" keyphrase was provided by both the author and the volunteer annotator. Output keyphrases that match with assigned keyphrases are presented in italic font.

Assigned keyphrases	Kea baseline	Our system
Neural network algorithm	Handover	Clusters
3G network	*Soft handover*	*Soft handover*
Visualization capability	3G	Data
Cluster analysis	Clusters	*3G network*
Self organizing map	*3G network*	Interesting clusters
Hierarchical clustering	Cell	Handover attempts
Key performance indicator of handover	Cell pairs	Method
Two-phase clustering algorithm	SHO	*Neural network*
Soft handover (2)	Active set	Measurements
Histograms	Handover measurements	Handover measurement
Decrease in computational complexity		
Mobility management		
Data mining		
neural networks		

Another problem area is in generating suitable long keyphrases (i.e., phrases with three words or more). Currently, these are rarely generated by the current methodology. In the sample text, no three-or-more word phrases are generated among in the ten outputs, although they make up 5 of the 14 manually assigned keyphrases in the gold standard set.

5   Conclusion

We have presented an improved feature set for the problem of keyphrase extraction in scientific publications. The set adds features for representing logical position of the keyphrase instances with respect to sections of the document, and features to model whether a candidate phrases is an acronym or abbreviation, two salient sources of keyphrases in scientific discourse. Applying the new features in Naïve Bayes model does have a significant improvement against the state-of-the-art baseline Kea [1].

In evaluating our work, we have also compiled a corpus of more than 200 scientific publications, with multiple keyphrase sets. Each publication was annotated by volunteers to provide additional keyphrase coverage aside from the set provided by the original author. Such coverage is essential to the evaluation of keyphrase extraction algorithms in terms of coverage and importance of individual keyphrases. We have made this corpus publicly available and we believe that it will be useful in future work on keyphrase extraction.

Our current work focuses on deployment, in which we apply this keyphrase extraction module automatically over a large set of freely available scientific publications found on the web (i.e., CiteSeer). We are interested in merging such an automated facility with social user tagging. Future work on the extraction algorithm itself will focus on generating longer, more descriptive keyphrases, a key weakness as discussed in our error analysis.

References

1. Frank, E., Paynter, G.W., Witten, H.I., Gutwin, C., Nevill-Manning, C.G.: Domain specific keyphrase extraction. In: Proceedings of the 16th International Joint Conference on Artificial Intelligence, pp. 668–673 (1999)
2. Kim, W., Wilbur, W.J.: Corpus-based statistical screening for content-bearing terms. J. Am. Soc. Inf. Sci. Technol. 52, 247–259 (2001)
3. Tomokiyo, T., Hurst, M.: A language model approach to keyphrase extraction. In: Proceedings of ACL Workshop on Multiword Expressions (2003)
4. Barker, K., Cornacchia, N.: Using noun phrase heads to extract document keyphrases. In: Proc. of the 13th Biennial Conf. of the Canadian Society on Computational Studies of Intelligence, pp. 40–52. Springer, Heidelberg (2000)
5. Turney, P.D.: Learning to extract keyphrases from text. Technical Report ERB-1057, National Research Council, Institute for Information Technology (1999)
6. Turney, P.D.: Coherent keyphrase extraction via web mining. In: IJCAI 2003. Proceedings of the Eighteenth International Joint Conference on Artificial Intelligence, pp. 434–439 (2003)
7. Steyvers, M., Griffiths, T.: Probabilistic topic models. In: Landauer, T., Mcnamara, D., Dennis, S., Kintsch, W. (eds.) Latent Semantic Analysis: A Road to Meaning, Laurence Erlbaum, Mahwah (2005)

8. Dumais, S.T., Platt, J., Hecherman, D., Sahami, M.: Inductive learning algorithms and representations for text categorization. In: CIKM. Proc. of 7th International Conference on Information and Knowledge Management, pp. 148–155 (1998)
9. Pouliquen, B., Steinberger, R., Ignat, C.: Automatic annotation of multilingual text collections with a conceptual thesaurus. In: BUG (2003)
10. Medelyan, O., Witten, I.H.: Thesaurus based automatic keyphrase indexing. In: Proceedings of the 6th ACM/IEEE-CS joint conference on Digital libraries, pp. 296–297. ACM Press, New York (2006)
11. Ratnaparkhi, A.: A maximum entropy part of speech tagger. In: Proc. ACL-SIGDAT Conference on Empirical Methods in Natural Language Processing, Philadelphia (1996)
12. Witten, I.H., Frank, E.: Data Mining: Practical machine learning tools and techniques, 2nd edn. Morgan Kaufmann, San Francisco (2005)
13. Nguyen, T.D.: Automatic keyphrase generation. Technical report, National University of Singapore (2007)
14. Jones, S., Paynter, G.W.: Human evaluation of Kea, an automatic keyphrasing system. In: ACM/IEEE Joint Conference on Digital Libraries, pp. 148–156 (2001)

Automated Template-Based Metadata Extraction Architecture

Paul Flynn, Li Zhou, Kurt Maly, Steven Zeil, and Mohammad Zubair

Department of Computer Science
Old Dominion University, Norfolk, VA. 23529
{pflynn,lzhou,maly,zeil,zubair}@cs.odu.edu

Abstract. This paper describes our efforts to develop a toolset and process for automated metadata extraction from large, diverse, and evolving document collections. A number of federal agencies, universities, laboratories, and companies are placing their collections online and making them searchable via metadata fields such as author, title, and publishing organization. Manually creating metadata for a large collection is an extremely time-consuming task, but is difficult to automate, particularly for collections consisting of documents with diverse layout and structure. Our automated process enables many more documents to be available online than would otherwise have been possible due to time and cost constraints. We describe our architecture and implementation and illustrate the effectiveness of the tool-set by providing experimental results on two major collections DTIC (Defense Technical Information Center) and NASA (National Aeronautics and Space Administration).

Keywords: Metadata, heterogeneous collections, automation.

1 Introduction

A number of federal agencies, universities, laboratories, and companies are placing their collections online and making them searchable via metadata fields such as author, title, and publishing organization. To enable this, every document in the collection must be catalogued using the metadata fields. A typical cataloguing process requires a human to view the document on the screen and identify the required metadata fields such as title, author, and publishing organization, and to enter these values in some online searchable database. Manually creating metadata for a large collection is an extremely time-consuming task. According to Chrystal [1], it would take about 60 employee-years to create metadata for 1 million documents. These enormous costs for manual metadata creation suggest a need for automated metadata extraction tools. The Library of Congress Cataloging Directorate recognized this problem [2] and sponsored a study, Automatic Metadata Generation Applications (AMeGA) [3], to identify challenges in automatic metadata creation.

Though time consuming, the task of identifying metadata fields by visually looking at the document is easy for a human. The visual cues in the formatting of the document along with accumulated knowledge and intelligence make it easy for a human to identify various metadata fields. Writing a computer program to automate this task is a

D.H.-L. Goh et al. (Eds.): ICADL 2007, LNCS 4822, pp. 327–336, 2007.
© Springer-Verlag Berlin Heidelberg 2007

research challenge. Researchers in the past have shown that it is possible to write programs to extract metadata automatically for a homogeneous collection (a collection consisting of documents with a common layout and structure). Unfortunately a number of federal organizations such as DTIC [4], GPO [5], and NASA [6] manage heterogeneous collections consisting of documents with diverse layout and structure, where these programs do not work well. Furthermore, even with the best possible automated procedures, numerous sources of error exist, including some that cannot be controlled, such as scanned documents with text obscured by smudges, signatures, or stamps. A commercially viable process for metadata extraction must remain robust in the presence of these external sources of error as well as in the face of the uncertainty that accompanies any attempts to automate "intelligent" behavior. How to reach the desired accuracy and robustness for a large and evolving diverse collection consisting of documents with different layout and structure is still a major research issue. We have developed and demonstrated a novel process for extracting metadata. Among the innovations is a two-part process that directly addresses the problem of coping with large heterogeneous collections by breaking the extraction problem into smaller, manageable pieces:

- A new document is classified, assigning it to a group of documents of similar layout. The goal is to group together documents whose title or other metadata-containing pages would appear similar when viewed (by humans) from several feet away.
- Associated with each class of document layouts is a template, a scripted description of how to associate blocks of text in the layout with metadata fields. For example, a template might state that the text set in the largest type font in the top-half of the first page is, in that layout, the document title.

We have tested our process and software against the DTIC collection which contains more than one million documents and adds tens of thousands of new documents each year. The documents are diverse, including scientific articles, slides from presentations, PhD theses, (entire) conference proceedings, promotional brochures, public laws, and acts of Congress. Contributions to DTIC come from a wide variety of organizations, each with their own in-house standards for layout and format, so, even among documents of similar kind, the layouts vary widely. Our tests resulted in an overall accuracy of 83% for documents with defined templates.

2   Metadata Extraction Approaches

Existing automated metadata extraction approaches can be divided into two main categories: learning systems and rule-based systems.

Learning techniques including SVM [7][8] and HMM [9] have been employed with promising results but to relatively homogeneous document sets. The investigators' own experiments with these techniques [10] suggest a significant decline in effectiveness as the heterogeneity of the collection increases. We believe that exposure of these learning systems to heterogeneous collections tends to dilute the internal probabilities that control their internal transitions. Evolution (changing characteristics over time, such as acquiring a new source of documents in an unfamiliar format) poses a difficulty for these techniques as well, as they necessarily exhibit significant inertia resisting changes to the internally acquired "knowledge" until a significant number of examples of the new characteristics have been encountered.

Rule-based systems [11][12][13] use programmed instructions to specify how to extract the information from targeted documents. With sufficiently powerful rule languages, such techniques are, almost by definition, capable of extracting quality metadata. Heterogeneity, however, can result in complex rule sets whose creation and testing can be very time-consuming [13]. Analogies to typical software complexity metrics [14] suggest that complexity will grow much more than linearly in the number of rules, in which case even a well-trained team of rule-writers will be hard-pressed to cope with changes in an evolving heterogeneous collection and maintain a conflict-free rule set.

Our own approach [10][15] can be seen as a variant of the rule-based approach, but we finesse the complexity induced by heterogeneity and evolution by first classifying documents by layout, then providing a template for each layout, so that templates are independent of one another and individually simple.

3   Architecture and Implementation

3.1   Overview of Architecture

Our template-based metadata extraction system is composed of commercial and public domain software in addition to components developed by our team. Figure 1 shows the complete process. Documents are input into the system in the form of PDF files, which may contain either text PDF or scanned images. Some documents may contain a Report Document Page (RDP), one of several standardized forms that is inserted into the document when the document is added to the collection. For the DTIC collection, more than 50% of the documents contain RDPs offering more than 20 metadata fields.

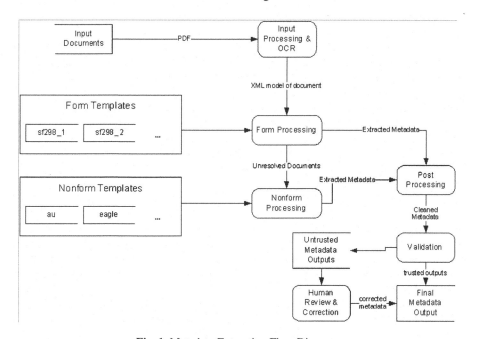

Fig. 1. Metadata Extraction Flow Diagram

The documents enter the input processing system where they are truncated, processed by an Optical Character Recognition (OCR) program and converted to a standardized XML format. The first extraction step is to search for and recognize any RDP forms present. Any documents without recognized forms enter the non-form extraction process. The non-form extraction process generates a candidate extraction solution from the templates available. After extraction, the metadata from both form and non-form processing enter the output processor. The output processor is comprised of two components: a post-processing module and a validation module. The post-processing module handles cleanup and normalization of the metadata. The final automated step of the process is the validation module which, using an array of deterministic and statistical tests, determines the acceptability of the extracted metadata. Any document that fails to meet the validation criteria is flagged for human review and correction.

3.2 Implementation

Input Processing. The source documents come into our system as PDF format files. These documents range from several pages to hundreds of pages in length. Our research into the collection has shown that the metadata we are interested in can typically be found in the first or last five pages of a document. Based on this observation, we use the program pdftk [16] to split the first and last five pages out of the document and into a new PDF document. This truncated PDF document is fed into a commercial optical character recognition (OCR) for conversion into an XML format. We have selected ScanSoft's OmniPage Pro as the OCR engine since it supports batch processing of PDF files with very good results. OmniPage saves the recognized file into a proprietary XML format which contains page layout as well as the recognized text.

The initial prototype of our extraction engine was based on the proprietary XML format used by OmniPage Pro version 14. However, by the time of the deployment of the initial prototype, the site was using OmniPage Pro version 15, which uses a different proprietary format that changed every XML tag except for the "word" tag and added dozens of new tags. Our form-based extraction engine is tightly coupled to the schema of the incoming XML documents, so supporting this new version of the OmniPage schema would require major recoding of the extraction engine, with the end result being another tight coupling to another proprietary schema. To forestall any future conflicts with schema changes, we decided to develop our own schema to decouple our project from proprietary schemas.

Independent Document Model (IDM). We based our new Independent Document Model (IDM) on the OmniPage 14 schema we already supported with our project. This step helped to minimize the re-coding cost for the extraction engine. The main structural elements are pages, regions, paragraphs, lines and words. The geometric boundaries of each of the structural elements are included as attributes. Style information such as font face, font size and font style, is recorded at the line and word levels. Alignment and line spacing are recorded at paragraph elements. Tables are composed of a sequence of cells that represent a virtual row-column table with each cell encoded with the upper-left coordinate and the row and column spans of the cell.

IDM documents are created by means of XSL 2.0 stylesheets. A different stylesheet is used for each type of source document. We have created stylesheets to support creation of IDM documents from either OmniPage 14 or 15 source documents. Similarly, additional OCR programs can be supported in the future by creation of XSL stylesheets to make the transformation.

Form Processing. Our experience with the DTIC collection has shown that about 50% of the documents contain an RDP form. The regular layout present in an RDP form makes it an attractive target for a template-based extraction process. In order to take advantage of the geometric relationships between fields in a form, we created an alternate version of our template language and extraction engine. The metadata fields are specified by a matching string and a set of rules indicating a positional relationship to one or more other fields (e.g., Figure 2). The number and layout of the fields for each different form constitute a unique signature for that form class. If a template describing form A is applied to a document containing form B, the resultant metadata returned will contain few if any fields. We have leveraged this property in the design of our extraction process.

Input processing finishes with IDM based documents exiting the input processor and entering the form processor. The processor is populated with a template developed for each version of RDP form found in the collection. We have found six different RDP forms within 9825 documents in the DTIC collection. The form processor runs the extraction process against the document using each of the templates and then selects the template, which returns the best results. If the form processor fails to match any template the document moves into the non-form extraction process described below. The extracted metadata is sent into the output processor.

```
        <field num="16->c"><line>c. THIS PAGE</line></field>
    </fixed>
    <extracted>
        <metadata name="ReportDate">
            <rule relation="belowof" field="1"/>
            <rule relation="aboveof" field="4|5a"/>
        </metadata>
```

Fig. 2. Form-based template fragment. The *(line)* elements in the *(field)* elements define string matching criteria. The *(rule)* elements defined for each *(metadata)* element defines the geometric placement.

Non-form Processing. As shown in Figure 1, documents without an RDP form enter the non-form processor. The documents are first transformed from IDM into another XML format called CleanML, which encodes the paragraphs and lines and their corresponding features (font size, style and alignment) into an XML structure. This simplified structure allows the extraction engine to repeatedly iterate over the content to apply the rules.

Template Construction. The non-form extraction engine also uses rule-based template extraction to locate and extract metadata. Each template contains a set of rules designed to extract metadata from a single class of similar documents. Figure 3 shows a template example. Each desired metadata item is described by a rule set designating the beginning and the end of the metadata. The rules are limited by features detectable at the line level resolution. We hope to address this deficiency in future versions. The first step in constructing a template is to identify a set of documents which share a structural or visual similarity. Once a class is selected, the template author determines the set of rules for each metadata tag by identifying the appropriate function to select the beginning and the end of the tag.

```
<structdef pagenumber="3" templateID="arl_1">
    <CorporateAuthor>
        <begin inclusive="current">
            <stringmatch case="no" loc="beginwith">Army
                Research</stringmatch>
        </begin>
        <end inclusive="before">
            <stringmatch case="no"
                loc="beginwith">ARL</stringmatch>
        </end>
```

Fig. 3. Non-form Template fragment

```
<val:validate collection="dtic">
    <val:sum>
        <val:field name="UnclassifiedTitle">
            <val:rescale
                function="0.499 -0.01 0.5 0.5 1.0 1.0">
            <val:average>
              <val:dictionary/>
              <val:length/>
            </val:average>
            </val:rescale>
        </val:field>
```

Fig. 4. Validation script fragment for DTIC collection. Each metadata field such as "*UnclassifiedTitle*" and "*PersonalAuthor*" is assigned a function for validation.

Non-form Classification. For purposes of our discussion we define a class as a group of documents from which the metadata can be extracted using the same template. The members of a class can be selected based on structural or visual similarity. The original design of our system used several different layout classification schemes in order to separate the incoming documents into the appropriate class for extraction [10][11]. As described later, we also created a validation system to flag suspicious data extracted by a template [17][18]. We found that by applying every available template to a document, we could use the validator as a post hoc classification system for selecting the proper template. This post hoc classification system is configured by creating a "validation script" (e.g., Figure 4), which defines a set of rules to be used for calculating a confidence value for individual fields as well as an overall confidence calculation. Figure 5 is an example of the validator output for the "alr_2" template. Table 1 shows the validation values for five of the eleven templates applied by the extraction system for the same file. (The other six templates did not produce any output for the file.) The best result, *alr_2*, differs from the next best, *alr_1*, by the extraction of an additional personal author. This is precisely the behavior and level of discrimination we desire in a classifier.

```
<metadata confidence="4.694">
   <UnclassifiedTitle confidence="0.891">Air Gun Launch
      Simulation Modeling and Finite Element Model
      Sensitivity Analysis</UnclassifiedTitle>
   <PersonalAuthor confidence="0.785">Mostafiz R.
      Chowdhury</PersonalAuthor>
   <PersonalAuthor confidence="0.713">Ala
      Tabiei</PersonalAuthor>
   <CorporateAuthor confidence="0.76">Army Research
      Laboratory Adelphi, MD 20783-1145</CorporateAuthor>
   <CorporateAuthor confidence="0.0"
      warning="CorporateAuthor: too many
         unknown words">Weapons and
      Materials Research Directorate, ARL</CorporateAuthor>
```

Fig. 5. Sample fragment of validator confidence values. In this example, we see that the second *CorporateAuthor* gives a low confidence score because of the existence of too many words not in the CorporateAuthor dictionary.

Table 1. Sample validator confidence values for a single file

Template	Total Confidence	Field Confidences			
		Unclassified Title	Personal Author	Corporate Author	Report Date
alr_2	4.694	0.891	0.785 0.713	0.760 0.000 0.546	1.000
alr_1	3.436	0.891	0.785	0.760 0.000	1.000
nsrp	1.000				1.000
rand	0.848	0.848	0.000		
nps_thesis	0.000		0.000		0.000

Output Processing. Referring back to the architecture diagram in Figure 1, the extracted metadata from both form and non-form processes enter output processing for post-processing cleanup and validation.

Post-processing. The post-processing step is designed to compensate for the inherent uncertainties involved in the OCR recognition and extraction process. We have designed a modularized post-processing system which can provide a variety of post-processing functions for each metadata field. For example, modules may be designed to parse multiple authors from a single personal or corporate author entry and to reformat date fields into a specific standard.

As an example of a post-processing module, we have one module that attempts to standardize acceptable field values in form processing and to overcome the potential for misrecognition by the OCR software. The module analyzes specific fields by comparing the extracted data to values in an authority file. The module compares these values via fuzzy string matching based on edit distance. Additionally, the post processor can match variable phases where the comparison is successful so long as every word in the authority file entry is contained in the extracted data. We generated the authority file by extracting field data from more than 9000 documents.

Validation. The final step in our process is the validation step. The primary purpose of this step is to determine whether or not to flag the extracted metadata for human review. We will be using the same validation engine as mentioned above in post hoc classification. This validation engine uses statistical models of previously extracted metadata in the collection along with dictionaries for names and specialized content to determine the norms for the collection. While the validator will use the same validation engine to assess individual field values, we do not anticipate using the same script used in the non-form post hoc classification system. At this point we have not yet integrated the final validation module into the implementation. We are currently experimenting to determine an appropriate script to use.

4   Experimental Results

For our experiments we downloaded 9825 documents from the DTIC collection and 728 from the NASA collection. The internal distribution between forms and non-form documents for the collections are 94% RDP forms for DTIC and 21% RDP for NASA. We conducted a series of experiments to evaluate the effectiveness of the extraction process.

4.1   Form Extraction Experiments

The large number of form documents involved prohibits inspecting every document during testing. As such, we randomly sampled 100 form documents from the DTIC collection distributed roughly along the same distribution of the collection. We examined

Table 2. Results for DTIC Form Extraction

Class	Samples	Recall	Precision
Citation_1	10	100%	100%
Sf298_1	30	91%	95%
Sf298_2	30	98%	99%
Sf298_3	10	68%	96%
Sf298_4	10	100%	100%
Control	10	96%	100%

each of the 100 documents and determined the accuracy of the extracted metadata. The results of this experiment are shown in Table 2. Note that the low recall found under the SF298_3 class was due to poor quality of the source documents and resulting OCR recognition.

4.2 Non-form Extraction Experiments

We conducted experiments to confirm the efficiency of the post hoc classification system and the ability to extract the metadata. To test the ability of the system to select the appropriate template for extraction, we manually classified the DTIC non-form documents into 37 separate classes with at least 5 members. We wrote templates for the 11 largest classes and tested the ability of the extractor to correctly identify the proper class. We achieved an 87% classification accuracy when compared to manual classification results.

The overall accuracy for the non-form extractor was 66% for DTIC and 64% for NASA. The lower value is mostly due to the fact that we have only written a limited number of templates. Assuming that we write all the necessary templates, we expect accuracy in the 90% range.

5 Conclusions and Future Work

We have described our two-stage approach to metadata extraction that extends previous research in metadata extraction to growing, large, and heterogeneous collections. The basic system has been implemented and applied to two major collections with near perfect for documents that contain an RDP form and approximately 65% accuracy for those without a form. Significant contributions of our approach are the post-processing and the validation concepts. In post-processing, we clean metadata via field- and collection-specific modules. In validation we first obtain a statistical model of the collection (done only once) and use this model to validate the output.

We still have to design and implement the human correction interface together with the module that will invoke human intervention based on scores obtained in the validation phase.

References

1. Crystal, A., Land, P.: Metadata and Search: Global Corporate Circle. In: DCMI 2003 Workshop, Seattle, Washington, USA (2003), http://dublincore.org/groups/corporate/Seattle/
2. Library of Congress, Bibliographic Control of Web Resources: A Library of Congress Action Plan, http://www.loc.gov/catdir/bibcontrol/actionplan.html
3. Greenburg, J., Spurgin, K., Crystal, A.: Final Report for the Automatic Metadata Genersation Applications (AMeGA) Project (2005), UNC School of Information and Library Science, http://ils.unc.edu/mrc/amega/
4. Defense Technical Information Center. Public Scientific and Technical Information Network (2007), http://stinet.dtic.mil/str/index.html
5. National Aeronautics and Space Administration. NASA Technical Reports Server (2007), http://ntrs.nasa.gov/search.jsp

6. U.S. Government Printing Office. A Strategic Vision for the 21st Century. Technical report (2004)
7. Han, H., Manavoglu, E., Zha, H., Tsioutsiouliklis, K., Giles, C.L., Zhang, X.: Rule-based word clustering for document metadata extraction. In: Preneel, B., Tavares, S. (eds.) SAC 2005. LNCS, vol. 3897, pp. 1049–1053. Springer, Heidelberg (2006)
8. Han, H., Giles, C.L., Manavoglu, E., Zha, H., Zhang, Z., Fox, E.A.: Automatic document metadata extraction using support vector machines. In: Proceedings of the 3rd ACM/IEEE-CS Joint Conference on Digital Libraries. International Conference on Digital Libraries, pp. 37–48. IEEE Computer Society Press, Washington, DC (2003)
9. Seymore, K., McCallum, A., Rosenfeld, R.: Learning hidden Markov model structure for information extraction. In: AAAI 1999. Workshop on Machine Learning for Information Extraction (1999)
10. Tang, J., Maly, K., Zeil, S., Zubair, M.: Automated Building of OAI Compliant Repository from Legacy Collection. In: ELPUB. Proceedings of the 10th International Conference on Electronic Publishing (June 2006)
11. Mao, S., Kim, J.W., Thoma, G.R.: A Dynamic Feature Generation System for Automated Metadata Extraction in Preservation of Digital Materials. In: Dial 2004. Proceedings of the First international Workshop on Document Image Analysis For Libraries, vol. 225, IEEE Computer Society, Los Alamitos (2004)
12. Bergmark, D.: Automatic Extraction of Reference Linking Information from Online Documents. CSTR 2000-1821 (November 2000)
13. Klink, S., Dengel, A., Kieninger, T.: Document structure analysis based on layout and textual features. In: Proc. of Fourth IAPR International Workshop on Document Analysis Systems, pp. 99–111 (2000)
14. Marciniak, J.J. (ed.): Encyclopedia of Software Engineering, pp. 131–165. John Wiley & Sons, New York (1994)
15. Tang, J.: Template-based Metadata Extraction for Heterogeneous Collections. PhD thesis, Old Dominion University (2006)
16. Steward, Sid, pdftk – the PDF toolkit (2007) http://www.accesspdf.com/pdftk/
17. Maly, K., Zeil, S., Zubair, M.: Exploiting Dynamic Validation for Document Layout Classification During Metadata Extraction (2007), http://dtic.cs.odu.edu/publications/validationreal07.doc
18. Maly, K., Zeil, S., Zubair, M., Amrou, A., Aazhar, A., Ratkal, N.: A Scriptable, Statistical Oracle for a Metadata Extraction System. In: First International Workshop on Software Test Evaluation (STEV 2007), Portland, OR (October 11/12, 2007), (to appear, 2007), http://dtic.cs.odu.edu/publications/stev07.pdf

Using Automatic Metadata Extraction to Build a Structured Syllabus Repository

Xiaoyan Yu[1], Manas Tungare[1], Weiguo Fan[1], Manuel Pérez-Quiñones[1],
Edward A. Fox[1], William Cameron[2], and Lillian Cassel[2]

[1] Virginia Tech, Blacksburg VA 24061, USA
{xiaoyany,manas,wfan,perez,fox}@vt.edu
http://syllabus.cs.vt.edu/
[2] Villanova University, Villanova PA 19085, USA
{william.cameron,lillian.cassel}@villanova.edu

Abstract. Syllabi are important documents created by instructors for students. Gathering syllabi that are freely available, and creating useful services on top of the collection, will yield a digital library of value for the educational community. However, gathering and building a repository of syllabi is complicated by the unstructured nature of syllabus representation and the lack of a unified vocabulary for syllabus construction. In this paper, we propose an intelligent approach to automatically annotate freely-available syllabi from the Web to benefit the educational community through supporting services such as semantic search. We discuss our detailed process for converting unstructured syllabi to structured representations through entity recognition, segmentation, and association. Our evaluation results demonstrate the effectiveness of our extractor and also suggest improvements. We hope our work will benefit not only users of our services but also people who are interested in building other genre-specific repositories.

1 Introduction

A course syllabus is the skeleton of a course. One of the first steps taken by an educator in planning a course is to construct a syllabus. Later, a syllabus can be improved by adapting information from other relevant syllabi. Typically, a syllabus sets forth the objectives of the course. It may assist students in selecting electives and help faculty identify courses with goals similar to their own. In addition, a life-long learner identifies the basic topics of a course and the popular textbooks by comparing syllabi from different universities. A syllabus is thus an essential component of the educational system.

Supporting activities like those mentioned above can be facilitated if metadata is extracted from syllabi. However, two obstacles hinder this, especially with respect to the syllabus genre. First, no metadata standard is specific to the syllabus genre, although markup schemes, such as IEEE LOM [1], exist for educational resources. Thus, while we are able to annotate a document as a syllabus by the LOM's resource type property, we are unable to annotate a piece of information inside a syllabus as a textbook using any of the available metadata standards. Second, it requires too much effort to manually annotate information inside syllabi, and no approach is available to automate the process of information extraction from the syllabus genre. Motivated by these two observations,

D.H.-L. Goh et al. (Eds.): ICADL 2007, LNCS 4822, pp. 337–346, 2007.

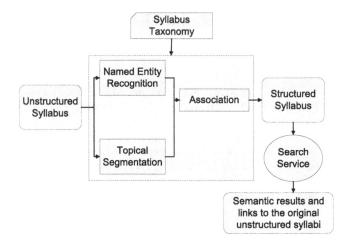

Fig. 1. Workflow from a unstructured syllabus to a structured syllabus

we propose a taxonomy and an extraction approach specific to the syllabus genre, build a structured syllabus digital library (DL) by extracting metadata from each syllabus, and support semantic search of the syllabi through the DL (and thus through the Semantic Web).

Figure 1 shows the flow of the transformation from an unstructured syllabus to a structured syllabus and then the retrieval of structured syllabi. Following our syllabus taxonomy (Section 2), semantic information can be extracted from a syllabus, which becomes part of the Semantic Web. The named entity recognition module identifies entities such as people and dates. The topical segmentation module identifies the boundary of a syllabus component such as a course description or a grading policy. Finally, the association module associates a list of syllabus properties with the segmented values, and stores them in the structured syllabus repository. These three modules work together for the information extraction task (Section 3). The search service (Section 4) indexes structured syllabi and provides semantic search results through both RDF[1] and links to the raw syllabi.

There are many other types of unstructured data on the Web; thus, success with our genre-specific structured repository suggests that there are opportunities to use such other data in similar innovative applications. We hope that our application of machine learning techniques to extract and obtain structured genre-specific data will encourage the creation of other similar systems.

2   Syllabus Taxonomy

Our syllabus taxonomy is designed to help reconcile different vocabularies for a syllabus used by different instructors. For example, instructors often start a course description with headings such as *'Description'*, *'Overview'*, or *'About the Course'*. Such variations

---

[1] http://www.w3.org/RDF/

make it difficult to reuse information from these syllabi. It also is very hard to locate a particular syllabus section because the section headings are not uniquely named. In order to facilitate processing of syllabi by different applications, we propose a syllabus taxonomy[2] and show the first level of the taxonomy in Table 1. Among these 15 properties, some are data types of a syllabus such as `title` (a course title) and `description` (a course description) while others are object types such as `teachingStaff` and `specificSchedule` that utilize other vocabularies at a deeper level. For example, a `courseCode` is defined as an abbreviation of the department offering the course and a number assigned to the course, while a `prerequisite` is composed of one or more `courseCode` objects. It also is worth noting that we define a `specificSchedule` as topics and specific dates to cover them, and a `generalSchedule` as semester, year, class time, and class location.

The taxonomy will help both our extraction of the list of property values from each syllabus, and our making the collection of structured syllabi available in RDF.

Table 1. First level of syllabus taxonomy

Data Type	affiliation, title, objective, description, courseWebsite
Object Type	assignment, resource, courseCode, teachingStaff, grading, specificSchedule, prerequisite, textbook, exam, generalSchedule

3   Information Extraction

Information extraction aims to extract structured knowledge, including entity relationships, from unstructured data. In our case, for example, we would extract relations such as an instance of the TEACH relation *"(Mary, Data Structure, Fall 2006)"* from a syllabus, *"(Mary teaches the Data Structure course in Fall 2006)"*. There are plenty of research studies, reviewed in [2], that have applied machine learning technology to the information extraction task. These approaches can be broadly divided into rule-based approaches such as Decision Tree, and statistics-based approaches such as Hidden Markov Model (HMM). The extraction task usually involves four major subtasks: segmentation, association, normalization, and deduplication [2]. For our extractor, the segmentation task includes mainly two steps – named entity recognition and topical segmentation – while the deduplication task is integrated into the association task. In addition, the normalization task, which puts extracted information into a standard format such as presenting *"3:00pm-4:00pm"* and *"15:00-16:00"* uniformly as *"15:00-16:00"* for the class time, will be performed in the future since it does not affect extraction accuracy.

Thompson *et al.* [3] have tried completing these tasks with an HMM approach on course syllabi for five properties: course code, title, instructor, date, and readings. They manually identified the five properties on 219 syllabi to train the HMM. However, it

---

[2] http://syllabus.cs.vt.edu/ontologies

would take us much more effort to label 15 properties for a large collection of unstructured syllabi. Therefore, we needed a method that is unsupervised, i.e., not requiring training data. In the following subsections, we explain our approach in detail.

3.1 Named Entity Recognition

Named Entity Recognition (NER), a sub-task of information extraction, can recognize entities such as persons, dates, locations, and organizations. An NER F_1 (a combination of the precision and the recall of recognition) of around 90% commonly has been achieved since the 7th Message Understanding Conference, MUC [3], in 1998. We therefore chose to base our named entity recognizer on a proven routine, ANNIE[4], part of the GATE natural language processing tool [4]. It has been successfully applied to many information extraction tasks such as in [5] and is easily embedded in other applications. Our recognizer also can recognize course codes by matching them to the pattern of two to five letters, followed by zero or more spaces, and then two to five digits.

3.2 Topical Segmentation

A course syllabus might describe many different aspects of the course such as topics to be covered, grading policies, and readings. Because such information is usually expressed in arbitrary sentences, NER is not applicable for that part of the extraction task. In order to extract such information, it is essential to find the boundaries indicating topic change and then to classify the content between identified boundaries into one of the syllabus data/object types. The first half falls in the topical segmentation task and the other half will be described in the next section. Much research work has already been done on topical segmentation. We chose C99 [6] because it does not require training data and has performance comparable to the supervised learning approach which requires training data [7]. C99 measures lexical cohesion to divide a document into pieces of topics. It requires a pre-defined list of preliminary blocks of a document. Each sentence in a document is usually regarded as a preliminary block. C99 calculates the cosine similarity between the blocks by stemming and removing stop words from each block. After the contrast enhancement of the similarity matrix, it partitions the matrix successively into segments.

C99 is not good, however, at identifying a short topic, which will be put into its neighboring segment. Therefore, we do not expect the segmenter to locate a segment with only a single syllabus property, but expect it not to split a syllabus property value into different segments. It also is critical to define a correct preliminary block which is the building block of a topical segment of C99. We defined a preliminary block at the sentence or the heading level. A heading is a sequence of words just before a syllabus property. It is usually short, and often occupies a line. At other times the heading and its contents are separated by the delimiter ':'. We first located possible headings and sentences. If two headings were found next to each other, the first one was treated as a preliminary block; otherwise a heading and the following sentence form a preliminary block in case they are partitioned into different segments.

---

[3] http://www-nlpir.nist.gov/related_projects/muc/

[4] http://www.aktors.org/technologies/annie/

```
Input: a people list (P), a date list (D), an organization list (O), a
location list (L), a course code list (C), a segment list and a property
pattern list (PP).
Output: a list of property names and extracted values, E.

Begin
1  For the first segment
2     If a code c in C falls into this segment
3     Then E← ('courseCode',c)
4           If the words following the code is a heading
5           Then E←('title', the words)
6        If an organization o in O falls into this segment
7     Then E←('courseAffiliation', o)
8     If an semester item d in D falls into this segment
9     Then E←('generalSchedule', d)
10 For each segment
11    If no entry of staff information is obtained
12    Then if a person p in P falls in this segment
13        Then if the teachingStaff pattern occurs before the occurrence of
                this person
14           Then E←('teachingStaff', ts)where Start_Pos(ts) = Start_Pos(p)
15               If there are more items in D and L falling in this segment
16               Then End_Pos(ts) = max(End_Pos(these items))
17               Else
18                   End_Pos(ts) = End_Pos(the segment)
19    If a URL in L falling in this segment contains the course code extracted
      already
20    Then E←('courseWebsite', the URL)
21    If the segment starts with a heading
22    Then for each pattern pp in PP
23           If pp occurs in the heading
24           Then E←(pn, the segment without the heading) where pn is the
                property name for the pattern pp.
25           Extraction is completed for this segment.
End
```

Fig. 2. The algorithm to associate topical segments and named entities with syllabus properties

Table 2. Heading Patterns for Syllabus Properties

Property	Regular Expression (Regex)
description	description|overview|abstract|summary|catalog|about the course
objective	objective|goal|rationale|purpose
assignment	assignment|homework|project
textbook	text|book|manual
prerequisite	prerequi
grading	grading
specificSchedule	lecture|topic|reading|schedule|content|outline
teachingStaff	instructor|lecturer|teacher|professor|head|coordinator|teaching assistant|grader
exam	exam|test
schedule	reference|reading|material|lecture[∧r]

3.3 Association

Given the topical segments and named entities of a syllabus, the final step is to associate them with the list of interesting syllabus properties. The algorithm for this final step is shown in Figure 2 and the details are explained below.

First of all, lines 1–9 in Figure 2 identify a course code, a semester, and a course affiliation (university and department) at the top of a syllabus, i.e., in the first segment. A course title is a heading and follows a course code. Second, lines 11–18 indicate information about teaching staff by a heading with keywords such as 'instructor', 'lecturer' and more in Table 2. It might include their names, email addresses, Website URLs, phone numbers, and office hours. They should fall in the same segment. Third, lines 19–20 identify a course Web site by looking for the course code inside. Finally, lines 21–25 look for other syllabus properties: each starts with the heading of the property and falls into a single topical segment. A heading is identified based on a list of keywords, as shown in Table 2. For example, a course description heading might contain 'description', 'overview', 'abstract', 'summary', 'catalog', or 'about the course'.

3.4 Evaluation

To evaluate the accuracy of the information extraction and conversion process, we randomly selected 60 out of over 700 syllabi manually identified from our potential syllabus collection [8], all in HTML format. The free text of each syllabus document (obtained by removing all HTML tags) was fed into our extractor.

One of the co-authors, an expert in the syllabus genre, judged the correctness of extraction manually by the following procedure: our judgment criterion was that a piece of information for a syllabus property is considered extracted correctly if it is identified at the correct starting position in the syllabus as obtained via manual inspection. It was considered acceptable to include extra information that did not affect the understanding of this piece of information. For example, we judged a course title that also contained semester information, as a positive extraction.

We calculated the F_1 on each property of interest, over the syllabi with this property. The F_1 is a widely accepted evaluation metric on information extraction tasks. It is a combination of precision and recall, expressed as $F_1 = 2*Precision*Recall/(Precision+Recall)$. Precision on a property is the ratio of the number of syllabi with the property correctly extracted over the total number of syllabi with the property extracted. Recall on a property is the ratio of the number of syllabi with this property correctly extracted over the total number of syllabi with this property. The higher the F_1 value, the better the extraction performance.

Our extractor is more effective on some properties than others. The performance on the more effective properties is shown in Figure 3. For example, we achieved high accu-

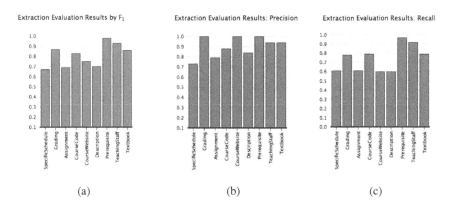

Fig. 3. Extraction evaluation results by F_1

racy on the `prerequisite` property at the F_1 value of 0.98 since this field usually starts with the heading keyword *'prerequisite'* and contains a course code. On the other hand, by examining the false extractions on the properties with low accuracy, we summarize our findings as follows.

- The heuristic rule to identify a course title, namely finding the heading next to a course code, is too specific to obtain high accuracy. Among the 60 syllabi we inspected, many have course titles and course codes separated by semester information.
- The extraction accuracy of a course title is also affected by that of a course code. Quite a few course codes do not match the pattern we defined. There is a larger variety of formats than we thought. For example, some course codes consist entirely of digits separated by a dot (such as '6136.123'), while some consist of two department abbreviations separated by a slash for two codes of the same course (such as 'CS/STAT 5984').
- The `resource` property is identified with high precision at 0.8, but low recall at 0.42, because it is misclassified as other properties such as `textbook`. For example, many readings are present under the textbook section without an additional heading. In addition, some resources such as required software for the course are hard to identify simply from the heading. The same reason causes the `schedule`, `objective`, and `courseAffiliation` properties to be extracted with very high precision but low recall.
- The accuracy on the `exam` property is low in terms of recall and precision, both at the F_1 value of nearly 0.5. It is mis-classified into `grading` sometimes, which leads to low recall. On the other hand, the low precision is because the exam time which belongs to the `specificSchedule` property is mis-classified into an `exam` property.

The evaluation results discussed above indicate challenges in the syllabus extraction task. First, there are many properties in a syllabus with varied presentations in varied syllabi. Trying to extract all of them at once will reduce the probability of obtaining high quality metadata on any of them. Therefore, we found it better to prioritize the few most important properties first and extract the rest later. Second, many properties' values contain long content, so the heading approach can only help in finding the starting position, not the ending position: the `schedule` property is the best example of this observation. We should use HTML tags to ascertain the structure of HTML documents. For example, schedules usually are included in an HTML table; we expect that if these tags are available during processing, the complete schedules can be extracted with high accuracy. This also will help extraction of information like textbooks, which are commonly presented in an HTML list. Creating an exhaustive set of patterns for all properties is a tedious and error-prone process. Thus, we started off with a smaller subset of patterns and properties.

4   Searching Syllabi

The availability of syllabi in a standard format with the appropriate metadata extracted from them makes several beneficial applications and services possible. We present one

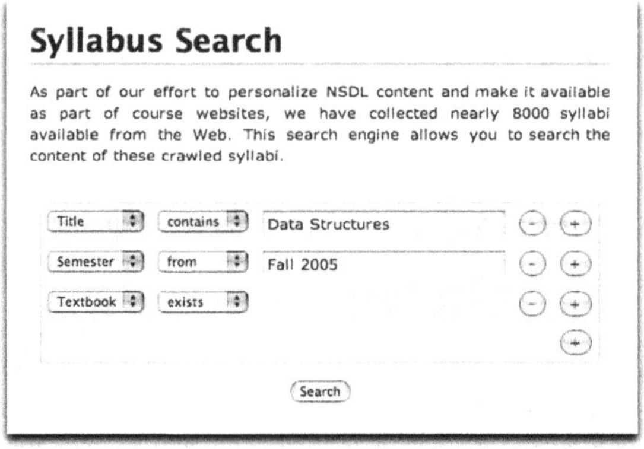

Fig. 4. Syllabus search engine interface: advanced search dialog

of these services, Semantic Search over syllabi, in detail below. Some others are discussed in [9].

We have provided a semantic search service over our structured syllabus repository. This is different from other general-purpose keyword search engines in that our search engine indexes a set of documents known with confidence to be syllabi, and provides extracted metadata to assist the user in various tasks.

For example, as shown in Figure 4, an instructor may query, from our advanced search dialog box, popular textbooks used in Data Structures courses since Fall 2005. The search results will highlight indicative keywords and also identified textbooks; there also will be a link to the original unstructured syllabus, and a link to the parsed syllabus in RDF format.

Our implementation is developed upon Lucene[5], a search engine development package. We index extracted metadata fields for each syllabus, and support basic search and advanced search functionalities. When a user types queries without specifying particular fields, our service searches all the indexed fields for desired syllabi. When the user specifies some constraints with the query through our advanced search dialog box, we only search in specific fields, which can find syllabi with greater accuracy. For example, only a syllabus with textbooks will be returned for the case shown in Figure 4.

Our semantic search service also would benefit agent-based systems and other semantic web applications. For example, an application is to list popular books in a variety of courses especially in computer science. It will obtain different lists of syllabi in RDF format by the same query as the instructors's but with different course titles and then for each list rank the textbooks by their occurrences in the list.

5   Related Work

There are a few ongoing research studies on collecting and making use of syllabi. The MIT OpenCourseWare project manually collects and publishes 1,400 MIT course

---
[5] http://lucene.apache.org/

syllabi in a uniform structure for public use. A lot of effort from experts and faculty is required in manual collecting approaches, which is the issue that our approach tries to address. Our previous work [10] helps with automating the syllabus acquisition process by identifying true syllabi from search results on the Web.

Some have addressed the problem of lack of standardization of syllabi. Along with a defined syllabus schema, SylViA [11] supports a nice interface to help faculty members construct their syllabi in a common format. More work has been done on defining the ontology or taxonomy of a variety of objects, such as the ontology of a learner, especially in a remote learning environment [12]. Our proposed syllabus taxonomy also describes the features of a course, such as the course instructor, textbooks, and topics to be covered. We will use these features to provide additional services such as recommending educational resources to students of a particular course.

In order to fulfill a general goal of the Semantic Web, annotation and semantic search systems have been successfully proposed for other genre (such as television and radio news [5]). Such systems vary in keeping with the different genre, due to their own characteristics and service objectives. To our knowledge, there is no specific annotation and semantic search system for the broad syllabus genre.

Much work has been done on metadata extraction from other genre such as academic papers. For example, Han *et al.* [13] described using Support Vector Machines for metadata extraction from a paper's header field.

6   Conclusions

In this paper, we proposed an intelligent approach to automatically annotate freely-available syllabi from the Internet and to benefit the education community through supporting services such as semantic search. We discussed our detailed process to automatically convert unstructured syllabi to structured data. Our work indicates that an unsupervised machine learning approach can lead to generally good metadata extraction results on syllabi, which are hard to label manually for a training data set. The challenges of extraction on the syllabus genre, along with suggestions for refinement, are discussed. We hope that the experience of our approach in building genre-specific structured repositories will encourage similar contributions for other genre, eventually leading to the creation of a true Semantic Web.

Acknowledgments

This work was funded by the National Science Foundation under DUE grant #0532825.

References

1. Hodgins, W., Duval, E.: Draft standard for learning technology - Learning Object Metadata - ISO/IEC 11404. Technical report (2002)
2. Mccallum, A.: Information extraction: Distilling structured data from unstructured text. ACM Queue 3(9) (November 2005)

3. Thompson, C.A., Smarr, J., Nguyen, H., Manning, C.: Finding educational resources on the web: Exploiting automatic extraction of metadata. In: Proc. ECML Workshop on Adaptive Text Extraction and Mining (2003)
4. Cunningham, H., Maynard, D., Bontcheva, K., Tablan, V.: Gate: A framework and graphical development environment for robust nlp tools and applications. In: Proceedings of the 40th Anniversary Meeting of the Association for Computational Linguistics (ACL 2002), Philadelphia (July 2002)
5. Dowman, M., Tablan, V., Cunningham, H., Popov, B.: Web-assisted annotation, semantic indexing and search of television and radio news. In: WWW 2005. Proceedings of the 14th international conference on World Wide Web, pp. 225–234. ACM Press, New York (2005)
6. Choi, F.Y.Y.: Advances in domain independent linear text segmentation. In: Proceedings of the first conference on North American chapter of the Association for Computational Linguistics, pp. 26–33. Morgan Kaufmann, San Francisco (2000)
7. Kehagias, A., Nicolaou, A., Petridis, V., Fragkou, P.: Text segmentation by product partition models and dynamic programming. Mathematical and Computer 39(2-3), 209–217 (2004)
8. Tungare, M., Yu, X., Cameron, W., Teng, G., Pérez-Quiñones, M., Fox, E., Fan, W., Cassel, L.: Towards a syllabus repository for computer science courses. In: SIGCSE 2007. Proceedings of the 38th Technical Symposium on Computer Science Education, vol. 39, pp. 55–59. ACM Press, New York, NY, USA (2007)
9. Tungare, M., Yu, X., Teng, G., P érez Quiñones, M., Fox, E., Fan, W., Cassel, L.: Towards a standardized representation of syllabi to facilitate sharing and personalization of digital library content. In: Proceedings of the 4th International Workshop on Applications of Semantic Web Technologies for E-Learning (SW-EL) (2006)
10. Yu, X., Tungare, M., Fan, W., Pérez-Quiñones, M., Fox, E.A., Cameron, W., Teng, G., Cassel, L.: Automatic syllabus classification. In: Proceedings of the Seventh ACM/IEEE-CS Joint Conference on Digital Libraries - JCDL 2007, pp. 440–441 (2007)
11. de Larios-Heiman, L., Cracraft, C.: (SylViA: The Syllabus Viewer Application)
12. Dolog: Reasoning and ontologies for personalized e-learning. Educational Technology and Society (2004)
13. Han, H., Giles, C.L., Manavoglu, E., Zha, H., Zhang, Z., Fox, E.A.: Automatic document metadata extraction using support vector machines. In: JCDL 2003. Proceedings of the 3rd ACM/IEEE-CS joint conference on Digital Libraries, Washington, DC, USA, pp. 37–48. IEEE Computer Society Press, Los Alamitos (2003)

Automatic Text Summarization in Engineering Information Management

Jiaming Zhan[1], Han Tong Loh[1], Ying Liu[2], and Aixin Sun[3]

[1] Department of Mechanical Engineering, National University of Singapore, Singapore 119077
{jiaming,mpelht}@nus.edu.sg
[2] Department of Industrial and Systems Engineering, The Hong Kong Polytechnic University,
Hung Hom, Kowloon, Hong Kong SAR, China
mfyliu@inet.polyu.edu.hk
[3] School of Computer Engineering, Nanyang Technological University, Singapore 639798
axsun@ntu.edu.sg

Abstract. In today's knowledge-intensive engineering environment, information management is an important and essential activity. However, existing researches of Engineering Information Management (EIM) mainly focused on numerical data such as computer models and process data. Textual data, especially the case of free texts, which constitute a significant part of engineering information, have been somewhat ignored, mainly due to their lack of structure and the noisy information contained in them. Since summarization is a process to distill important information from source documents and at the same time remove irrelevant and redundant information, it could address the obstacles for handling textual data in EIM. Moreover, text summarization could address the increasing demand to integrate information from multiple documents and reduce the time in acquiring useful information from massive textual data in the engineering domain. This paper discusses in detail the need to apply text summarization in EIM and introduces a case study in summarizing multiple online customer reviews.

Keywords: Automatic Text Summarization, Engineering Information Management, Online Customer Reviews.

1 Textual Information Within Engineering Domain

Information management is an important and essential activity in today's knowledge-intensive engineering environment. Information lies at the core of a modern engineering environment, comprising not only numerical data but also textual data. Effective and efficient information management is one of the key factors by which industrial and engineering performance can be greatly improved [1].

However, most existing Engineering Information Management (EIM) applications focus on the handling of numerical data. Textual data, such as technical papers, patent documents, e-mails and customer reviews, which constitute a significant part of engineering information, have been somewhat ignored. There are probably three major reasons for this lack of attention:

D.H.-L. Goh et al. (Eds.): ICADL 2007, LNCS 4822, pp. 347–350, 2007.

- Numerical data are well structured and organized in databases, making them relatively easy to handle. In comparison, textual data are usually stored as unstructured free texts or semi-structured data so that there is a greater level of difficulty in handling textual databases.
- Compared to the clean and purified numerical data, textual data contain a lot of noisy and redundant information.
- Most existing EIM applications have focused on design and manufacturing phases in which numerical information dominates.

As with numerical data, textual data offer a wealth of information in engineering activities, especially with the explosive growth of enterprise Intranet and the Internet [2, 3]. There has been an increasing demand of advanced techniques to reduce the time in acquiring useful information and knowledge from massive textual data, commonly appearing in technical papers, patent documents, reports, white papers, e-mails, Web pages, and notes from call centers [4].

2   The Need of Text Summarization Within EIM

Because of the rich information involved in textual data, how to utilize and how to discover knowledge from them effectively and efficiently has become a concern. However, only a few studies have been reported on textual information management within engineering domain, due to the aforementioned limitations. These existing studies can be divided into two major areas: information indexing & searching [5] and automatic text classification [6]. On the other hand, another important issue, i.e. integrating information from multiple textual sources and extracting useful information to fulfill users' requirements, has not yet been covered by previous studies.

Due to the current overload of engineering information, even with the powerful classification and searching tools, users often encounter a huge amount of retrieved documents for any given topic. Users have to screen these documents manually, which often takes a lot of time, until they satisfactorily identify documents relevant to their specific purposes. In such context, a summarization system, which can integrate the information from retrieved documents and facilitate the searching process, is much needed. The retrieved documents, regarding the same query, must share much common information which is interesting to readers. Besides, in some documents there must exist some unique information. Therefore, the summarization system should be able to integrate the common and unique information from all documents. At the same time, this summarization system should be able to exclude the redundant and noisy information across the documents.

Summarization is a process to distill the most important information from source documents and at the same time remove irrelevant and redundant information. Moreover, the output of a summarization system would be a well structured text compared to the unstructured source documents. Therefore, automatic text summarization could probably address the aforementioned limitations for handling textual information in EIM.

The first implementation of automatic text summarization can be traced back to 1950s [7]. During the last decade, there has been increasing interest with Multi-Document Summarization (MDS), as an outcome of the capability to collect large sets of documents online [8, 9]. The most popular MDS approach is clustering-summarization which separates a document set into non-overlapping clusters of documents and summarizes each cluster. However, when applied to the real-world engineering document sets, the number of clusters is difficult to determine, and moreover, topics often overlap with each other and are not perfectly distributed in non-overlapping clusters of documents [3].

3 Case Study: Summarizing Online Customer Reviews

This case study aims to investigate the domain of customer reviews, which constitute a typical kind of documents used in EIM. Some work has been reported dealing with the vast amount of customer reviews [10, 11]. All these work focused on opinion mining which was to discover the reviewers' orientations, whether positive or negative, regarding various features of a product, e.g. weight of a laptop and picture quality of a digital camera. However, opinion mining is not enough to cover all the important information from customer reviews and there is a desire to apply summarization techniques to identify the significant topics from multiple customer reviews [3].

In this case study, we propose a summarization approach based on the topical structure, which consists of a list of significant topics that are extracted from a document set [3]. The summarization performance was compared with the approaches of opinion mining and clustering-summarization. The data sets used in the experiment included five sets from Hu's corpus [10] and three sets from Amazon.com. These document sets were moderate-sized with 40 to 100 documents per set. The compression ratio of summarization, i.e. the length ratio of summary to original text, was set to 10%. Summarization performance was evaluated according to users' responsiveness. Human assessors were required to give a score for each summary based on its content and coverage of important topics in the review set. The score was an integer between 1 and 5, with 1 being the least responsive and 5 being the most responsive. In order to reduce bias in the evaluation, three human assessors from different backgrounds joined the scoring process. For one set, all the peer summaries were evaluated by the same human assessor so that the hypothesis testing (paired t-test) could be performed to compare the peer summaries.

Table 1 shows the average responsiveness scores of opinion mining, clustering-summarization and our approach based on all the review sets. Table 2 presents the results of paired t-test between our approach and other methods. It could be found that our approach based on topical structure performed significantly better than other peer methods.

Table 1. Average responsiveness scores

	Responsiveness score
Opinion mining	2.9
Clustering-summarization	2.3
Our approach	4.3

Table 2. Hypothesis testing (paired t-test)

Null hypothesis (H0): There is no difference between the two methods. Alternative hypothesis (H1): The first method outperforms the second one.	
	P-value
Our approach vs. opinion mining	$1.91 \times 10\text{-}3$
Our approach vs. clustering-summarization	$2.43 \times 10\text{-}4$

4 Conclusion

This paper reviews the existing EIM applications, with the focus on textual information management, and addresses the need to apply automatic text summarization in EIM. Moreover, a case study to summarize multiple online customer reviews is introduced. This work might enrich the research of EIM since it examined the textual information which has been somewhat ignored in the previous studies. In the future, we will apply the summarization techniques to other types of documents in engineering domain, such as technical papers and patent documents.

References

1. Hicks, B.J., Culley, S.J., McMahon, C.A.: A study of issues relating to information management across engineering SMEs. International Journal of Information Management 26(4), 267–289 (2006)
2. Liu, Y.: A concept-based text classification system for manufacturing information retrieval. Ph.D. Thesis, National University of Singapore (2005)
3. Zhan, J., Loh, H.T., Liu, Y.: Automatic summarization of online customer reviews. In: Proceedings of the 3rd International Conference on WEBIST, Barcelona, Spain (2007)
4. Blumberg, R., Atre, S.: The problem with unstructured data. DM Review (2003)
5. Fong, A.C.M., Hui, S.C.: An intelligent online machine fault diagnosis system. Computing and Control Engineering Journal 12(5), 217–223 (2001)
6. Menon, R., Loh, H.T., Keerthi, S.S., Brombacher, A.C., Leong, C.: The needs and benefits of applying textual data mining within the product development process. Quality and Reliability Engineering International 20(1), 1–15 (2004)
7. Luhn, H.P.: The automatic creation of literature abstracts. IBM Journal of Research and Development 2(2), 159–165 (1958)
8. Mani, I., Bloedorn, E.: Summarizing similarities and differences among related documents. Information Retrieval 1(1-2), 35–67 (1999)
9. Radev, D.R., Jing, H., Styś, M., Tam, D.: Centroid-based summarization of multiple documents. Information Processing and Management 40(6), 919–938 (2004)
10. Hu, M., Liu, B.: Mining and summarizing customer reviews. In: Proceedings of the 10th ACM SIGKDD, Seattle, WA, pp. 168–177 (2004)
11. Popescu, A.-M., Etzioni, O.: Extracting product features and opinions from reviews. In: Proceedings of HLT/EMNLP 2005, Vancouver, Canada, pp. 339–346 (2005)

The PENG System: Integrating Push and Pull for Information Access

Mark Baillie[1], Gloria Bordogna[2], Fabio Crestani[3], Monica Landoni[3], and Gabriella Pasi[4]

[1] Dept. Computer Information Sciences, University of Strathclyde, Glasgow, UK
[2] CNR IDPA, Dalmine, Bergamo, Italy
[3] Faculty of Informatics, University of Lugano, Lugano, Switzerland
[4] DISCO, University Milano-Bicocca, Milano, Italy

Abstract. This paper describes the PENG project that integrates personalized push and pull technologies to access relevant information. PENG integrates several key tasks, including personalized filtering, retrieval, and presentation of multimedia news, into a single system. In this paper we provide an overview of PENG, describing our approach to constructing a dedicated retrieval and content management system for a specific user group. We also report critically on the results of a user and task based evaluation.

1 Introduction

News professionals, such as Radio, TV and Newsprint journalists and editors, now have at their disposal a large and varied collection of digital information resources. News Agencies such as ANSA, Reuters and AP can, for example, provide live feeds of breaking stories directly into a newsroom. Journalists can also search and browse a variety of online news archives, digital libraries and web repositories when researching and compiling a report. However, to utilise this wealth of digital information, it is expected that the busy news professional is proficient in a number of systems or interfaces. The aim of PENG, an EC funded Specific Targeted Research (STREP) Project[1], was to address the issue of news content management allowing the news professional to access information under a single interface. PENG integrates several key tasks, including personalised filtering, retrieval, and presentation of multimedia news, into a single system. In this paper we provide an overview of PENG, describing our approach to constructing a dedicated retrieval and content management system for a specific user group. We show how detailed knowledge of a user group and the information tasks they perform has been used to inform the design of retrieval and filtering system components.

---

[1] More information at `http://www.peng-project.org/`

D.H.-L. Goh et al. (Eds.): ICADL 2007, LNCS 4822, pp. 351–360, 2007.

2   The PENG System

The design of the PENG system was motivated by a user study undertaken as part of the project, investigating the work practices of European journalists and editors from different news mediums i.e. TV, Radio and Print. The study investigated how journalists searched, verified, processed and then used information gathered from a wide variety of electronic resources, and in particular how they exploited current systems to support these daily work tasks. Although we cannot report here in full the results of this study, several important requirements came out of this study. Firstly, it was highlighted that journalists require a high level of control over the operation of a system due to the fear of missing important information. Ideally, they wish to view all potentially relevant documents across a number of (disparate) news archives and information resource providers, in contrast to what is offered by current systems.

A common theme highlighted during the study was that journalists use a range of criteria when gathering information for a task. These criteria are not static but constantly changing as the journalist and environment changed. For example, across the journalists surveyed, documents were judged by the accuracy of their contents, the reliability and verification of the information source, the accessibility of the information (in terms of speed, cost, etc.), the timeliness of the information, and also the proximity of the information to the journalist (i.e. local news concerning local issues). These findings mirrored previous studies that have highlighted the dynamic and multidimensional nature of relevance, where many factors beyond topicality and aboutness influence how a user assesses information [1,2,3].

One important criteria in particular was the notion of *trust*: the interviewed declared how important it was to identify the original source of the document in order to determine the accuracy of its content. This was considered a vital step when assessing information, and reflected what has previously been cited as one of the key elements of journalism: the verification of a news source [4]. Given the nature of information retrieved from the web, where the original publisher and source of content can be difficult to identify, and where many documents do not go through a strict refereeing or editorial process, the journalist has to be vigilant [5]. To address this issue of source verification, journalists often restrict their search for information to a number of (trusted) resources. Therefore, which resources the journalist searched for information were of particular importance (i.e. news archives, digital libraries or web resources). For example, web search engines were often used because of ease of use and speed, while internal databases were used for checking personal details of sources. Overall, the type(s) of resource a journalist would access at any given time was dependent on the journalist and their individual tasks and needs.

With regards to the requirements analysis, three important phases were identified during the daily workflow of a journalist; information *push, pull* and *presentation*. To address these three phases, PENG combines information filtering, searching and presentation within a unified framework which also provides support for personalisation.

PENG has a typical client server architecture. There are five main components on the server:

Filtering module: It provides support for the filtering of incoming stream of news from various resources reporting breaking stories.

Distributed Information Retrieval (DIR) module: It provides support for the search of multiple disparate third party information resources.

Common database manager module: It co-ordinates communication and functionality across the system, also maintaining both databases.

Document database and indexes: It provides a central repository for documents or other information gathered through filtering, as well as important results required for the DIR (such as query history), allowing data to persist over time.

User profile database: It manages a persistent profile for each user stored in this database.

User interface: It provides support for all the functionality of push, pull and presentation of the system, in addition to several others aimed at facilitating the user information gathering, organisation and composition tasks.

This architecture was designed to provide the following advantages (i) a single document representation which can be used consistently by all modules in PENG, (ii) a single user profile representation used by all PENG modules, and (iii) a single method of access to both the information artefacts (e.g. documents) irrespective of whether they are returned by filtering or retrieval. We are now going to describe the functionality of the two main modules of the PENG system, the filtering and DIR components, responsible for the push and the pull of information. Due to space limitations the other modules of the system, though important, will not be presented here.

3   The Information Filtering Module

The filtering component implements the push phase of the PENG system. From the requirements analysis, journalists were found to be "fearfull" of a filtering system which may cause them to miss important information. This was one of the prime motivations for the design of the filtering component, which goes beyond the functionality of what are normally termed filters. In particular, the filtering component was designed to (i) organise the incoming new feeds pushed directly into the PENG system through the use of (fuzzy) clustering, (ii) organise the personalised information of each user by filtering information with respect to each individual users interests, and (iii) rank relevant information by using a variety of criteria alongside relevance, such as timeliness and novelty. With this motivation, the internal architecture of the filtering module was divided into four main sub-modules: 1) Gathering, that receives or actively gathers new material from pushed external information news feeds such as Reuters and ANSA; 2) Clustering, that periodically identify topically related groups of recently arrived documents, thus providing an overview of the current scenario of recent news;

3) Filtering new documents or clusters to individual interests within each user profile by applying personalized multicriteria evaluating content based relevance of news, actuality and novelty of news and trust of the news sources; 4) User monitoring of user actions (carried out in the PENG user interface) and records these actions for later use by the training sub-system. The two most important sub-components are described in the following.

The *personalised filtering system* can filter either individual documents or clusters of documents (i.e. category-based filtering) to user interests [6]. Each PENG user may have a number of individual interests. Each interest is defined by the user, either by example or explicitly by typing a textual description. These user interests provide the second method of organization of the data within PENG, providing a view of the incoming stream of filtered data personalized to each user, and each user interest. Many traditional filtering systems carry out a hard classification of the input stream of documents, classifying each document as it arrives, as either relevant or not relevant to a (user) profile. From the PENG requirements gathering, we posit that this is exactly what journalists do not want. Because of this, the filtering model used in PENG applies a multi-criteria decision for each individual user interest, thus reflecting the multiple criteria used by journalists to assess relevant information. The following measures are computed and used by the filter: *aboutness*: it is a usual measure of the content-based similarity of a new document to the interests of the user; *coverage*: it is a measure of the inclusion of the user interests in the contents of the latest news; *novelty*: it is a measure of the new information offered to the user by an incoming document; *reliability*: it is a measure related to the user trust in the resource from which the news is coming; *timeliness*: it is a measure of the usefulness of a news item to the user-specified time-window. The personalised filtering can be split into two main stages. The first stage computes the relevance judgment of a news to the user interest. This stage first applies a pre-filtering phase based on the consideration of the trust score specified in the users profile. Then the relevance score is generated by combining aboutness and coverage. The value so obtained can then be thresholded to determine the final relevance, i.e. the selection condition of the document to a user interest. If the document is deemed relevant to the interest it will pass to the second stage, i.e. a merging stage which inserts a topically relevant document into the existing result list, generating a number of different scores allowing the result list to the re-ranked by different criteria. The relevance from the previous stage can be combined with either the novelty, trust or timeliness values, to produce different ranking the output of the filter customisable to the users. Such a scheme can be considered as the maintenance of a single ranked list over time, where the job of the filtering system is to place each new document in the ranked list, relative to the other existing documents. In other words, the filtering system must now determine not just whether the document is relevant, but also how is this document relevant relatively to existing results.

A *fuzzy clustering module* operates periodically, currently once every day but potentially every few minutes, to automatically generate a set of clusters which characterise the incoming stream of news. The clustering is not based on a preset classification, and so may vary from day to day, depending on the content of the daily news. This is intended to provide journalists with an overview of the current "news landscape", easing the identification of relevant breaking news stories, and also providing a means to classifying individual documents within this daily structure. The output of the algorithm is a fuzzy hierarchy of the news reflecting the nature of news, which may deal with multiple topics. The algorithm computes a membership degree score (between $[0,1]$) for each item (news) to each generated fuzzy cluster allowing the documents to be ranked within a cluster, easily supporting flexible filtering strategies such as the selection of the top ranked news within a cluster of interest.

The generated fuzzy hierarchy represents the topics at different levels of granularity, from the most specific ones corresponding to the clusters of the lowest hierarchical level (the deepest level in the tree structure representing the hierarchy), to the most general ones, corresponding with the clusters of the top level. Since topics may overlap one another, the hierarchy is fuzzy allowing each cluster of a level to belong with distinct degrees to each cluster in the next upper level. To generate such a fuzzy hierarchy, we have defined a fuzzy agglomerative clustering algorithm based on the recursive application of the Fuzzy C-means algorithm (FCM). The algorithm works bottom up in building the levels of the fuzzy hierarchy. Once the centroids of the clusters in a level of the hierarchy are generated, the FCM is re-applied to group the newly identified centroids into new fuzzy clusters of the next upper level. In this way, each level contains fuzzy clusters that reflect topics homogeneous with respect to their specificity (or granularity), so that, in going up the hierarchy, more general topics are identified [7]. The algorithm can also operate an updating of a generated hierarchy of clusters with new news arriving on the stream. This incremental modality can eventually add new clusters when the content of new news is too different from that represented in current clusters of the hierarchy. More specific characteristics of the clustering algorithm are the automatic estimation of the number of clusters to generate, the efficient management of sparse vectors of documents features, and the use of a cosine similarity [7]. Finally, when the clusters are identified their labelling takes place with the aim of summarizing the main contents of the most representative news of the clusters. The summarization criteria identify the index terms with highest share among the top ranked news of the cluster and with the highest discrimination power among all the clusters. The balance of these two criteria makes it possible to generate unique labels for the overlapping fuzzy clusters.

4   The Distributed Information Retrieval Module

The pull phase allows the journalist to find background context on breaking news stories, deepen their knowledge of the story and/or assist during the compilation of news reports or articles. In PENG we facilitate search across a wide range of remote and local information resources. To enable search across a number of distributed resources within an integrated framework, we had to address four main research problems: automatic query generation and refinement, resource description acquisition, resource selection and data fusion [8].

The input to the retrieval component can be either from standard ad-hoc querying or pushed news documents explicitly selected to "deepen" the topic by the journalist. In the former case, the journalist enters keywords into the system when searching for information. In the latter case, automatic queries are formulated from pushed documents selected from the filtering. This process minimises the workload for the journalist, by extracting query terms based on term importance, and also provides a longer query than is typically submitted to a search interface potentially providing better retrieval accuracy. If available, queries are also expanded using terms from the journalist's user interest determined by the filtering module during the push phase. The refined query is then used to search distributed collections available to PENG.

We investigated a number of solutions for *automatically generating queries* from pushed news feeds. We investigated the use of representative and discriminative terms for query expansion, which has been found to be an effective technique for query expansion in centralised retrieval. The assumption for using query expansion with representative and/or discriminative terms is: user's with little topic familiarity in the topic, representative terms for the topic will be able to locate documents that are very general e.g. overview documents. In comparison, discriminative terms can be used to find detailed documents about a topic, for those user's with previous knowledge of that subject area. To extract either discriminative and representative terms, a topic language model is formed from the pushed news documents. The Kullback-Leibler Divergence measure is then applied to determine a term's contribution to the topic model [9]. Terms in the topic model are then ranked according to how representative or discriminative they are, and then used as a input query to the DIR module. For those users with low familiarity of the topic, the top ranked representative terms are used, while a user with high topic familiarity the set of discriminative terms used. Across an exploratory analysis of using both sets of queries for various user contexts, within the 2005 HARD track of TREC, it was discovered that using discriminative queries provided improved retrieval accuracy when compared to other query expansion techniques for users of varying topical knowledge. As a result, we have adopted this approach for generating automatic queries from all pushed news documents.

Journalists interact with a variety of resources and an integrated system must search across resources for a single information need. This means that we must obtain a *description of the resource to be searched*, an important stage because the perceived quality of such representations will impact on resource selection accuracy and ultimately retrieval performance. PENG uses Query-based Sampling (QBS) for the acquisition of resource description information [8]. Our approach is based on measuring the Predictive Likelihood (PL) of the journalist's information needs given the estimated resource description. This provides an indication of the description quality and indicates when a sufficiently good representation of the resource has been obtained [10]. Integrating PL as part of the QBS algorithm, performance was improved both in terms of efficiency and effectiveness when compared to currently adopted threshold based stopping method, minimising overheads while maintaining performance. Our approach is fundamentally different to existing work which measure the quality of an estimate against the actual resource. This requires full collection knowledge which is not readily available except in an artificial environments and is not realistic for journalists who are searching actual information resources. PL requires that only a set of queries are available for evaluating each resource description. In PENG, we mine the journalists query logs to obtain queries that are representative to the typical information needs of the journalists. Past queries are stored in the user profiles database, alongside a record of the meta-data of the current documents searched and queried by each user (stored in the document database).

Finally, the goal of *resource selection* is to search only those collections that hold relevant documents given a query request. In PENG, we rank collections by combining two evidence sources (using simple weighted averages): (1) an estimation of collection relevance with respect to a query using CORI [8], and (2) a user specified trust score for each resource. Trust scores are an estimate of the quality of information held in each resource. Applying *trust* addresses a key concern of journalists who often use such criteria when researching a story. To illustrate, using trust alongside relevance, a digital library of refereed academic articles can be given more importance than a collection of unpublished web articles even though the resource has been given a higher relevance score, thus in turn reflecting the current users needs. After ranking the collections the top k ranked are searched by asking for a decreasing number of documents form each collection based on the position in the ranking. The returned document results are then fused using the CORI algorithm.

5 Evaluation

The evaluation of the prototype system followed a task-based and user-oriented methodology in the context of a formative design evaluation framework. The evaluation involved 9 professional journalists, and 13 postgraduate students of journalism. Professional journalists, considered in this context expert users, were asked to complete forms describing typical information filtering and information search tasks carried out during their daily work activities. These forms gathered, over a period of 3 months, contained information about the nature of each task, the way it was carried out using any system available to the journalists, and the information found to be relevant for the task. During the same period material from the newswires and information repositories the journalists typically accessed was logged and copied into a separate storage. The documents reflected the nature of the tasks defined, resulting in a mixture of various multimedia and also multilingual documents (English, Italian, German and French) to reflect the working environment for this sample of users. Information extracted from the forms was used to design specific information filtering and information search tasks that both students and professional journalists carried out using the prototype.

To simulate appropriately the information filtering task, newswires and other pushed information was delivered in a time delayed fashion. The information available through the search was also the information that was originally available to the journalist at time $t - k$, as the users were then required to carry out at time t with the prototype the same filtering and search tasks performed by the experts at time $t - k$ with different systems. Users were then asked to judge the performance of PENG in relation to a number of evaluation qualitative dimensions pertaining to ease of use, learnability, satisfaction, likeability, and general attitude to the system and compare that with the systems they used during their everyday work. Quantitative data was also collected related to rate of task completion, time to carry out tasks, error and recovery rates. Additional information related to the performance in these tasks was also collected using the "think aloud" technique, direct observation and interviews[2]. Each user involved in the evaluation experiments was asked to carry out one filtering and one search task per session for a total of 3 sessions spread out over a period of two months.

The results in general indicated that the *students* found search tasks intuitive and were comfortable with using the PENG prototype. When considering the prototype against the system they would have normally used, Google, the overall usability of the prototype was comparable and helped them retrieve relevant documents with 75 % finding the system easy to use and browse. Approximately 50% of the user group believed that the prototype helped in completing their tasks faster; general consensus being that the ability to search simultaneously a lot of various news sources and resource was an advantage over a generic search engine. In particular, a cited advantage of the prototype was the importance given to news agencies and resources used for the region of Europe the students lived (i.e. Switzerland). This was potentially an indication of the trust model of the resource selection algorithm placing more weight on the geographical proximity of some resources in comparison to others. Also, the ability to search local repositories (both personal and shared in the PENG system) was considered a useful feature. One limitation of the prototype during the pull phase, however, was that the accuracy of the retrieval results varied across tasks. In particular the search for named entities such as people or specific places was often variable. A number of students asked specifically for Boolean operators to be available while formulating a query. Possibly as some queries were not returning documents with the people or places expected result, the users did not feel in control, hence the request for more advanced search features. Users were confused by the ability to search multiple language resources. In terms of the push phase, the students found overall the filtering tasks complex or even too complex for them to cope with the extra complications of not really understanding initially the meaning and implications of filtering. While interacting with the PENG retrieval module

---

[2] The interviewers did not belong to the design and development team in order to avoid any bias.

students could relate to their previous experience with search engines, they found it hard to deal with the filtering component as they were lacking of previous experience in this area. This was probably one of the underlying factors that influenced the less positive scores to the usability of PENG filtering component.

In comparison to the student responses, the *professional journalists* on the whole felt PENG was an obstacle for completing tasks. Professional journalists expect a high quality of service, in particular radio news journalists who are very dependent on the speed that they can find relevant information. As a result of these high expectations, the speed of the search was a negative comment even for a prototype version of the system. While the professionals agreed with the feature of searching agencies, repositories, personal archives being an advantage in the prototype, there was also emphasis that this should be extended to include local newspapers' and local authorities' archives. Even with a prototype system, the variable accuracy and speed made the journalists very sceptical about using the system in future to perform new tasks. Also, a general observation was the lack of obvious Boolean search operators traditionally used for search in library style systems but nowadays substituted by natural language interfaces. This could be justified as above for students with the users' need to feel more in control and possibly relate to users' previous extensive experience with more traditional interfaces for searching. A total of 71% negative responses of users not feeling satisfied about task completion indicates how journalists did consider filtering as an hindrance more than a useful functionality. As filtering is in nature passive and transparent to users it was difficult for them to understand how the system could be used to accomplish their tasks. Indeed the opinions expressed by journalist were inevitably biased by their everyday experience using Open Media, a professional tool described by them as extremely effective and flexible. Overall filtering proved not popular nor used as journalists had no frame of reference and lack of trust that it was not hiding news.

In general the opinions expressed by journalists and students were in line. If anything the journalists were more confident in their choice of scores, especially in negative terms, than the students. They were also more experienced in using a variety of sources and tools for finding relevant information, in particular they tended to compare PENG to the professional tool in use. Journalists were quite critical about PENG usability from the very beginning when performing search tasks and complained they could have performed better without PENG.

It is without doubt that being compared versus a professionally designed and engineered tool such as Open Media did not help PENG, a tool still in its prototype version. Even more so as the final experiments had to be rushed in order to take place on time for the completion of the project. The lack of time unfortunately affected particularly the filtering as it is the most time consuming of the modules and resulted in journalists not being able to perform properly some of the proposed tasks, as commented by the group locally in charge of running the evaluation experiments. In particular the filtering suffered from the confusion and lack of understanding in both user groups of what filtering really was for, that resulted in confused/mistaken expectations. This had emerged and was reported earlier on, while collecting task descriptions, and these findings confirmed that both user groups, students and journalists, felt equally confused when defining and later on performing filtering tasks. While search engines have educated users on retrieval tasks, filtering systems are not yet as popular and effective in educating users.

6   Conclusions

We have highlighted one solution to the information access and seeking problems that journalists currently face. PENG is an initial attempt at modelling and integrating the push, pull and presentation phases of a journalists workflow. In this study real users and in particular professional journalists were involved and this allowed us to have very valuable feedback as opposed to the tradition student lab based approach used in literature, where students are conveniently involved in evaluation of retrieval and filtering systems even if they are not necessarily representative in terms of genuine needs, skills and motivations of the final users. Indeed involving busy professionals added a level of complexity both in practical and conceptual terms.

References

1. Barry, C.L.: User-defined relevance criteria: an exploratory study. J. Am. Soc. Inf. Sci. (JASIS) 45, 149–159 (1994)
2. Schamber, L., Eisenberg, M., Nilan, M.S.: A re-examination of relevance: toward a dynamic, situational definition. Inf. Process. Manage. 26, 755–776 (1990)
3. Ruthven, I.: Integrating approaches to relevance. New Directions in Cognitive Information Retrieval 19, 61–80 (2005)
4. Kovach, B., Rosenstiel, T.: The Elements of Journalism. Random House (2001)
5. Pharo, N., Jarvelin, K.: Irrational searchers and ir-rational researchers. Journal of the American Society for Information Science and Technology 57, 222–232 (2005)
6. Bordogna, G., Pagani, M., Pasi, G., Villa, R.: A Flexible News Filtering Model Exploiting a Hierarchical Fuzzy Categorization. In: Larsen, H.L., Pasi, G., Ortiz-Arroyo, D., Andreasen, T., Christiansen, H. (eds.) FQAS 2006. LNCS (LNAI), vol. 4027, Springer, Heidelberg (2006)
7. Bordogna, G., Pagani, M., Pasi, G., Invernizzi, F., Antoniolli, L.: An Incremental Hierarchical Fuzzy Clustering Algorithm Supporting News Filtering. In: IPMU. Information Processing and Management of Uncertainty in Knowledge-Based Systems (2006)
8. Callan, J.P.: Distributed information retrieval. In: Advances in information retrieval, pp. 127–150. Kluwer Academic Publishers, Dordrecht (2000)
9. Baillie, M., Azzopardi, L., Crestani, F.: An evaluation of resource description quality measures. In: ACM SAC 2006, ACM, New York (2006)
10. Baillie, M., Azzopardi, L., Crestani, F.: Adaptive query-based sampling of distributed collections. In: Crestani, F., Ferragina, P., Sanderson, M. (eds.) SPIRE 2006. LNCS, vol. 4209, Springer, Heidelberg (2006)

BrowsReader: A System for Realizing a New Children's Reading Environment in a Library

Jia Liu, Makoto Nakashima, and Tetsuro Ito

Dept. of Computer Science and Intelligent Systems, Oita University
700 Dannoharu, Oita-shi, Oita 870-1192, Japan
{wenyao,nakasima,ito}@csis.oita-u.ac.jp

Abstract. This paper proposes a system for realizing a new children's reading environment in a library. Conventional libraries establish children's rooms for supporting children's activities of reading printed picture books. The collections there, however, are often biased in content and limited in number, and children are usually not satisfied with them. When the entire collection becomes large, it occurs that younger children cannot effectively search that collection. Similar search problems also exist in children's digital libraries. We have designed a system called Brows-Reader with the purpose of realizing a new environment, where children can browse in a large virtual bookshelf and can easily find the digitized and/or printed picture books that they are interested in. A user study was conducted to test the effectiveness of the system.

Keywords: Children's digital reading room, children's digital library, picture books, browsing interface.

1 Introduction

Picture books play an important role as textbooks for children. By reading stories and seeing pictures children can see an unknown world, improve their levels of imagination and expression, and learn the words and wording [3]. Conventional libraries, especially public libraries, arrange printed picture books in the bookshelves in their children's reading rooms or corners. The collections there, however, are often biased in content and limited in number, and cannot fully satisfy the children's needs. In addition, it is usually hard for children to read digitized versions of old picture books and newly published digitized picture books. When a library collects a large quantity of picture books, children would face the unfamiliar keyword-based index search different from the bookshelf browsing.

Some of these problems have been alleviated by realizing children's digital libraries on the Web, e.g., International Library of Children's Literature in the National Diet Library of Japan [8], and International Children's Digital Library created in Maryland University [7]. When utilizing digital libraries children are still required to input a keyword or click a category symbol in order to access to the desired digitized books, in spite of the difficulty that younger children cannot deal with keywords and categories very well [6]. They also have to trace

D.H.-L. Goh et al. (Eds.): ICADL 2007, LNCS 4822, pp. 361–371, 2007.
© Springer-Verlag Berlin Heidelberg 2007

Fig. 1. An image of a children's digital reading room

the links to get to the desired picture books. Further they will not be able to read the printed versions as in a conventional library.

We here try to solve such various problems by formalizing a new reading environment, a children's digital reading room, where a table with a built-in touch display [14] and a personal computer installed with the proposed BrowsReader is placed in a conventional reading room as shown in Fig. 1. BrowsReader, a system to assist *a group of children as well as a single child* in book-finding and book-reading activities, consists of Bookshelf-Browser and Book-Reader. The main features of the activities in this room involve:

(1) Children can easily find the books of their interest from a large quantity of intermixed collections of digitized/printed picture books.
(2) Children can read both digitized and printed picture books with the same interface on the table.

For (1), Bookshelf-Browser that assists children to browse the titles/author-names and the front cover images of digitized/printed picture books, is designed with the consideration of the cognitive ability of younger children as to the use of keywords and categories. Although children start to become interested in characters from around 3 years old, they could not correctly spell words until 11. The vocabulary of younger children is not rich enough and the categories have not been well formed [1][11][12]. By using Bookshelf-Browser children can narrow a large index of the title/author-names to be browsed by incrementally inputting a character, and a large virtual bookshelf showing the front cover images by touching a bookcase of the bookshelf. For (2), Book-Reader together with Bookshelf-Browser is specifically designed for children so that they can read, with the same interface, the just found digitized picture book together with any other previously found books.

BrowsReader was evaluated by establishing a temporary children's digital reading room in a playground during the 2007 Children's Event at Oita University. The collection consisted of the digitized versions of 162 printed picture books. Total 111 children, ages from 2 to 13, visited the room with their friends and families. All the children could launch Book-Reader via Bookshelf-Browser even though they had never learned how to use BrowsReader before. About a quarter of the children utilized the character input function. From the answers to the questionnaire, BrowsReader was shown to be interesting, useful, and easy to use even for younger children.

2   Design Policy

BrowsReader consists of Bookshelf-Browser accompanied by Book-Reader. The former is responsible for managing the primary and the secondary information of digitized picture books and the secondary information of printed books. The latter is mainly responsible for flipping the pages of digitized picture books. Digitized picture books as well as printed ones are read on a table with a touch display and a BrowsReader installed on a personal computer by page flipping.

2.1   Primary and Secondary Information

The primary information is about the digitized/printed picture books themselves, and the secondary information here is extracted from the primary information as to the bibliographic attributes "title" and "author." The secondary information functions as an index to the primary information for managing a large quantity of picture books. For adults there would be no problem in distinguishing and making use of the two kinds of information. For children, however, it would be difficult to make connections between the two [9].

2.2   BrowsReader

Fig. 2 illustrates the processing of the information in BrowsReader. With Bookshelf-Browser children can find books of their interest by browsing in the virtual bookshelf and/or the index. The virtual bookshelf showing the front cover

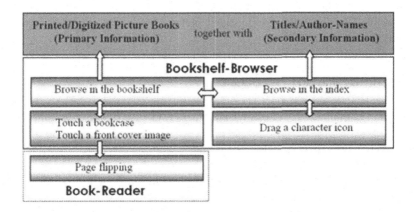

Fig. 2. Information processing in BrowsReader

images of digitized/printed picture books in the form of arranged bookcases is browsed in as seen in the real bookshelf browsing, and the index listing the titles and author-names as seen in the Keyword-in-Context index browsing. For easy browsing the bookshelf can be enlarged by touching a bookcase, and the index by dragging a character icon. To better help children understand the relationship between the primary information and the secondary information, the left and right parts of Bookshelf-Browser in Fig. 2 should be processed seamlessly, and the results be displayed promptly. Bookshelf-Browser is designed to do exactly this.

The front cover images of the books in the virtual bookshelf are intended to be treated as the primary information. The found digitized books can be read by just touching their cover images and flipping the following pages as if they were printed versions. Book-Reader, designed to be activated by Bookshelf-Browser, assists children in page flipping. When the touched image is for a printed book, children can find the book quickly from the real bookshelf based on the information from Bookshelf-Browser. On the table they can read both digitized and printed picture books together by page flipping.

The way of processing the information in a conventional children's reading room can be seen in the left part of Fig. 2, where the primary information is fully utilized. Children find the printed picture books by browsing in the bookshelves, pick them up, and then read them by flipping through the pages on a table. When the entire collection increases in size, children will have difficulty in searching it just by the primary information. A search system for processing the secondary information may help them in this case. This system, however, requires children to input keywords, a not easy task for younger children to do. Children who find books of their interest also need to go to the bookshelves to check the front covers to be certain they have got the right ones. In this environment, the primary information and the secondary information are processed separately, rendering it hard for children to make connections between the two.

In children's digital libraries, the processing of the two kinds of information is still separated even though the primary information is digitized. Children have to know and input keywords or category symbols before starting any search. The accesses to the primary information and the secondary information are independent. Besides, only the digitized picture books are available to read, no the printed versions.

3   Implementation Process

This section explains how Bookshelf-Browser and Book-Reader are implemented.

3.1   Bookshelf-Browser

The virtual bookshelf and the index are unified in Bookshelf-Browser. Children can browse in the bookshelf and in the index freely without noticing the difference. The bookshelf and the index, respectively, are enlarged by touching a bookcase and by dragging a character icon. The user's character input filters the

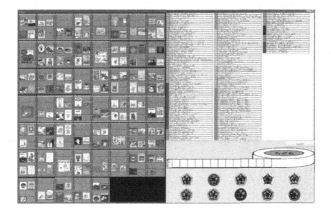

Fig. 3. Bookshelf-Browser

results to contain only the titles and author-names with the input character(s). The enlargements of the bookshelf and the index are synchronized such that when a part of the bookshelf (or the index) is enlarged its corresponding part of the index (or the bookshelf) becomes enlarged as well. This synchronization helps children relate the picture books to their titles/author-names in a visible way. Fig. 3 shows Bookshelf-Browser, where the bookshelf with 162 book cover images on the left and the index with input character icons on the right are displayed. The white ribbon-shaped paper on the middle right is prepared for displaying the history of the inputted characters.

The bookshelf is organized after arranging the books in the lexicographical order of the titles and clustering them into bookcases in fixed size. Each bookcase forms a unit for enlargement. The arrangement can be done as to an attribute, e.g., "author," "cover color," and "subject" other than "title," and its result corresponds to a simultaneous menu [4][5] for the employed attribute. When the "subject" is employed, consecutive bookcases store the images of the books with the similar categories. This makes it much easier for children to browse the books based on categories, even if they may not have enough knowledge about them.

The index called Character-String-in-Context index, devised as a sibling of the KWIC index, is made for listing the titles and author-names with the input character(s) in the middle of each string. The CSIC index is enlarged in font size just by dragging every input character icon. No enter-key is needed. The input character information is materialized in the CSIC index and in the white ribbon-shaped paper. Children with inadequate spelling skills can freely change the sequence of the input characters in the ribbon-shaped paper, and accordingly refashion the CSIC index. By this way children can manage the secondary information without pre-condition of having knowledge about keywords.[1]

---

[1] BrowsReader works as an enhanced search system in a library when children input a keyword instead of a character. The CSIC index in this case works as one of the KWIC index as seen in the outputs of a search engine, e.g. Google [2].

3.2 Book-Reader

Book-Reader, which works cooperatively with Bookshelf-Browser, is a viewer to read digitized books in a way similar to read printed ones. Its main function is page flipping with three options: page-by-page reading, scan reading, and bulk reading. Other functions include graffiti and magnifying, by which children can enjoy reading with imagination. Book-Reader is automatically started when the front cover image of a digitized picture book is touched. The foreground in Fig. 4 shows two digitized picture books (which were on the bookshelf before having been touched) being read.

4 An Example of Utilizing BrowsReader

Fig. 4 shows an example situation of the utilization of BrowsReader. A child, who has imagined about " うさぎ (rabbit)," can find a book by dragging an icon for a character, e.g., ' さ,' that occurred in his/her mind, by browsing in the enlarged CSIC index or the enlarged bookshelf and then by checking the details of the front cover images in some bookcases. The books found can be read with his/her friends, where they together or one after another flip the pages by dragging the displayed images. In Fig. 4, another book about " うさぎ " which had been under reading is displayed behind the reading one.

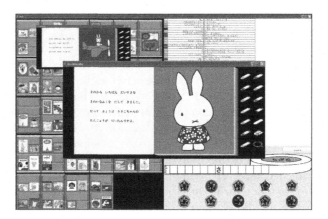

Fig. 4. Utilization of BrowsReader

5 Usability Experiment

We conducted a user study to evaluate the usability of BrowsReader by establishing a temporary children's digital reading room in a playground during the 2007 Children's Event hosted at Oita University. All participated children could visit the room freely. Though some experiment under the controlled situation could be performed, we adopted such a study in order not to bear unpredictable influence on the children's reading activities in the future. The collection consisted of the digitized versions of 162 printed picture books. A 30-inch (2560×1600 pixel resolution) display controlled by a personal computer with an Intel® Core^{TM}2 Duo CPU working under Windows XP was utilized to implement our prototype of BrowsReader.

We fixed several parameters on our prototype before we conducted the user study. Nine university students took part in this preliminary experiment. For Bookshelf-Browser, after examining how effective and easy the interesting books could be found while browsing in the bookshelf, it was decided that each bookcase was to have 4 books. In [5] it is mentioned that a character icon not smaller than 96×96 pixels in size would be desirable for a 30-inch display, but we set the size to be 64×64 so that total 50 input character icons could be displayed, and we grouped 5 icons together into a big soccer-ball icon whose size was at least 96×96 pixels. Any character could be inputted by touching a soccer-ball icon and then dragging the corresponding character icon. The size of each bookcase was at least 96×96 pixels.

For Book-Reader each book was displayed so that the font size of a character was at least 15 pt (which was decided by examining the above 162 picture books). The speed of page flipping was designed to be 1 page per second based on experience from preliminary experiment; but a faster speed was also set for smaller books. The pixel size of the icons for graffiti and magnifying was set to be at least 96×96.

5.1   Participants and Logged Data

Fig. 5 shows the environment for the user study. Two tables were prepared, where each display was put on a table though it should be fit in the center of a table top in a real environment. Total 111 children (ages 2-13) took part in the study (10:00 a.m.-15:00 p.m.) together with their friends and families. There were 16 kindergartners (ages 2-5), 68 lower graders (ages 6-9) and 27 higher graders (ages 10-13) from elementary schools. They did not know anything about BrowsReader before they came to the event. All the children were delighted to touch the bookcases/characters, to browse in the bookshelf/index, and to read the digitized picture books.

A logging system was utilized to record touch operations by the children for the statistical survey. A questionnaire for the children was also prepared.

5.2   Tasks and Results

The children were asked to use BrowsReader just as they pleased. University students would assist them only when they had problems. All the children launched Book-Reader via Bookshelf-Browser, and most of them were able to flip pages. After using BrowsReader, each of the children was interviewed in a casual conversation and requested to fill in the questionnaire shown in Table 1.

Fig. 5. An experimental environment

Table 1. Questions in the questionnaire

No.	Question
Q1	Are you happy when you used it?
Q2	Did you read a book when you used it?
Q3	Did you use it with others together?
Q4	Would you like to use it again?
Q5	Do you often go to a library?

About Browsing: Fig. 6 shows the results about browsing derived from the logged information, where gray, black, and white bars are for kindergarteners, lower graders, and higher graders, respectively (The similar bars are used in the following figures). Most children still found the books via the browsing of the primary information; but there were about 15% to 30% of them used mainly the secondary information to find the books. The children in higher school grades preferred to use primary information when browsing. The reason is that they are more used to read printed picture books. Before the study, we were uncertain about how many children could handle the CSIC index browsing. The result turns out that it worked fairly well when displaying the index together with the bookshelf.

About character input: Even the kindergartners and the lower graders tried to input a character as shown in Fig. 7. They often input a part of their names or some known characters. The results of the study show the kindergartners and the lower grader students were more likely to input characters than the higher grader students. This result, which seemingly contradicts to the discussion in [12] where category symbol input is recommended, shows some rationale about preparing the character input for searching the picture books.

About Reading: Page flipping was rather popular among the kindergartners as shown in Fig. 8. The kindergartners liked page flipping, which matches the findings in [12]. However, all children rarely used the magnifying. This fact shows

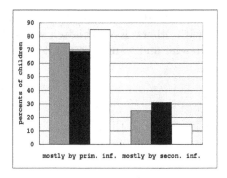

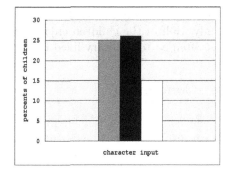

Fig. 6. Statistics as to browsing   **Fig. 7.** Statistics as to character input

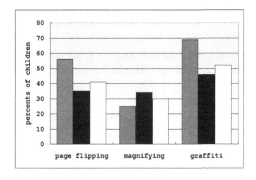

Fig. 8. Percents of children who used different functions in Book-Reader

the pertinence of the display size of the books. Also, many children, especially the kindergartners, preferred to write graffiti, which matches the finding in [13]. Because graffiti is not allowed in a printed picture book in a conventional library, this added function favors the use of the children's digital reading room.

About the questionnaire: According to the statistics of the questionnaire in Table 2, BrowsReader was said to be interesting, useful, and easy to use. For question 1, one child returned the negative answer "there is no my favorite book."

Table 2. Percents of positive answers

No.	kindergartners	lower graders	higher graders	average
Q1	100	98.5	100	99.5
Q2	81.3	89.7	96.3	89.1
Q3	81.3	32.4	29.6	50.3
Q4	93.8	94.1	100	96.0
Q5	75.0	75.0	66.7	72.2

This might be due to the fact that there were only 162 books prepared in this study. Half of the children enjoyed reading with others. Even children who did not often go to a library liked to use BrowsReader and wanted to use it again.

6   Conclusion

A new children's reading environment is suggested by settling tables, for which touch displays and BrowsReader installed personal computers are set up, in a conventional children's room in a library. A child there can enjoy with his/her friends or family by finding the interesting books from a large quantity of digitized/printed books, and by reading the just found books together with any other previously found books. The effectiveness of BrowsReader was also shown in the user study.

The 30-inch screen utilized here can display up to about 500 front cover images each of which is distinguishable from the others without enlargement. Many conventional children's reading rooms, however, stores more than 10,000 printed picture books. We plan to evaluate BrowsReader for a collection of 10,000 to 20,000 picture books, and then set it in the children's reading corner of Oita University library for the future usability tests.

Though a display was placed on a table in the user study, it should be fit in the center of a table. Such an arrangement will be realized by adopting the research findings in [14], and/or by employing an emerging device seen in [10].

References

1. Borgman, C.L., et al.: Children's Searching Behavior on Browsing and Keyword Online Catalogs: The Science Library Catalog Project. Journal of the American Society for Information Science 46, 663–684 (1995)
2. Google, http://www.google.co.jp/
3. Guidelines for Children's Libraries Services, http://www.ifla.org/VII/s10/pubs/ChildrensGuidelines.pdf
4. Hochheiser, H., Shneiderman, B.: Performance Benefits of Simultaneous Over Sequential Menus as Task Complexity Increases. Int'l Journal of Human-Computer Interaction 12(2), 173–192 (2000)
5. Hutchinson, H.B., et al.: How Do I Find Blue Books About Dogs? The Errors and Frustration of Young Digital Library Users. In: CD-ROM. Proc. of 11th Int'l Conference on Human-Computer Interaction (2005)
6. Hutchinson, H.B., et al.: The Evolution of the International Children's Digital Library Searching and Browsing Interface. Interaction Design and Children, 105–112 (2006)
7. International Children's Digital Library, http://icdlbooks.org/
8. International Library of Children's Literature in the National Diet Library of Japan, http://kodomo4.kodomo.go.jp/web/ippangz/html/TOP.html
9. Kuhlthau, C.: Meeting the Information Needs of Children and Young Adults: Basing Library Media Programs on Developmental States. Journal of Youth Services in Libraries, 51–57 (1988)

10. Microsoft Surface: A 30-inch Diagonal Display Table, `http://www.microsoft.com/surface/`
11. The National Institute for Japanese Language: Reading and Writing Ability in Pre-school Children (in Japanese). Tokyo Shoseki, Tokyo (1972)
12. Reuter, K., Druin, A.: Bringing Together Children and Books: An Initial Descriptive Study of Children's Book Searching and Selection Behavior in a Digital Library. In: Proc. of ASIS&T, 339–348 (2004)
13. Sakamoto, T., Suzuki, S.: Computer is Good for Younger children! - Verified Computer Effort in the Term of Small Children (in Japanese). Sunchoh Publishing, Tokyo (1997)
14. Scott, S., Sheelagh, C.: Interacting with Digital Tabletops. IEEE CG&A 26, 24–27 (2006)

Desktop Search Engine Visualisation and Evaluation

Schubert Foo and Douglas Hendry

Division of Information Studies, School of Communication and Information
Nanyang Technological University, Singapore 637718
{assfoo,hend0007}@ntu.edu.sg

Abstract. This work investigates the potential of applying a suite of visualisation for query processing in a desktop search environment. While each of these visualizations may not be totally new on its own, we have attempted to add value to each one by endowing it with useful features, and to seamless integrate them to allow easy switching of views, thereby providing the novelty in this work to create a potentially useful means to process search results and carry out query refinements and exploration. These visualisations include a List View, Tree View, Map View, Bubble View, Tile View and Cloud View. A first evaluation was undertaken by 94 M.Sc. participants to gauge the system's potential usefulness and to detect usability issues with its interface and graphical presentations. The evaluation results were encouraging and showed that these views to be both effective and useful on the whole, and support the research premise that a combination of integrated visualisations will result in a more effective search tool.

Keywords: Query result processing, query reformulation, tree view, map view, bubble view, tile view, cloud view, evaluation, search engine, user interface.

1 Introduction

Internet search and information seeking is predominated by large search engines such as Google, Yahoo, MSN and AOL (more commonly referred to as GYMA) where textual search results are largely displayed as text-based URLs and a few lines of document contents to provide the context to the URLs. These established search engines' strength lies mainly in their very sophisticated search and ranking algorithms but lack somewhat in innovation in terms of their interfaces which are traditional text-based. However, we begin to see emergence of online search engines that have started to offer more graphical interfaces to assist users such as Grokker and Ujiko.

In contrast to the developments of online tools to search the Web, desktop search tools for searching increasing large volumes of documents held on local computers hard drives, have been slower to develop. While search tools are now being incorporated into the latest desktop operating systems, such as Spotlight in Apple's OSX and Vista in Microsoft Windows, their search result visualisations have taken the traditional display of textual output. With result lists becoming increasingly longer, the challenge is to improve efficiency and effectiveness of search and result selection. In this respect, the metaphor that "a picture paints a thousand words" neatly encapsulates

D.H.-L. Goh et al. (Eds.): ICADL 2007, LNCS 4822, pp. 372–382, 2007.

the concept that well presented graphical views can convey large amounts of complex information in a simple and easy to understand manner. It is therefore not surprising that graphical visualisations have been employed in search engines to assist users.

This work reports on a number of visualisations that were developed for desktop searching. While each of the individual visualisation might not be new by itself, we believe that the seamless integration of these views and value-added functionality in them are novel to assist in the results review, selection and query refinement. A set of first evaluations was carried out to determine the usefulness of the developed views and to identify areas for further development. The work therefore aims to contribute towards desktop searching research through the provision of a suite of integrated views, eliciting and confirming the characteristics (applicability, usefulness, limitations) of the views, supported by evaluation results.

2 Related Work

In order to overcome the limitations of text-based result lists, a number of researchers have developed a variety of 2D and 3D graphical visualisations in order to allow the user to explore and understand the results of their query. PFNET was an early to attempt to introduce visualisation through a network based tool that used a thesaurus to create associated document networks based on the query results [1]. Users could browse results as a network and view individual documents.

Envision [2] is a tool that was built as an alternative to the query-document similarity ranking. The tool allowed the searcher to graphically display the search results from a bibliographic IRS using an X-Y graph. Different attributes such as author, year, document type, number of citations and relevance can be plotted on the X and Y-axes and the result documents are represented as icons in the main plot area.

GRIDL uses a grid-based approach to present large volumes of digital library search results using categorical and hierarchical axes (hieraxes) to simplify the display [3]. The documents in a search result are clustered together based on metadata. These hierarchical categories are then used as the axes for a grid-based plot that displays the relative number of documents at the intersection of each axis attributes. Users can then drill down these hierarchies to explore the search results. Other examples of visualisation tools include the Visual Information Browsing Environment (VIBE) that presents search results as a 2D map [4], Periscope which is a system for adaptive 3D visualisation of web search results [5], and YAVI that uses a 3D information space to display a vector-space model output of search results [6].

To support query reformulation, visualisations usually present terms that are related to the query terms in use. These related terms often come from a controlled dictionary, thesaurus or other system metadata. The user can review these new terms and use them to modify their query. One such tool is the AquaBrowser Library [7] which shows a visual word cloud that suggests words similar or related to the users query terms.

Whilst some evaluation studies have reported mixed results [8] many have found positive support that the visualisations have aided user performance [9, 10]. Even in cases where performance has not improved users often report better satisfaction with tools incorporating visualizations [11]. Visualisations seem to be particularly effective where the complexity of the task and volumes of data are at their highest [12, 13]. They also seem to work well when they are kept as simple as possible [14].

3 System Design

DSE is a Java-based lightweight desktop search engine developed to index and search content on a desktop computer. The indexing and searching sub-systems were designed based on traditional IRS principles and incorporates stop word removal, stemming and is based on Boolean logic. DSE uses a plug-in architecture for supporting different file types that include text, HTML, RTF, XML, MS Word, MS Excel and PDF.

The design of the user interface was based on the following research premises:

- Visualisations can assist users to search for documents [10, 13]
- Different visualisations can be used to support different elements of the searching process (results review and query reformulation)
- Different graphical techniques can be used to assist users to visualise different kinds of information
- Visualisations work best when they are kept simple [14].

The search engine GUI has a plug-in view architecture that allows different views to be created independent of the searching mechanism. Six views were constructed for use and evaluation: List View, Tree View, Map View, Bubble View, Tile View and Cloud View. These will be elaborated later in the paper.

4 Evaluation Methodology

After development, a first evaluation was carried out to determine the usefulness of the views, particularly to find out if the availability of such views was useful and which of them would be worth developing further. As such, the work reported here is intended as a proof of concept and to have a first gauge of usability and usefulness of the views.

The evaluation was carried out through a user survey based a questionnaire comprising 51 questions to gather user opinion about the visualisations and their usefulness. The evaluation was split into five tasks – each required the user to perform a search in support of a given information need and the participants worked through these in sequence. For each task they were given five minutes to interact with an interface to find the most relevant documents to satisfy the information need. Following this, they were asked to complete a series of questions pertaining to the interface used.

Prior to the evaluation, a handout describing the search engine and views was provided to the evaluators, followed by a briefing and demonstration. Evaluators were also invited to download the search engine to familiarise themselves with it and the various views. Each evaluator had a minimum of one week of familiarity prior to the evaluation, and each spent 45 to 60 minutes in completing the evaluation.

Since the purpose of this evaluation was to review the visualisation aspects of the search tool, it was important to remove other factors that could influence user responses. Therefore, for each task, the set of search terms to be used was pre-specified.

We used a small set of 30 documents for evaluation since DSE had the functionality to show the top *n* documents in the display, thereby restricting the volume of results displayed even if a larger collection was indexed and searched. We expect future evaluations to be more comprehensive and extended to test for scalability to larger document collections, and to test how the visualisations will lose their effectiveness and usefulness when the displays are increasingly populated with more search results.

The documents used in the evaluation covered topics on information retrieval systems, the World Wide Web, indexing and programming languages. A portion of the ERIC thesaurus was used to create a hierarchical folder structure and these documents were then stored into the folders based upon their categorisation in ERIC.

The post-graduate students of the Nanyang Technological University M.Sc. Information Studies class of 2006/07 participated in the evaluation. There were 94 participants – 57% female, 43% were male. 98% of participants rated themselves having Satisfactory (or better) computer skills, while 76% rated themselves having Satisfactory (or better) searching skills.

5   Search Engine Views and Evaluation Results

This section presents the various views constructed in DSE for presenting search results from the queries. It highlights the characteristics of each display and reports on the salient findings of the evaluation.

5.1   List View

The List View (Fig 1) is the classic search results view. It contains a list of files that, based on Boolean logic, match the users query. Each file name is shown along with the number of "hits" from the query. A hit is defined as one occurrence of one query term in the file contents. The files are listed in descending order of the total number of hits.

Toolbar   Result List   Search Control   Paging Control   Show "n" documents   Highlighted Search Terms

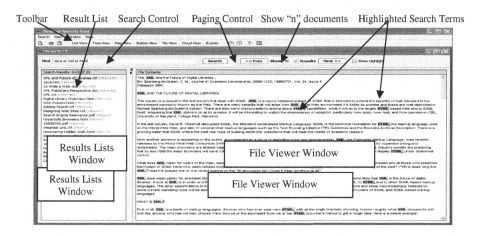

Fig. 1. List View

This view also contains a File Viewer window that will display the textual contents of a selected file in order to allow quick review of the file contents. Matching query terms are highlighted in different colours to aid the user identify where these terms occur in the document. Any non-textual content, such as images, etc. in the actual document are not displayed. Similarly, some of the source document formatting will be lost as only line and paragraph breaks are preserved in the extraction process.

Double clicking on a selected file displays the original document in a separate window. Such a view, although elementary, is simple, intuitive, provides clarity and a quick preview of the documents. Unsurprisingly, most evaluation participants gave strong support to this view as both easy to use (89% who Agree or Strongly Agree) and useful (86%) in reviewing the results. The evaluators liked the highlighting of the search terms in the file viewer and clear indication of the number of hits per result file, and suggested improvements related to more flexible sorting of the results and more document and result information to be made easily available. These results confirm that the basic design and operation of the desktop search engine is effective and useful.

5.2 Tree View

The Tree View is similar to the List View (Fig 2) except the result files are organised based on their underlying folder structure. For each file in the results list, all of its parent folders are added to the folder hierarchy (avoiding duplicates). The Result files are then added into the tree at the appropriate folder for their physical location.

Hierarchical Folder   Result Files per Folder with Hit Count Structure

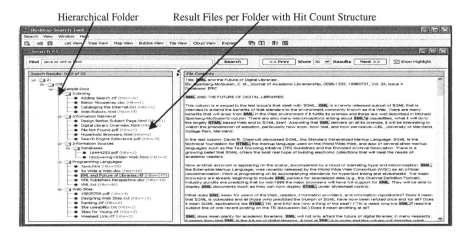

Fig. 2. Tree View

This view is very similar to the Microsoft Windows Explorer view. However, only files that match the query string are displayed and only the parent folders of these files are included in the tree. The purpose of this view is to use the physical file structure as part of the results display. If users have taken the time to organise their documents into meaningful folders and hierarchies then this information may be useful when reviewing results. This view is particularly suited for thesaurus or taxonomy based folder organisations where documents are stored in the respective nodes of this organisation scheme.

As such, related documents would already have been assessed and organised into folder hierarchies that will help users to quickly zoom into documents of interest.

With the familiarity of Windows Explorer, participants strongly indicated that this view was easy to use (93%) and useful (91%) in reviewing the results. They found the view clear and obvious. 87% of them acknowledged that if they had organise their documents logically in folders, then this view would be especially useful for them. This confirms the design premise that the user's folder structure is a useful aid to present search results, as well as a means to logically organise information in thesaurus/taxonomy-like structures that can support browsing as well as searching.

5.3 Map View

The Map View (Fig 3) provides an overview of the relationship between the query terms and the result files. Each query term is depicted as a blue rectangle and each result file as a green ellipse. Lines link related query terms and results files. These are annotated (in red) with the number of occurrences of the query term in the result file.

The view can be zoomed and rotated and individual shapes can be moved around on screen to obtain views that are more legible and avoid cluttering. If the mouse is moved over a query term it will display a popup window that lists all the result files that contain the query term along with their respective number of hits (not shown in Fig 3). Similarly, if the mouse is moved over a result file then a popup window will display all the query terms found in this file with their respective hit counts (Fig. 3).

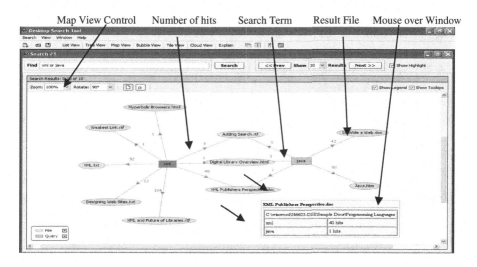

Fig. 3. Map View

This view shows how individual query terms affect the results and which files contain one or more query terms. This bird's eye view can be used to detect problems in the query specification if the required results are not as expected. It will clearly show the relative influence of each query term in producing the result files and therefore help the user in deciding whether the query needs to be reformulated and how to do so.

The evaluation found that slightly over half of the evaluators (51%) agreed or strongly agreed that the Map View was useful in reviewing their query results and reformulating their query. The distribution of responses for ease of use and usefulness are very similar. Qualitative comment analysis indicated that the most useful aspect noted by the evaluators (35%) was the ability to see an overview of the relationship between the query terms and the results files. This was the design premise for the Map View – to provide a clear overview of the query and its effect on matched results. However, the view can become very crowded for complex Boolean queries with a large number of items displayed resulting in overlapping of the graphic objects. A significant number of evaluators (36%) indicated that this caused confusion.

5.4 Bubble View

Boolean logic systems make it difficult to judge the relevance of a result file. The total number of hits alone is not necessarily a good guide to relevance especially when document length is taken into consideration. Therefore, it is desirable to normalise this measure to take into account document size. In this work, a hit density is calculated as the number of hits per 1,000 searchable terms (non stopwords) in the document.

The intention of the Bubble View (Fig 4) is to help the user better assess the relevance of different documents. The axes of the graph are the number of hits and the calculated hit density. These measures are used to distribute the documents along each axis as they provide good document discrimination in order to achieve a better visualisation. The diameter of the bubble is determined by the number of query terms present in the result file and its colour represents its file type.

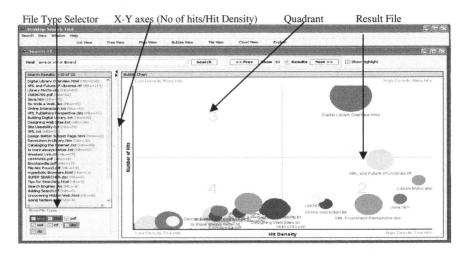

Fig. 4. Bubble View

Quadrant 1 is expected to contain the most relevant documents as both the number and density of hits is greatest. Correspondingly, quadrant 4 will be expected to contain the least relevant documents, as both the hit count and density are smallest. The display suggests that documents should be explored in priority according to the Quadrant numbers. The view therefore attempts to provide an overview of document relevance for a given query and aids the review of documents most likely to be relevant to the query.

The evaluation results show that 46% of the evaluators found this view useful in reviewing their query results. The majority found the position (65%) and size (59%) of the bubbles gave them useful information, which supports the concept of this view as a means to convey several dimensions about the relevance of the results documents.

The comments analysis showed that useful features were the ability to get a quick and easy overview of the relevancy of the results and the ability to see the hit density. The major confusion factors related to the display of a large number of result documents where the titles overlap and become unreadable and the display was found to be very cluttered and messy. Suggestions for potential improvement relate mainly to improving the layout to increase clarity and for help on how to interpret the view.

5.5 Tile View

The Tile View (Fig 5) presents each result file as a coloured tile using a Treemap. A Treemap is "a space-constrained visualization of hierarchical structures" [15]. The size of each tile is determined by a measure such as Total Number of Hits, File Size, and Hit Density (Hits per 1,000 searchable terms). Using the control panel, the user can change the measure used to determine the size of a tile. As before, the colour of a tile is determined by its file type and the display can be restricted to certain file types.

The Tile View can optionally include the folder hierarchy of the results files (not shown in figure). In this variant, all the result files in a specific folder are grouped together in a "super tile". Each folder is enclosed within a tile representing its parent so that the entire folder structure of the results files can be displayed. This is an alternative display of the tree view but with value-added information in the tiles. The purpose of the Tile View is to allow users to review the results visually and judge their relevance based on different criteria with larger tiles denoting the most relevant documents.

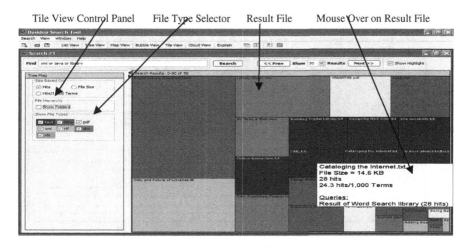

Fig. 5. Tile View

The results of the evaluation of the Tile View show over half the evaluators (59%) agreed or strongly agreed that the Tile View was useful in reviewing their results. Over two thirds found the tiles to be obvious and easy to understand (68%) and the ability to use different criteria to control their sizing was found to be useful (69%). This supports the design objective for this view to easily support the use of different criteria for judging the relevance of the results documents.

The ability to group files by folders also received strong support with 75% of evaluators agreeing or strongly agreeing that this was useful. The comments analysis indicated that useful features were the ability to change tile size based on different criteria, the ability to group files by folder and the use of colour to distinguish file types.

5.6   Cloud View

The Cloud View (Fig 6) is adapted from the Tag Clouds popular on social networking sites such as Flickr. A Tag Cloud is a weighted list which contains the most popular tags used on that site and the relative popularity of each tag is indicated by changing its font size. It is thus easy to see the most popular tags. The Cloud View creates a Word Cloud based on the (indexable) content of the result files. The file contents are examined and stop words and non-indexable terms are removed. The words are then stemmed and a simple term count of the documents contents. The top 300 terms are then displayed in a Word Cloud as they represent the most common indexable terms.

Only files selected in the Results List (in the left hand window) have their contents included in the Word Cloud. If the selection of files is changed, the Word Cloud is dynamically refreshed with information based on the new selection of files.

When the user clicks on a word in the Word Cloud a popup menu appears offering the choice to expand (OR), restrict (AND) or exclude (NOT) the word from the current query or to create a new search (NEW) using the selected word (Fig 6).

Selectable Result List   Word Cloud (Alphabetically arranged)   Common word (larger fonts)   Query refinement

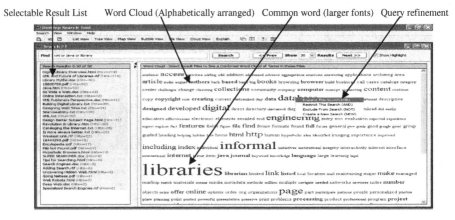

Fig. 6. Cloud View

The purpose of the Cloud View is to provide uses with the most common *n* words (300 in this instance) found in the selected files in the result lists thereby providing an idea of the contents of the result files (i.e. basically a concordancer) and information on potential words that can be used in the query refinement process.

The evaluation results for the Cloud View showed that nearly two thirds of the evaluators found the Cloud View useful in reformulating their query (63%) and easy to use (61%). However, the distribution profile for Question 37 (usefulness of Cloud View) is different with a bi-polar distribution, with a peak for Disagree and Agree. This implies that the evaluators were split into two groups, a conclusion strongly supported by a review of the comments. Those evaluators who scored the usefulness of the Cloud View very low (Strongly disagree or Disagree) reported a lot of confusion as to the contents of the Cloud. In other words, they did not find the view useful because they did not understand what it does. Those who did rated it highly. This implies that some users had not seen this type of visualisation before nor understood its potential.

6   Conclusion

A lightweight desktop search engine, DSE, along with six integrated views was developed to study the effectiveness and usefulness of them to aid the processing of query results and query refinement. The evaluation carried out on 94 participants indicates these visualisations useful and easy to use on the whole. They generally felt that it would help them find their desired results quicker. In particular, the Tree View and Cloud View were rated highly by the evaluators. The Tree View takes advantage of the users own defined hierarchies (their folder structures) to present the search results in a format that significant numbers of the evaluators found useful. The Cloud View, while novel, posed difficulty for some evaluators who did not understand it. Those who did rated the view to be very useful. Our work indicates that the provision of a suite of tightly linked yet different visualisations has the potential to increase the usefulness and ease of use for result processing and query refinement for desktop searching which is very much at its infancy stage of development in contrast to its more established online search engine counterpart. Our next stage of work is to improve the current views in light of the evaluation findings in preparation for the scalability and longitudinal tests for a series of increasingly larger result sets of documents.

References

1. Fowler, R.H., Fowler, W.A.L., Wilson, B.A.: Integrating query thesaurus, and documents through a common visual representation. In: Proceedings of the 14th annual international ACM SIGIR conference on Research and development in information retrieval, ACM Press, New York (1991)
2. Nowell, L.T., et al.: Visualizing search results: some alternatives to query-document similarity. In: Proceedings of the 19th annual international ACM SIGIR conference on Research and development in information retrieval, ACM Press, New York (1996)

3. Shneiderman, B., et al.: Visualizing digital library search results with categorical and hierarchical axes. In: Proceedings of the fifth ACM conference on Digital libraries, ACM Press, New York (2000)

4. Koshman, S.: Testing user interaction with a prototype visualization-based information retrieval system. Journal of the American Society for Information Science and Technology 56(8), 824 (2005)

5. Wojciech, W., Krzysztof, W., Wojciech, C.: Periscope: a system for adaptive 3D visualization of search results. In: Wojciech, W. (ed.) Proceedings of the ninth international conference on 3D Web technology, ACM Press, New York (2004)

6. Newby, G.B.: Empirical study of a 3D visualization for information retrieval tasks. Journal of Intelligent Information Systems 18(1), 31–53 (2002)

7. Kaizer, J., Hodge, A.: AquaBrowser Library: Search, Discover, Refine. Library Hi Tech News 22(10), 9–12 (2005)

8. Heo, M., Hirtle, S.C.: An empirical comparison of visualization tools to assist information retrieval on the Web. Journal of the American Society for Information Science and Technology 52(8), 666 (2001)

9. Sebrechts, M.M., et al.: Visualization of search results: a comparative evaluation of text, 2D, and 3D interfaces. In: Proceedings of SIGIR: International Conference on R&D in Information Retrieval, vol. 22, pp. 3–10 (1999)

10. Veerasamy, A., Heikes, R.: Effectiveness of a graphical display of retrieval results. In: Proceedings of the 20th annual international ACM SIGIR conference on Research and development in information retrieval, ACM Press, New York (1997)

11. Heflin, J., et al.: WebTOC: Evaluation of a Hierarchical Browsing Interface for the World Wide Web (1997), http://www.otal.umd.edu/SHORE/bs11/

12. Morse, E., Lewis, M., Olsen, K.A.: Testing visual information retrieval methodologies case study: comparative analysis of textual, icon, graphical, and "spring" displays. J. Am. Soc. Inf. Sci. Technol. 53(1), 28–40 (2002)

13. Wingyan, C., Chen, H., Nunamaker Jr, J.F.: A Visual Framework for Knowledge Discovery on the Web: An Empirical Study of Business Intelligence Exploration. Journal of Management Information Systems 21(4), 57–84 (2005)

14. Chen, C., Yu, Y.: Empirical studies of information visualization: A meta-analysis. International. Journal of Human-Computer Studies 53(5), 851–866 (2000)

15. Bederson, B.B., Shneiderman, B., Wattenberg, M.: Ordered and Quantum Treemaps: Making Effective Use of 2D Space to Display Hierarchies. ACM Transactions on Graphics (TOG) 21(4), 833–854 (2002)

Digital Libraries and Digitised Maps: An Early Overview of the DIGMAP Project

José Borbinha[1], Gilberto Pedrosa[1], João Gil [1], Bruno Martins[1], Nuno Freire[2], Milena Dobreva[3], and Alberto Wyttenbach[4]

[1] IST – Instituto Superior Técnico, Av. Rovisco Pais, 1049-001 Lisboa, Portugal
jlb@ist.utl.pt, gilberto.pedrosa@ist.utl.pt, jg7l@netcabo.pt,
bruno.martins@tagus.ist.utl.pt
[2] BNP – Biblioteca Nacional de Portugal, Campo Grande, 1749-081 Lisboa, Portugal
nmaf@bn.pt
[3] IMI – Institute of Mathematics and Informatics, 1113 Sofia, Bulgaria
milena.dobreva@gmail.com
[4] UPM – ETSI Topografía, Geodesia y Cartografía, 28031 Madrid, Spain
a.fernandez@topografia.upm.es

Abstract. DIGMAP is a project to find solutions for digital libraries scenarios focused on digitised historical maps. The main service will reuse metadata from European national libraries and other relevant third party metadata sources to provide discovery and access to contents. This will also include a proof of concept of a scenario of reusing and enriching these metadata by automatic processes that will try to extract relevant indexing information from the images of the digitised maps, as well as from any kind of associated text.

1 Introduction

DIGMAP[1] is a project co-funded by the European Community programme eContent-plus[2]. DIGMAP stands for "Discovering our Past World with Digitized Historical Maps", but could also stand for "digging on maps"! Historical maps have important details that make them very difficult (and therefore expensive) to describe, classify and index in "hand-made" descriptive metadata structures. With these assumptions in mind, we propose to develop a modular software solution to enable the development of flexible services to support the description and indexing, as also the searching and browsing in collections of digitized historical maps. Maps have a special characteristic: they always represent a limited physical space in the real world. Therefore, we propose to develop technology for services to make it easy to classify, index, search and browse them by their geographic boundaries, with the support of multilingual geographic thesauri or gazetteers, as also by their decorative and stylistic details.

This paper will proceed by describing the main DIGMAP use cases, followed by a description of the architecture of the system.

---

[1] http://www.digmap.eu
[2] http://ec.europa.eu/information_society/activities/econtentplus/index_en.htm

D.H.-L. Goh et al. (Eds.): ICADL 2007, LNCS 4822, pp. 383–386, 2007.
© Springer-Verlag Berlin Heidelberg 2007

2 DIGMAP Use Cases

The final results of the project will consist in a set of services available in the Internet, according to the main use cases presented in the Figure 1.

A **User** is an anonymous person, a **Registered User** is an actor that has been recognized by the system, and an **External System** is a machine. A **Cataloguer** will be able to update the metadata, while a **Geographer** is a specialist who manages Thesaurus data types. The **Administrator** will manage all the services and infrastructure.

The users will be able to access resources in two ways: **Searching Resources** (simple and complex searches in the metadata) or **Browsing Resources** (browsing in indexes built from the metadata). It will also be possible for users to propose resources for inclusion in the system, through the use case **Proposing Resources**. Through **Asking Expert** it will be possible to submit questions to a **DIGMAP Forum**, to be answered by experts with specific knowledge in the area. The specialists will answer and manage those questions through the use case **Dispatching Questions**.

There will be three different data management use cases in DIGMAP. **Updating Catalogue** will feed the Catalogue with metadata from remote data providers or through a local cataloguing interface. At any time, it also might be need to care about **Updating Authorities**, to assure a better browsing and search. Also with the purpose of supporting those cases we have the case **Updating Gazetteer**.

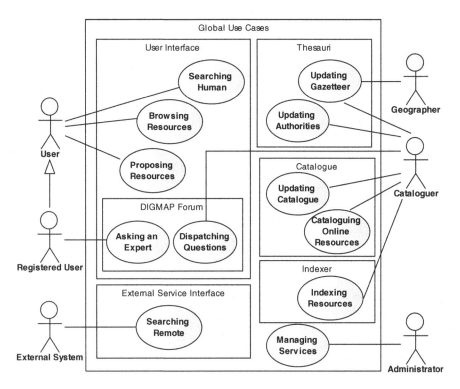

Fig. 1. DIGMAP Use Cases

3 DIGMAP Architecture

All the software solutions produced in DIGMAP will be based on standard and open data models and will be released as open-source, so the results will be useful for local digital libraries of maps, as standalone systems, or as interoperable components for wider and distributed systems (as for example portal management systems).

A **Resource** is an information object relevant for the scope of the service which is registered in the system by descriptive and indexing metadata. Examples are maps, books or web sites. When a Resource is a map, it must be possible to register its geographic information (geographic boundaries, scales, details of sub-maps inside a wider generic map, etc.).

The **Catalogue** is the service responsible by the management of the records describing the Resources. It must support the definition of collections, which are ways to group the Resources according to any desired criteria (type or genre of the resource, the source or provenance of the record, etc.). It will be possible to register Resources through a local user interface, or importing records through OAI-PMH[3]. The local user interface of the Catalogue must make it possible to edit any existing record. The Catalogue must be able to maintain the descriptions of authorities and of the maps in multiple metadata formats, especially in UNIMARC[4] MARC 21[5] and Dublin Core[6].

The **Thesauri** is the generic component of the system that registers ancillary information for the purpose of indexing, searching and browsing for Resources. It will be developed as a multilingual semantic information system, to manage and provide access to information relating to concepts, geographic coordinates, names of places, areas and persons, and related historical events (with dates or time intervals). The Thesauri will be made up of two major sub-systems (not detailed in the diagram): the **Gazetteer** (as defined by the OGC[7]) and the **Authority File** (to maintain and provide easy access to the author's information and identification and disambiguating from similar and duplicate authors).

The **User Interface** is how the users will interact with the system. Besides the traditional OPAC interface, the DIGMAP service will offer a browsing environment for human users exploring paradigms inspired by Google Maps[8], Virtual Earth[9], Time-Map[10], etc. Also, special specific functions will be provided, such as timelines and other indexes for powerful but easy and pragmatic retrieving.

The **External Service Interface** will provide services for external services by Z39.50[11], SRW/SRU[12] and OAI-PMH.

---

[3] OAI-PMH –http://www.openarchives.org
[4] UNIMARC Forum – http://www.unimarc.net/
[5] MARC Standards – http://www.loc.gov/marc/
[6] Dublin Core Metadata Initiative – http://dublincore.org/
[7] Open Geospatial Consortium – http://www.opengeospatial.org
[8] Google Maps - http://maps.google.com
[9] MSN Virtual Earth - http://virtualearth.msn.com
[10] TimeMap Open Source Consortium - http://www.timemap.net
[11] Z39.50 Maintenance Agency - http://www.loc.gov/z3950/agency
[12] SRW – Search/Retrieve Web Service - http://www.loc.gov/z3950/agency/zing/srw

The Indexer presents a specific challenge. Its purpose is to provide support to the automatic indexing of maps. That indexing can be done in multiple perspectives: geographic (the bounding area represented by the map); semantic features (graphical scales, ornaments, etc.); technical features (histogram, etc.). For that, several algorithms have been considered, based on ideas from [1], [2], [3], [4], [5] and others. Concerning the applicability to old maps, the most effective automatic methods appear to be geometry-based vectorization and text classification. These approaches will be seriously considered for further image processing work in the DIGMAP project.

Concluding, any DIGMAP map processing software should bear in mind the benefits and costs of introducing human interaction in its workflow.

4 Conclusions

The project started on October 2006, and will run for 24 months. It is not a pure research project, but an attempt to develop innovative services reusing existing software and metadata. It will make a proof of concept reusing and enriching the contents from the National Library of Portugal (BNP), the Royal Library of Belgium (KBR/BRB), the National Library of Italy in Florence (BNCF), and the National Library of Estonia (NLE). In the future we expect it to be complemented with contents and references from other libraries, archives and information sources, namely from other European national libraries members of TEL – The European Library[13] (in case of success, a ultimate goal of DIGMAP will be to become a service fully integrated in TEL - in this sense the project is fully aligned with the "European Digital Library" vision, as expressed in the "i2010 digital libraries" initiative of the European Commission).

References

1. Arbeláez, P.: Boundary Extraction in Natural Images Using Ultrametric Contour Maps. In: POCV 2006. Proceedings 5th IEEE Workshop on Perceptual Organization in Computer Vision, New York, USA (June 2006)
2. Frischknecht, S., Kanani, E.: Automatic Interpretation of Scanned Topographic Maps: A Raster-Based Approach. In: Tombre, K., Chhabra, A.K. (eds.) GREC 1997. LNCS, vol. 1389, pp. 207–220. Springer, Heidelberg (1998)
3. Ganesan, A.: Integration of Surveying and Cadastral GIS: From Field-to-Fabric & Land Records-to-Fabric. In: 22nd ESRI User Conference, Redlands, CA, USA (2002)
4. Levachkine, S.: Raster to vector conversion of color cartographic maps. In: Lladós, J., Kwon, Y.-B. (eds.) GREC 2003. LNCS, vol. 3088, pp. 49–60. Springer, Heidelberg (2004)
5. Liu, W., Dori, D.: Genericity in Graphics Recognition Algorithms, Graphics Recognition: Algorithms and Systems. In: Tombre, K., Chhabra, A.K. (eds.) GREC 1997. LNCS, vol. 1389, pp. 9–21. Springer, Heidelberg (1998)

---

[13] The European Library – http://www.theeuropeanlibrary.org/

SMS – Its Use in the Digital Library

Ailsa Parker

Whitireia Community Polytechnic Library, Porirua, New Zealand
ailsa.parker@whitireia.ac.nz

Abstract. SMS or short messaging service is a form of text messaging used extensively throughout the world. It is a cheap and mobile form of communication but there is limited research into its library use. Using Internet search techniques and content analysis, this research investigated how libraries use SMS. Fifty libraries with English language websites were found to be using SMS and were divided evenly between academic and public, with two national libraries. They spread over fourteen countries with the United Kingdom having the most libraries using the technology. Usage was mainly in the circulation area, particularly reserves. Some libraries offered reference services. Cost and complexity varied, with some libraries offering examples of possible future use. Suggestions are made as to the importance of libraries helping each other in implementing SMS.

1 Introduction

Libraries are always looking for new ways of communicating with users and with the growth in mobile phone ownership, a new avenue has opened up. Many businesses now use SMS for communicating with clients - SMS, or short messaging service, being defined as "a service similar to paging for sending up to 160-characters-long messages to mobile phones." [1] Software is available which sends SMS messages from a PC to a mobile phone and it can be integrated into library systems. However, in this stage of early adoption, it is difficult to establish the extent and type of library use. Content analysis has previously been used to determine library weblog use [2] and this technique can be used to establish SMS usage in libraries.

2 Library Use of SMS

This survey focuses on English language library literature and websites and a small body of literature is available about using SMS in libraries. In 2001, the Helsinki University of Technology introduced Liblet which let the library communicate with users via mobile phone. It linked to the Voyager Library System and reservations, renewals and even payments were possible. Users were required to register and costs were shared, depending on whether the texts were library or student generated. [3] At Curtin University of Technology in Perth, Giles and Grey-Smith in 2005 described their SMS reference service. Limited staff training was required, the service was heavily promoted and usage was steady. Most queries were explanatory. [4] In an American context, South Eastern Louisiana University Library introduced an SMS reference service in 2005. Set messages were used, librarians required limited training and questions were short and factual. It was

D.H.-L. Goh et al. (Eds.): ICADL 2007, LNCS 4822, pp. 387–390, 2007.
© Springer-Verlag Berlin Heidelberg 2007

concluded that the service might receive limited use and needed different marketing. [5] In 2006, Herman described a SMS Reference at the Southbank Institute of Technology Library in Melbourne. MessageNet software met cost constraints, allowed for a unique number and interfaced with Microsoft Outlook. Protocols were developed for texts.[6] Also in 2006, Monash, another Australian university, began using MessageNet to overcome reservation pickup problems. In 2005, 35% of intra-library loans were not collected. A SMS trial found quicker pick-up times. [7] In Malaysia, university librarians in 2006 found that all students in a library survey had a mobile phone. Perceptions of SMS use by the library were positive with renewals service rating more highly than reference. The researchers found limited examples of SMS library use but they concluded that libraries should begin introducing SMS services as soon as possible. [8]

3 Methodology

Laurel Clyde is a well-known researcher into library websites and associated technologies. In 2004, using Internet search engines and directories and content analysis techniques, she found weblogs on library sites in only three countries – the United States of America (USA), Canada and the United Kingdom (UK).[2] A similar multi-step process was used in the current study to investigate library use of SMS.

1. Data was collected between June 20 and 26, 2007.
2. English language library websites were identified using google keyword searches including sms, "text messaging," "mobile phones" and library or libraries.
3. Libraries using SMS were then identified, the relevant page printed out and then examined.
4. Categories were established and named. These were country, type, registration, charges, privacy, homepage link and special name.
5. TypeS of use were also established: a. Circulation (outward messages) b. Queries and reference (inward messages).
6. Numbers in each category were then counted and results analysed.

4 Results

Fifty libraries used SMS. There was an equal number of academic and public libraries and two national libraries. Spread over fourteen countries, most were in the United Kingdom, where six public libraries used it for circulation (reserves or overdues) and four for inward queries and reference. Australia had eight libraries, six academic and two public. Three of the academic libraries used MessageNet, where students could text into the library. Finland had most libraries using it in Scandanavia, with a group of libraries using Liblet which allowed for inward and outward communication.

The most advanced use of SMS was at Seoul National University where the mobile phone used a downloadable library card for entry, borrowing and circulation. [9]

Registration was required mainly because mobile numbers were not on file. Charges usually clarified that users paid for texts. Three of the four libraries mentioning privacy were in the United States. Home page links often had special names. Most popular were SMS with a suffix e.g. SMS SCU or a group term like Mobile Services. Variations on txt were used for libraries with reference services e.g. Txt ur Library!

Use was mostly for sending outward circulation messages and ranged from multiple functions to many libraries using it just for reserve notices. Inward queries and reference services were obviously less used.

Table 1. Location and types of the fifty libraries using SMS

Countries	No	Academic	Public	National
UK	10	1	9	
Australia	8	6	2	
Finland	8	3	4	1
Denmark	6		6	
USA	4	2	2	
Korea	3	3		
Norway	3	2	1	
Hong Kong	2	2		
Sweden	2	2		
Israel	1	1		
New Zealand	1	1		
Singapore	1			1
Thailand	1	1		

Table 2. Characteristics of SMS website pages

Categories	Number of library website pages
Registration	21
Charges	16
Privacy	4
Home Page	14

Table 3. Use of SMS

	Number of library website pages
Circulation (Outward)	40
Queries and reference (Inward)	12

5 Conclusion

Given the popularity of the mobile phone, the fact that only fifty English language sites with SMS services were found demonstrates that its potential is not being realised. Nonetheless, individual libraries, regardless of location, are trying to be innovative. In addition, geographic clusters indicate that libraries are either learning off each other or that local SMS technologies are available. Obviously, the cost and complexity of the technology varies and Seoul National Library represents the future. In contrast, smaller libraries in isolation are still trying to provide a service, which could just be from mobile to mobile or from a free internet site. The dominant use of SMS in circulation is also interesting in that it demonstrates, given the 160-character constraint, the best use of the technology in libraries. It is also apparent that libraries are developing protocols for its use. In whatever form chosen, however, libraries need to start collecting mobile phone numbers if they plan to implement an SMS service.

A similar survey in five years time will undoubtedly be very different, especially when SMS is integrated into library systems. In the meantime, libraries using SMS should be informing others about software, use, popularity, problems, constraints and costs. If such information is shared, it is possible that, regardless of budget, geography and technology, SMSLIB or TXT2LIB will be a common feature on library websites.

References

1. Kajan, E.: Information technology encyclopedia and acronyms. Springer, Berlin (2002)
2. Clyde, L.: Library Weblogs. Lib. Man. Emerald database 25, 183–189 (2004)
3. Pasanen, I., Muhonen, A.: Library in Your Pocket (2001), http://www.iatul.org/conference/proceedings/vol12/papers/Muhonen.pdf
4. Giles, N., Grey-Smith, S.: Txting Librarians @ Curtin. (2005) http://conferences.alia.org.au/online2005/papers/a12.pdf
5. Hill, J.B.: Text a Librarian: Integrating Reference by SMS into Digital Reference (2005), http://data.webjunction.org/wj/documents/12542.pdf
6. Herman, S.: SMS Reference: Keeping Up with Your Clients (2006), http://conferences.alia.org.au/alia2006/Papers/Sonia_Herman.pdf
7. MessageNet: Monash University (2006), http://www.messagenet.com.au/caseStudies/casestudies_MonashUniLibrary.pdf3
8. Karim, N.S.A., Darus, S.H., Hussin, R.: Mobile Phone Applications in Academic Library Services: A Students' Feedback Survey. Campus - Wide Information Systems 23, 35–51 (2006)
9. Seoul National University Library (2007), http://library.snu.ac.kr/Eng/StaticView.jsp?page=MobileIDCard

Understanding Topic Influence Based on Module Network

Jinlong Wang[1], Congfu Xu[2], Dou Shen[3], Guojing Luo[2], and Xueyu Geng[4]

[1] School of Computer Engineering, Qingdao Technological University
Qingdao, 266033, China
WangJinlong@gmail.com
[2] Institute of Artificial Intelligence, Zhejiang University
Hangzhou, 310027, China
xucongfu@cs.zju.edu.cn
[3] Microsoft adCenter Labs, Redmond WA 98052
doushen@microsoft.com
[4] Institute of Geotechnical Engineering Research, Zhejiang University
Hangzhou, 310027, China

Abstract. Topic detection and analysis is very important to understand academic document collections. By further modeling the influence among the topics, we can understand the evolution of research topics better. This problem has attracted much attention recently. Different from the existing works, this paper proposes a solution which discovers hidden topics as well as the relative change of their intensity as a first step and then uses them to construct a module network. Through this way, we can produce a generalization module among different topics. In order to eliminate the instability of topic intensity for analyzing topic changes, we adopt the piece-wise linear representation so that we can model the topic influence accurately. Some experiments on real data sets validate the effectiveness of our proposed method.

1   Introduction

Topic analysis over academic document collections is beneficial to researchers since it can help researchers find out hot topics at a certain stage and the topic evolution patterns. It can even discover how topic changes affect the researchers' actions and vise versa.

Some recent works [1, 2] have focused on discovering the relationships among the topics. In [1], the authors work on the problem of discovering evolutionary transitions. When two themes have a smaller evolution distance, measured with the KL-Divergence, the themes are claimed to be closer to each other. However, this measurement only reflects the similarity between two themes, and cannot make multiple topics simultaneously. In [2], the authors propose a method for discovering dependency relationships among the topics in a collection of documents shared in social networks. This paper's hypothesis is that one topic evolves into another topic through the interaction between the corresponding social actors with different topics in the latent social network. Based on the hypothesis,

D.H.-L. Goh et al. (Eds.): ICADL 2007, LNCS 4822, pp. 391–399, 2007.

the authors can compute the transition probability among topics and rank the authors to see who dominate the topic transition. However, this relation is only a transition between topics, it computes the quantitative realtionship of different topics through the co-authors. The Markov transition graph among topics is obtained by computing the transition between topics respectively, which is not a global graph. In [3], we use the dynamical bayesian network to model the research field development with a global graph, but it needs input to be sequential data, and it only reflects the relations among fields, not the topics.

Different from previous methods, in this paper, we attempt to generalize the topic's interactional relations reflected by topic intensity, which reflects the topic development and trend. We use the relative change of topic intensity over time to build a module network for topic interrelation generalization. As a generative model, the module network [4] can partition the variables into modules and learn the dependency structure of each module. The variables in each module share the same parents in the network and the same conditional probability distribution. By feeding a set of variables and the maximum module number, this method can generate a module network automatically. [4] applies this method to cluster stock sectors based on the stock price change, which generalizes better than the traditional bayesian network. Especially, the learned module network provides some important insights for stock sector based on the stock price. For example, the stock of high-tech service companies are in the same module, and have the same parent of manufacturer. Essentially, the topic intensity with time is similar to the stock data. Thus, we can use the module network to learn topic relation for a better understanding of the latent influence among the topics.

The rest of the paper is organized as follows. Section 2 introduces the related works. Section 3 presents our method. Section 4 describes the experimental setup and the results. The last section concludes the paper.

2 Related Works

2.1 Document Content Analysis

A variety of statistical approaches have been proposed to model a document, such as LSA (Latent Semantic Analysis) [5], pLSI (Probabilistic Latent Semantic Indexing) [6], LDA (Latent dirichlet allocation) [7] *etc.* In recent years, LDA models, as effective approach for generating topics, receives more and more attentions. The model uses the Dirichlet distribution to model the distribution of topics for each document. Each word is considered sampled from a multinomial distribution over words specific to topics. LDA models are well-defined generative models and generalize easily to new documents without overfitting. Some recent work has been concerned with temporal documents [8, 9].

2.2 Module Network

As a generative model, a module network [4] partitions the variables into modules, and studies the dependent relation among the modules. A module network can be viewed simply as a Bayesian network [10] in which variables in the same module share parents and parameters. This provides a good representation for understanding the data. When learning module network, given a domain of random variables $X = \{X_1, \cdots, X_n\}$, we can obtain K modules $M_1, \cdots, M_K$. A module network consists of two components. The first defines a template probabilistic model for each module; all of the variables assigned to the module will share this probabilistic model. The second component is a module assignment function that assigns each variable to one of the K modules.

Fig. 1 represents an example of module network, the rectangle denotes the module, and the variables in one module share the same conditional probability table, parameter and same parent node, but may have different descendants. For example, the three variable $\{X_2, X_3, X_4\}$ in M_2 have the same parent X_1, but only the variable X_3 has descendants, which are X_6, X_7 in module M_4. Learning module network includes an iterative approach consisting of two steps: module assignment search step (assign variables to modules) and structure search step (learn the network structure).

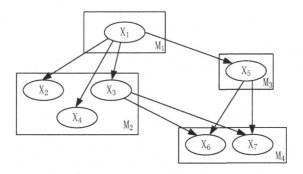

Fig. 1. The module network example

3   Methods

In our method, we take a document collection with time stamps as the input and proceed with three steps to build the module network: (1) discover the topics/themes in the collection as well as the change of their intensity; (2) process the themes strength with piece-wise linear analysis; (3) construct module networks. The module network supplies a graph reflecting topic relations. Based on the graph, we can analyze influences among topics.

3.1   Obtaining Topic Intensity Time Series

As the first step, we extract the topics from the document collection using a LDA model [7], and then obtain the topic dynamics referring to the series of topic with various strength of probability over time. The process of the method is shown in Fig. 2.

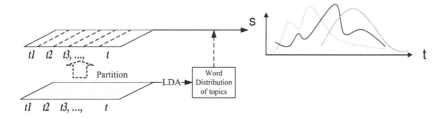

Fig. 2. Obtaining topic intensity time series

For a certain year, the strength of a topic is calculated as the normalized sum of all the probabilities of this topic inferred from all documents in that year. We partition the documents by year. For each year, all of the words are assigned to their most likely topic. The fraction of words assigned to each topic for a given year is then calculated for each of the topics and each year. These provide relative topic popularity in the document collection, and present the trend of topics over time.

3.2   Piece-Wise Linear Representation

With the topic intensity induced in Section 3.1, we can directly use the relative changes between successive years as the input of module network building. However, the induced intensity changes are not stable and may contain much noise due to many reasons such as the fluctuation of the number of documents each year and the statistical error in LDA model. These instability and noise will debase the module network's generalization performance. Actually, comparing with the concrete intensity of a topic at a certain time, the general trend is more important in our analysis. Therefore, we use the piece-wise linear segmentation [11], an effective method for trend representation of time series to eliminate the fluctuation in our problem. The piece-wise linear segmentation attempts to model the data as sequences of straight lines, which has been proved to be effective for data compression and noise filtering. In our problem, for each topic, we set its intensity every-year with the responding slope which is mapped close to the range of [-1, 1]. Fig. 3 is the topic intensity (red line) of SVM topic discovered in the above way and its piece-wise linear representation (blue line). As the input of the module network, it also ensures that the topic's change consistent in a small range of time. In this way, the module network can generate an ordered time sections which better accord with reality.

3.3   Constructing the Module Network

In step 3, we construct a module network. We treat each topic as a variable, and each instance to correspond to a year, where the value of the variable is the result processed with piece-wise linear representation, and the range is in [-1, 1]. Taking these data and the maximum module number as input, we can obtain a high scoring module network based on the scoring function in [4]. Same

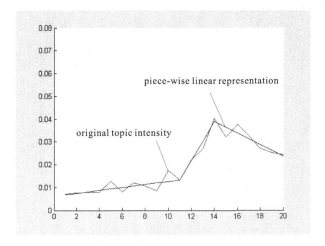

Fig. 3. The topic intensity of SVM and its piece-wise linear representation

as the Bayesian network, given a scoring function over network, we can find a high scoring module network. We encourage the readers to refer to [4] for more information about module network.

4 Experiments

We use the NIPS (Neural Information Processing Systems) conference data to explain our approach. The NIPS dataset consists of the full text of the 20 years of proceedings from 1987 to 2006. In our experiments, we use the abstracts extracting from the all NIPS papers data download from the website[1]. In order to improve the accuracy of abstract extraction from pdf/ps/djvu format to text, we use different methods toward different time literature data. For the papers in 1987-1999, we use Roweis' raw data, which were corrected errors by hand on yann's djvutotxt data. For papers of 2000-2001, we use the DjVuLibre 3.5.17-1 to transform the file format of djvu to plain text, and the result is much better than using the other two formats. And for the remaining 977 papers without djvu files, we use the text save function of Acrobat7.0 for text export, which is less OCR errors then using PDF2TXT tools. For better discovering the topics, we filter some phases, such as "the", "a", *etc.* by a stopwordlist. At the same time, we deleted two letters' words except some domain words, such as "SVM", "HMM", "KNN", "RNN", "AI", "KL", *etc.* We also deleted words that appeared less than five times in the whole collection, for these words are mostly generated by OCR errors. Finally we extract 3102 abstracts and total 4517 distinct words from the collection.

With the LDA topic model, we extract 50 topics and then obtain each topic's intensity time series as shown in Section 3.1. We select 9 significant topics

---

[1] http://nips.cc/

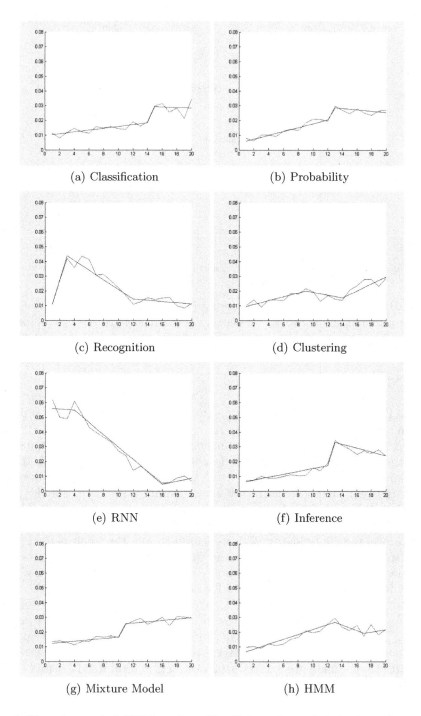

(a) Classification

(b) Probability

(c) Recognition

(d) Clustering

(e) RNN

(f) Inference

(g) Mixture Model

(h) HMM

Fig. 4. The other topics' (SVM result as Fig. 3) intensity and their piece-wise representation

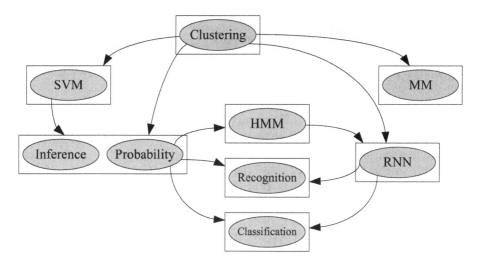

Fig. 5. The module network with 9 topics

Table 1. Conditional probabilistic table

(a) CPT of SVM node

	Clustering(Up)	Clustering(Down)
SVM(Up)	0.533	1
SVM(Down)	0.467	0

(b) CPT of Prob node

	SVM(Up) Clustering(Down)	SVM(Down) Clustering(Up)
Pro(Up)	0.8	0
Pro(Down)	0.2	1

Mixture Model (MM), Clustering, Recurrent Neural Network (RNN), SVM, HMM, Classification, Inference, Probability, Recognition, that are frequently referred to machine learning in the NIPS.

With the piece-wise linear segmentation, the topic intensity can be presented better with trend on a macroscopical view. Since we have only 20 years' data, the more the segmentation number is, the harder to see the trend. In our experiments, we vary the segment number from 2 to 3, 4 and 5. We find that the result is best for understanding with the 3 segmentation. The result is as Fig. 4.

Using the 9 topics' intensity data to construct the module network, and the maximum module number is set as 9, we obtain the network with 6 modules as shown in Fig. 5. The topic Inference and Probability are clustered together. The figure shows the topic relation/influence according to the arrowhead. For example, the topic Clustering influences the topic SVM, MM, RNN and Prob (Inference and Probability). The topic RNN influences Recognition and

Fig. 6. The four topics intensity description (more red, more rise; more green, more fall)

Classification, and the topic Probability is related with the topic Classification, HMM and Recognition because it introduces new theory methods to these methods and applications. These results are consistent with our understanding about this field. In the following, we show the detailed relation with a part of CPT (conditional probability table) as Table 1.

We take the module SVM and Prob (Inference, Probability) as examples. As Table 1(a), when Clustering (Clu) rises (U), the influence is not obvious; when Clustering (Clu) falls (D), the influence is negative. Through the detailed description of topics (Clustering, SVM, Inference, Probability) trend in Fig. 6, we can find that, in the beginning, with the development of NIPS, the four machine learning methods increase synchronization, but in recent years, with the steady of development, the influence is obvious. Fig. 6 shows that recently SVM is positively correlated with Prob topic (Inference and Probability), but Clustering is negatively correlated with them, which is consistent with Table 1(b). And recently, when Clustering rises, SVM falls, which complements the result in Table 1(a).

5   Conclusion

This paper investigated the problem of topic influence based on the topic intensity evolution and used the topic intensity time series to construct the module network for understanding the topic influence. Future work includes improving the accuracy of topic intensity by virtue of topic tracking technology, introducing the semantic information of topic, considering the time series information in module network building.

Acknowledgements

This work was supported by the National Natural Science Foundation of P.R.China (No.60402010) and Zheiang Provincial Natural Science Foundation of P.R.China (Y105250).

References

[1] Mei, Q.Z., Zhai, C.X.: Discovering evolutionary theme patterns from text: an exploration of temporal text mining. In: KDD 2005. Proceeding of the eleventh ACM SIGKDD international conference on Knowledge discovery in data mining, pp. 198–207. ACM Press, New York (2005)

[2] Zhou, D., Ji, X., Zha, H., Giles, C.L.: Topic evolution and social interactions: how authors effect research. In: CIKM 2006. Proceedings of the fifteenth ACM international conference on Information and knowledge management, pp. 248–257. ACM Press, New York (2006)

[3] Wang, J.L., Xu, C.F., Li, G., Dai, Z.W., Luo, G.J.: Understanding research field evolving and trend with dynamic bayesian networks. In: PAKDD 2007. Proceedings of the eleventh Pacific-Asia Conference on Knowledge Discovery and Data Mining, pp. 320–331. Springer, Heidelberg (2007)

[4] Segal, E., Pe'er, D., Regev, A., Koller, D., Friedman, N.: Learning module networks. Journal of Machine Learning Research 6, 557–588 (2005)

[5] Landauer, T., Foltz, P., Laham, D.: Introduction to latent semantic analysis. Discourse Processes 25, 259–284 (1998)

[6] Hofmann, T.: Probabilistic latent semantic indexing. In: SIGIR 1999. Proceedings of the twenty-second annual international ACM SIGIR conference on Research and development in information retrieval, pp. 50–57. ACM Press, New York (1999)

[7] Blei, D.M., Ng, A.Y., Jordan, M.I.: Latent dirichlet aladdress. Journal of Machine Learning Research 3, 993–1022 (2003)

[8] Wang, X.R., McCallum, A.: Topics over time: a non-markov continuous-time model of topical trends. In: KDD 2006. Proceedings of the twelfth ACM SIGKDD international conference on Knowledge discovery and data mining, pp. 424–433. ACM Press, New York (2006)

[9] Blei, D.M., Lafferty, J.D.: Dynamic topic models. In: ICML 2006. Proceedings of the twenty-third international conference on Machine learning, pp. 113–120. ACM Press, New York (2006)

[10] Pearl, J.: Probabilistic reasoning in intelligent systems: networks of plausible inference. Morgan Kaufmann Publishers Inc, San Francisco (1988)

[11] Keogh, E.J., Pazzani, M.J.: An enhanced representation of time series which allows fast and accurate classification, clustering and relevance feedback. In: KDD 1998. Proceedings of the fourth ACM SIGKDD international conference on Knowledge discovery and data mining, pp. 239–243. AAAI Press, New York (1998)

Development of Indian Agricultural Research Ontology: Semantic Rich Relations Based Information Retrieval System for Vidyanidhi Digital Library

M.A. Angrosh and Shalini R. Urs

International School of Information Management
University of Mysore, Manasagangotri Mysore, India
angrosh@isim.ac.in, shalini@isim.ac.in

Abstract. Digital Libraries represent semantically rich collections of digital documents. Ontology-based information retrieval systems capture semantic relations for providing value added information services. Deviating from the regular approach of developing ontologies on the basis of domain knowledge, the present paper puts forward a novel method for developing ontologies from the semantic information available in the titles of digital documents. Such an approach gathers significance due to its simplicity in ontology development process. To examine the same, the study considered the case of Agricultural Electronic Theses and Dissertations (ETDs) present in Vidyanidhi Digital Library. The study resulted in the development of Indian Agricultural Research domain ontology, which was used for developing ontology-based information retrieval system. This paper while describing the methodology followed for developing the ontology presents the technical details of the developed system.

Keywords: Indian Agricultural Research Ontology, Ontology, Web Ontology Language, Semantic Web, Vidyanidhi Digital Library.

1 Introduction

The field of Information Retrieval is a central area of research in Digital Libraries. Information Retrieval (IR) is a process of finding all relevant documents from a document collection, satisfying user information need [1]. Unfortunately, currently employed information retrieval mechanisms suffer from various limitations. Issues such as information overload, rapid technological developments, fluctuating user trends and behaviour call for better IR mechanisms. The emerging Semantic Web technologies such as ontologies promise knowledge-based systems capable of performing crucial tasks of information retrieval and extraction [2]. Domain ontology based systems supporting navigation and querying facilities form ideal information retrieval systems for digital libraries. The reasoning and querying capabilities offer valuable search strategies for digital libraries.

Ontology-based IR systems mainly rely on domain ontologies, which are further extended for information retrieval. Development of domain ontologies is not only costly [3], time consuming and cumbersome, but also leads to difficulties particularly

D.H.-L. Goh et al. (Eds.): ICADL 2007, LNCS 4822, pp. 400–409, 2007.

at the time of mapping document instances to the ontology. Further, the high volume of knowledge represented in the ontology may not be used in its entirety. Thus, instead of representing the entire knowledge into domain ontology and mapping documents to the ontology, it would be appropriate to develop ontology, based on the available documents and deploy the same for information retrieval. This would also facilitate in the easy mapping of documents to the ontology. The present paper puts forward a simple yet powerful method for developing ontologies from the information present in the titles of electronic documents, resulting in effective information retrieval systems for Digital Libraries. This paper is the result of the study carried at Vidyanidhi Digital Library (VDL). The study focused on developing ontology from the information available in the titles of Agricultural Electronic Theses and Dissertations (ETDs) present in VDL, resulting in 'Indian Agricultural Research' ontology. This was mapped to Agricultural ETDs for developing a knowledge base, which was used for information retrieval.

The paper is organized as follows. In Section 2, we discuss the related work. Section 3 details the process of building Indian Agricultural Research ontology. Section 4 brings out the relation of the developed ontology with Description Logics. Section 5 outlines the technical details of the developed ontology-based information retrieval system. While Section 6 describes various search options provided by the system, Section 7 brings out the importance of such systems for Digital Libraries. We conclude with a brief summary and our plans for future research in Section 7.

2 Related Work

Efforts are underway for developing ontology-based information retrieval systems for agriculture domain. The Food and Agriculture Organization of the United Nations (FAO) has made commendable contribution through the development of AGROVOC, a multilingual, structured and controlled vocabulary, which covers the terminology of all subject fields in agriculture, forestry, fisheries, food and related domains [4]. An attempt to convert the AGROVOC thesaurus into RDFS ontology was carried out during 2003 [5]. The FAO has also initiated the Agricultural Ontology Service (AOS), a reference tool that structures and standardizes agricultural terminology in multiple languages [6]. The AOS provides terms, definition and relationship components for sharing among associated partners and increasing the functionality for indexing and retrieving of resources. AGROVOC is being extensively used across the globe for developing multilingual agricultural thesaurus. Such thesaurus is being used for developing Semantic Web technologies based agricultural information systems. In the Indian scenario, the DEAL project [7] is currently in progress for developing ontology and a metadata system that supports the knowledge archiving & retrieval reuse in Indian Agriculture and rural livelihood domain. DEAL also proposes to develop a multilingual agricultural thesaurus based on AGROVOC. Angrosh and Urs [8] have successfully developed a prototype ontology-based information retrieval system for a specific case of Agri-Pest domain.

Most of the approaches for developing information systems employing AGROVOC focus on multilingual features and lay less emphasis on semantic relations. For instance Liang et al. [9] put forward a schema for mapping Chinese Agricultural Thesaurus to

FAO's AGROVOC. However, the crucial task of representing rich semantic relations of agricultural theses is absent. Though Sini et al. [10] have developed an ontology-based navigation system in the domain of food, nutrition and agriculture, the scope is limited to bibliographic metadata. Though such knowledge models facilitate semantic browsing, the real value of ontologies is obtained through use of rich semantic relations derived beyond bibliographic metadata relations. Thus, we present here a different methodology for developing ontologies based on the information available in the titles of the documents. We also show that the proposed methodology can be used for developing ontology-based information retrieval system.

3   Indian Agricultural Research Ontology

The field of Indian Agriculture is a cascade of many interacting effects between plant and environmental factors. Socio-economic factors such as timely-credit, price support, availability of critical inputs, crop insurance etc. play an important role in crop production. Post-harvest techniques such as storage and preservation mechanisms are also equally important. Thus the domain of Indian agriculture is a blend of various inter-disciplinary subjects and research in this domain interweaves these disciplines. Further, these research issues are specific to particular geographical regions. Thus, the Indian Agricultural Research ontology besides representing these interdisciplinary subject characteristics should also represent the geographical knowledge. Use of agricultural vocabularies such as AGROVOC would add value to the ontology. A broad framework of Indian Agricultural Research system mapping agricultural eTheses is shown in Figure 1.

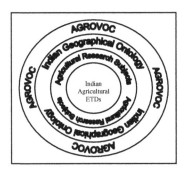

Fig. 1. Framework of Indian Agricultural Research Ontology

At a fundamental level, ontologies capture static domain knowledge in a generic way and provide a commonly agreed understanding of that domain, which may be reused and shared across applications and groups [11]. Thus, the primary task in developing ontology based information system for agricultural eTheses is to develop a shared understanding of the agricultural domain, which captures knowledge represented in agricultural ETDs. The data pertinent to agricultural ETDs in Vidyanidhi is of two types viz., full-text theses and bibliographic metadata. Vidyanidhi Digital Library and E-Scholarship Portal is set up at the University of Mysore for catering to the research needs of the scholarly community in India [12]. Vidyanidhi currently hosts a full-text database of 6000 doctoral theses of various subjects and a bibliographic metadata database of more than 1 lakh eTheses records. There were nearly 200 full-text theses and 2800 agricultural eTheses records in the domain of agriculture at Vidyanidhi. The study focused on developing an ontology, using the limited semantic information available in the titles of agricultural records. The methodology followed for developing the ontology is as follows:

- Identify all agricultural eTheses titles in the Vidyanidhi Digital Library
- Identify all possible keywords in these titles. Keywords are primarily those terms that are used by a user for searching information.
- Identify and define classes and subclasses to which these terms belong to
- Define relations binding individuals of different classes.

Each of the 2800 agricultural eTheses titles was carefully analyzed to identify the possible keywords present in these titles. Upon identification of keywords, we identified the different classes and subclasses to which these keywords would belong. The classes and subclasses relationships were identified using agricultural handbooks and subject classification systems [13]. Table 1 shows a sample database of the identified keywords, classes and subclasses and relationships between keywords of different classes.

Table 1. Keywords, Classes, Subclasses and Relations identified in Agricultural eTheses

Sl. No.	Title	Class & Individuals	Class & Individuals	Class & Individuals	Geographical Class
1	Capital formation in arid agriculture: a study of resource conservation and reclamation measures applied to arid agriculture in Andhra Pradesh	Types of Agriculture (C) → Arid Agriculture (I)	Natural Resource Mgmt. (C) → Resource Utilization (I)	Agribusiness (c) → Agriculture Finance (c) → Capital Formation (I)	Indian States → Andhra Pradesh (I)
		Capital Formation inRelationTo Resource Utilization inRelationTo Arid Agriculture inRelationTo Andhra Pradesh			
2	Employment in Indian agriculture: Analytical and policy issue	Agricultural Economics → Agricultural Labour → Agricultural Employment (I)	Agricultural Policy → Agricultural Policy Issues		
		Agricultural Employment isRelated To Agricultural Policy			
3	Pattern of investment in agriculture in Orissa during the plan period	Agricultural Economics → Agricultural Investments → Pattern of Investments (I)	Agricultural Policy → Five Year Plans		Indian States → Orissa
		Agricultural Investments inRelationTo Five Year Plans inRelationTo Orissa			
4	Economic impact of central sector scheme women in agriculture on farm women in Maharashtra State	Agricultural Labour → Women Labour (I)	Agricultural Policy → Central Sector Schemes (I)		Indian States → Maharashtra
		Agricultural Women Labour inRelationTo Central Sector Schemes inRelationTo Maharashtra			

The agricultural research ontology was implemented in the OWL Web Ontology Language, a W3C recommendation for defining and instantiating Web Ontologies [14]. The Protégé-OWL editor was used for developing the ontology [15]. Figure 2 shows the screenshot of the ontology developed in Protégé-OWL Editor.

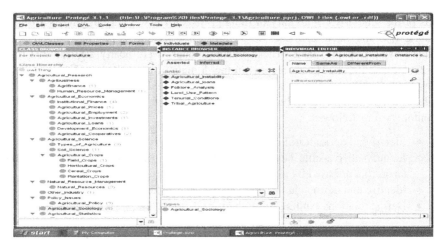

Fig. 2. Agricultural Research Ontology developed in Protégé-OWL Editor

A schematic representation of a part of Indian Agricultural Research ontology mapped with AGROVOC as is shown in Figure 3. These mappings provide rich valuable search points for retrieving agricultural ETDs.

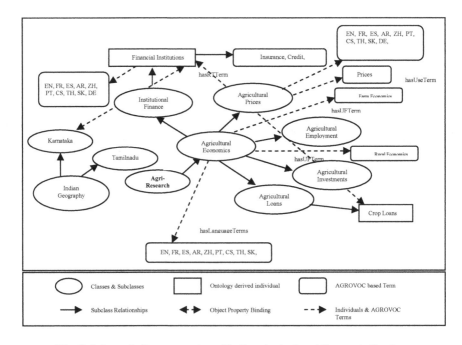

Fig. 3. Schematic Representation of Indian Agricultural Research Ontology

4 Description Logics Based Formalisms

The knowledge structures obtained above evolve into Description Logics (DLs) based knowledge representation (KR) formalisms. Primarily, DLs formalisms represent knowledge of an application domain by first defining the relevant concepts of the domain (its terminology) and then using these concepts to specify properties of objects and individuals occurring in the domain [16]. Description Logics based knowledge representation formalisms provide strong support for reasoning services, allowing inference of implicitly represented knowledge from the knowledge that is explicitly contained in the knowledge base. The subsumption relationships resulting in a hierarchical structure of classes and subclasses of the agricultural research domain can be used for designing value added information services. The classification and binding of individuals through relative object properties facilitate in deriving explicit knowledge about individuals associated with various classes.

The semantics of concept description in DLs is defined by the notion of interpretations, wherein an interpretation I consists of a non-empty set Δ^I (the domain of interpretation) and an interpretation function, which assigns to every atomic concept A a set $A^I \subseteq \Delta^I$ and to every atomic role R a binary relation $\subseteq \Delta^I \times \Delta^I$

$$\text{Interpretation } I = (\Delta^I, .^I)$$

DLs knowledge base typically comprises of two components viz. a "TBox" and an "ABox". While the TBox contains intensional knowledge in the form of a terminology, built through declarations that describe general properties of concepts, the ABox contains extensional knowledge – also referred to as assertional knowledge. Assertional knowledge refers to the knowledge that is specific to the individuals of the domain of discourse.

The Agricultural Research domain ontology developed in the study falls in line with DLs, with the keywords forming the assertional knowledge and the classes and subclasses forming the terminology of the domain. Further, the study, while conceptualizing the vocabulary of the knowledge base in terms of concepts and roles, maintained the important assumptions about DL terminologies, which included:

- allowance of only one definition for a concept.
- acyclic characteristics of the definitions – in the sense that concepts are neither defined in terms of themselves nor in terms of other concepts that indirectly refer to them

DLs based reasoners are employed for drawing inferences from the knowledge representation derived above. We used Pellet, a capable OWL-DL reasoner with acceptable to very good performance [17] for deriving inferences. The following section outlines the technical details of the developed system.

5 Technical Details of the System

The OWL-DL Agricultural Research ontology was used for developing ontology-based information retrieval system for Indian agricultural eTheses in Vidyanidhi Digital Library. The architecture of the system is as shown in Figure 4.

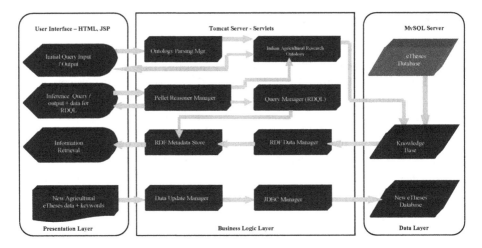

Fig. 4. Architecture of ontology-based IRS for Vidyanidhi Digital Library

5.1 Presentation Layer

The presentation layer mainly used HTML and JSP pages for creating user-friendly interfaces. The user interfaces are divided into two categories viz., Information Retrieval and Information Updating interfaces.

5.2 Business Logic Layer

The business logic layer has the following components:

5.2.1 Ontology Parsing Manager
The system uses Jena, an open source Java based API, developed by HP Labs for handling semantic web information model and languages [18]. The Jena2 ontology API is used to parse OWL for deriving class and subclass relationships and listing individuals.

5.2.2 Pellet Reasoner Manager
This module uses Pellet - a OWL-DL reasoner [17] to reason about individuals in the ontology. The reasoner is employed to retrieve related individuals connected by object-property relationships in the ontology. The retrieved individuals form the input for the Query Manager for retrieving respective agricultural eTheses.

5.2.3 Query Manager
The Query Manager is responsible for retrieving eTheses from an RDF data model, mapping agricultural eTheses and keywords defined in the ontology. The module uses Jena's Resource Description Query Language (RDQL) specific API function calls for querying the data model. The SQL-like syntax of RDQL is proved to be an effective way of querying an RDF data model [19].

5.2.4 RDF Data Manager
RDF Data Manager is primarily responsible for creating Resource Description Framework (RDF) Metadata store of agricultural eTheses, mapped with Indian Agricultural Research knowledge base. Jena's RDF API is used for representation of

models, resources, properties, literals, statements and other key concepts of RDF [20]. Table 2 shows RDF metadata capturing ontology defined keywords and Agrovoc vocabulary terms for a sample record.

Table 2. RDF metadata capturing keywords and Agrovoc vocabulary terms for a sample record

```
<rdf:RDF
xmlns:rss="http://purl.org/rss/1.0/"
xmlns:jms="http://jena.hpl.hp.com/2003/08/jms#"
xmlns:rdf="http://www.w3.org/1999/02/22-rdf-syntax-ns#"
xmlns:rdfs="http://www.w3.org/2000/01/rdf-schema#"
xmlns:owl="http://www.w3.org/2002/07/owl#"
xmlns:j.0="http://localhost:8080/vidyanidhi/india/agriresearch.owl#"
xmlns:ns3="http://localhost:8080/vidyanidhi/india/agriresearch.owl"
xmlns:daml="http://www.daml.org/2001/03/daml+oil#"
xmlns:ns2="http://purl.org/dc/elements/1.1/"
xml:base="http://localhost:8080/publication/documentowl" >
<rdf:Description rdf:about="http://localhost:8080/vidyanidhi/org/india/ETD1.htm">
    <j.0:hasKeyword1>Arid Agriculture</j.0:hasKeyword1>
    <ns2:title>Capital formation in arid agriculture: a study of resource conservation and reclamation
measures applied to arid agriculture</ns2:title>
    <j.0:hasKeyword2>Resource Utilization</j.0:hasKeyword2>
    <j.0:hasKeyword3>Capital Formation</j.0:hasKeyword3>
    <j.0:hasAgrovocBT1>Climate Zones</j.0:hasAgrovocBT1>
    <j.0:hasAgrovocNT1>Deserts</j.0:hasAgrovocNT1>
    <j.0:hasAgrovocRT1>Minimum Tillage</j.0:hasAgrovocRT1>
    <j.0:hasAgrovocRT2>Scrublands</j.0:hasAgrovocRT2>
    <j.0:hasTitle>Capital formation in arid agriculture: a study of resource conservation and reclamation
measures applied to arid agriculture</j.0:hasTitle>
    <j.0:hasAgrovocRT3>Dryland Management</j.0:hasAgrovocRT3>
    <ns2:creator>Jodha, Narpat Singh</ns2:creator>
    <ns2:contributor>Bardhan, Pranab</ns2:contributor>
    <ns2:contributor>Choudhury, Mrinal Dutta</ns2:contributor>
    <ns2:language>English</ns2:language>
    <ns2:degreeGrantor>University of Delhi</ns2:degreeGrantor>
    <ns2:year>University of Delhi</ns2:year>
    <j.0:hasAgrovocUF1>Drylands</j.0:hasAgrovocUF1>
    <j.0:hasAgrovocBT5>Resource Management</j.0:hasAgrovocBT5>
    <j.0:hasAgrovocNT5>Soil Conservation</j.0:hasAgrovocNT5>
    <j.0:hasAgrovocRT5>Natural resources</j.0:hasAgrovocRT5>
    <j.0:hasAgrovocRT6>Sustainable Development</j.0:hasAgrovocRT6>
    <ns2:description>The theses describes capital formation in arid agriculture</ns2:description>
    <rdf:type rdf:resource="http://localhost:8080/vidyanidhi/india/agriresearch.owl#hasIdentifier"/>
  </rdf:Description>
</rdf:RDF>
```

5.2.5 Data Update Manager

The Data Update Manager is responsible for adding new eTheses and ontology related keywords to the system.

5.3 Data Layer

The system employs MySQL database for creating the data backend, comprising of eTheses database and knowledge base created with the ontology.

6 Ontology-Based Information Search

The ontology-based information system facilitated the following search options:

6.1 Simple Search

The simple search option facilitates in performing search on terms present in the ontology, metadata of agricultural records and AGROVOC vocabulary. The generic search option retrieves all records related to a specific term by using the rich semantic relations binding metadata records with the ontology and the AGROVOC vocabulary.

6.2 Taxonomic View

A taxonomic view of the agricultural research domain is presented to the user. The user is provided with the option of browsing the subject hierarchy and view the instances (or keywords) of a specific class and retrieve records related to a specific keyword. This facility facilitated in confining user's search to a specific term.

6.3 Query Building Mechanisms

The system also facilitates in extending the taxonomic based information search and retrieval for query building mechanisms. The Pellet Reasoner is employed for developing such query mechanisms. For example, consider a specific individual, say 'Rice Crops' of the class 'Agricultural Crops' being searched. The system notifies the user that the individual 'Rice Crops' is related to Classes such as 'Agricultural Statistics', 'Agricultural Economics' etc. Further, if the user is interested in say, 'Agricultural Production', the system lists out various keywords such as 'Plant Development', 'Disease Control', etc., giving an option for the user to choose from the popup list. Furthermore, on specifying a specific keyword from the related class, the system notifies the available geographical entities related to the keyword. The user is again given the option for selecting the desired geographical location. Thus, this process results in building a chain of related keywords and retrieves records based on a specific chain, resulting in 'query building mechanisms'.

7 Implications of the Present Work

The process of ontology building primarily experiences difficulties in providing a significant coverage of the domain. It is also equally important to foster at the same time, the conciseness of the model by determining the meaningful and consistent generalizations [21]. The approach adopted in the present study overcomes these problems to a great extent. The development process of the ontology is relatively simpler and manageable. Further, while achieving the conciseness of the knowledge model, the identification of classes and subclasses provides a significant coverage of the domain. The ontology can be easily extended through addition of more keywords, classes and relationships. The ontology framework developed for Indian Agricultural Research domain is generic and is extendable for other domains in the Digital Library.

8 Conclusions and Future Work

In the present study, we have reported the development of an Indian Agricultural Research ontology based on the information available from the titles of agricultural eTheses. This was utilized for developing ontology-based information retrieval system for Vidyanidhi Digital Library. The work carried out in the present study is a novel effort for developing ontology-based information retrieval systems for digital libraries. The evolved framework and methodology can be extended for other domains as well. The use of AGROVOC increased the robustness of the system. Similar vocabularies in other domains can be used in the system. In future, we plan to target at extending the system for other domains in the Digital Library. This would result in the development of Indian Research ontology comprising various sub-domains. We also propose to investigate the use of vocabularies for different domains. Employing the same, we aim at developing a robust ontology-based information system covering all domains for Vidyanidhi Digital Library.

References

1. Baeza-Yates, R., Ribeiro-Neto, B.: Modern Information Retrieval. Addison Wesley Longman Publishing (1999)
2. Guarino, N.: Semantic Matching: Formal Ontological Distinctions for Information Organization, Extraction and Integration. In: International Summer School - SCIE-97 on Information Extraction: A Multidisciplinary Approach to an Emeging Information Technology, pp. 139–170 (1997)
3. Ratsch, E., Schultz, J., Saric, J., Cimiano, P., Wittig, U., Reyle, U., Rojas, I.: Developing a Protein Interactions Ontology. Comparative and Functional Genomics 4(1), 85–89 (2003)
4. AGROVOC Thesaurus, FAO (2007), http://www.fao.org/aims/ag_intro.htm
5. Applied Ontologies in FAO.: FAO (2007), http://www.fao.org/aims/onto_domains.jsp
6. AGROVOC Concept Server.: FAO (2007), http://www.fao.org/aims/aos.jsp
7. Digital Ecosystem for Agriculture and Rural Livelihood Project.: (DEAL), Indian Institute of Technology Kanpur (2007), http://emandi.mla.iitk.ac.in/deal/
8. Angrosh, M.A., Urs, S.R.: Ontology-driven Knowledge Management Systems for Digital Libraries: Towards creating semantic metadata based information services. In: Proceedings of National Seminar on Knowledge Representation and Information Retrieval, Paper:N. Document Research & Training Centre, ISI, Bangalore (March 22-24, 2006)
9. Liang, A., Sini, M., Chun, C., Sijing, L., Wenlin, L., Chunpei, H., Keizer, J.: The Mapping Schema from Chinese Agricultural Thesaurus to AGROVOC, FAO (2000), ftp://ftp.fao.org/docrep/fao/008/af241e/af241e00.pdf
10. Sini, M., Salokhe, G., Pardy, C., Albert, J., Keizer, J., Katz, S.: Ontology-based Navigation of Biliographic Metadata: Example from the Food, Nutrition and Agriculture Journal, FAO (2000), ftp://ftp.fao.org/docrep/fao/009/ah765e/ah765e00.pdf
11. Castano, S., Ferrara, A., Montanelli, S.: Dynamic Knowledge Discovery in Open, Distributed and Multi-Ontology Systems: Techniques and Applications. In: Web Semantics and Ontology, ch. 5, Idea Group Publishing (2005)
12. Vidyanidhi Digital Library and E-Scholarshop Portal.: University of Mysore (2007), www.vidyanidhi.org.in
13. Indian Council for Agricultural Research.: Handbook of Agriculture. Indian Council of Agricultural Research, New Delhi (2006)
14. Web Ontology Language (OWL), W3C (2004), http://www.w3.org/2004/OWL/
15. Protégé-OWL (ed.): Stanford Medical Informatics (2007), http://protege.stanford.edu/overview/protege-owl.html
16. Baader, F., Knutt, W.: Basic Description Logics. In: Baader, F., Calvanese, D., McGuinness, D.L., Nardi, D., Patel-Schneider, P.F. (eds.) The Description Logic Handbook: Theory, implementation and applications, pp. 47–100. Cambridge University Press, Cambridge (2003)
17. Sirin, E., Parsia, B., Grau, B.C., Kalyanpur, A., Katz, Y.: Pellet: A practical OWL-DL reasoner. Web Semantics: Science, Services and Agents on the World Wide Web 5, 51–53 (2007)
18. McBridge, B.: Jena: A semantic web toolkit. IEEE Internet Computing. November-December, 55–59 (2002)
19. Powers, S.: Practical RDF. O'Reilly (2003)
20. Min, W., Jianping, D., Yang, X., Xenxing, X.: The Research on the Jena-based Web Page Ontology Extraction and Processing. In: SKG 2005. Proceedings of the First International Conference on Semantics, Knowledge and Grid, IEEE Computer Society, Los Alamitos (2006)
21. Cimiano, P.: Ontology Learning and Population from Text: Algorithms, evaluation and applications. Springer, New York (2006)

Organizing News Archives by Near-Duplicate Copy Detection in Digital Libraries

Hung-Chi Chang[1] and Jenq-Haur Wang[2]

[1] Institute of Information Science
Academia Sinica, Taiwan
[2] Department of Computer Science and Information Engineering
National Taipei University of Technology, Taiwan
hungchi@iis.sinica.edu.tw, jhwang@csie.ntut.edu.tw

Abstract. There are huge numbers of documents in digital libraries. How to effectively organize these documents so that humans can easily browse or reference is a challenging task. Existing classification methods and chronological or geographical ordering only provide partial views of the news articles. The relationships among news articles might not be easily grasped. In this paper, we propose a near-duplicate copy detection approach to organizing news archives in digital libraries. Conventional copy detection methods use word-level features which could be time-consuming and not robust to term substitutions. In this paper, we propose a sentence-level statistics-based approach to detect near-duplicate documents, which is language independent, simple but effective. It's orthogonal to and can be used to complement word-based approaches. Also it's insensitive to actual page layout of articles. The experimental results showed the high efficiency and good accuracy of the proposed approach in detecting near-duplicates in news archives.

Keywords: Near-duplicate document copy detection, sentence-level features, news archive organization.

1 Introduction

There are huge numbers of documents in digital libraries, for example, academic papers, news archives, and historic documents, to name a few. How to effectively organize these documents so that humans can easily browse or reference is a challenging task. Several existing methods might help better organizing them. For articles in news archives, they might be related to specific named entities such as people, event, time, location, and objects. Named entity recognition or extraction methods can extract these named entities with good accuracy. Text classification or clustering approaches could be used to effectively group news articles with similar semantic content into thematic topics. However, these classification methods in different dimensions only provide partial views of the news articles. For example, the relationships among news articles might not be easily grasped using these methods. Chronological ordering of news articles might organize according to their time of writing or archiving. Geographical information such as the place of writing or archiving provides another good

D.H.-L. Goh et al. (Eds.): ICADL 2007, LNCS 4822, pp. 410–419, 2007.

source of clues. However, time and location information may not be readily available if not explicitly specified in the news articles or their metadata. They are hard to pinpoint without further analysis of their relationships with other articles that have explicit time/location information.

In this paper, we propose a different approach to organizing news archives in digital libraries. First, near-duplicate copy detection techniques are applied to articles in news archives. Features are extracted and similarity among documents is estimated. Then, the candidate lists of near-duplicates in the archive form several clusters of documents which might be combined with above-mentioned classification methods as an alternative way to organize them.

Conventional document copy detection methods deal with the problem in two general word-based approaches. First, exact or "almost identical" copies can be detected with *document fingerprinting* approaches. Second, similar or "relevant" documents can be identified using *information retrieval* approaches based on bag-of-word frequencies. In the growing applications in Web community such as blogs, partial or "subset" copies are more common than exact copies of whole documents. Conventional near-duplicate document copy detection methods are usually based on different types of content editing behaviors. For example, several subcategories of near-duplicate copy detection are defined [12] based on the editing styles such as block adding/deletion, minor change, minor change and block edit, block reordering, bag-of-word similarity, and exact copy. In this paper, we intend to focus on efficiently detecting near-duplicates based on the editing results of documents.

Several issues have to be addressed. First, conventional word-based features cannot be directly applied since the word features in an extracted text segment might be quite different from those in a target document. Second, efficiency is critical in real applications. Word-based approach usually requires comparison of word n-grams in source document with all candidates in the target documents to find all possible matches, which is time-consuming. Third, most word-based approaches try to speed up the matching process by using "hashed n-grams" or shingles [2]. However, the accuracy of detection could be significantly influenced if some terms were substituted in the copied segments.

In this paper, we propose an effective sentence-level approach that exploits features beyond word-level features. We try to somehow capture part of the writing style above the word semantics. The most representative sentence-level statistics are extracted from the documents, the *sentence length*. It's the most primitive syntactic information about a document which could vary in different people's writing styles. One important advantage of the approach is its independence of the specific language used in writing. Since only the sentence delimiter will be used in determining the boundaries, it's insensitive to actual article layout. Also, it's both efficient and effective in detecting partial copies of text segments. The promising experimental results showed high accuracy of detection and good efficiency of our proposed approach.

The rest of the paper is organized as follows. Section 2 lists some related work. In Section 3, the details of the proposed approach are illustrated. Section 4 presents the experiments and potential applications of the approach. In Section 5, we list the conclusions.

2 Related Work

The research field of document copy detection has received much attention thanks to many popular applications on the Web, for example, duplicate Web page detection [5] and removal in Web search engines, document versioning and plagiarism detection [6] in digital libraries, and duplicate document removal in databases [1][9]. Conventionally, there are two general approaches to near-duplicate document copy detection: document fingerprinting and bag-of-word similarity. Document fingerprinting approaches [4] can detect almost-identical documents with hash-value based checksum or "fingerprint". Earlier copy detection systems such as SCAM [9] and COPS [1] are some examples. Bag-of-word similarity approaches try to identify relevant documents with similar distribution of word frequencies.

New applications in Web community such as weblogs and wikis raise new challenges to the existing approaches. First, there are more partial or "subset" copies than exact copies of the whole document. Copy detection techniques dealing with whole document matching might not be directly applicable to the partial copy cases. There are similar efforts in detecting documents with "intermediate-level similarity" between the semantic relevance and syntactically identical. For example, Shulman [10] targeted at the task of near-duplicate detection in notice and public comment rulemaking. Yang and Callan [12] proposed an instance-level constrained clustering approach that incorporates additional information such as document attributes and content structure into the clustering process to form near-duplicate clusters.

3 The Proposed Approach

The basic idea of the proposed approach is to use sentence-level features, i.e. *sentence length*, as an alternative to word-based features. The system architecture of the main functional blocks is illustrated in Fig. 1.

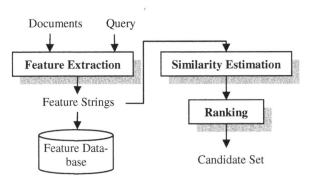

Fig. 1. Architecture of the main functional blocks

The source data is first converted into a *feature string*, a numeric sequence representing the number of words in sentences, by the *feature extraction* module. Some pre-processing steps such as HTML markup removal and stopword removal might be applied first. The *similarity estimation* module then extracts *feature vectors* from the feature string and indexes them into the feature database which records the (Feature vector, Document ID) pairs.

For a given query, the same feature extraction procedure is used to generate a query feature string. The feature vectors extracted from the querying feature string are then used to search the feature database. A *ranking* scheme would then be applied to order the candidate set of query results by their similarity scores.

3.1 Feature Extraction

In feature extraction module, text-based sentences are determined by the pre-defined delimiter set. A string of text would be considered a single sentence if and only if it satisfies both conditions: (1) the string is between two arbitrary symbols defined in the delimiter set; and (2) it does not contain any symbol defined in the delimiter set. For example, punctuation marks such as "." and "!" are included in the delimiter set. Delimiters should be chosen carefully, because it would influence the result of conversion and also the quality of subsequent functions. The output of feature extraction would be a numeric sequence, the *feature string* for the document. Further processing would be done on the feature string instead of the original text content.

3.2 Similarity Estimation

In order to detect partial document copies, a sliding window is used to extract feature vectors, the sub-sequences from the original feature strings, for similarity estimation. There are two major parameters: the window size WS and the sliding step width SW. Window size defines the constant length of the current feature vectors that we're dealing with, and sliding step width defines the number of elements to slide through at each step. It is referred as jumping window if SW > 1 [13]. For static jumping, SW is constant during indexing. For example, for feature string [0123456789...] (with WS=4 and SW=2), the first two extracted feature vectors would be [0123] and [2345] respectively. The effect of varying window size and sliding step width is listed in Fig. 2. We would verify these effects in later section.

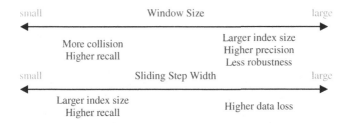

Fig. 2. The effect of window size and sliding step width

Considering the case when the feature strings are [0123456789...] and [1234567890...] respectively. It should be identified as near-duplicate for the matched segment [123456789]. However, in previous example of static jumping (where WS=4 and SW=2), we will not be able to find any match between the feature vectors extracted from the two feature strings, since none of the respective features vectors [0123], [2345], [4567], ..., and [1234], [3456], [5678], ..., will match. To avoid such situation due to the jumping in feature string, we only applied static jumping window in indexing stage to reduce the database size in our implementation.

Instead of indexing the extracted feature vector in each sliding step, another scheme called *dynamic jumping* might be applied to decrease the number of indexed records and to reduce the storage space usage. The idea of dynamic jumping is described as follows: If the leading *SW* elements (E_n, E_{n+1}, ..., E_{n+SW-1}) at the current window position are equal to the *SW* elements at the previous window (E_{n-SW}, E_{n-SW+1}, ..., E_{n-1}), that is, the *SW* elements in front of the current window, and also equal to the *SW* elements at the next window (E_{n+SW}, E_{n+SW+1}, ..., $E_{n+2SW-1}$), then *discard* the current window since the repeated pattern has been extracted in the adjacent vectors.

When determining if two numeric sequences are near-duplicates, we set a *minimum matching threshold t*, which controls the minimum number of contiguously matched sub-sequences before being considered as near-duplicates. The *minimum matching length* could be expressed as follows:

$$Length_{MIN\_Match} = WS + SW * (t - 1). \tag{1}$$

When $t=1$, we're only considering the effect of window size for determining if two sequences are near-duplicates. Assume that there are two configurations with different windows sizes $WS_L > WS_S$, step widths SW_L and SW_S, thresholds t_L and t_S, respectively. If we fix the minimum matching length, then we have:

$$Length_{MIN\_Match} = WS_S + SW_S * (t_S - 1) = WS_L + SW_L * (t_L - 1). \tag{2}$$

With the constraint of fixed index size, we would get that $SW_L > SW_S$. The relationship among these arguments is shown as Fig. 3.

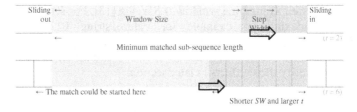

Fig. 3. The relationship among parameters. The top half shows the case of larger window size WS_L while the bottom half shows the smaller windows size WS_S.

From equation (2), we could see that a configuration with smaller window size is more flexible and might achieve higher recall than the one with larger window size because shorter *SW* implies that the matched sub-sequence or feature vector could start at more points. The case of smaller windows size also preserves more details about original feature string because of shorter *SW*. However it requires more computation because $t_S > t_L$.

3.3 Ranking Scheme

After similarity estimation, there could be more than one document matching feature vectors extracted from a given query Q. We assume that the more matches of a document by a feature vector in Q, the more relevant the query is. Then, the similarity score of a document to a query Q would be calculated based on its weighted sum of the number of matched feature vectors. That is, a document is more likely to contain duplicated content with the query if it has a higher score. Finally, we could rank the documents in the candidate set according to their scores.

4 Experiments

The experiments are divided into two parts. First, the effects and relationship among various parameter configurations were evaluated. Then, the performance of different matching and ranking schemes were compared. All experiments were carried out on the same desktop PC with Intel Pentium 4 3.0 GHz CPU and 2 GB DDR2 RAM.

4.1 Parameter Fine-Tuning

The first experiment was conducted to evaluate the effects and relationship among parameter configurations. We first collected the proceedings of WWW and SIGIR conferences from 2004 to 2006, a total of 1,090 papers in PDF format. These papers were converted to text files to form the *source data set*. Then, we selected the 27 nominated best paper award candidates and extracted their introduction and conclusion sections into a single query file as the *test query*. For each selected paper, this step could be regarded as the copying and editing behaviors; while the query file could be considered as a near-duplicate of each selected paper. We tested our proposed approach using 15 different parameter configurations as listed in Table 1.

Table 1. Various parameter configurations and their index statistics

Notation	WS	SW	DJ	Index number	Index time (sec.)
WS08SW1	8	1	No	545,254	2.233
WS08SW1wDJ	8	1	Yes	508,468	2.096
WS08SW2	8	2	No	273,442	0.984
WS16SW1	16	1	No	536,642	2.475
WS16SW1wDJ	16	1	Yes	501,422	2.310
WS16SW2	16	2	No	269,135	0.951
WS16SW4	16	4	No	135,386	0.404
WS32SW1	32	1	No	519,426	3.132
WS32SW1wDJ	32	1	Yes	486,835	2.716
WS32SW4	32	4	No	131,082	0.479
WS32SW8	32	8	No	66,338	0.209
WS64SW1	64	1	No	485,134	3.457
WS64SW1wDJ	64	1	Yes	456,995	3.121
WS64SW8	64	8	No	62,039	0.310
WS64SW16	64	16	No	31,827	0.139

Note that, for each *WS* setting, we only applied the *dynamic jumping* (denoted as *DJ* in the table) scheme to the case *SW*=1 for comparison. Since the purpose of these experiments was to evaluate the effect of various parameter configurations, no ranking were done here. The matched documents by any feature vector extracted from the query file would be added to query result. As shown in Table 1, we could see that the dynamic jumping scheme reduced the number of indexed feature vectors by around 7% and index time by around 10% of the original in indexing stage. Under these configurations, we tested the query performance in terms of the efficiency and effectiveness as shown in Tables 2 and 3.

Table 2. Search time in querying stage

Configuration	Search time (sec.)	Configuration	Search time (sec.)
WS8SW1	6.469	WS32SW1	0.344
WS8SW1wDJ	1.766	WS32SW1wDJ	0.297
WS8SW2	3.204	WS32SW4	0.141
		WS32SW8	0.078
WS16SW1	0.219	WS64SW1	0.125
WS16SW1wDJ	0.172	WS64SW1wDJ	0.125
WS16SW2	0.156	WS64SW8	0.078
WS16SW4	0.109	WS64SW16	0.062

Table 3. Query result of various configurations

Configuration	Recog.	Total	Recall	Precision	F1
WS8SW1	26	570	0.963	0.046	0.087
WS8SW1wDJ	26	353	0.963	0.074	0.137
WS8SW2	26	459	0.963	0.057	0.107
WS16SW1	24	24	0.889	1.0	**0.941**
WS16SW1wDJ	24	24	0.889	1.0	**0.941**
WS16SW2	24	24	0.889	1.0	**0.941**
WS16SW4	23	23	0.852	1.0	0.920
WS32SW1	12	12	0.444	1.0	0.615
WS32SW1wDJ	12	12	0.444	1.0	0.615
WS32SW4	12	12	0.444	1.0	0.615
WS32SW8	12	12	0.444	1.0	0.615
WS64SW1	1	1	0.037	1.0	0.071
WS64SW1wDJ	1	1	0.037	1.0	0.071
WS64SW8	1	1	0.037	1.0	0.071
WS64SW16	1	1	0.037	1.0	0.071

We use the term *recognition* to denote the number of selected papers found in query result. We could see that the dynamic jumping scheme improved the efficiency without losing the effectiveness of the query result. We could also find that recall dropped dramatically as the window size increased. Nevertheless, precision reached 1.0 when window size is larger than or equal to 16. This observation is encouraging that the proposed approach is discriminative enough with low cost.

Note that one of the selected papers could not be matched by all queries. We checked this paper and found that the introduction and conclusion of the poster paper are too short. There are also problems for converting PDF to text file, such as unexpected hyphenation to break single words into two. This requires manual processing before further experiments can be done.

4.2 Evaluation on Similarity Estimation and Ranking

When we adopted a smaller window size (WS=8), the precision was quite low as in Table 3. The reason could be lots of false positives incurred by matching smaller fragment of numeric string. Thus, we further applied our ranking scheme to improve the precision. As shown in Table 4, we simply compared the top 27 of the ranked list with the original result without ranking. It could be found that our ranking scheme helped to achieve high precision without sacrificing much recall.

Table 4. Query performance with and without ranking

Configuration	Recog.	Total	Recall	Precision	F1
WS8SW1wDJ	26	353	0.963	0.074	0.137
WS8SW1wDJ, w/ ranking, top 27	25	27	0.926	0.926	0.926

The second experiment was conducted on a larger data set to evaluate our similarity estimation and ranking schemes. The source data set is shown in Table 5. We selected six English document collections from NTCIR-4 Test Collections for CLIR task [7]. 30 documents from each collection, that is, a total of 180 documents, were randomly selected as the test queries. Then we manually composed the query file in the same manner as the previous experiments. At most 10 paragraphs per document (about 6 paragraphs per document on the average) were copied. Note that documents in this dataset (news articles), were much shorter in general and more varied in length than the previous one (academic papers).

Table 5. Source data set

Document collection	# of doc	Size (MB)
EIRB010 (Taiwan News and Chinatimes English News, 1998 - 1999)	10,204	24.6
Hong Kong Standard (1998 – 1999)	96,856	253.2
Korea Times (1998 – 1999)	19,599	55.9
Korea Times (2000 – 2001)	30,530	81.1
Mainichi Daily News (1998 – 1999)	12,723	33.3
Mainichi Daily News (2000 – 2001)	12,155	26.3
Total	182,067	474.4

The feature database in this experiment was stored on disk instead of in memory because it's much larger in size. The query data set was indexed with two configurations: WS8SW1 ($WS=8$ and $SW=1$) and WS16SW2 ($WS=16$ and $SW=2$). To concentrate our attention on the performance of ranking scheme, dynamic jumping was disabled in both configurations. Summary of index statistics for these two configurations is listed in Table 6. It took much longer to index because of the extra time requirement introduced by disk access. The performance comparison among configurations with various *minimum matching threshold t* is shown in Table 7.

Table 6. Summary of index statistics

Metrics	WS8SW1	WS16SW2
Index number	5,356,152	2,195,642
Index time (second)	2,100.49	274.58

Table 7. Performance comparison among configurations with various thresholds t

Configuration	Time	Recog.	Total	R	P	F1
WS8SW1, t=1	0.156	178	230	**0.989**	0.774	0.868
WS8SW1, t=3	0.172	175	198	0.972	0.884	0.926
WS8SW1, t=5	0.188	163	168	0.906	0.970	**0.937**
WS8SW1, t=9	0.188	143	146	0.794	0.979	0.877
WS16SW2, t=1	0.094	139	141	0.772	**0.986**	0.866

R: Recall, P: Precision

For different thresholds t, recall decreased and precision increased as t increased. Comparing the case t=1 of WS16SW2 with the case t=9 of WS8SW1 in Table 7, which is an example of Fig. 3, we verified that smaller windows size was more flexible and achieved higher recall than larger ones under the assumptions of invariant minimum matching length and constant index size. However, it would take longer time for query. There is a tradeoff between efficiency and efficacy.

4.3 Possible Applications

The motivation of this work is to detect quotation without explicit reference to the original news articles. It could be applied in other situations. For example, when a blogger is posting a new article on the blog site, the system checks the submitted content against the protected archive. If any matched article is found, notification would be sent to the posting user and the authors of matched candidates.

In addition to triggering the detection by a query document, for some applications, we have only one data set, for example, the collection of Web pages. The proposed approach could be easily modified to handle such case. Instead of extracting and indexing feature vectors from feature string of each document, similarity scores among documents would be calculated. Then, documents could be clustered into groups based on the similarity scores and clustering policy.

Link analysis has been proven effective for many Web applications, such as Web page ranking and classification. However, explicit hyperlinks do not exist among news articles. One possible solution is to introduce an alternative resource of links among articles, that is, implicit links [8][11]. Under the assumption that documents with the same duplicated content could be regarded relevant to each other, our proposed approach could be modified to identify this kind of implicit links.

5   Conclusions

In this paper, we propose a sentence-level statistics-based near-duplicate copy detection approach to organizing news archives in digital libraries. The proposed approach is language-independent, simple but effective. It is orthogonal to and can be used to complement conventional word-based approaches. The experimental results show the potential of its effectiveness and efficiency in near-duplicate document copy detection for news archive. One possible limitation of the approach is that it could not effectively handle the well-structured syntactic paragraphs with repetitive or regular sentence length pattern, for example, Tang poetry or lyrics, since their sentence lengths carry not much information about the document content. Further investigation is needed to exploit such special application in copy detection.

References

1. Brin, S., Davis, J., Garcia-Molina, H.: Copy Detection Mechanisms for Digital Documents. In: ACM SIGMOD International Conference on Management of Data, pp. 398–409 (1995)
2. Broder, A., Glassman, S., Manasse, M., Zweig, G.: Syntactic Clustering of the Web. In: 6th International World Wide Web Conference, pp. 393–404 (1997)
3. Charikar, M.S.: Similarity Estimation Techniques from Rounding Algorithms. In: 34th Annual ACM Symposium on Theory of Computing, pp. 380–388 (2002)
4. Heintze, N.: Scalable Document Fingerprinting. In: Proceedings of the 2nd USENIX workshop on Electronic Commerce, pp. 191–200 (1996)
5. Henzinger, M.: Finding Near-Duplicate Web Pages: A Large-Scale Evaluation of Algorithms. In: Proceedings of SIGIR 2006, pp. 284–291 (2006)
6. Hoad, T.C., Zobel, J.: Methods for identifying versioned and plagiarized documents. Journal of the American Society for Information Science and Technology 54(3), 203–215 (2003)
7. NTCIR (NII Test Collection for IR Systems) Project, http://research.nii.ac.jp/ntcir/
8. Shen, D., Sun, J.T., Tang, Q., Chen, Z.: A Comparison of Implicit and Explicit Links for Web Page Classification. In Proceedings of WWW 2006, pp. 643–650 (2006)
9. Shivakumar, N., Garcia-Molina, H.: SCAM: a copy detection mechanism for digital documents. In: Proceedings of International Conference on Theory and Practice of Digital Libraries (1995)
10. Shulman, S.: E-Rulemaking: Issues in Current Research and Practice. International Journal of Public Administration 28, 621–641 (2005)
11. Xu, G., Ma, W.Y.: Building Implicit Links from Content for Forum Search. In: Proceedings of SIGIR 2006, pp. 300–307 (2006)
12. Yang, H., Callan, J.: Near-Duplicate Detection by Instance-level Constrained Clustering. In: Proceedings of SIGIR 2006, pp. 421–428 (2006)
13. Zhu, Y., Shasha, D.: StatStream: Statistical Monitoring of Thousands of Data Streams in Real Time. In: Proceedings of the 28th ACM VLDB International Conference on Very Large Data Base, pp. 358–369 (2002)

An Efficient Dictionary Mechanism Based on Double-Byte

Lei Yang[1], Jian-Yun Shang[1], and Yan-Ping Zhao[2]

[1] Dept. of Computer Science, Beijing Institute of Technology
[2] School of Management and Economics, Beijing Institute of Technology
Beijing 100081, P.R. China
jeffy2008@gmail.com, shangjia@bit.edu.cn, zhaoyp@bit.edu.cn

Abstract. Dictionary is an efficient management of large sets of distinct strings in memory. It has significant influence on Natural Language Process, Information Retrieval and other areas. In this paper, we propose an efficient dictionary mechanism, which is suitable for Double-Byte coding languages. Compared with other five popular dictionary mechanisms, this mechanism performs the best of all. It improves the search performance greatly and reduces the complexity of the construction and maintenance of the dictionary. It can be well applied in large-scale and real-time processing systems. Since Unicode is a typical double-byte code which can represents all kinds of characters in the world, this dictionary will be applicable for multi-language dictionaries.

Keywords: Dictionary, Double-Byte, Information Retrieve, multi-language.

1 Introduction

Dictionaries play an important part in improving efficiency of information retrieval, text filtration, semantic analysis, Chinese word segmentation and other areas. The pioneers have done a lot of work in studying dictionary mechanism and proposed many efficient dictionary mechanisms[1]-[7]. Summarizing the characteristics of the previous mechanisms, we design an efficient dictionary mechanism, the Double-Byte, with experiments to show the efficiency of our algorithm. The time complexity of the algorithm is O(n) and the space complexity of that is O(C^n), n being the length of a word and C being a constant. This mechanism is suitable for various double-byte languages and mixed multi-languages without extra modifications. Therefore, it has broad prospects in future.

2 Related Work

There has been a lot of work such as Binary-Seek-by-Character[3], TRIE indexing tree[1][4], Binary-Seek-by-Word mechanism[2][3], Double-Array TRIE[4] and Double Coding[5][6]. Binary-Seek-by-Word consists of a binary search after the location of the first character; TRIE indexing tree comprises multiple chains to save dictionary and a

D.H.-L. Goh et al. (Eds.): ICADL 2007, LNCS 4822, pp. 420–423, 2007.
© Springer-Verlag Berlin Heidelberg 2007

Hash table for location; Binary-Seek-by-Character combines the previous two mechanisms; Double array TRIE introduces the DFA(Definite Finite Automata), and Double Coding exploits the serial code of Chinese words to obtain higher efficiency.

3 Double-Byte Based Dictionary Mechanism

3.1 Motivation

There are 6768 popular characters in Chinese and are coded using GB2312[8] which is a double-byte code and the same as Unicode. If we treat each character as a node in tree, then each node may connect to at least 6768 nodes(out-node). This will waste a lot of space. But we find that each double-byte character can be split into two bytes and the range of each byte in Chinese Area Code is about ($\sqrt[2]{6768}$) = 82. If we divide each character into two bytes and use two nodes to represent one character in Chinese tree, we can cut down the number of out-nodes from 6768 to 82 and this will make Chinese tree similar to English tree.

3.2 Dictionary Mechanism Based on Double-Byte

With a group of Chinese words, we define our dictionary as follows:

(1) We split each character of these words into two bytes.
(2) Construct a tree with each byte being a node in the tree.
(3) Define a hash function for each node in the tree for state transition.

   a) The range of each byte of a character is from 0xa1 to 0xfe. We change the range into 0-94 by subtracting 0xa1.

   b) The index of the Hash function is the decimal value of each byte subtracting 0xa1.

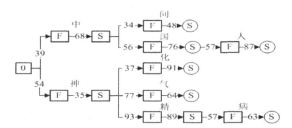

Fig. 1. Double-Byte based dictionary tree

In figure 1, node $\boxed{F}$ represents the first byte of a character and $\boxed{S}$ represents the second byte and $\widehat{s}$ represents the final state.
(4) The last node of each branch is the final state. A node in middle of a branch represents any word is set from the first to the second state.

4   Experiments and Discussions

To evaluate the performance, we conduct experiments on comparing the mechanisms mentioned previously. Experiments are run in the same environment (Windows XP, Intel(R) Pentium(R) 4 CPU 1.80GHz, 512MB memory).

4.1   Experiments Data

Our data comes from half-year's corpus of People's Daily, containing 49500 Chinese words that include single-character, double-character, three-character and four-character words with a frequency affixed to each word. The total frequency of 2595 single-character words is 2062393 by adding up each single-character word's frequency, the total frequency of 31838 double-character words is 2860757, the total frequency of 7858 three-character words is 165044 and 7209 four-character words have the total frequency of 62829. Totally, there are 5151023 words in the corpus.

4.2   Dynamic Performance

We use these algorithms to search all 5151023 words in the half-year's corpus and count the number of different operations defined in Table 1, then divide the number of each operation by 5151023 to get the average value. Table 1 gives the results.

Table 1. The Dynamic Performance (by average number of operations per word)

NO.	Numerical Computation	String-Length Computation	Reading Array	String Comparison
(1) Double-Byte	23.18	0	10.6	0
(2) Binary-Seek-by-Character	38.244	0	8.268	4.268
(3) Binary-Seek-by-Word	46.24	0	14.48	8.232
(4) Double-Array TRIE	30.44	1	1.857	0

From Table 1, we can see that algorithm (1) performs the best on numerical computation which is the most important factor. Actually, most of algorithm (1)'s numerical operation is assigning but not computation. In only one case algorithm (4) is better than algorithm (1) on reading array but worse in other aspects. Algorithm (2) and (3) increases the complexity because of the extra string comparison operation.

4.3   Search Performance

We compare the search performance of the new algorithm with other three typical algorithms. Table 2 below shows the results of the time consumption of each algorithm by searching 5151023 words (average time per word in ms).

Table 2. Time Consumption of the Four Algorithms

Algorithm	Time(s)
Double-Byte	0.150
Binary-Seek-by-Characters	0.688
Binary-Seek-by-Word	1.282
Double-Array TRIE	0.484

We can see that algorithm Double-Byte performs the best of all. It improves search performance by nearly 10 times which has gone far beyond the other three algorithms.

5 Conclusion

This paper proposes an efficient dictionary mechanism whose performance and efficiency meet the requirement of large-scale and massive processing systems with highly real-time requirements. Furthermore, this mechanism is suitable to other double-byte coding languages, and wide application aspects.

Acknowledgments. This research is supported by the National Science Foundation of China, the project code: 70471064, and the National Innovation Base in Philosophy and Social Science, the project of National Defense Science and Technology, of the second phase of "985 Project", code 107008200400024.

References

1. Aoe, J.: An Efficient Digital Search Algorithm by Using a Double-Array Structure. IEEE Transactions on Software Engineering (9) (1989)
2. Li, X., Yang, W., Chen, G.: PATRICIA-tree based Dictionary Mechanism for Chinese Word Segmentation. Journal of Chinese Information Processing (2001.03)
3. Sun, M., Zuo, Z., Huang, C.: An Experimental Study on Dictionary Mechanism for Chinese Word Segmentation. Journal of Chinese Information Processing 14(1) (2000)
4. Karoonboonyanan, T.: An Implementation of Double-Array TRIE, http://linux.thai.net/~thep/datrie/datrie.html
5. Li, J., Zhou, Q., Chen, Z.-s.: A Study on Fast Algorithm for Chinese Dictionary Lookup. Journal of Chinese Information Processing 20(5), 31–39 (2003.4)
6. Li, J., Zhou, Q. , Chen, Z.-s.: A Study on Rapid Algorithm for Chinese Dictionary Query. In: Proceedings of Large-Scale Information Retrieval and Content Security. Beijing, 9, pp. 380–390 (2005)
7. Morrison, D.: PATRICIA2Pratrical Algorithm to Retrieve Information Coded in Alphanumeric. JACM (15) (1968)
8. GB 2312-1980. Code of chinese graphic character set for information interchange, Primary set, http://www.csres.com/detail/1417.html

Content-Based Language Learning in a Digital Library

Shaoqun Wu and Ian H. Witten

Department of Computer Science, University of Waikato
Hamilton, New Zealand
{shaoqun,ihw}@cs.waikato.ac.nz

Abstract. Digital libraries have untapped potential for supporting language teaching and learning. This paper describes a new scheme for automating topic-specific language learning using a specially built digital library. Three exercises of different types are generated automatically from the library content: one that learners undertake individually, one in which learners collaborate in pairs, and one in which a group of learners compete. The system aims to foster content-based language learning, which greatly increases students' motivation, fosters long-term recollection, and can be culturally situated in appropriate ways.

1 Introduction

Digital libraries have untapped potential for supporting language learning and teaching. They include an unprecedented supply of authentic linguistic material in the form of top-quality prose. They make language material easily accessible through purposeful searching and browsing. They include rich metadata that can support interesting linguistic exercises. They provide a safe and controlled learning environment. Socially-oriented library software can support collaborative activities that strengthen and enrich the students' learning experience. Exercise content can be focused on a particular subject. Last but not least, digital libraries can be distributed to people who lack the opportunity to attend traditional classroom lessons.

We are extending the Greenstone digital library software [8] and its metadata extraction tools to support language learning activities [9]. This paper describes a project in which articles on a chosen topic were harvested to create a collection that supports language learning activities. The material came from Wikipedia, and we derived metadata appropriate to language learning by mining its structured format and richly linked hypertext using standard natural language processing tools.

We implemented three vocabulary learning activities that offer challenging exercises within a particular domain utilizing the above metadata. The exercises are automatically generated from the digital library content; in some, learners select material using Greenstone's standard search and retrieval facilities. One exercise is done by individual learners; in another they collaborate in pairs; and in the third a group of learners compete. Together these exercises provide a learning environment in which students can improve their topic-specific vocabulary knowledge—we chose the example domain of *business*. The exercises have the following unique features:

- They draw students' attention to the salient vocabulary of a particular topic.
- They help students learn vocabulary from context.

D.H.-L. Goh et al. (Eds.): ICADL 2007, LNCS 4822, pp. 424–433, 2007.

- They increase the students' encounters with relevant topic-related vocabulary.
- Collaborative learning helps sustain learning motivation and interests.

2 Digital Libraries in Language Learning

Digital libraries have an important role to play in language education. They provide genre-specific, focused material that is carefully selected and organized. By exploring the authenticate material that digital libraries provide, learners are exposed to contemporary language usage. Subject-specific collections provide the opportunity to encounter key terms and grammatical constructions that rarely occur in general texts. For example, Fuentes [4] reports that students' knowledge of business language is greatly enriched by basing learning on a corpus of business reports and product reviews. Digital libraries of multimedia can provide a rich and coherent learning context, which aids retention and reinforces learners' knowledge of language. They can promote culturally situated learning by working with collections that introduce the target language's people, history, environment, art, literature, and music.

Digital libraries can provide a safe learning community in which teachers share thoughts, tips and lesson plans, and organize collaborative task-based, content-based language projects; and learners meet their peers, exchange learning ideas, and engage in competitive or collaborative tasks. Pedagogically tuned search and browse facilities can meet the special needs of individual learners and teachers without bogging them down in fruitless tangential explorations. Earlier [9] we developed eight activities that are automatically generated from digital library content and utilize the search and retrieval facilities to illustrate new ways of supporting language study.

Supporting language learning with digital libraries is particularly relevant in developing countries where the ability to speak another language can make the difference between poverty and success. Language education traditionally takes place in classrooms, and many students are denied the opportunity because of scarce resources. Although the Internet is beginning to invade everyday life—and the classroom—even in developing countries, many people living in remote areas are still deprived of this learning facility. However, stand-alone digital libraries packed with unprecedented amounts of language material can reach out to those in areas that technology has forgotten and give them the means to escape the poverty trap.

3 Content-Based Language Learning in Digital Libraries

Content-based language learning refers to two separate ideas: learning a subject (such as business or mathematics) through the medium of a foreign language, and learning a foreign language by studying a particular subject [3]. Although a relatively new research area it has promising educational implications. First, it makes language learning more interesting and motivational. For example, young children would enjoy reading science fiction or simple encyclopedia articles using vocabulary learned in the science classroom. Second, it helps those who need to upgrade their language knowledge in a particular domain but are hampered by time constraints or the inability to find suitable courses to further their career. Third, being subject-specific makes it natural and meaningful to introduce the culture of the target language.

Digital libraries, like traditional ones, can make a central contribution to education. The genre-specific nature of many collections matches the content-based language learning paradigm. Such collections provide a vast body of samples of authentic language use. Language is often topic-specific: "different genres may exhibit very different pattern in the use of both lexis and grammar." [7] Learners supported by a digital library can acquire knowledge of a subject and at the same time improve their linguistic ability. Activities can help them to notice linguistic features, give them the opportunity to acquire a core vocabulary and expressions relevant to that particular subject, and practice as much as they like.

4   Building a Topic-Related Collection

Wikipedia is a massive and constantly growing online encyclopedia with a unique "anyone can edit" philosophy. At the time of writing, its English version contains 1.8 million articles covering a very wide range of topics. Traditional encyclopedias and their new multimedia electronic counterparts are a valuable language learning resource that can enrich a learners' knowledge of the target language. However, their cost often discourages use. Wikipedia, which is online and completely free, opens up new opportunities for supporting language study in innovative and creative ways.

Figure 1 shows a typical article. It comprises the entry's title (in this case *business*), an accompanying picture, a definition, and a brief description. The sections that follow include detailed explanations and introduce related topics. One of the most striking features is the wealth of hyperlinks that are scattered through the articles, putting related material right at the learner's fingertips. Moreover, the hyperlinks are labeled by short phrases called the link's "anchor text." These manually assigned key-phrases have great pedagogical value.

4.1   Selecting the Articles

We built a small digital library focused on the topic *business*. This term led directly to the Wikipedia page in Figure 1. All hyperlinks from this article to other Wikipedia articles were followed, resulting in about one hundred business-related articles.

In order to accomplish this we used the WikipediaMiner toolkit [10]. Wikipedia provides a standard procedure for exporting its entire database of articles, and WikipediaMiner connects to this database. The user starts by submitting a query term, in this case *business*. WikipediaMiner returns a list of matching articles. More than one article would be returned in the case of an ambiguous query. At this stage, the user must select one of more articles that correspond to the desired senses of the word.

We chose the article shown in Figure 1, and specified that we are also interested all articles to which that article links. WikipediaMiner locates the articles and downloads their full text (which is not in the database) from the web.

Fig. 1. Typical Wikipedia article

4.2 Generating Metadata

A set of pedagogically useful metadata is extracted from each article, to support the vocabulary exercises described below. It is extracted with the assistance of OpenNLP, a collection of software tools for natural language processing that perform sentence detection, tokenization, part-of-speech-tagging, chunking and parsing.

The metadata we extract from each article comprises:

- the term that describes the article, and its definition
- the illustrative picture (as shown in Figure 1), if any, and its caption
- key-phrases relating to the article
- the number of key-phrases and paragraphs in each section of the article.

The first sentence of the article usually contains the relevant *term and its definition*. The sentence is located using OpenNLP's sentence detector and parsed into linguistic phrases. Then the opening noun phrase and its following verb phrase are identified by string pattern matching, taking into account the possibility of an intervening adjectival phrase. The noun is marked as a term, and the remainder is marked as the definition.

For example, the first sentence of the page in Figure 1 reads

In economics, business is the social science of managing people to organize and maintain collective productivity toward accomplishing particular creative and productive goals, usually to generate profit.

The net result is to remove the initial qualifier, identify *business* as the relevant term, and return the rest of the sentence as its definition:

business is the social science of managing people [continues as above].

This simple heuristic procedure works well. It works even when the term defined differs slightly from the article name, and when it is accompanied by a qualifier. In our example, terms and definitions were successfully extracted from 80 of the 100 articles.

The procedure failed on the remaining 20 because the page structure was ill-formed. In one case out of 80 the extraction procedure yielded an ill-formed definition.

The page's *picture and caption* are extracted by seeking a particular configuration of HTML tags, including the *<img>* tag that signals an image.

As mentioned earlier, *key-phrases* are the anchor text of hyperlinks in the article, and thus are easily located. Care is taken to ignore the links used for navigation and special functions such as *search* and *edit*. Different sections of a Wikipedia article normally provide supplementary or complementary information about the topic: this makes them good sources of focused coherent text. The number of key-phrases and paragraphs in each section provide some indication of its pedagogical character and are calculated by counting the relevant HTML tags.

4.3 Building the Collection

The Wikipedia articles extracted using WikipediaMiner were fed into Greenstone to build a digital library collection. The automatic metadata extraction described above was built into a special Greenstone plugin which processed each article and associated derived metadata with it.

The final collection seeded from the term *business* contained about 100 articles each with four kinds of extracted metadata: the term and its definition, a picture and caption, several key-phrases, and the number of paragraphs and key-phrases in each section. Of course, Wikipedia content is by no means definitive: it has been widely criticized as a reliable information source. However, it provides useful supplementary reading material for studying the subject.

5   Using the Collection for Language Learning

Three learning activities, *Match-term-and-definition*, *Fill-in-the-blanks* and *Predict-words*, provide systematic study of vocabulary. They make use of the material in the digital library, augmented with pedagogically valuable metadata. They pull out the vocabulary that is important to the subject in question, namely *business*, and help students notice the salient language features of that particular subject.

The importance of explicit vocabulary learning has been widely recognized. "The brightest of students will not be able to recall and use new words without repeated meaningful contact with them."[2] Reading topic-related articles engenders some degree of familiarity with particular words. But follow-up learning activities that use the same material, presented through various kinds of individual and interactive exercises, maximize the chance of the word being retained over the long term.

The exercises we describe are automatically generated from the content of the articles and the extracted metadata, and do not place any extra burden on language example of *business*. Their design has been guided by the psychological conditions teachers. They work for any topic, and are not in any way specific to our running for the retention and ultimate mastery of a word [6]:

- Noticing: giving an attention to an item.
- Retrieval: giving a word form and retrieving its meaning or vice versus.
- Creative and generative use: using the word in a new context.

Noticing plays a central role in every activity. *Match-term-and-definition* asks learners to associate a word form with its meaning: Nation's *retrieval* task. *Fill-in-the-blanks* promotes deeper reflection by asking learners to guess the meaning of a word from its context. *Predict-words* facilitates *creative and generative use* of a word by asking learners to predict what words might appear in a text.

The exercises reinforce learning through repetition: subsequent encounters of a particular word help strengthen and enrich previous knowledge [5]. Moreover, the last two activities engage learners in collaborative activity, making the vocabulary learning less daunting and more enjoyable and effective. They embody a "chat" facility, creating an environment in which learners can practice communication skills by discussing with peers, seeking help, and negotiating tasks.

Exercise material for some activities comes from individual documents; for others it comes from the whole collection. *Match-term-and-definition* is a collection level activity. It uses the term and definition metadata associated with each article. Students reach the exercise from a link that is placed on the digital library collection's home page, which in Greenstone typically contains information about the provenance of the collection as a whole, and information on how to use it. In the case of these language learning collections, the home page briefly describes the searching and browsing functionality, and introduces the language activities. The other two activities, *Fill-in-the-blanks* and *Predict-words*, work at the document level. They use the content of a particular article, along with its metadata, as the exercise material. Students reach these exercises from links that are placed on the digital library page that displays each document. They reach the document in the normal way, by searching or browsing.

The language learning exercises are built on the top of the Greenstone run-time software. Like it, they follow the client-server model. The interface is explicitly designed to be multilingual—this is a particular strength of Greenstone. To add a new interface language it is necessary to create a "resource bundle" for that language and drop it into the appropriate folder. The Greenstone Translator's Interface [1] provides interactive tools for creating and maintaining language interfaces. The language learning activities are implemented using JavaScript and Ajax technology.

5.1 Matching Terms with Their Definition

In "matching" activities, learners must find two items that are related in some way, such as having similar or opposite meanings. These activities can easily cater for learners at different levels, and are widely used for vocabulary study. The *Match-term-and-definition* exercise asks the learner to match terms with their definitions. Terms are words or phrases, and definitions are sentences whose subject is missing.

Figure 2 illustrates the interface to this activity. This exercise comprises five pairs of terms and definitions, scrambled. The learner drags a term on the left side into a definition box on the right. The system formats the result by highlighting the term. In Figure 2, the learner has matched *International trade* (second row) and *Agriculture* (third row) with their definitions, and has yet to match *Expertise*, *Defecation* and *Jurisdiction*.

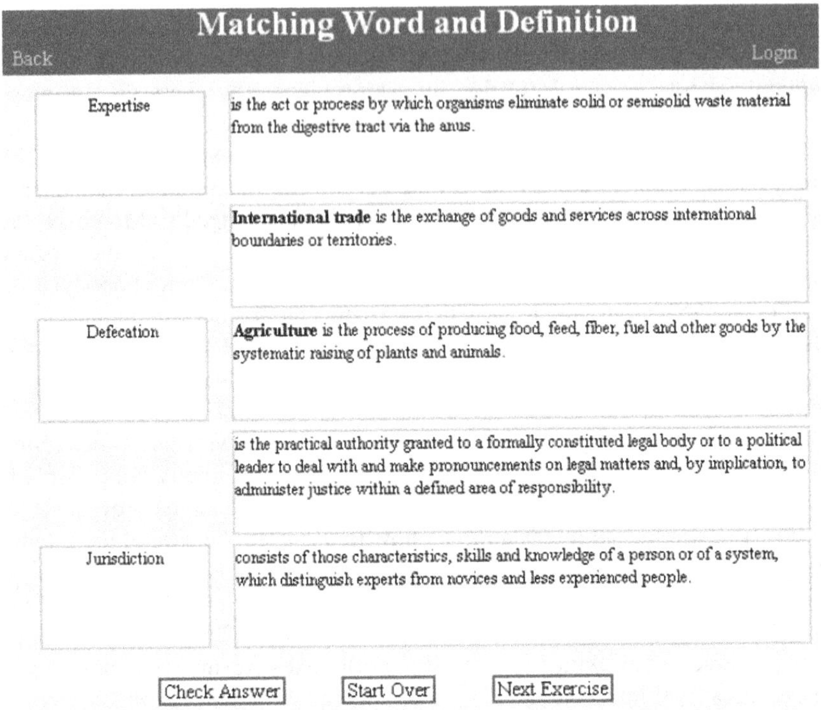

Fig. 2. The Match-*term-and-definition* activity

At any stage the learner can use the buttons at the bottom to check the answers (this causes incorrect ones to be colored red), start this exercise over again (this reinitializes the boxes), or move to the next exercise. The scheme is designed for motivated learners, and for exercising rather than testing: it does not enforce the completion of one exercise before allowing the next to begin. At the beginning of the exercise the activity server retrieves five phrases and their associated definitions from the collection at random, using the metadata computed as described earlier.

5.2 Filling in the Blanks

Fill-in-the-blanks exercises are created by cutting target words or phrases out of a sentence or article and having students fill them in. Human tutors can judge free word choices, but in computer-assisted environments the excised words or phrases are invariably displayed so that a student's choice can be checked automatically. Our implementation of this activity is novel because pairs of learners must work together to complete the task. The necessary information is split: each student holds part of it and in order to succeed they must cooperate. Such tasks are known to have great pedagogical value: they encourage negotiation of meaning, promote implicit or explicit corrective feedback between learners, and improve performance by stimulating modifications to the output [6]. In this activity, each learner has only half the missing words, and in order to achieve the goal of filling them in they must exchange information verbally using a built-in "chat" system.

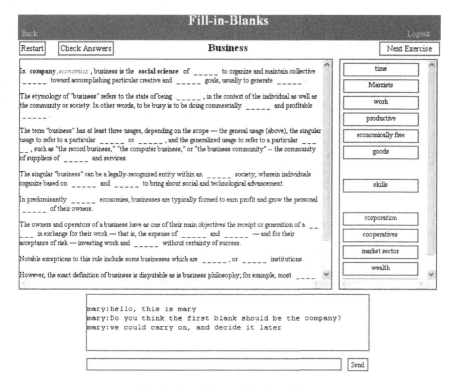

Fig. 3. The fill-in-the-blanks activity

Learners are paired based on the article they have selected. An exercise is constructed using the text of a particular section and the key-phrases that occur within it. Not all sections provide suitable material: some have too few key-phrases; others have too many paragraphs. The key-phrase and paragraph count metadata is used to select sections of modest length that contain a high density of key-phrases.

Figure 3 shows the interface. The key-phrases in the text are replaced with dashed lines. Each learner is given half of them and can drag them and drop them into the gaps. A move can be undone by clicking the dropped word. Each learner sees the moves their partner makes, in real time. In Figure 3, this learner has filled in the first blank with the words *company* and *social science*. Of the word that follows, *economics* appear in a different color and font style because it was filled in by the partner. The buttons at the top are only displayed to one of the two learners. That learner can check the answers at any time and control the progress of the activity, restarting it or passing on to the next exercise.

5.3 Predicting Words

Given an article's topic, students compete to predict words they think will occur in it. This traditional pre-reading activity is often played in a classroom to stimulate interest and facilitate comprehension before students begin reading. It can also be used to

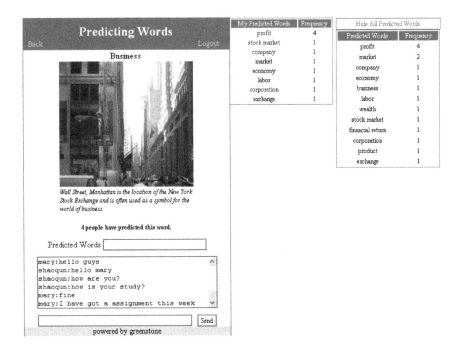

Fig. 4. *The* Predicting *Words* activity

brainstorm suitable vocabulary for a forthcoming essay, or serve as a retrospective activity where learners recall and review a list of expressions and collocations that are important for accurately expressing the ideas relevant to the topic learners.

We have implemented the exercise in its traditional form to provide a collaborative learning environment in which learners help each other by sharing information and exchanging ideas. This is a document level activity: learners who have chosen the same article are grouped together.

Figure 4 shows the interface. The title, picture, and caption are presented to convey the context of the original article. Each learner enters predicted words in the *Predicted words* field and can "chat" on the system to other learners. On the right hand side, the *My Predicted Words* table shows the words that this learner has predicted, colored blue if they occur in the key-phrase list. The table is ordered by the number of participants who have chosen that word. In the Figure, this student has predicted eight words, four of which appear in the article's key-phrase list.

The positions and counts change dynamically to reflect work by other learners. For example, *profit* has been predicted by four students. The *All Predicted Words* table, which learners can hide, lists the predictions of other participants.

6   Conclusions

This paper has described a scheme for supporting content-based language learning with a digital library. First, an instructor decides on a topic and interacts with the

WikipediaMiner system to retrieve a suitable selection of related articles. This is done by issuing a command-line statement which could easily be automated within a web-based form. Next, the articles are built into a digital library collection, using the Greenstone software augmented with a purpose-built plugin that extracts language learning metadata. These operations are independent of whatever topic was chosen for the collection.

Once the collection has been constructed, language learners interact through the ordinary Greenstone interface. Special facilities have been built into the run-time system to present three types of exercise. Learners access these by clicking on links in the collection. For the *Match-term-and-definition* exercise, which involves the whole collection, a link is placed on the collection's home page. Learners access the other exercises, *Fill-in-the-blanks* and *Predict-words*, once they have reached a particular document in the collection through links that are presented alongside the document.

The three exercise types illustrate different modes of interaction. *Match-term-and-definition* is done by individual students, working independently. *Fill-in-the-blanks* shares the necessary information between pairs of students who must cooperate in order to solve the exercise. In *Predict-words* all students who choose the same article see the same information, and compete in an informal manner to guess as many words as possible. In the last two cases the different displays are updated simultaneously, and users are able to communicate with each other through a "chat" panel.

These exercises have been devised and implemented, but not yet field tested. Interested readers can access them at *http://www.nzdl.org/language_learning*. Our next step is to test them in an actual classroom environment and a self-study setting.

References

1. Bainbridge, D., Edgar, K.D., McPherson, J.R., Witten, I.H.: Managing change in a digital library system with many interface languages. In: Koch, T., Sølvberg, I.T. (eds.) ECDL 2003. LNCS, vol. 2769, pp. 350–361. Springer, Heidelberg (2003)
2. Conzett, J.: Integrating collocation into a reading and writing course. In: Michael, L. (ed.) Teaching Collocation, LTP, England, pp. 70–87 (2000)
3. Darn, S.: Content and Language Integrated Learning. British Council Teaching English (2006), http://www.teachingenglish.org.uk/think/methodology/clil.shtml
4. Fuentes, C.A.: The use of corpora and IT in a comparative evaluation approach to oral business English. ReCALL 15(2), 189–201 (2003)
5. Nation, I.S.P.: Learning vocabulary in another language. Cambridge Univ. Press, Cambridge (2001)
6. Pica, T., Kang, H.S., Sauro, S.: Information Gap tasks. SSLA 28, 301–338 (2006)
7. Sinclair, J.: Corpus, concordance, collocation. Oxford University Press, Oxford (1991)
8. Witten, I.H., Bainbridge, D.: How to build a digital library. Morgan Kaufmann, San Francisco (2003)
9. Wu, S., Witten, I.H.: Towards a digital library for language learning. In: Proc European Conf on Digital Libraries, Alicante, Spain, pp. 341–352 (2006)
10. Milne, D.: WikipediaMiner (2007), http://wikipedia-miner.sourceforge.net

Further Development of a Digital Library Curriculum: Evaluation Approaches and New Tools

Seungwon Yang[1], Barbara M. Wildemuth[2], Seonho Kim[1], Uma Murthy[1],
Jeffrey P. Pomerantz[2], Sanghee Oh[2], and Edward A. Fox[1]

[1] Department of Computer Science, Virginia Tech,
Blacksburg, VA 24061 U.S.A.
{seungwon,shk,umurthy,fox}@vt.edu
Tel.: +1 540-231-5113
[2] School of Information and Library Science, University of North Carolina – Chapel Hill,
Chapel Hill, NC 27599-3360 U.S.A.
wildem@ils.unc.edu, {jpom,shoh}@email.unc.edu
Tel.: +1 919-962-8366

Abstract. This paper is a follow-up to our ICADL 2006 paper, reporting on our progress over the past year in developing a digital library curriculum. It presents and describes the current curriculum framework, which now includes ten modules and 41 sub-modules. It provides an overview of the curriculum development lifecycle, and our progress through that lifecycle. In particular, it reports on our evaluation of the modules that have been drafted. It concludes with a description of two new technologies – Superimposed Information (SI) to help resource presentation in a module and Visual User model Data Mining (VUDM) to help long-term module upgrade by visualizing the user community and its trends.

Keywords: Digital library, curriculum, education, evaluation, superimposed information, community visualization.

1 Introduction

Our Digital Library (DL) Curriculum Development Project[1] [3] is now in its second year (of three years of funding). Since describing the project at ICADL 2006 in Kyoto, Japan, our curriculum framework has continued to evolve, based on analyses of the literature and course syllabi in information and library science and in computer science. In addition, we have developed draft versions of seven sub-modules, and have conducted a pilot test of our formative evaluation procedures.

In section 2, we show the recently-updated DL curriculum framework, which has ten core modules, which in turn have 41 sub-modules. Section 3 presents our curriculum development lifecycle, and discusses our progress through that lifecycle. In particular, it describes our plans for expert review of draft modules, already begun at a meeting held during the Joint Conference on Digital Libraries, June 2007, and our

---

[1] Funded by US NSF Grants IIS-0535057 to VT and IIS-0535060 to UNC-CH for 2006-2008.

plans for field testing of the revised modules. In section 4, we introduce two software tools – our Superimposed Information (SI) tool [16] and our user community/trend visualization tool, i.e., Visual User model Data Mining (VUDM) [10]. They will be used to enhance the presentation of papers in a module (SI tool) and to help long-term upgrading of the DL modules by visualizing user trends over time (VUDM tool).

2   Evolution of the DL Curriculum Framework

New technologies are emerging in every area of computer science such as databases, human-computer interaction, web-based technologies, multimedia, hypermedia, search algorithms, security, etc. Considering that a digital library (DL) integrates most of those technologies under a unified system, the topics in the DL area are changing as well.

The model presented in Fig. 1 is the fourth and current version of a DL curriculum framework developed for this project.[2] Initially, our work was based on four sources. The first was an analysis of the Computing Curriculum 2001 (CC2001), recommendations for undergraduate curricula and course content in computer science. CC2001 recommends that Digital Libraries be included as one component of education on Information Management. Second, we built our initial framework on the 5S theoretical framework [5,6,7]. This framework specifies streams, structures, spaces, scenarios, and societies as the core components of digital libraries and, thus, necessary components of a curriculum on digital libraries. Third, our framework was influenced by the results of a survey of digital librarians, concerning the skills and knowledge required to manage a digital library [2]. Fourth, we based our initial framework on our own experience in teaching courses on digital libraries.

Since then, the project team has been refining the curriculum framework through a series of analyses and classification tasks. First, we analyzed the recent literature in digital libraries [14], by classifying the papers presented at JCDL and its predecessor conferences or published in *D-Lib Magazine* over the past decade. The core modules of our curriculum framework were used as the basis for the classification of 1064 papers; in addition, when a paper could not be easily classified, it initiated a discussion of possible revision of our framework. Next, we similarly analyzed syllabi from 40 digital library courses offered by 29 U.S. programs in information and library science [15]. Finally, we analyzed syllabi from digital libraries courses offered in computer science (CS) programs in the U.S. [13]. Only five courses focusing on digital libraries were identified from an examination of the websites of 296 CS programs. The readings from these five courses were added to our analysis. During each analysis, there were two types of gaps between the analytical findings and the curriculum framework that were investigated. First, if very few papers were classified into a module in the framework, we discussed whether the topics could be merged together or one of them should be removed from the framework. Second, if an unusually high number of readings were classified into the same module, we discussed the possibility of splitting that module.

---

[2] Earlier versions are available on the project's Module Development web page, http://curric. dlib.vt.edu/modDev/modDev.html

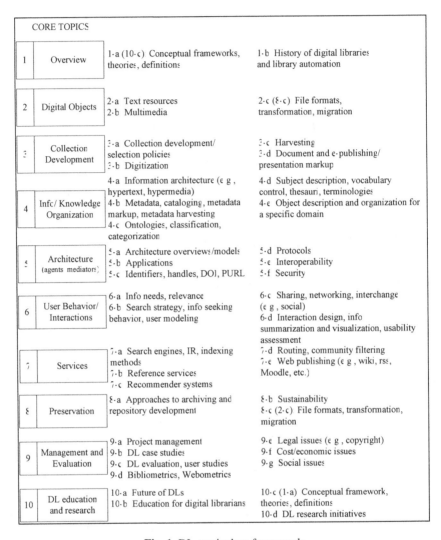

Fig. 1. DL curriculum framework

Based on these discussions, several changes were made in the curriculum framework, to bring it to the form found in Fig. 1. First, the order of the two core modules, '2: Digital object' and '3: Collection development', were reversed to be more natural (moving from considerations of a single digital object to a group of them). Second, two new sub-modules were introduced: '1-b: History of digital library and library automation' and '4-a: Information architecture (e.g., hypertext, hypermedia)'.

While we believe that the current framework is stable enough to serve as the basis for development of draft modules, it is likely that it will continue to evolve. In particular, it is important to remember that this framework's development was based on present and past teaching practices and published resources. Therefore, the current framework does not yet include emerging topics in the DL field (except in 10-a and 10-d). As evidence of their importance is identified, however, those topics will be incorporated into future versions of the module framework.

3 Digital Library Curriculum Development Lifecycle

Fig. 2 shows the development process in a spiral lifecycle, as suggested by Boehm's [1] model of the system development lifecycle. Currently, we are focused on three stages: 'Design modules', 'Evaluate via inspection', and 'Feedback on strengths & weaknesses.' In Fig. 2, these three are marked in yellow.

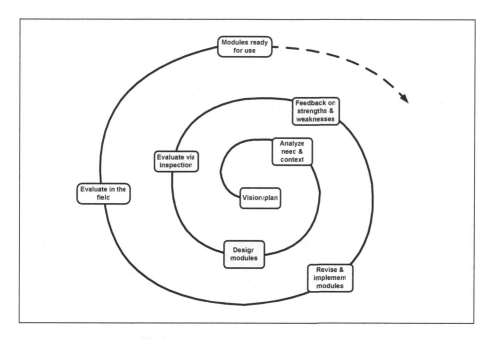

Fig. 2. DL curriculum development lifecycle

For the 'Design modules' step, in this iteration through the cycle, we developed seven draft modules. These modules were selected to cover some of the high-priority areas of the curriculum, as suggested by earlier analyses [2,13,14]. They are:

- 1-b: History of digital libraries and library automation. The origin of the DL research agenda, DLI, DLI-2, NSDL, and the origin of other long-term DL projects still extant.
- 5-a: Architecture overviews/models. Different types of DL architectures and models such as operational, technical, systems, component-based, federated, distributed, and service-oriented architectures.
- 5-b: Applications. Different types of DL systems, such as repository-based systems (e.g., DSpace, EPrints, Greenstone, FEDORA), metadata-based

systems (e.g., CiteSeer), and component-based systems (e.g., Open Digital Libraries – ODL – and WS-ODL).

- 6-a: Information needs, relevance. Aspects of a user's experience of an information need and how that experience might affect the user's interactions with the DL; relevance judgments and their relationship to the user's information needs.
- 6-b: Search strategy, information seeking behavior, user modeling. The fundamental concepts, definitions, and theoretical models of online information seeking behaviors, as they apply to digital libraries; user behaviors that have been identified in empirical studies of digital libraries.
- 7-b: Reference services. Services for meeting different types of user information needs addressed to digital libraries, including human-mediated reference, automated information retrieval (IR), and question answering (QA) services.
- 9-c: DL evaluation, user studies. Methods for evaluating the outcomes, impacts, or benefits of a digital library, including cost/benefit analyses.

Each module has the same structure following the module template[3]. Key components of each module include its learning objectives, the prerequisite knowledge required and its relationship to other modules, the body of knowledge to be covered, assignments and learning activities, and readings and other resources supporting the module.

The seven modules listed above are now ready for the next step in the curriculum development lifecycle: evaluation via inspection by experts. To pilot test our procedures for these evaluations, the project team convened a meeting during the 2007 Joint Conference on Digital Libraries (JCDL). It included members of our Advisory Board, participants in the JCDL Doctoral Consortium, and other members of the community with particular interest in curriculum evaluation. We first provided an overview of our progress in developing modules, and reviewed the curriculum development lifecycle. The participants then worked in pairs to evaluate/review a module (each pair selected a module of particular interest to them). They were guided in their evaluation by an evaluation form provided by the project team (see Fig. 3), and developed based on the work of Diamond [4], Grunert [8], and Wiggins and McTighe [17].

The final portion of the meeting was spent discussing the evaluation process itself, and how it should be implemented more widely. We received several suggestions on the module development and evaluation process. The first is to prioritize the modules, so that the most critical modules are developed first. To prioritize them, we will conduct a study of the distribution of the topics appearing in the table of contents of various DL textbooks, in the articles of DL-related magazines and journals, and in the DL class syllabi. The second suggestion is to define the scope of each module and to specify the dependencies and relationships among the modules more consistently. We will undertake this effort in the near future and will conclude it before developing an additional set of modules. Third, the group encouraged us to include more international scholars and doctoral students in the development process, so that the modules would be usable outside the U.S.

---

[3] http://curric.dlib.vt.edu/DLcurric/moduleTemplate.html

---

1. **Objectives:**
   Guiding question: Are the objectives appropriate for the topic?
   Specific questions:
   - Are the objectives observable?
   - Will students be able to achieve the objectives, given the content in the body of knowledge?
2. **Body of knowledge:**
   Guiding question: Does the module address all areas of the topic that need to be addressed?
   Specific questions:
   - Will the body of knowledge enable students to achieve the objectives?
   - Are there any topics that you think are critical to add to the body of knowledge?
   - Are there any topics that you would remove from the body of knowledge?
3. **Readings:**
   Guiding question: Are the readings the best and most appropriate for the topic?
   Specific questions:
   - Are there any readings that you think are critical to add to the list?
   - Are there any readings on the list that you would remove?
4. **Learning Activities:**
   Guiding question: Are the activities appropriate for the topic?
   Specific questions:
   - Will students be able to accomplish the activities, given the content in the body of knowledge?
   - Will the activities enable students to achieve the objectives?
   - Can you suggest any other learning activities that may be appropriate for this module?
5. **Level of Effort and Prerequisites:**
   Guiding question: Is it feasible to teach the module as it is currently constructed?
   Specific questions:
   - Is the level of effort required in class appropriate to the scope of the body of knowledge? Prior to class?
   - Is the prerequisite knowledge required sufficient for students to comprehend the body of knowledge?

Fig. 3. DL module formative evaluation form

As we continue the evaluation process, evaluators will be identified through individual nominations, review of the membership of the American Society for Information Science & Technology (ASIST) Special Interest Group on Digital Libraries (SIG DL), review of the attendance list for JCDL 2007, and other sources. To support the involvement in the evaluation process of a wide variety of experts in DLs, we will establish a password-protected wiki. Evaluators will be invited (or may volunteer) to evaluate a particular module of interest to them, and will be supplied with the wiki password. The current version of the module will be available for their review; a separate discussion page for each module will be established, so that people may comment on the module. As changes to the module are suggested and consensus is reached about the need for those changes, a revised version of the module will be provided and the wiki-based discussion will be edited to contain only those issues not yet resolved. In this way, any number of evaluators may be involved in discussing and suggesting improvements for each module. The project team will closely monitor these discussions to provide support and to finalize the module draft when the discussion has reached a conclusion.

When each module has been evaluated, the feedback received will be incorporated into its design. The next step is to field test the module. Again, volunteers will be recruited to implement particular modules in their regular classes. These classes may include those focused on digital libraries, or other classes in which a particular module would be useful. We will check the implementation of each module by tracking which portions of the module were used as proposed and which were modified, and we will capture the actual assignments completed in connection with the module. At the completion of the module in each class, we will interview the instructor and survey the students. The instructor will be asked the same types of questions that are included on the expert review evaluation form (see Fig. 3); the students will be asked about the ways in which the module (its content, readings, and assignments) affected their learning. The results from these evaluations will be available via our project website, as well as being published in more formal venues, so that potential users of each module can adapt the module as appropriate for a particular situation. At that point, we will have reached the end of our curriculum development lifecycle, with 'Modules ready for use'.

4 Future Work

We plan to incorporate two new technologies. The Superimposed Information (SI) technology will be used to enhance module resource presentation by displaying only the relevant portion of a resource. Therefore, it will help students save time studying the module resources assigned to them. Once all the modules are developed and deployed, Visual User model Data Mining tool (VUDM) will be used to visualize the module user groups and analyze the usage trend over time for further module update.

4.1 Superimposed Information (SI) Technology

In many educational tasks, there is a need to be specific about a reference – to operate at a finer level of granularity than a complete document. For example, a user may want to work with a definition in a paper, a section in a book, or part of an image or a short clip in an audio/video document. Current annotation and knowledge management tools support this functionality to some extent. However, they provide limited support for working with information at sub-document granularity across heterogeneous formats. Using SI technology, users may lay new information over existing or *base information*, typically to highlight, annotate, elaborate, select, collect, organize, connect, or reuse information elements. This functionality enables the user to work with information at sub-document granularity *in situ* (in its original form and context) [11]. These tools employ *marks*, which are references to selected regions within base information (of text or multimedia content) [12]. Fig. 4 shows how a mark may be used as part of the list of resources for the DL application module, and the resolved mark highlighting the selection that describes DSpace. In addition, this mark may be used in any place where a URI[4] may be used, such as in a module web page, as part of a concept map[5] about a module, etc. For details on work done on SI by the authors and their collaborators, please refer to [16].

We believe that using superimposed tools in the DL curriculum project has advantages. They allow users to select and work with sub-document information, while simultaneously retaining links to the original information context. This can be beneficial in providing greater specificity of reference in multiple ways:

- In module development: Using superimposed tools to describe resources in modules will help in focusing on part of a resource while still providing links to the complete documents.
- In module usage: Students and instructors can customize/personalize their use of modules in various assignments and projects by being specific about a part of a resource.

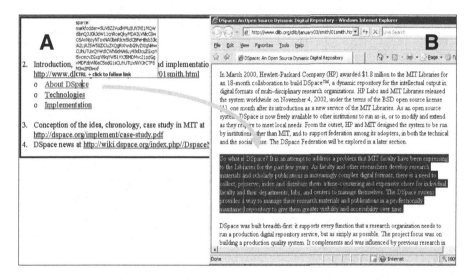

Fig. 4. A) Mark (represented as a URI) used in a module; B) Mark highlighting the desired selection that describes DSpace in an article

4.2 User Community/Trend Visualization Tool – VUDM

Visual User model Data Mining tool, VUDM [10], was designed to visualize users, user communities, and usage trends of complex information system, e.g., to analyze the DL curriculum module usage. VUDM visualizes user communities based on the

---

[4] Short for *Uniform Resource Identifier,* the generic term for all types of names and addresses that refers to objects on the World Wide Web.
[5] A concept map is a graphical knowledge representation tool, where the nodes represent concepts and the links represent relationships between concepts.

long-term history of usages of each user, instead of an explicitly-entered user profile. A web-based curriculum module server will be built to collect the usage history of each curriculum module for all users.

Fig. 5 illustrates VUDM visualizing the change in usage trends of curriculum modules for three consecutive weeks. Module number is displayed for each user group. By observing the spirals and their module number each week, we can see the changes of the user groups' interests (e.g., a group might become bigger, smaller or disappear). This information will be useful for understanding the evolution of DL curriculum modules. Further, VUDM is able to visualize the trends of positive or negative rankings entered by module users through the distribution web site. This trend indicates controversial, erroneous, and out-dated modules, and is useful for module upgrade if analyzed along with users' comments from the online forum.

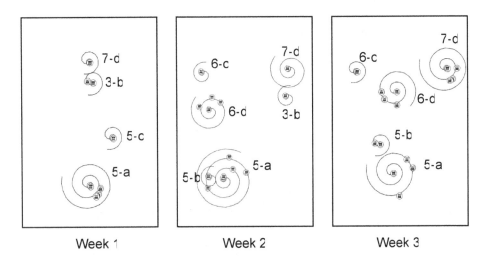

Fig. 5. Visualization of DL module usage trends for three weeks

5   Conclusion

The collaborative Virginia Tech - University of North Carolina DL curriculum development project is in the second of its three years. We have developed several curricular modules and have pilot tested our preliminary evaluation methods. In this paper, we presented the upgraded DL module framework, the curriculum development lifecycle, the draft module evaluation form, and two technologies that will support the presentation of the resources in the module (SI tool) and will help the project group to understand the module users' needs by visualizing the user community topics and trend changes (VUDM tool). We hope that the international DL community will become actively involved in this interdisciplinary effort, and that its results will improve the education of DL professionals.

References

1. Boehm, B.W.: A spiral model of software development and enhancement. IEEE Computer 21, 5, 61–72 (1988)
2. Choi, Y., Rasmussen, E.: What is needed to educate future digital librarians: A study of current practice and staffing patterns in academic and research libraries. D-Lib Magazine 12, 9
3. DL Curriculum Project Group: Collaborative Research: Curriculum Development for Digital Library Education (2006), http://curric.dlib.vt.edu/
4. Diamond, R.M.: Designing & Assessing Courses & Curricula: A Practical Guide. Jossey-Bass, San Francisco (1998)
5. Gonçalves, M.A.: Streams, Structures, Spaces, Scenarios, and Societies (5S): A Formal Digital Library Framework and Its Applications. Ph.D. dissertation. Computer Science Dept., Virginia Tech, Blacksburg, VA (2004), http://scholar.lib.vt.edu/theses/available/etd-12052004-135923/unrestricted/MarcosDissertation.pdf
6. Gonçalves, M.A., Fox, E.A.: 5SL - A language for declarative specification and generation of digital libraries. In: Proceedings of JCDL 2002, Second ACM / IEEE-CS Joint Conference on Digital Libraries, pp. 263–272 (2002)
7. Gonçalves, M., Fox, E., Watson, L., Kipp, N.: Streams, structures, spaces, scenarios, societies (5S): A formal model for digital libraries. ACM Transactions on Information Systems 22, 270–312 (2004)
8. Grunert, J.: The Course Syllabus. Anker Publishing Company, Inc., Bolton, MA (1997)
9. Gustafson, K.L., Branch, R.M.: Survey of Instructional Development Models, 3rd edn. ERIC Clearinghouse on Information Resources, Syracuse University, Syracuse, NY (1997)
10. Kim, S., Lele, S., Ramalingam, S., Fox, E.A.: Visualizing user communities and usage trends of digital libraries based on user tracking information. In: Sugimoto, S., Hunter, J., Rauber, A., Morishima, A. (eds.) ICADL 2006. LNCS, vol. 4312, pp. 111–120. Springer, Heidelberg (2006)
11. Maier, D., Delcambre, L.: Superimposed information for the Internet. In: WebDB Workshop, Philadelphia, PA, June 3-4, pp. 1–9 (1999)
12. Murthy, S., Maier, D., Delcambre, L., Bowers, S.: Putting integrated information into context: Superimposing conceptual models with SPARCE. In: First Asia-Pacific Conference of Conceptual Modeling, Denedin, New Zealand, pp. 71–80 (2004)
13. Pomerantz, J., Oh, S., Wildemuth, B.M., Yang, S., Fox, E.A.: Digital library education in computer science programs. In: Proceedings of the ACM/IEEE-CS Joint Conference on Digital Libraries, Vancouver, pp. 177–178 (2007)
14. Pomerantz, J., Oh, S., Yang, S., Fox, E. A., Wildemuth, B.: The core: Digital library education in library and information science programs. D-Lib Magazine, 12, 11 (2006), http://dlib.org/dlib/november06/pomerantz/11pomerantz.html
15. Pomerantz, J., Wildemuth, B., Fox, E. A., Yang, S.: Curriculum development for digital libraries. In: Proceedings of the 6th ACM/IEEE-CS Joint Conference on Digital Libraries, pp. 175–184 (2006)
16. Murthy, U., Fox, E.A.: Sidewalk at Virginia Tech. 2006. Digital Library Research Laboratory, Computer Science Dept., Virginia Tech, Blacksburg, VA, http://si.dlib.vt.edu/
17. Wiggins, G., McTighe, J.: Understanding by Design, 2nd edn. Association for Supervision and Curriculum Development, Alexandria, VA (2005)

Managing Offline Educational Web Contents with Search Engine Tools

Choochart Haruechaiyasak[1], Chatchawal Sangkeettrakarn[1],
and Wittawat Jitkrittum[2]

[1] Human Language Technology Laboratory (HLT),
National Electronics and Computer Technology Center (NECTEC)
Thailand Science Park, Klong Luang, Pathumthani 12120, Thailand
{choochart.haruechaiyasak,chatchawal.sangkeettrakarn}@nectec.or.th
[2] School of Information and Computer Technology (ICT),
Sirindhorn International Institute of Technology (SIIT),
Thammasat University, Bangkadi Campus, Pathumthani 12000, Thailand

Abstract. In this paper, we describe our ongoing project to help alleviate the digital divide problem among high schools in rural areas of Thailand. The idea is to select, organize, index and distribute useful educational Web contents to schools where the Internet connection is not available. These Web contents can be used by teachers and students to enhance the teaching and learning for many class subjects. We have collaborated with a group of teachers from different high schools in order to gather the requirements for designing our software tools. One of the challenging issues is the variation in computer hardwares and network configuration found in different schools. Some shools have PCs connected to the school's server via the Local Area Network (LAN). While some other schools have low-performance PCs without any network connection. To support both cases, we provide two solutions via two different search engine tools. These tools support content administrators, e.g., teachers, with the features to organize and index the contents. The tools also provide general users with the features to browse and search for needed information. Since the contents and index are locally stored on hard disk or some removable media such as CD-ROM, the Internet connection is not needed.

Keywords: Web Content Management, Search Engine Tools, Educational Web Contents.

1 Introduction

With its exponential growth rate, the World Wide Web is now well recognized as the world's largest online information resource. Today there are many Web sites which provide excellent educational contents. Some of the interesting examples include BBC's Learning [5], MIT's OpenCourseWare [11], Wikipedia [12] and Google Book Search's Library Project [7]. BBC's Learning is a Web portal that provides links to various learning subjects for different audience ranging

D.H.-L. Goh et al. (Eds.): ICADL 2007, LNCS 4822, pp. 444–453, 2007.

from kids to adults. MIT's OpenCourseWare Web site provides a free and open educational resource (OER) for educators, students, and self-learners around the world. The contents are course materials such as presentation slides and publications used for teaching many different classes at MIT. Wikipedia is a free well-known online encyclopedia. The outstanding feature of Wikipedia is the use of collaborative concept in which each user is allowed to edit and share the definitions and contents of the posted articles. Currently there are over *1,600,000* articles available in English and many in other languages. Google's Library Project has collaborated with many universities in order to scan and export contents of many books into digitized format. Users are then able to search and retrieve the book contents which match their interests. Besides these examples, there are many other Web sites which provide great information and knowledge and are publicly available.

These information and knowledge resources are however only accessible by users who could connnect to the Internet. Today, the problem known as *digital divide* still exists among people who live in rural or remote areas, especially of developing countries. The digital divide problem can generally be described as the lack of computer equipments, network infrastructure, and knowledge in IT. Therefore, our main goal is to bridge this gap for the users who have difficulties in accessing the Internet. In this paper, we focus on high schools in rural areas of Thailand. Most of these schools do not have the Internet connection. Some schools do have the Internet connection via satellite communication. However the available bandwidth is very limited, thus making the Internet usage impractical. To use the bandwidth efficiently, the teachers often save and share useful Web contents by using a Web browser or some downloading software. However, these software tools do not provide enough functions to help users organize, index and search the contents.

To design some suitable solutions, we have discussed with a group of teachers from many high schools in different regions of Thailand. One of the challenging issues that we discovered is the variation in computer hardwares and network configuration found in different schools. Some schools have low-performance PCs with limited hard disk capacity. These computers are often donated by some universities and organizations. Some fortunate shools have PCs connected to the server via the Local Area Network (LAN) configuration. While some other schools have individual PCs without any form of network connection. To support these cases, we provide two solutions via two different search engine tools: *Sansarn Look!* and *Sansarn Offline*. Sansarn Look! was designed as a Web-based application and is therefore suitable for the client-server model. Sansarn Look! allows Web content and its index to be stored on the server. Users can use a Web browser to browse and search for the content via LAN. On the other hand, Sansarn Offline was designed as a stand-alone application and is therefore suitable for stand-alone PCs without any network connection. Web content and its index are stored on removable media such as CD-ROM. Users can retrieve the content through the interface of the program.

Other important design issue is to provide and distribute the software tools to interested users without any licensing fee. Therefore, our tools are based on the open-source software concept. With the thorough survey of the open-source Information Retrieval (IR) libraries, we found *Lucene* to be a suitable choice for implementing our search engine tools. Lucene is one of the most widely used IR library currently maintained under the Apache project [10]. Lucene provides the core indexing and searching functions which are very efficient and scalable [1]. The key success to the high scalability is the intelligent algorithm in managing the inverted index files between the memory and the secondary storage. The users may set the number of index segments and their sizes according to the system specification in order to optimize the indexing and searching performance. Lucene library, however, only comes with the support of English language. To extend the library to support other languages, developers must add language-specific analysis package into the library. To make Sansarn Look! and Sansarn Offline support both English and Thai texts, we have developed a Thai-language analysis package, *ThaiAnalyzer*, which is integrated into the Lucene package. Integrating and extending Lucene library is very simple and easy, since its creator has designed the framework via the object-oriented programming concept of Java [2]. This object-oriented feature is another reason that makes Lucene a very attractive choice for implementing our tools.

To demonstrate our search engine tools, we have asked a group of volunteer teachers to help select, organize and index some class-related Web contents. The contents and its index will be distributed to high schools in rural areas where the Internet connection is slow or not available. Students can search for useful information from either server's hard disk or some CD-ROMs as if they are connected to the Internet. After the tools and the contents are distributed, we will collect the feedbacks from the teachers and students in order to improve the features for the next version of the tools.

The remainder of this paper is organized as follows. In next section, we explains a case study of designing solutions to support different hardware and network configurations in high schools. Section 3 gives details on Lucene library and the implementation of ThaiAnalyzer package. Section 4 and 5 gives the discussion on the system design and architecture of Sansarn Look! and Sansarn Offline, respectively. Section 6 gives the conclusions and future work.

2   A Case Study of Managing Educational Web Contents for High Schools

One potential application of our search engine tools is to manage educational Web contents and distribute them to schools which have slow or no Internet connection. As mentioned earlier, we have found from the survey and discussion with many teachers that the hardware and network configuration is extremely varied among different schools. Therefore, we have designed our solutions to support all different use cases.

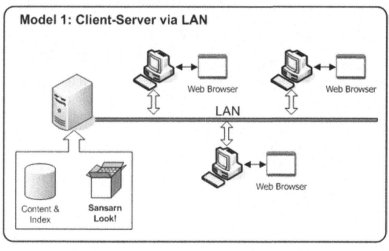

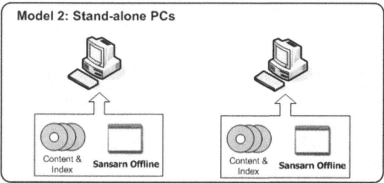

Fig. 1. Hardware and network configuration models

Figure 1 illustrates two solutions according to different hardware and network configuration models. The first configuration model is a typical client-server via LAN. In this model, the administrator installs Sansarn Look! on the server and uses the tool to collect and index the contents. Users on the client PCs can use a Web browser to connect to the server via the *localhost* URL. A user can perform the full-text search by entering query on the Web browser interface. The query is sent to the local server where the search result is compiled and returned to the user. This model offers scalability, sharability and ease of maintenance. Since contents are organized and indexed on the hard disk of the server, they can easily be updated and shared among many users at the same time.

The second configuration model is the stand-alone PCs. Each PC must install Sansarn Offline program. The contents with the index must be recorded and distributed via removable media such as CD-ROM. This model offers great reliability since the query and search results are performed on local PC. Another advantage is that the response time depends only on the I/O activity of the computer which is generally more reliable than the network.

We have asked a group of volunteer teachers to select and organize the Web contents according to the class materials taught in high schools. Typical learning contents are composed of Web pages with some image and other multimedia files such as Macromedia Flash. In Thailand, classes are organized into different subjects such as Thai literatures, English language, physics, biology and chemistry. Web contents often offer better learning approach for students because they are more dynamic than materials available from textbooks. In some contents related to physics, an animation simulating projectile movement can be replayed to help the students understand the concept better. Therefore, it is quite obvious that by offering these offline educational contents to high schools, both teachers and students can benifit more from teaching and learning the class subjects than using only textbooks.

3   Lucene Library with ThaiAnalyzer Package

Lucene was designed and implemented based on the object-oriented programming framework of Java. Figure 2 illustrates the *Lucene* API which is organized into the following four packages.

- *Document*: In Lucene, a text is organized as a document which represents a collection of fields. Before a text document could be indexed, it must be created by constructing a new instance of *Document* class. The seach results are also returned in the form of *Documents*.
- *Index*: Index contains many different classes responsible for indexing process. *IndexWriter* is the main class whose task is to create a new index and add documents into an existing index.
- *Search*: Search contains classes related to search operations. *IndexSearcher* is the main class whose task is to search within the given index directory for matching documents.
- *Analysis*: Analysis contains classes for performing the text processing. Different *Analyzer* classes are provided for various specific tasks such as stopword removal and stemming which are suitable only for English language. To make the analysis applicable to Thai language, we developed *ThaiAnalyzer* package whose basic task is to tokenize Thai texts into a set of words [3].

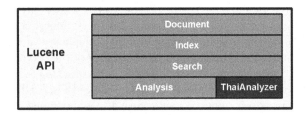

Fig. 2. Lucene API with ThaiAnalyzer package

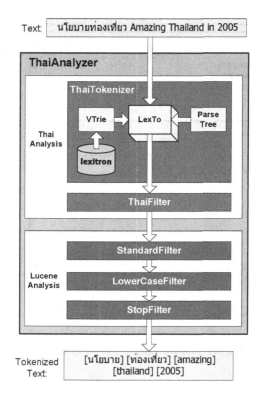

Fig. 3. ThaiAnalyzer package

Figure 3 illustrates the work flow within the ThaiAnalyzer package. ThaiAnalyzer consists of two main analysis packages: Lucene and Thai. Lucene analysis package is originally provided with the library. Thai analysis package is implemented for tokenizing Thai texts. *ThaiTokenizer* is the main class in Thai analysis package which performs the Thai-text tokenization. Our tokenizing algorithm is based on the use of a dictionary, i.e., *LEXiTRON*[9]. To improve the speed of dictionary look-up, words from dictionary are stored by using the *trie* data structure. During the tokenizing process, a text is segmented by looking up the dictionary set stored in the trie. Given a text, *ThaiTokenizer* will perform tokenizing process on Thai texts. English segments and other special characters will be filtered via *ThaiFilter* class. Those English texts and special characters will be further handled by *StandardFilter*, *LowerCaseFilter* and *StopFilter*. LowerCaseFilter will normalize the English token to lower-case characters while StopFilter will remove English stopwords to improve the space efficiency.

4   Sansarn Look! for Client-Server Via LAN Configuration

Sansarn Look! was designed as a Web-based application and is therefore suitable for the client-server model. Sansarn Look! allows Web content and its index to

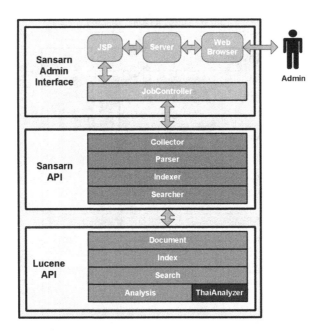

Fig. 4. Sansarn Look!: system architecture

be stored on the server. Users can use a Web browser to browse and search for the content via LAN. Figure 4 illustrates the system architecture of *Sansarn Look!*. The system is composed of three layers: Lucene API (LAPI), Sansarn API (SAPI) and Sansarn Admin Interface (SADI). The source codes in this project is written in Java which provides the platform-independent feature that we need. LAPI is the layer which provides two fundamental IR functions: indexing and searhing. The details of LAPI were given in the previous section.

On top of LAPI is the SAPI which provides the integrated packages by making use of LAPI layer. SAPI is composed of the following four packages.

- *Collector*: To provide flexibility, the *Collector* package contains various classes for collecting documents from either remote servers or local file system. For remote documents, i.e., Web pages, the collector functions exactly as a Web crawler.
- *Parser*: This package contains classes for removing HTML tags or pre-specified XML tags.
- *Indexer*: This package contains classes which perform indexing functions. Indexer package contain higher-level classes which make use of the *Index* package from LAPI.
- *Seacher*: This package contains classes which perform searching functions. *Searcher* contain higher-level classes which make use of the *Search* package from LAPI.

The topmost layer of the platform is the SADI which is composed of necessary components and interfacing functions between users and the system. *JobController* provides classes for managing all related processes of the system. Our system is run on a server called *JBoss* which is an open-source, well-designed and well-implemented technology [8]. To provide an easy-to-use interface, we adopt the *Java Server Page (JSP)* technology which could be integrated seamlessly with any Web browser. Therefore, once installed, users may use a Web browser to control all processes of the system. Sansarn Look! is available for free download from our Web site at *http://sansarn.com/look*.

5   Sansarn Offline for Stand-Alone PCs Configuration

Sansarn Offline was designed as a stand-alone application and is therefore suitable for stand-alone PCs without any network connection [4]. The process of using Sansarn Offline consists of two main tasks: content administration and content retrieval. For the first task, content administrator, i.e., teachers, will use the tool to organize and index Web contents into some predefined categories. The contents and its index are then recorded into some removable media such as CD-ROM. The second task involves user entering search query through the interface. The tool then looks up from the index and returning search results to the user. Since the contents and their indexes are stored on the removable media, the retrieval process can be done without the Internet connection. Another advantage is that the tool could yield search results which are more focused than the ones obtained by using regular search engines. This is because the content administrator would select, filter and collect only high quality Web contents which match the users' information need.

Figure 5 illustrates the overall architecture of *Sansarn Offline*. The system consists of two functions: *content administration* and *content retrieval*. For the first function, there are four modules which support the content administration function: Collector, Parser, Indexer and Categorizer. The first three modules are the same as used in Sansarn Look!. The categorizer module contains classes which perform automatic categorization of collected contents into predefined subjects or categories. The outputs from the first phase are data contents with the index which are then recorded into some removable media.

The second *content retrieval* function consists of two modules which support the retrieval of contents for the users. The searcher module is the same as used in Sansarn Look!. Search Interface module provides a GUI-based interface which connects to the searcher module in order to pass the query from the users and to return the search results back to the users.

Sansarn Offline is implemented in Java by using Eclipse tool [6]. Using Java offers several advantages including platform-independent and open-source development. Sansarn Offline allows users to browse according to the organized subjects and to perform the full-text search on the contents. The returned search results show the snippets with the highlighted terms similar to typical search engines. User can click on the link to view the full page content. Our tool also

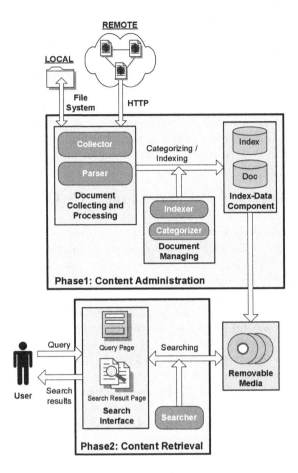

Fig. 5. Sansarn Offline: system architecture

allows the users to set some configuration such as the color used to highlight the terms in the results and number of results shown per page. More details of Sansarn Offline can be found from our Web site at *http://sansarn.com/offline*.

6   Conclusions and Future Work

We proposed two search engine tools: *Sansarn Look!* and *Sansarn Offline* to support managing educational Web contents among high schools with different hardware and network configurations. Both softwares are freely avaiable and open-source which were developed by using *Lucene* IR (Information Retrieval) library. To support the retrieval of Thai texts, we developed a Thai-language analysis package called *ThaiAnalyzer* whose task is to segment Thai written texts into word tokens. Sansarn Look! was designed as a Web-based application and is therefore suitable for the client-server model. Sansarn Offline was designed as a stand-alone application and is therefore suitable for stand-alone PCs without any network connection. Both tools allow searching for contents stored on either server's hard disks or from the removable media, therefore the retrieval process could be done offline or without the Internet connection.

Our future work includes the plan to select, index and distribute more useful copyright-free Web contents. One of the potential contents is the Wikipedia Selection for Schools [13] This Selection is about the size of a *15* volume encyclopaedia with *24,000* pictures, and articles on *4,625* topics. The articles have been cleaned up and checked for suitability for children. However, the current version of the content can only be accessed by browsing on subject index and title word index. By using our tools, users can perform the full-text search function on the contents.

Acknowledgement. This work is part of an ongoing project, Multimedia Search System (VG5001), which is supported by the Knowledge Engineering program, under the National Science and Technology Development Agency (NSTDA), the Ministry of Science and Technology, Thailand.

References

1. Cutting, D., Pedersen, J.: Optimization for dynamic inverted index maintenance. In: Proc. of the 13th Int. ACM SIGIR conf., pp. 405–411 (1989)
2. Cutting, D., Pedersen, J., Halvorsen, P.: An object-oriented architecture for text retrieval. Proc. of Intelligent Text and Image Handling (RIAO 91), 285–298 (1991)
3. Haruechaiyasak, C., et al.: Sansarn Look!: A Platform for Developing Thai-Language Information Retrieval Systems. In: Proc. of the 21st International Technical Conference on Circuits/Systems, Computers and Communications (ITC-CSCC 2006), pp. 85–88 (2006)
4. Haruechaiyasak, C., Sangkeettrakarn, C.: Sansarn Offline: A Search Engine Tool for Managing, Archiving and Retrieving Offline Web Contents. In: Proc. of the Int. Conf. on Electrical Engineering/Electronics, Computer, Telecommunications and Information Technology, pp. 1034–1037 (2007)
5. BBC - Learning, http://www.bbc.co.uk/learning/
6. Eclipse - An Open Development Platform, http://www.eclipse.org
7. Google's Book Search: Library Project - An enhanced card catalog of the world's books, http://books.google.com/googleprint/library.html
8. JBoss Enterprise Middleware Suite, http://www.jboss.com/ products/index
9. LEXiTRON Thai-English Dictionary, http://lexitron.nectec. or.th
10. Overview - Apache Lucene, http://lucene.apache.org/java/docs/ index.html
11. MIT OpenCourseWare, http://ocw.mit.edu
12. Wikipedia: The Free Encyclopedia, http://wikipedia.org
13. Wikipedia Selection for Schools, http://schools-wikipedia.org

Presentation Lag Reduction by Scheduling Media Objects for Auto-assembled Multimedia Presentations from Educational Digital Libraries

Feng-Cheng Lin[1], Chien-Yen Lai[1], Pai-Hsun Chen[2], and Jen-Shin Hong[1,*]

[1] Department of Computer Engineering, National Chi-Nan University, Taiwan
[2] Department of Computer Engineering, Nan-Kai Institute of Technology, Taiwan
jshong@ncnu.edu.tw

Abstract. This study investigates the optimization of the ordering of retrieved media objects from educational multimedia repositories for a continuously-played presentation such that the total presentation lag through a slow network is minimized. We present a computation-efficient NEH-based heuristic algorithm that can obtain a near-optimal solution with minimal computation time. A simulation experiment shows the average gaps between the idle rate of heuristic solutions and randomly generated sequences are around 26.3%. The results indicate the proposed heuristic solution can significantly reduce the presentation lag as compared to a random ordering approach that is commonly applied in conventional multimedia repositories.

Keywords: Multimedia presentation, repository, scheduling, flowshop.

1 Introduction

Over the last decade, there is a tremendous growth in education resource repositories of learning materials. Typically, a query to a specific educational learning material repository often retrieved a bag of relevant multimedia items. To present the retrieved media objects, while many repositories typically support interfaces for a user to "click and play" the interested items one by one for downloading and presentations, there are also interests that aim to dynamically compose the media items selected into a continuously-played TV-like multimedia presentation. A continuously-played presentation particularly suits well for hand-held portable devices with which the input interfaces are usually less easy to operate than a mouse. In a dynamically generated presentation, often, the multimedia servers randomly or chronologically push the content to the users. Since there is a delay between the points when the user clicks on a presentation and when it is played, the user often experiences certain idleness while waiting for the transmission of the media objects. To present an auto-assembled multimedia document online, an important concern for optimally scheduling the objects is the "total latency" experienced by the user during a presentation. The presentation latency is particular important when the multimedia

---

* Corresponding author.

D.H.-L. Goh et al. (Eds.): ICADL 2007, LNCS 4822, pp. 454–457, 2007.

presentation is delivered through a low-bandwidth communication channel, which might be caused by bandwidth limitation in the backbone network, a busy server, or a client with a slow "last-mile" connection. To cope with such a delay issue, a commonly used strategy is by the "streaming" technology. However, in a slow network environment where the media data consumption rate is higher than the data transmission rate, there are still intermittent idle durations during a streaming-based presentation. Besides, while the streaming technology has been widely and successfully applied in commercial web sites, its applicability in real-life application in campuses is somehow more restricted. In reality, many teachers do not have knowledge and tools to author and distribute streaming-based multimedia objects.

The media objects in a typical educational repository cover a wide spectrum of modalities with different formats including text, images, audio, video, and vector graphics. For a same amount of data transmitted, the anticipated presentation time for different media types actually differs drastically. How to order them in a prefetch-enabled continuously-played presentation to reduce the total presentation latency in a bandwidth-limited environment is interesting to investigate. This paper explores techniques for optimizing the ordering of media items to provide better Quality of Service (QoS). The sequence optimization techniques can be integrated into a server for dynamically assigning the order of the selected media items to be delivered. In the following, the problem will be formulated and solved using techniques adapted from conventional studies in operational researches.

2 A Heuristic Solution for Approximating the Optimal Sequence

In principle, the media scheduling problem addressed here can be mapped to a "two-machine flowshop problem" [1-2] that aims to optimize a scheduling of sequential jobs to be processed on two machines. The objective is to minimize the completion time of all jobs. Table 1 gives a mapping of the parameters between the media scheduling problem and the two-machine flowshop problem. In this study, we present a computation efficient heuristic algorithm that can be used to generate feasible near-optimal approximate solutions. The performance of the heuristic approach will be evaluated empirically using simulations. In the following, we describe the heuristic approach for solving this sequence optimization problem with a limited buffer size.

Table 1. Correspondence between media sequence problem and two-machine flowshop problem

Sequence optimization problem	Two-machine flowshop problem	Notation
A set of n media items	A set of n jobs	M
A media item	A job	m_i
Server	First machine	$server$
Client	Second machine	$client$
Transmission time for a media item	Processing time on the first machine	a_i
Playback time for a media item	Processing time on the second machine	b_i
Completion time for a media item	Completion time for a job	c_i
Presentation span	Makespan (maximum c_i of M)	C_{max}

2.1 NEH-Based Approach for Real Time Optimization

A recent study [3-4] compared 25 well-known heuristic approaches for obtaining near-optimal solutions for flowshop problems and showed that the NEH approach frequently outperforms other approaches on several benchmark problem sets. Furthermore, the NEH approach is rather straight forward and easy to implement. Therefore, we have applied the NEH heuristic solution for the media sequence optimization problem. We outline the major steps of the NEH heuristic as follows.

Step 1:   For each m_i, compute $c_i=a_i+b_i$. Sort the media jobs by non-increasing c_i.
Step 2:   Take the first two media items in the sorted M, and order the two items such that the partial completion time of the first two items is minimized.
Step 3:   Insert the 3 rd jobs into the partial schedule in the previous step. Since there are two existing jobs in step 2, there are three possible positions for the insertion of the 3rd job. Among these 3 positions, select the position which minimizes the partial completion time under the given buffer size.
Step 4:   For $k=4$ to n do
          Insert the k-th job into the previous partial schedule. There are k possible positions for this insertion. Select the position which minimizes the partial completion time under the given buffer size.
Step 5:   Output the final sequence from step 4.

2.2 Performance Evaluations for the NEH Heuristic Solutions

This section presents the computational experiments designed to evaluate the effectiveness and efficiency of the proposed NEH-based heuristic approximations for searching near-optimal media sequences. For each job set, the processing times a_i and b_i were randomly generated numbers ranging between 1 and 100. The transmission bit rate is 160KB/s. The simulation codes were written in C++ language, and the experiments were performed on an IBM xSeries x206m computer. We conducted a series of computational experiments with different problem sizes, from 15 to 50 media items. The problem generation procedure yields the test problems that possibly encompass a wide variety of real life scenarios of online multimedia applications. For each experiment, 50 different media sets are used. Three different buffer size constraints, 16,000 KB, 22,400KB, and 30,720 KB respectively, were given for the experiments. A typical computation time using the NEH algorithm for problems with 20 media objects is less than 0.0005 seconds of CPU time. For each media set, the idle rate is defined as the ratio of the total idle time and the total transmission time of the media items, that is,

$$Idle\_rate = \frac{C_{max} - \sum b_i}{\sum a_i} * 100\% \qquad (1)$$

An *Idle_rate* close to 1 refers to a case where the media items are badly ordered such that the playbacks are mostly halted during the downloading time. Table 2 lists the *Idle_rate*s of the NEH heuristic solutions and randomly generated sequences in problems with different number of media items. The average *Idle_rate*s of the NEH solutions are 45.2%, 14.4%, and 6.3% for problems with 16,000KB, 22,400KB, and 30,720KB buffer constraints respectively. The average *Idle_rate*s of the random (RAN) are 75.0%, 47.6%, and 23.0% respectively. The average gaps between the *Idle_rates* of NEH and RAN are around 26.3%. These results indicate the NEH solutions significantly reduce the presentation lags as compared to random sequences.

Table 2. Computation results on the *Idle_rate* (%)

n	Buffer	16,000KB	22,400KB	30,720KB	n	Buffer	16,000KB	22,400KB	30,720KB
15	NEH	49.9	18.8	9.2	35	NEH	42.5	13.0	5.9
	RAN	76.1	49.7	26.1		RAN	73.7	46.5	22.5
20	NEH	46.3	15.5	6.4	40	NEH	43.0	11.9	5.2
	RAN	74.9	46.3	25.3		RAN	74.8	44.9	22.3
25	NEH	45.9	16.4	7.5	45	NEH	46.1	12.9	4.9
	RAN	74.8	46.6	23.8		RAN	76.3	46.8	21.0
30	NEH	46.6	15.7	8.0	50	NEH	40.9	10.8	3.6
	RAN	76.0	48.1	23.4		RAN	73.4	44.1	19.6

3 Conclusions

This study investigated the optimization of the sequences of retrieved media objects such that the total presentation lag of a continuously-played multimedia presentation through a slow network is possibly minimized. Aiming for the real time online applications, we present a computation-efficient NEH-based heuristic algorithm that can obtain a near-optimal solution with minimal computation time. Overall speaking, the average gap between the idle rates of heuristic solutions and randomly generated sequences is around 26.3%. These simulation results indicate that the NEH-based heuristic solutions can significantly reduce the presentation lags as compared to a random ordering approach which is commonly applied in conventional multimedia applications.

Acknowledgments. This work is supported by the National Science Council of Taiwan, under grant NSC-94-2422-H-260-002.

References

1. Pinedo, M.: Scheduling Theory, Algorithm, and Systems, 2nd edn. Prentice Hall, New Jersey (2002)
2. Graham, R.L., Lawler, E.L., Lenstra, J.K., Kan, R., A.H.G.: Optimization and approximation in deterministic sequencing and scheduling: a survey. Annals of Discrete Mathematics 5(1), 287–326 (1979)
3. Nawaz, M., Enscore, E., Ham, I.: A heuristic algorithm for the m-machine, n-job flow shop sequencing problem. OMEGA, International Journal of Management Science 11, 91–95 (1983)
4. Ruiz, R., Maroto, C.: A comprehensive review and evaluation of permutation flowshop heuristic. European Journal of Operational Research 165, 479–494 (2005)

Metadata and Organizational Structures in Personal Photograph Digital Libraries

Sally Jo Cunningham and Masood Masoodian

Computer Science Department, University of Waikato, Hamilton, New Zealand
{sallyjo,masood}@cs.waikato.ac.nz

Abstract. We examine the 'native' metadata and organizational structures that individuals create for their personal photo digital libraries, by analyzing the behavior of photo collectors as recorded in 37 autoethnographies and ethnographies. The findings confirm several common assumptions about how people organize their photos that have been the basis for features in earlier photo digital libraries—that photos are commonly organized by time, event, and location, and that collection owners create very little metadata manually. We discuss alternate sources of metadata that arise as a consequence of sharing photos, and consider additional features for photo digital libraries that may be useful in supporting searching and browsing of personal collections.

Keywords: Personal digital libraries, photo metadata, user studies.

1 Introduction

Digital camera users take more photos than did film camera owners: digital cameras free users from the expense of developing film and creating prints, and offer instant feedback on each shot. Digital photos can be inexpensively stored as well—on CDROM, DVD, hard drives, and online archives. We can afford to create personal photo archives that document our lives, our hobbies, and indeed any passing interest.

Locating an existing photo in our collection and maintaining a sense of the collection's contents, however, may not be straightforward. Which PC did we store Uncle Bob's photo on—or was it on the hard drive that failed? How can we easily search our meagerly labeled photo directories to find my uni graduation shots? And this photo of a party: is this an event I attended, or did someone else send it to me?

A number of novel searching, browsing, and collection management features have been prototyped to address one or more of the above situations (Section 2). In this paper, we explore the ways that people currently organize their digital photo collections. Our work is based on a large-scale ethnographic investigation of personal photo management (37 participants, over 150 pages of observational summaries; Section 3). We examine analyze of the metadata and photo collection organization schemes created by participants, and suggest additional metadata that might be useful to these users (Section 4). We conclude (Section 5) by identifying photo management behavior that can suggest additional software support for personal photo collection digital libraries.

D.H.-L. Goh et al. (Eds.): ICADL 2007, LNCS 4822, pp. 458–467, 2007.
© Springer-Verlag Berlin Heidelberg 2007

2 Previous Work

Earlier work on personal photo digital libraries has focused on using a commonsense understanding of how and when people take photos to inform the design of novel features specific to personal photo collections. For example, the insights that photos are taken in a time-linear order and that people tend to take several snapshots of the same event or person in a brief period of time, have led to experimental systems using photo timestamps as a basis for browsing structures [7], using time and image content to automatically cluster photos into 'events' [2], or using GPS location information to identify geographically proximate photos [11]. Since these are personal photos, the collection owner is expected to recognize the images—and so browsing, rather than searching, is anticipated to be the primary means of locating specific photos in the collection. A number of systems have been developed to enhance the layout and selection of representative thumbnails for browsing (eg, [7]). The insight that an individual tends to take multiple photos of the same people and places over time can be exploited by using time and location metadata in previous, annotated photos to develop metadata for new, un-captioned photos [9].

However, personal photo collections can rapidly become too large for browsing to be practicable as the main access method. Digital cameras enable their owners to take more photos than was feasible with film cameras, since the cost of film and processing are eliminated, and printing is optional. Content-based searching relies on image processing and pattern recognition to retrieve images based on similarity to a query specification (often in the form of a sample image); query features are usually limited to color, shape, texture, and spatial relations between image components [13]. These features rely on the user's ability to describe the contents of the desired image—and so content-based searching is most useful for locating the photos that could also be recognized while browsing.

Annotation tools based on manually created metadata are less commonly presented in the digital libraries research literature—possibly because of another assumption about personal photo-taking habits, that photos are rarely labeled by their owner. This assumption is confirmed by recent studies ([12], [4], [8]).

Public image collections usually rely on text-based rather than content-based searching, where the search metadata may be manually created or (more commonly) leveraged from text associated with the image (for example, a figure title, or document contents proximate to the image). Image management tools created for large scale, public photo libraries may be expected to be less relevant to the needs of the individual, managing a personal collection. The users of a public collection are relatively unfamiliar with its contents, whereas the photographer is the creator and curator of the individual collection—and so the user needs and requirements for the personal collection will be different from those of the general-purpose photo libraries.

3 Data Collection

Data for this study was gathered through a project assigned to undergraduate students in an upper level university human-computer interaction course. The students' goal was to design and prototype a shared, online photo collection—essentially a digital library of personal photographs. The students based their designs on ethnographic investigations into the photo taking and sharing habits of themselves and at least one friend. These investigations were summarized, and these summaries are analyzed in this present paper. In total, the students conducted 18 personal ethnographies and 19 observations/interviews of another person, and the summaries come to over 150 pages (Table 1). To preserve anonymity, each participant is referred to with a randomly assigned letter of the alphabet.

The students first performed a 'personal ethnography' or autoethnography [3], examining their own photograph collections and photo-taking behavior. The students then performed a similar ethnographic observation and interview of a friend. In these investigations, the students describe their (or their informant's) photo collection's contents, when and under what circumstances they take new photos, how those photos are organized, metadata associated with the photos, and the ways in which photos are exchanged (e.g., by showing off photo albums, posting photos online, or sending photos via mobile phone). 'Organization' of a photo or collection was construed as broadly as possible, so as to capture as many aspects as possible for that behavior; in these investigations organization included creating physical and digital albums, storing photos on CD, tossing print photos in a drawer, and so forth. In this work, we focus on the behaviors associated with managing digital photo collections, rather than the now-historic management techniques for print photos.

An overview of photo organization behaviors, as described in the ethnographies and autoethnographies, is presented in Section 4. Grounded theory methods [14] were used to analyze the students' summaries of their interviews and observations. We analyze the summarized descriptions of photo taking, sharing, and storage/organization as reported in the ethnographies, rather than the students' own analyses and suggestions for photo digital library features for their projects.

Table 1. Gender and nationality of participants

	Male	Female	National Origin	Count
	22	15	NZ	17
			China	16
			Iran	2
			Korea	2

4 'Native' Metadata and Organizational Structures

In this section, we describe the types of metadata and organizational structures that are reported in the ethnographies.

Fig. 1. (a) Raw ingredients. (b) 'Steamed chicken on balsam pear'.

4.1 Group Labels and Individual Photo Labels

The most striking feature of the manually created photo metadata is that it is primarily applied to groups of photos, rather than to individual photos. People tend to take multiple shots, rather than individual snaps, and these groups of photos are downloaded and stored together. The folder that a group of photos are stored in is given a descriptive name (for example, "Wellington"). The filename of individual photos may be left at the default (eg, "Image12.JPG", "P80080003.JPG"), or the file may be renamed to be slightly more evocative (eg, "View 1", "View 2", etc. for photos of scenery taken on a trip; Participant I).

Many of the folder labels are 'events' in the sense used by Cooper et al [2]: a group of photos that are temporally clustered together (and that are temporally distant from other clusters), and that depict the same activity or location. A temporal cluster can be relatively tightly grouped (eg, "Wedding") or can include extended periods (eg, "Wellington trip"). Occasionally hierarchies of folders will be created (for example, to group together different days of a lengthy trip). Hierarchies very rarely are more than two folders deep.

Folders that are not event-related may simply be a group of photos that are downloaded from the camera at the same time (and so represent a number of un-labeled groups). These 'miscellaneous' folders might not be manually labeled, and may simply retain the default label of the photo management system. For these photos, the photo timestamps or the folder creation date may be the main access point.

Location-related labels may be records of a specific trip (and so represent an event, as well as a specific place), or may be (possibly temporally distant) images of a place that holds significance to the collection owner. Examples of location labels are "Home town" or "XXX University".

More rarely, a folder may be labeled with a theme that represents a hobby or special interest of the collection owner. These can be deeply idiosyncratic. Participant E's informant, for example, is a hobbyist cook; she takes photos of the raw ingredients and the final dish when she creates a meal "she could be proud of" (Figure 1a, b), with the cooking – related photos stored together across different dates.

Affect, or the emotional impact of a photo, also appears as a category for organizing photos. Only two emotion–related categories appear in the ethnographies:

sentimental (associated with close family members or romantically themed); and humorous or quirky. Sentimental photos are usually grouped with other photos associated with the individual (eg, "my mother"). Humorous photos may be part of a themed collection: for example, Participant K's informant "takes photos of 'engrish' – amusing mistakes in the use of the English language by foreign language users". Funny photos are also frequently reported as being shared with others by email or SMS, accompanied by a brief description of the comical aspect of the image ("hey look at this hilarious picture of my cat sleeping on my dog"; Participant M).

4.2 Time

Participants almost uniformly reported that they appreciated the association of a timestamp with a photo, and used the timestamps to browse along their personal timeline of activities. Without a timestamp, it can be difficult to distinguish between similar events (eg, is this a photo of last week's party, or last month's?). It can be annoying, however, to have the timestamp appear on the photo itself ("because it can make a really good picture look bad with the red/white date in the bottom of the screen"; Participant Q). The preference is for having the timestamp—and indeed, all metadata—viewable separately from the photo, so that the image is not spoiled.

A timestamp indicates when a photo was snapped. A second aspect of time is duration—how long it's been since that photo was taken. While raw timestamps are useful in sorting photos for browsing, an indication of duration may be useful in appreciating an individual photo. Participant G, for example, notes that for him the timestamp "serves dual functions, first of all, it tells me when I took this picture [of his mother] and secondly, the most of all, it also reminds me how long I have been away from home or my dear loving mother".

4.3 Verbosity of Metadata

Photo labels tend to be brief—a word or phrase, perhaps simply a few characters in a filename, rather than an exhaustive description of the people, places, events and so forth appearing in the image. Exceptions occur, with some photos given much more extensive metadata. Some photos may be selected to form a sort of visual, personal diary; these selected images are given more detailed captions ("I want to know exactly which Christmas that was, what club that concert was in, and so on"; Participant K). Photos that are intended to be shared are more likely to have manual descriptions, and these descriptions tend to be more detailed than those of photos retained for personal viewing only. The intention is to provide an explanation and interpretation of the photo for the viewer, ("if the photo shows people who were doing something and it isn't clear for the person who will see that photo, I add a brief explanation" (Participant L); "so that other people who happen to look at my photos know who the people in them are, or what we're doing, or where we are" (Participant Q)). At times this more detailed metadata can be critical: Participant K's informant, for example, posted on her website a photo in which it appears that several people are beating up one of their friends, and so "in the description, she explained that the fight was only a pretend one".

Photos shared via email or SMS usually have an accompanying message that explains the significance of the image to the photo taker. Even very brief messages can contain a surprising amount of contextual information. For example, the SMS message accompanying a photo of shelves of books ("This is one spot of the Waikato University's library! I read very hard there!" Participant T) includes the location (*Waikato University, library*) and the activity (*read*). Email messages tend to have much lengthier, and much richer, descriptions of attached photos.

While participants recognized that they infrequently recorded manual metadata on print photos, they also felt that they recorded longer captions on physical prints. Likely explanations for these phenomena are that captions on physical prints are usually created prior to archiving the print in a formal album, or to giving the print to someone else. The ease of writing on the back of prints—keeping the metadata bound to the photo, but not spoiling the image—may also be a factor.

Why is manually created metadata for digital photos usually so limited—or non-existent? One reason, it seems, is that photo takers tend to over-estimate their ability to remember the context surrounding their snapping a picture: a typical explanation for a lack of captioning is that "I did not note down when and where I took the pictures since I thought that it would not be necessary and I could remember" (Participant I). Several of the participants were shocked to realize while creating their autoethnographies that their memories, even of relatively recent events, were fallible:

> As I have just looked through my photo collection I am aware that I cannot remember what is happening in many of the photos. I also have no idea when most of the photos were taken, I can only guess—because of the chronological nature of the organisation of the photos and how old that people look in the photos. (Participant J)

Another impediment to the manual creation of metadata is the disjunct between creating a photo and adding it to the photo collection; a photo cannot be captioned until it is transferred from the digital camera to the photo storage system, and at that point the relatively large number of photos being transferred makes creating metadata seem a daunting task. The exception is the labeling of photos taken on a mobile; the user is generally taking only one or a few photos at a time, the mobile has a built-in keyboard or keypad for recording metadata, and the point of creating the photo is often to send it to someone else (with a brief message that can serve as a caption).

And, of course, it can simply be difficult to motivate oneself to put effort into organizing and providing metadata for a photo collection, when the payoff for this effort will be in increased ease locating particular photos in the distant future ("Also, I am too lazy to write down some information about those photos beside them [T]"). This behavior is hardly restricted to managing photo collections; most users have poorly maintained, badly organized, and cryptically labeled file storage structures on their personal computers [10]. An effective photo metadata scheme, then, would make it easy to create labels at the most convenient time for the user—perhaps at the time the photo is being taken—and would also be forgiving enough to automatically create metadata (for example, by image content analysis).

Fig. 2. (a) Photo taken in Guild Wars, 'as a tourist' at Perdition Rock (Participant K). (b) 'An action photo' of Participant K at Augery Rock, in Guild Wars.

4.4 Technical Metadata

Serious photographers may wish to have a record of the technical details of their photos: camera model, exposure time, environmental and lighting conditions, etc. These details may be held for personal use, or may be associated with photos that are intended for distribution, as they may be useful in critiquing the photos (Participant B notes that his informant believes that "Sharing photos [and their associated technical metadata] can improve his photography skill"). Note that some of these details could be automatically harvested from the camera, as a shot is being taken.

Photo filesize can be useful in deciding how to transfer a photo, or group of photos, to another person—can they be sent as an attachment, or will they break the mailer? Can an image be sent via PXT? Is my online photo account reaching its size limits?

Photo resolution is useful in selecting photos for printing or display as computer 'wallpaper'. Users may not be technically savvy enough to predict what resolution is appropriate for different uses. This understanding isn't critical for selecting wallpaper—the user can easily experiment with different background images—but it can be expensive and disappointing to print A3 copies of low-resolution photos.

Participants reported occasional use of Photoshop or other photo processing software to eliminate minor imperfections (for example, removing red eye), or to improve the appearance or impact of an image (by cropping, altering contrast, or removing people or objects extraneous to the focus of the photo). Serious photographers would want to document their processing of the raw photos.

'Photoshopping' is also used to radically alter photos, generally for creative or humorous effect. Sepia tinting is particularly popular, to make photos appear old, and converting to black and white makes an image appear more 'artistic' (Participant C). Swapping heads on bodies, including celebrities in photos of oneself and friends, or adding objects ("classic example is you can stamp someone's cheeks with lipstick marks"; Participant D) are stock techniques in the creating of comic photos. Rarely, an image may be 'photoshopped' for more serious reasons; Participant T regretted having missed his graduation ceremony, and so he added himself to his copy of the official photograph of the graduating group. While many of these alterations are so outrageous or conspicuous as to be easily detected by the viewer, subtle changes to content can be difficult to spot. Again, a record of modifications would be appropriate, for example to aid in maintaining an accurate record of one's activities.

Some photos may be of entirely virtual composition. Online communities offer the opportunity for online experiences that are compelling enough to capture; the 'photos' can be snapped from a first person point of view (Figure 2a) or third person (Figure 2b). These images can be shared within the game or other role-playing environment, and can also be mixed in with digital photo collections in the 'real' world ("I even used both of the images as my desktop wallpapers"; Participant K).

4.5 Heterogeneous Sources and Heterogeneous Storage Destinations

Although the photo collections described in the ethnographies were 'personal' (in the sense of belonging to one individual), one striking aspect of the collections was that the owner did not necessarily take the photos. Photos are frequently shared [1]— people create photo CDs as gifts, email photos of family events to relatives too far away to attend, and use mobile phones to snap and send quirky shots to friends. This situation suggests that browsing may not be sufficient for locating photos taken by others, since the collection owner will be less familiar with these images.

The participants reported that their digital collections tend to be distributed: some photos are on CDs or DVDs, the majority on a hard drive in a photo organizer, a few on their mobile, others in an online archive (perhaps for sharing with overseas friends and relatives), still more photos held within their email system as attachments. Finding a particular photo, then, involves first remembering how and where it is stored, and then recalling the particulars of that system's interface.

But a distributed (and duplicated) photo collection also has can be an advantage: Participant W's informant describes an exceptionally bad experience: '...where their harddrive suffered a melt down. Nothing was recoverable without paying an exorbient [sic] fee to professionals, subsequently many photos were lost. They now backup the images onto CD from time to time to safeguard against a reoccurrence." A physically distributed photo collection is at present a practical necessity; storing a photo collection across several media or storage systems provides a backup in case of hardware failure. Additionally, most participants recognized that storing photos in more than one software system allows the collection owner to separate out photos for personal viewing only from photos that are share-able with others (for example, by placing some photos in an online archive).

Given these arguments for the continuance of distributed collections across multiple photo organization systems, a meta-organizing system would be welcome. One such prototype, developed at Hewlett-Packard Laboratories, supports searching of photos across both local and remote computers, and maintains links between multiple copies of a photo [5].

5 Conclusions

Earlier research on personal photo collection digital libraries is based on assumptions about how people 'naturally' organize their photo collections: by time, by event, and by location. This paper provides evidence that these assumptions are valid; the students and their informants did indeed report that significant portions of their personal photo collections were organized along these lines.

For digital photo collections, these organizing principles are frequently enforced by the file storage structures. Participants tend to label groups of digital photos, rather than individual photos. This behavior suggests that the most practical tools for locating photos within a collection will be based on browsing, rather than searching; the emphasis should be in identifying the group(s) most likely to include the desired image, and then aiding the user in efficiently browsing them.

It may be possible to automatically harvest some types of useful metadata. The timestamp for a photo allows the user to temporally order the collection for browsing. Different ways of recording the timestamp—for example, as the length of time since the photo was taken—may be helpful in appreciating a specific photo. Photo enthusiasts may desire technical details on camera settings and post-processing. Coupling a digital camera with a GPS could help the user identify photos of a single location that are scattered through time [11], although using GPS data to create human-readable location labels is still challenging. When we share photos by email or SMS, the email or SMS itself is a potentially rich source of metadata. These messages could supply keywords for searching, and the structure of the message itself can also be useful (for example, to record the sender and receiver of the photo).

The highest quality (in the sense of the most personally meaningful) metadata is that which the photo collection owner manually creates. Unfortunately, our participants rarely added more than a few terms to an aggregation of photos, primarily as a folder label. Very few individual photos were annotated. This behavior is not a sign of atypical laziness in our participants; other studies have noted a similar reluctance to formally organize and annotate their photos ([4], [12]). Given these behaviors, it appears that an annotation tool will be used only if it requires very little effort on the part of the user—if, for example, the user can apply metadata to variably sized groups of photos, rather than having to label each photo individually.

Participants recognized that they needed descriptive metadata for their photos both to locate specific images and to appreciate the photos whose contexts were being lost to imperfect human memory. Since photos may be given to others, there is also a concern that the recipients may not store descriptions of photos. But participants are reluctant to record captions visibly on the photo, as this detracts from the image. Ideally, the metadata would be separate from, but bound to, a photo, so that the metadata would 'follow' the photo as the photo is copied or passed along to others ("'captions should be allowed to stick with the picture" [S]).

When photos are grouped into labeled folders, the clustering principles are broad, and primarily based around events ("Wedding", "Birthday"), places ("Wellington"), or people ("Mother", "My Girlfriend"). But a classification scheme based solely on a photo's location limits the access points unnecessarily; a photo of a girlfriend in Wellington could logically fit into two folders, but will be physically stored in only one. The ability to easily (and we must stress "easily", as very few people appear to possess the discipline or desire to consistently catalog their photos) create virtual folders would allow the user to place a given image into multiple browsing categories.

Creating additional virtual folders (or more conventionally, 'albums') may be a better metaphor than providing textual tags, as albums are visually structured to suggest browsing rather than searching—and browsing groups of related photos is by far the more frequent activity for a personal photo collection.

In the short term, the paucity of metadata does not pose a problem for managing photo collection. We can rely on our memory of the approximate time the photo was taken to locate it, and we are also confident that we can remember the significant details of people, places, and activities depicted. In the long term, however, our memory inevitably fails us—and we end up with a hard drive full of unidentifiable images. At that point, we are grateful for any metadata that helps us recall the context of our photos, to recover our chronicle of our life. Any tool that encourages us to record photo metadata, or that automatically collects metadata, will be appreciated.

References

1. Adams, A., Cunningham, S.J., Masoodian, M.: Sharing, Privacy and Trust Issues for Photo Collections. Working Paper 01/2007, Computer Science, Waikato Uni. (2007)
2. Cooper, M., Foote, J., Girgesohn, A., Wilcox, L.: Temporal event clustering for digital photo collections. ACM Transactions On Multimedia Computing, Communications, and Applications 1/3 (2005)
3. Cunningham, S.J., Jones, M.: Autoethnography: a tool for practice and education. In: Procs. of the 6th New Zealand Int. Conf. on Computer-Human Interaction, pp. 1–8 (2005)
4. Frohlich, D., Kuchinsky, A., Pering, C., Don, A., Ariss, S.: Requirements for Photoware. In: Proc. CSCW 2002, pp. 166–175
5. Gargi, U., Deng, Y., Tretter, D.R.: Managing and Searching Personal Photo Collections. HP Labs Technical Report HPL-2002-67, p. 9 (2002)
6. Glaser, B., Strauss, A.: The Discovery of Grounded Theory, Aldine, Chicago IL (1967)
7. Graham, A., Garcia-Molina, H., Paepcke, A., Winograd, T.: Image and cultural digital libraries: Time as essence for photo browsing through personal digital libraries. In: Proceedings of the ACM/IEEE Joint Conference on Digital Libraries 2002, pp. 326–335 (2002)
8. Mills, T., Pye, D., Sinclair, D., Wood, K.: ShoeBox: A Digital Photo Management System. AT&T Laboratories, Cambridge (2000)
9. Naamen, M., Yeh, R.B., Garcia-Molena, H., Paepcke, A.: Leveraging context to resolve identity in photo albums. In: Proceedings of the ACM/IEEE Joint Conference on Digital Libraries 2005, pp. 178–187 (2005)
10. Nardi, B.A.: A small matter of programming: perspectives on end user programming. MIT Press, Cambridge, Mass (1993)
11. O'Hare, N., Gurrin, C., Lee, H., Murphy, N., Smeaton, A.F., Jones, G.J.F.: My Digital Photos: Where and When? In: Proceedings of Multimedia 2005, pp. 261–262 (2005)
12. Rodden, K., Wood, K.R.: How do people manage their digital photographs? In: Proceedings of CHI 2003, pp. 409–416 (2003)
13. Smeulders, A.W.M.: Content-Base Image Retrieval at the End of the Early Years. IEEE Transactions on Pattern Analysis and Machine Intelligence 22/12 (December 2000)
14. Strauss, A., Corbin, J.: Basics of Qualitative Research: Grounded Theory Procedures and Techniques. Sage, Newbury Park (1990)

Building a Directory for the Underdeveloped Web: An Experiment on the Arabic Medical Web Directory

Wingyan Chung[1] and Hsinchun Chen[2]

[1] Department of Operations and Management Information Systems, Leavey School of
Business, Santa Clara University, Santa Clara, CA 95053, USA
[2] Department of Management Information Systems, Eller College of Management,
The University of Arizona, AZ 85721, USA
wchung@scu.edu, hchen@eller.arizona.edu

Abstract. Despite significant growth of the Web in recent years, some portions
of the Web remain largely underdeveloped, as shown in a lack of high quality
content and functionality. An example is the Arabic Web, in which a lack of
well-structured Web directories has limited users' ability to browse for Arabic
resources. In this research, we proposed an approach to building Web directories for the underdeveloped Web and developed a proof-of-concept prototype
called Arabic Medical (AMed) Web Directory that supports browsing of over
5,000 Arabic medical Web sites and pages organized in a hierarchical structure.
We conducted an experiment involving Arab subjects and found that AMed directory significantly outperformed a benchmark Arabic Web directory in terms
of browsing effectiveness and user ratings. This research thus contributes to
developing a useful Web directory for organizing information of the Arabic
medical domain and to better understanding of supporting browsing on the
underdeveloped Web.

Keywords: Browsing, information seeking, Arabic, medical domain, meta-
searching, Web directory, underdeveloped Web, user study, experiment.

1 Introduction

Internet usage has been growing rapidly in recent years, especially in many developing countries and regions where more and more people are getting access to the Internet. For example, between 2000 and 2007, the online populations of the Middle East
grew by 491.4% [1]. However, the functionality and quality of content in the Web
sites of these regions often lack behind the growth of their user base. Users are challenged to browse over a large number of Web sites without well-designed Web directories. Here we define the "underdeveloped Web" as the portion of the World Wide
Web in which the growth of its usage far outpaces the growths of its content and functionality, thus users' needs for Web resources are largely not satisfied. An example of
the underdeveloped Web is the Arabic Web, which consists of Web sites from the
Middle-Eastern regions and is used by over 19 million Arabic-speaking people. Despite the rapid growth in Arabic Web usage, a lack of well-structured Web directories
has limited users' ability to browse the Arabic Web. Although some search engines

D.H.-L. Goh et al. (Eds.): ICADL 2007, LNCS 4822, pp. 468–477, 2007.

support searching in Arabic, the relatively little content of the Arabic Web (compared with other languages such as English) makes it difficult for search engines to provide a comprehensive indexing of Arabic content. The unique characteristics of the Arabic language further complicate the problems.

To address the needs, we developed an approach to organizing information on the underdeveloped Web by using a combination of manual and automatic methods. Based on the approach, we developed the Arabic Medical (AMed) Web Directory as a proof-of-concept prototype that facilitates browsing of Arabic medical information on the Web. To understand the usability and effectiveness of the AMed Web directory, we conducted an experiment involving Arab subjects to compare AMed directory with a benchmark Web directory. In the following, we review prior work on Web directory development, describe our work on developing the AMed Web Directory to address the needs, report findings of an experiment on studying the effectiveness and usability of the directory, and finally conclude our work and outline future directions.

2 Literature Review

Research into Web searching and browsing on the non-English Web has been growing in recent years [e.g., 2, 3, 4]. While many efforts are related to Web searching, relatively little attention has been paid on Web browsing, an activity that users often perform when seeking information. Web directory is often the starting point of users' Web browsing and different approaches are used to develop Web directories. In this section, we review previous research on browsing and on Web directory development. We also review Web resources in Arabic, the language we chose to study browsing on the underdeveloped Web.

2.1 Web Browsing

Browsing is a major activity that users frequently engage when seeking information on the Web. It has been defined as an exploratory information seeking process characterized by the absence of planning, with a view to forming a mental model of the content being browsed [5]. In exploratory browsing, a user first transforms his general information need into a problem. He then articulates his needs as search terms or hyperlinks that appear on the system interface, searches using those terms or explores hyperlinks using browse supports such as Web directories, and finally evaluates the results by scanning through them. As the Internet evolves as a major information-seeking platform for many developing regions, supporting Web browsing by using Web directories have become increasingly important. Consequently, the development of these directories has drawn attention from researchers and practitioners.

2.2 Web Directory Development

Previous work in developing Web directories falls into two categories: (1) Extensive manual identification and categorization of Web resources; and (2) Automatic construction of directories using machine learning or Web mining techniques. We review the work done in these two categories below.

Manual identification and categorization have been used in various domains, ranging from general search engines to domain-specific Web portals. The Open Directory Project, also known as Directory Mozilla (DMOZ) (http://dmoz.org), is constructed and maintained by a large, global community of volunteer editors. With 71,053 human editors, it lists more than 5,199,707 sites classified into over 590,000 categories. The rationale of DMOZ is to use extensive human work to combat growth of human-created Web resources, which often grow with the size of online population. Currently, DMOZ powers the core directory services of many search engines, including Netscape Search, AOL Search, Google, Lycos, HotBot, and DirectHit. There are several other Web directories developed by paid or volunteer editors. The Yahoo! Directory (http://dir.yahoo.com/) is built and maintained by a team of paid editors who organize Web sites into categories and subcategories. The Librarian's Index to the Internet (LII, http://lii.org/) provides a searchable, annotated subject directory of more than 12,000 Internet resources selected and evaluated by librarians for their usefulness to users of public libraries. The UMLS Semantic network is one of three UMLS Knowledge Sources being developed by the National Library of Medicine (http://www.nlm.nih.gov/pubs/factsheets/umlssemn.html). Based on 134 semantic types and 54 links, the network provides a categorization of all concepts represented in the UMLS Metathesaurus and represents important relationships in the biomedical domain. The construction of the above-mentioned Web directories relies heavily on expert participation and their domain knowledge. Moreover, the quality of the directory constructed by this method depends highly on the volunteer editors' domain knowledge, which usually varies from person to person.

Beside manual methods, automatic approaches to constructing directory and ontology have been proposed in previous research. For example, Sato and Sato developed an automated editing system that generates a Web directory from a given category word without human intervention [6]. Chuang and Chien propose a query-categorization approach to facilitate the construction of Web directories [7]. To automatically generate Web directory and identify directory labels, a self-organizing map approach was proposed that built up the relationships among Web pages and extracted category labels [8]. Chen and colleagues proposed a self-organizing approach to Internet search and categorization [9]. Stamou et al. developed an approach to automatically assigning Web pages to a directory framework based on the linguistic information on the Web textual data [10]. In general, these automatic approaches yielded high efficiency at the expense of accuracy. The typically unsatisfactory categorization accuracy shows the deficiency of these approaches in constructing Web directories, especially on the underdeveloped Web where the usage is growing rapidly nowadays.

2.3 Arabic Web Resources

The Arabic Web serves as a good example of the underdeveloped Web. Arabic is spoken by more than 284 million people in about 22 Middle-Eastern and North African countries. Although Arabic is the fifth most frequently spoken language in the world, the Arabic Web is still in its infancy, constituting less than 1% of the total Web content and having a low 2.2% penetration rate [11]. The cross-regional use of Arabic and the exponential growth of Arabic Web [12] nevertheless have highlighted the necessity of supporting better Web browsing. Here we review several Arabic Web portals to understand their support on Web browsing.

Ajeeb (http://www.ajeeb.com/) is a bilingual Web portal (English/Arabic) launched in 2000 by Sakhr Software Company. Its database contains over one million searchable Arabic Web pages, which can be translated to English using the online version of Sakhr's machine-translation software. In addition, Ajeeb has a multilingual dictionary and is known for its large Web directory, "Dalil Ajeeb," which the company claims is the world's largest online Arabic directory. Albawaba (http://www.albawaba.com/) is a consumer portal offering comprehensive services including news, sports, entertainment, e-mail, and online chatting. The portal supports searching for both Arabic and English pages and the results are classified according to language and relevancy. Albawaba also provides metasearching of other search engines (Google, Yahoo, Excite, Alltheweb, Dogpile) and a comprehensive directory of all Arab countries. Launched in 2000, UAE-based Albahhar (http://www.albahhar.com/) provides a wide range of online services such as searching, news, online chatting, and entertainment. The portal searches its 1.25 million Arabic Web pages and provides Arabic speakers a wide range of other online services like news, chat, and entertainment. Based in New Hampshire, Ayna (http://www.ayna.com/) is a Web portal providing an Arabic Web directory, an Arabic search engine, and other services such as a bilingual (English/Arabic) email system, chat, greeting cards, personal homepage hosting, and personal commercial classifieds. In July 2001, Ayna had over 700,000 registered users and provided access to more than 25 million pages per month. Due to Ayna's popularity, Alexa Research ranks it among the top three leading Web sites in the Arab World.

3 The Arabic Medical Web Directory

As shown in the review, previous approaches to building information directory have several limitations. On the one hand, manual approaches typically introduce biases due to limited knowledge of the group of directory editors. The fact that many Web pages are generated dynamically also makes this approach not scalable to the rapid growth of the underdeveloped Web. On the other hand, automatic approaches lack precision in identifying category terms and organizing items inside the directory. Previous efforts relying on such approaches typically exploit limited information sources, thus the quality of the resulting directory is limited. The lack of expert knowledge in many of these approaches also creates problems in the usability of the directory created. On the underdeveloped Web (such as the Arabic Web), a lack of comprehensive Web directory further aggravate the problems. To our knowledge, no previous attempt has been made to develop an approach to developing Web directories for the underdeveloped Web.

To address the research gaps, we have developed a generic approach to facilitating Web directory development for the underdeveloped Web. Our approach tried to overcome problems found in previous research by combining human knowledge and machine efficiency, while incorporating various information sources to ensure a high quality of content. Searching multiple high-quality search engines, which has been shown to provide higher quality of the results than relying on only a few search engines [13], and manual filtering of results are the main components. In the following,

Table 1. Summary statistics of the Arabic Medical Web Directory

Statistics	AMed Directory
Total number of categories	232
Total number of Web pages	5,107
Average number of pages per categories	22.1
Maximum depth	5

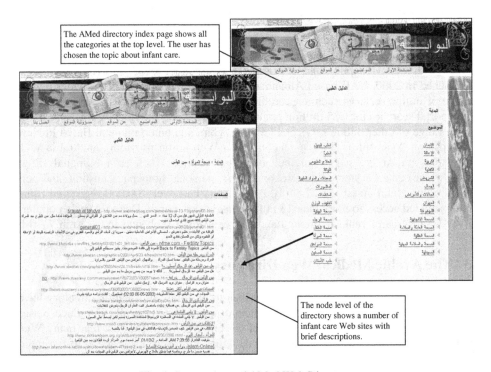

Fig. 1. Screen shots of AMed Web Directory

we explain our approach in the context of building the Arabic Medical Intelligence (AMed) Web directories. We chose the Arabic Web as our research test bed because of its high growth rate yet limited development in its resources. Providing timely and accurate Web directory of a domain that most Arabic people concern is very important; and the Arabic medical domain serves as a good example in this regard.

In the first step, we identified an existing Web directory as the base directory and modified its category labels as queries for meta-searching. We used the DMOZ directory as the base directory because of its comprehensiveness in the English medical domain. We removed 46 nodes from the original 356 nodes of its medical subdirectory, leaving 310 nodes in the directory. Then, 11 nodes were manually added by including cultural specific items such as Islamic medicine, resulting in a 321-node Arabic medical directory framework.

In the second step, we filled in the directory framework (obtained from the first step) with items obtained by automatic meta-searching. By sending queries to multiple search engines and collating the set of top-ranked results from each search engine, meta-searching can greatly reduce bias in search results and improve coverage. To fill in the Arabic medical directory, we used six major search engines (Ba7th, Arabmedmag, Google, Ayna, Sehha, and ArabVista) as meta-searchers and the 321 category labels of the framework (from step 1) as input queries. To our knowledge, these meta-searchers provide the richest Arabic resources on the Web. After running an automatic meta-search program adapted to the Arabic language, we obtained 8,040 unique URLs related to 292 category labels (non-empty nodes) out of the 321 nodes. The maximum depth of the resulting directory was 5.

In the third step, we manually filtered out non-relevant items and added necessary items. We followed a number of heuristics to filter and enhance the directories. URLs were removed if they were not relevant to the topic or they were not related to the domain (i.e., Arabic medical domain) being considered. Empty nodes were removed. Sub-topics of deleted nodes were removed as well. Web sites that contained too few links and pages (typically fewer than 10) were removed. Duplicated category labels were consolidated into one label. The statistics of the two resulting directories are shown in Table 1. Figure 1 show screen shots of the AMed Web directory.

4 Experimental Results and Discussion

In this section, we describe an experiment to evaluate the usability of the AMed Web directory and report the experimental findings. We selected Albawaba (http://www.albawaba.com/) as the benchmark directory to compare against AMed directory because of Albawaba's high stability and reliability. To our knowledge, Albawaba's medical Web directory provides the most comprehensive listing of medical topics and Web sites in Arabic. We designed scenario-based browse tasks consistent with Text Retrieval Conference standards [14] to evaluate the performance of the two directories. For example, a browse task related to prevention and treatment of cancer was: "Find articles about healthy diet and cancer prevention." In each task, the subject used the directory to find addresses (represented by URL links) of relevant Web sites or pages.

The subjects who voluntarily participated in the experiment were native Arabic speakers who could understand the content of the Web directories. Seven subjects participated in the experiment and each subject performed five different browse tasks using each of the two directories, making a sample size of $n = 35$ (5×7). In the half-hour experiment, we introduced the two directories (our directory and the benchmark directory) to each subject. Each subject worked on five tasks in the first section (using one directory) and five different tasks in the second section (using another directory), thus preventing learning effect in their performance. The order in which the directories were used was randomly assigned to avoid bias due to their sequence. We also randomly assigned the two sets of five tasks to evaluate the two directories. Because each subject used the two directories, a repeated-measure design was used in the experiment.

After using a Web directory, the subject filled in a post-section questionnaire about his rating and comments on the directory. The experimenter recorded all verbal comments or behavioral observations that were later analyzed using protocol analysis [15]. Upon finishing the study, the subject also filled in a post-study questionnaire to compare the two directories and to provide their demographic information, which was kept confidential in accordance with the Institutional Review Board Guidebook [16].

We measured the effectiveness of using a directory by precision, recall, and F value as shown in the formulas below. Serving in this research as a medical expert, a final-year Arabic medical student in a Middle East country graded the subjects' answers to calculate the values of the measures. The expert also verified all the experimental tasks to be appropriate for the experiment.

$$Precision = \frac{Number\ of\ relevant\ URLs\ identified\ by\ the\ subject}{Number\ of\ all\ URLs\ identified\ by\ the\ subject}$$

$$Recall = \frac{Number\ of\ relevant\ URLs\ identified\ by\ the\ subject}{Number\ of\ relevant\ URLs\ identified\ by\ the\ expert}$$

$$F\ value = \frac{2 \times Recall \times Precision}{Recall + Precision}$$

The hypotheses we tested are as follows:

H1: The AMed Web Directory achieves a significantly higher precision than the benchmark Web directory.

H2: The AMed Web Directory achieves a significantly higher recall than the benchmark Web directory.

H3: The AMed Web Directory achieves a significantly higher F value than the benchmark Web directory.

H4: The AMed Web Directory achieves a significantly better rating on its helpfulness to browsing than the benchmark Web directory.

H5: The AMed Web Directory achieves a significantly better rating on user satisfaction than the benchmark Web directory.

Table 2 summarizes the results of testing the five hypotheses. The AMed Web Directory (AMedDir) achieved a significantly better effectiveness than Albawaba, as shown in the significantly higher precision, recall, and F value. We believe that these favorable results were due to AMedDir's well-organized information hierarchy and comprehensive coverage of Arabic medical resources on the Web. The approach we used to create the AMedDir has incorporated the preciseness of manual methods and the efficiency of automatic methods, thus giving the benefits of high effectiveness and broad content coverage. Subjects were able to obtain answers to their tasks from AMedDir effectively. Therefore, hypotheses H1, H2, and H3 were supported.

In addition to the favorable results on effectiveness, AMedDir achieved significantly better ratings on helpfulness to browsing and on user satisfaction than Albawaba. On a 7-point Likert scale where "1" refers to the most favorable option, AMedDir obtained on average 2.14 on helpfulness to browsing and 2.43 on user satisfaction, while Albawaba obtained ratings of 5.57 and 5.86 respectively. We believe that the wide margins between the two directories' ratings were due to AMedDir's high effectiveness in supporting subjects' task performance and relatively cleaner interface, and Albawaba's problems in browsing support. Some subjects complained that Albawaba's collection was too small and had inadequate organization of information, thus providing insufficient results or non-relevant results. Based on the significant differences in subjects' ratings, we conclude that hypotheses H4 and H5 were supported.

Table 2. Results of hypothesis testing

Hypothesis	AMedDir		Albawaba		p-value	Result
	Mean	**SD**	**Mean**	**SD**		
H1: Precision	**0.74**	0.44	0.09	0.28	0.000	Supported
H2: Recall	**0.32**	0.29	0.06	0.20	0.000	Supported
H3: F value	**0.43**	0.30	0.07	0.23	0.000	Supported
H4: Helpfulness to browsing*	**2.10**	1.07	5.57	0.54	0.000	Supported
H5: Satisfaction*	**2.43**	1.27	5.86	0.69	0.000	Supported

* The subject rating was based on a 7-point Likert scale, where "1" refers to the most favorable option.

The favorable experimental results bring about several implications. First, the development of AMedDir not only helped support browsing, but also demonstrated an effective way to improve organization of vast and growing amounts of information on the underdeveloped Web. The hierarchical structure helped organize information effectively, facilitating concept categorization and topic classification. Second, the experimental findings revealed the need for better Arabic Web directories for browsing. While the Arabic Web is growing much faster than other parts of the Web, Arabic users are not getting the same level of services that many non-Arabic Web users currently enjoy. In our review, we found that many Arabic portals are unstable and contain limited information. This study highlights the need for researchers and practitioners to enrich the Arabic Web with better content and functionality. Third, the review and findings from this research contribute to the growing body of research on non-English Web browsing, which becomes increasingly important as it demonstrates a strong potential of growth in the coming decades. Rapidly emerging issues such as browsing Web directories, information organization, and directory development were addressed in this research.

5 Conclusions

While the developments of many parts of the World Wide Web have matured over the past decade, there are still some portions of the Web that are largely underdeveloped. This underdeveloped Web is often characterized by a lack of high quality content and functionality, despite having a rapid growth in usage. These problems reveal the needs for better organization and presentation of information on many Web sites of the underdeveloped Web. In this research, we developed an approach to building Web directories for the underdeveloped Web and used the approach to construct the Arabic Medical Web Directory (AMedDir) that supports browsing the Arabic medical domain. Findings from an experiment involving Arab subjects show that AMedDir significantly outperformed a benchmark Web directory in terms of effectiveness and subject ratings. We conclude that the Arabic Medical Web Directory has high effectiveness and usability for supporting browsing the Arabic Web. This research thus contributes to developing a proof-of-concept prototype for organizing information of the Arabic medical domain and to better understanding of supporting browsing in the underdeveloped Web.

The research was limited in a number of ways. First, the AMed directory was a research prototype and hence lacked the scalability of commercial Web portals. Second, scarce literature on browsing the underdeveloped Web made it difficult for us to perform a more comprehensive review of the field. Third, we were limited by our ability to recruit more Arab subjects for the experiment.

Future directions of this research include developing Web directories for other parts of the underdeveloped Web and studying their impacts on browsing. For example, Spanish is the language of many Hispanic communities and Latin American countries whose populations are growing rapidly. Yet many Web sites in Spanish still lack browsing support and information and can be improved by introducing better Web directories or other functionality. Another direction is to develop automatic techniques to generate Web directory framework and to collect information on the Web. A third direction is to enhance human filtering and classification process to increase the precision and accuracy. These efforts will yield more useful and effective Web directories and enhance the browsing experience of many people in the world.

Acknowledgements

This research was partly supported by NSF Knowledge Discovery and Dissemination (KDD) program #9983304 (June 2003 – March 2004 and October 2003 – March 2004) and Santa Clara University. We thank the subjects and expert who participated in the experiment and the members of the system development team.

References

[1] Miniwatts International.: Internet Usage Statistics - The Big Picture (2007), http://www.internetworldstats.com/stats.htm

[2] Chung, W.: Studying information seeking in the non-English Web: An experiment on a Spanish business Web portal. International Journal of Human-Computer Studies 64, 811–829 (2006)

[3] Chung, W., Zhang, Y., Huang, Z., Wang, G., Ong, T.-H., Chen, H.: Internet searching and browsing in a multilingual world: An experiment on the Chinese Business Intelligence Portal (CBizPort). Journal of the American Society for Information Science and Technology 55, 818–831 (2004)

[4] Spink, A., Ozmutlu, S., Ozmutlu, H.C., Jansen, B.J.: U.S. versus European Web Searching Trends. SIGIR Forum 36 (2002)

[5] Chung, W., Chen, H., Nunamaker, J.F.: A visual framework for knowledge discovery on the Web. Journal of Management Information Systems 21, 57–84 (2005)

[6] Sato, S., Sato, M.: Automatic generation of Web directories for specific categories. In: Proceedings of the AAAI Workshop on Intelligent Information Systems, Orlando, FL (1999)

[7] Chuang, S.-L., Chien, L.-F.: Enriching Web taxonomies through subject categorization of query terms from search engine logs. Decision Support Systems 35, 113–127 (2003)

[8] Yang, H.-C., Lee, C.-H.: A text mining approach on automatic generation of Web directories and hierarchies. In: Proceedings of the IEEE/WIC International Conference on Web Intelligence, Halifax, Canada (2003)

[9] Chen, H., Schuffels, C., Orwig, R.: Internet categorization and search: a self-organizing approach. Journal of Visual Communication and Image Representation 7, 88–102 (1996)

[10] Stamou, S., Krikos, V., Kokosis, P., Ntoulas, A., Christodoulakis, D.: Web directory construction using lexical chains. In: Proceedings of the 10th International Conference on Applications of Natural Language to Information Systems (2005)

[11] Abbi, R.: Internet in the Arab World, UNESCO Observatory on the Information Society, p. 3 (2002)

[12] Norton, L.: The Expanding Universe: Internet Adoption in the Arab Region. World Markets Research Centre, Report (2001)

[13] Mowshowitz, A., Kawaguchi, A.: Bias on the Web. Communications of the ACM 45, 56–60 (2002)

[14] Voorhees, E., Harman, D.: Overview of the Sixth Text Retrieval Conference (TREC-6), In: NIST Special Publication 500-240: The Sixth Text Retrieval Conference (TREC-6), Gaithersburg, MD, USA (1997)

[15] Ericsson, K.A., Simon, H.A.: Protocol analysis: verbal reports as data. MIT Press, Cambridge (1993)

[16] Penslar, R.L.: Institutional Review Board Guidebook,Office for Human Research Protection, U.S. Department of Health and Human Services (2006), http://www.hhs.gov/ohrp/irb/irb_guidebook.htm

Recommending Scientific Literatures in a Collaborative Tagging Environment*

Ping Yin, Ming Zhang, and Xiaoming Li

School of Electronics Engineering and Computer Science
Peking University, Beijing, China
pkufranky@gmail.com, mzhang@net.pku.edu.cn, lxm@pku.edu.cn

Abstract. Recently, collaborative tagging has become popular in the web2.0 world. Tags can be helpful if used for the recommendation since they reflect characteristic content features of the resources. However, there are few researches which introduce tags into the recommendation. This paper proposes a tag-based recommendation framework for scientific literatures which models the user interests with tags and literature keywords. A hybrid recommendation algorithm is then applied which is similar to the user-user collaborative filtering algorithm except that the user similarity is measured based on the vector model of user keywords other than the rating matrix, and that the rating is not from the user but represented as user-item similarity computed with the dot-product-based similarity instead of the cosine-based similarity. Experiments show that our tag-based algorithm is better than the baseline algorithm and the extension of user model and dot-product-based similarity computation are also helpful.

1 Introduction

Collaborative recommendation and content-based recommendation are widely used in recommendation systems. Due to advantages and flaws of both technologies, it's a hot research to combine them to achieve better results [1, 2].

In recent years, collaborative tagging [3] becomes more and more popular. Tags can reflect both user's opinion and content features of resources. The utilization of the tag content for recommendation is worthy of a further research.

This paper focuses on scientific literature recommendation in a collaborative environment, considering both collaborative tags and content information.

There is much work related to ours. Digital libraries such as ACM[1] list similar papers in the form of text search. CiteSeer[2] provides content-based and citation-based recommendations. McNee etc. generate recommendations by mapping the web of citations between papers into the CF user-item rating matrix [4, 5].

The remainder of the paper is organized as follows. Section 2 describes the key steps for scientific literature recommendation in a collaborative tagging environment. Section 3 experimentally evaluates the algorithm. Section 4 summarizes this paper.

---

* This work is supported by the National Natural Science Foundation of China under Grant No. 90412010, HP Labs China under "On line course organization".

[1] http://www.acm.org/dl
[2] http://citeseer.ist.psu.edu

D.H.-L. Goh et al. (Eds.): ICADL 2007, LNCS 4822, pp. 478–481, 2007.
© Springer-Verlag Berlin Heidelberg 2007

2 Recommending Scientific Literatures in a Collaborative Tagging Environment

This paper proposes a hybrid recommendation algorithm similar to the user-user CF algorithm for a collaborative tagging environment. The difference lies in the user interest modeling, the user similarity computation and the user rating simulation.

2.1 The Representation of User Interest and Literature

The user's interest keywords have three sources: the user tags of literatures, keywords of the tagged literatures and their citations. To distinguish the importance of these three sources, different weights are assigned to them respectively. Then, the keywords frequencies are used to form an m-dimension user interest vector as follows.

$$U =< u_1,...,u_m >$$ (1)

Here u_i denotes the weighted word frequency of the ith keyword.

Similarly, the model of a single literature consists of its keywords, keywords of its citations and all users' tags on it.

$$D =< d_1,...,d_m >$$ (2)

Here d_i denotes the relative weighted frequency of the ith keyword summed to one.

2.2 The Computation of User Interest Degree

The user rating is simulated by user interest degree which is not directly from the user, but measured by similarity between vectors of the user interest and the literature.

The formula for interest degree is as follows, where dot-product-based similarity is used instead of cosine similarity since the length of user interest vector is meaningful.

$$R(U,D) = \sum_{i=1}^{m} u_i d_i$$ (3)

2.3 The Computation of User Similarity and Prediction

Once the set of most similar users is isolated with the correlation-based similarity [6], the adjusted weighted sum approach is used to obtain prediction [7].

Formally, we can denote the prediction P_{ui} as

$$P_{ui} = \overline{R_u} + \frac{\sum_{n \in NSet} sim(u,n) \times (R_{ni} - \overline{R_n})}{\sum_{n \in NSet} |sim(u,n)|}$$ (4)

Here NSet denotes the nearest neighbor set, $sim(u,n)$ denotes similarity between user u and n. R_{ni} denotes user n's rate on item i, $\overline{R_u}$ denote the average rating of user u.

3 Experimental Evaluation

The dataset is adapted from citeulike[3] including ten thousands literatures with tags.

We use the 4-fold cross validation and the "All but one" scheme [5]. One literature is removed randomly from the tagged literatures of each user in the test dataset, and then the modified test dataset is merged into the training dataset. Then a top-10 recommendation is run on the whole dataset.

The hit percentage [5] is used to express this expectation that the removed literature can be recommended.

$$hit - percentage = hitcount / |testset| \tag{5}$$

Here *hitcount* denotes the number of successful recommendations and |*testset*| denotes the size of the testset, that is, the number of recommendations made.

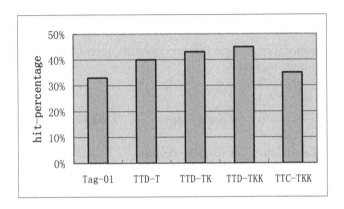

Fig. 1. Comparison of different algorithms for top-10 recommendation

All the algorithms which are used in the experiment are list below and figure 1 gives the result of all algorithms.

Tag-01: The baseline experiment which uses the user-user collaborative filtering algorithm. The rate is 0 or 1 according to whether the user has tagged the item.

Tag-text-dotproduct-T (TTD-T): User model and literature model are both represented as tag frequency vector. Dot-product-based similarity is used for the computation of user interest degree.

Tag-text-dotproduct-TK (TTD-TK): Almost the same with TTD-T except extending the user and literature model by literature keywords.

Tag-text-dotproduct-TKK (TTD-TKK): Almost the same with TTD-T except extending the user and literature model by keywords of the literature and the literature's citations.

---

[3] http://www.citeulike.org

Tag-text-cosine-TKK (TTC-TKK): Almost the same with TTD-TKK except that cosine-based similarity is used instead of dot-product-based similarity.

4   Conclusions and Acknowledgments

As the experiment shows, our tag-based algorithm is better than the baseline algorithm. The extension of user model with literature keywords and dot-product-based similarity computation also help to achieve better results. The prototype is now available under PKUSpace[4] [8].

This work is partially supported by NSCF Grant (60573166) as well as Network Key Lab Grant of Guang Dong Province.

References

1. Burke, R.: Hybrid Recommender Systems: Survey and Experiments. User Modeling and User-Adapted Interaction 12(4), 331–370 (2002)
2. Balabanovic, M., Shoham, Y.: Fab: Content-Based, Collaborative Recommendation. Communications of the ACM 40(3), 66–72 (1997)
3. Golder, S.A., Huberman, B.A.: Usage Patterns of Collaborative Tagging systems. Journal of Information Science 32(2), 198–208 (2006)
4. Torres, R., McNee, S.M., Abel, M., Konstan, J.A., Riedl, J.: Enhancing Digital Libraries with TechLens+. In: Proc. of the 2004 Joint ACM/IEEE Conference on Digital Libraries, pp. 228–236 (2004)
5. McNee, S.M., Albert, I., Cosley, D., Gopalkrishnan, P., Lam, S.K., Rashid, A.M., Konstan, J.A., Riedl, J.: On the Recommending of Citations for Research Papers. In: CSCW 2002: Proceedings of the 2002
6. Sarwa, B.M., Karypis, G., Konstan, J., Riedl, J.: Analysis of Recommendation Algorithms for E-commerce [R]. In: ACM Conference on Electronic Commerce, pp. 158–167 (2000)
7. Pan, H.Y., Lin, H.F., Zhao, J.: Collaborative Filtering Algorithm Based on Matrix Partition and Interest Variance. Journal of the China Society for Scientific and Technical Information 25(1), 49–54 (2006)
8. Zhang, M., Yang, D.Q., Deng, Z.H., Feng, Y., Wang, W.Q., Zhao, P.X., Wu, S., Wang, S.A., Tang, S.W.: PKUSpace: A Collaborative Platform for Scientific Researching. In: Liu, W., Shi, Y., Li, Q. (eds.) ICWL 2004. LNCS, vol. 3143, pp. 120–127. Springer, Heidelberg (2004)

---

[4] http://fusion.grids.cn:8080/PKUSpace/home.jsp

Bridging Community Resource Gateways by Linking Community Taxonomies

Wonsook Lee, Mitsuharu Nagamori, and Shigeo Sugimoto

University of Tsukuba, Tsukuba 1-2, Ibaraki, Japan
{wonsook,nagamori,sugimoto}@slis.tsukuba.ac.jp

Abstract. Many communities provide Web resource directories to help users find useful resources in the community. A typical example is a resource directory in a homepage of a local government. Crosswalk of the directories of neighboring communities is a crucial function for users to collect useful resources from the communities. However, an appropriate scheme bridging the community directories is required. This paper proposes a few mapping schemes to connect community directories and compares them by applying them to the resource directories of three local governments - Tokyo and Hokkaido in Japan and Chungcheongnam-do in Korea. The mapping schemes use National Diet Library Subject Heading (NDLSH) and/or Nippon Decimal Classification (NDC) as a switching language. Evaluation of the proposed schemes shows their advantages and limitations.

Keywords: Community Taxonomy, Crosswalk of Web Directories, Switching Language, Subject Headings, Knowledge Organization.

1 Introduction

There are many Web sites which provide a directory of useful Web resources in a specific domain for a specific community. Each of the directories is useful for the community to find valuable resources in their domain. A nice example is a resource directory of a local government - prefectures and cities – which provides a categorized list of useful resources for their community members. A crosswalk of resource directories of neighboring communities is a crucial function to collect resources from the communities. Those community resource directories are usually not huge. Their organizing taxonomies are much smaller than those of comprehensive resource directories – some tens to hundreds of terms. It is not straightforward, however, to crosswalk the resource directories of neighboring communities because of the difference of their taxonomies. In this paper, we propose a few schemes for mapping between the community resource directories and compare them.

2 Related Works

Jens-Erik proposed to use Dewey Decimal Classification (DDC) as a switching language among several taxonomies [1]. There are a few collaborative projects to merge resource directories in specific domains, e.g. Renardus project and Resource Discovery Network [2][3]. These projects use comprehensive conventional classification schemes

D.H.-L. Goh et al. (Eds.): ICADL 2007, LNCS 4822, pp. 482–486, 2007.

to merge the subject directories. Community resource directories have different features from comprehensive directories. In our previous work, we examined the characteristics of subject vocabularies of a community oriented directory of Web resources provided by Okayama Prefecture Library in Japan [4]. The library uses three taxonomies to classify resources – NDC, a taxonomy for children, and a taxonomy for the resources provided by the prefecture government. NDC is a comprehensive classification scheme and a major standard for Japanese libraries. On the other hand, both of the two local taxonomies have 300 to 400 categories. They each have a quite rich set of terms in specific subjects in accordance with their purpose.

3   Connecting Community Taxonomies

3.1   An Underlying Model

As shown in Fig. 1, there are two general schemes to connect taxonomies – one-to-one mapping and mapping via a hub taxonomy called *switching language*. It is obvious that mapping via a switching language is advantageous in terms of

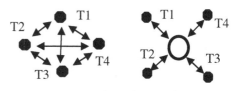

a. one-to-one mapping   b. mapping via a hub

Fig. 1. Mapping between Taxonomies

complexity. The mapping schemes proposed in the next section are based on this switching language model. The vocabularies used for the switching language in this paper are NDC and NDLSH. NDLSH is one of the authoritative subject headings in Japan and is maintained by the National Diet Library of Japan. These two authoritative vocabularies are primarily designed for Japanese resources but the sections used in this research are mostly not specific to Japanese contents.

3.2   Community Taxonomy Mapping Models

The four mapping schemes (MS1-4) shown below are evaluated for community taxonomy mapping in this paper. Each MS is explained as a procedure to obtain a mapping function. The taxonomies have a hierarchical structure of category terms.

MS1: All resources are assigned one or more NDLSH terms in advance.

1. For each leaf category of taxonomy A and B, create a test set of resources by randomly choosing resources classified into the category.
2. Create a NDLSH term list for every category by gathering NDLSH terms assigned to the resources of the test set of the category.
3. For every leaf category pair of A and B, create an intersection of their NDLSH term sets. Neglect the pairs whose intersection is empty.
4. Calculate an inter-category connectivity value of all of the leaf category pairs using (1) and (2).
5. Choose category pairs which have inter-category connectivity values above a heuristic threshold value. The set of category pairs obtained is a mapping function between A and B.

A connectivity value CV_{cn} from a category term c to an NDLSH term n and an inter-category connectivity value ICV_{th} between category t and h connected via an NDLSH term n are defined as follows.

$$CVcn = Rcn \, / \, Rn \tag{1}$$

$$ICVth = \sum CVtn \times CVhn \tag{2}$$

where R_{cn} is the number of resources classified in category c and assigned term n, and R_n is the number of all resources classified in category c.

MS2: By definition, every NDLSH term is associated with one or more NDC terms. MS2 uses the NDC in addition to the NDLSH terms.

1. Same operations as Step 1 and 2 of MS1.
2. For each leaf category of A and B, create a frequency list of NDC terms by accumulating every NDC term that is assigned to the NDLSH terms of the resources included in the test set of the category.
3. For each leaf category of A and B, find the most frequently used NDC term(s) and call it (or them) the surrogate(s) of the category.
4. For every category of A, find one or more categories of B which have the same surrogate and make a category pair. The set of category pairs obtained is the mapping function from A to B.

MS3: In MS3, all resources are assigned one or more NDC terms (up to five).

1. Find the most frequently used NDC term(s) for each category of A and B. Use the most frequently assigned NDC term(s) as the surrogate of the category.
2. Find source- and target-category pairs which have the identical NDC term(s) as their surrogate. The set of pairs created is the mapping function from A to B.

MS4: MS4 uses a direct mapping from a category to NDLSH terms.

1. Assign one or more NDLSH terms to each category of A and B. If a category is a compound term, split the term into simple terms and assign an NDLSH term(s) to each of the simple terms; for example, split "School Education and Life Long Education" into "School Education" and "Life Long Education".
2. Create a set of category pairs by coupling an A category and a B category that are assigned to at least one identical NDLSH term. The set of pairs created is the mapping function between A and B.

4   Experiments – Mapping Directories Between Local Governments

We applied all mapping schemes to resource directories of Tokyo, Hokkaido in Japan and Chungcheongnam-do in Korea. The directory of Tokyo has two layers and 56 leaf categories. The Hokkaido directory has two layers and 47 leaf categories. The Chungcheongnam-do directory has three layers and 118 leaf categories.

Table 1. Precision and Recall values of Community Taxonomy Mappings

	MS1		MS2		MS3		MS4	
	Precision	Recall	Precision	Recall	Precision	Recall	Precision	Recall
T to H	0.60	0.49	0.33	0.38	0.48	0.49	0.73	0.61
H to T	0.40	0.34	0.28	0.37	0.35	0.48	0.71	0.78
C to T	0.57	0.45	0.18	0.34	0.30	0.48	0.57	0.59
C to H	0.28	0.57	0.23	0.45	0.17	0.41	0.50	0.73

Precision =(number of correct mappings by MSx) / (number of all mappings by MSx)
Recall=(number of correct mappings by MSx) / (number of manual mappings)
T: Tokyo, H: Hokkaido, C: Chungcheongnam-do

In this experiment, we selected up to five resources from each category in MS1, MS2 and MS3. In the case that the number of resources under a category was smaller than five, we used all of the resources. We applied all three mapping schemes to the pairs of the directories. Assignments of NDLSH terms to the resources and the categories were manually done. We used 193, 218 and 149 resources of Tokyo, Hokkaido and Chungcheongnam-do, respectively. Table 1 summarizes the recall and precision values. This evaluation is performed by comparing the mapping functions created by the four mapping schemes with the answer set that were created by manual mapping of the categories between the taxonomies.

The recall and precision scores of MS4 are generally better than others. Among the four schemes, MS4 is the only scheme which directly maps a category term to another. These scores are calculated based on category-to-category mappings. Mapping schemes other than MS4 use subject terms of resources to create the mappings. Therefore, we consider that those schemes could be extended to mapping by sub-leaf-category which is implicitly formed by a group of resources or to mapping by single resource. These issues are left for our future work.

5   Conclusion

Community people know better and more deeply about community resources than Google. Crosswalking community resource directories is an essential function for end-users trying to find valuable resources. Taxonomies used in the community gateways are semi-controlled but not designed for interoperability with neighboring communities. Because of the nature of the human-created taxonomies, sharing the categories of community gateways is hard even when the taxonomies are not large. The mapping schemes and evaluation results discussed in this paper give us useful clues about how to create a crosswalk function for community gateways.

References

1. Mai, J.-E.: The Future of General Classification. Cataloging & Classification Quarterly 37 (1/2), 3–12 (2003)
2. Renardus, http://renardus.lub.lu.se/

3. Intute, http://www.intute.ac.uk/
4. Lee, W., Sugimoto, S.: Toward Core Subject Vocabularies for Community-oriented Subject Gateways. In: DC-2005: International Conference on Dublin Core and Metadata Applications, pp.15–24, Madrid (2005)

Functional Requirements for Subject Authority Records (FRSAR): A Conceptual Model of Aboutness

Maja Žumer[1], Athena Salaba[2], and Marcia Lei Zeng[2]

[1] University of Ljubljana, Slovenia
University of Ljubljana, Faculty of Arts Askerceva 2, 1000 Ljubljana Slovenia
{Maja.Zumer}@nuk.uni-lj.si
[2] Kent State University, Kent, Ohio, USA
314 Library, PO Box 5190, Kent, Ohio 44224, USA
{asalaba,mzeng}@kent.edu

Abstract. Provides a brief overview of the activities of the IFLA Functional Requirements for Subject Authority Records (FRSAR) working group. Introduces the group's terms of reference and the work completed so far, including definitions of user tasks and subject entities. Discusses the development of the entity-relationship conceptual model of subject entities in the bibliographic universe.

Keywords: FRSAR, IFLA, subject access, knowledge organization systems (KOS), conceptual model, subject authority data, subject authority records.

1 Background

IFLA's Working Group on the Functional Requirements for Bibliographic Records (FRBR)[1] developed an entity-relationship conceptual model of the bibliographic universe [1]. The working group began with identifying essential user tasks and then defining the relevant entities, their attributes, and relationships. The basic elements of the FRBR model are the result of a logical analysis undertaken by members of the Study Group of the data typically reflected in bibliographic records. The entities are divided into three groups:

- Group 1 consists of four entities that are the product of intellectual or artistic endeavors: *work, expression, manifestation*, and *item*.
- Group 2 consists of entities that are actors, those who are responsible for the intellectual or artistic content, the physical production and dissemination, or the custodianship, of Group 1 entities: *person, corporate body*
- Group 3 consists of entities that serve as subjects of intellectual or artistic endeavor.

---

[1] IFLA WG on Functional Requirements for Bibliographic Records (FRBR), http://www.ifla.org/ VII/s13/wgfrbr/finalreport.htm

D.H.-L. Goh et al. (Eds.): ICADL 2007, LNCS 4822, pp. 487–492, 2007.

We can illustrate Group 1 entities using the example of *The Da Vinci Code* by Dan Brown. *Items* are individual copies of the book (for instance, my personal copy *vs.* a copy owned by the Slovenian National Library with call number 601547, etc.). A particular publication is called a *manifestation* by FRBR (e.g., the edition published by Bantam in 2003). However, books are more than just physical objects and the intellectual content is really our main focus. For instance, one may be interested in the original text, in the Slovenian translation, or in an abridged version. These are all *expressions* of the same work. Therefore, the Bantam 2003 edition is a *manifestation* that contains Brown's original text (an *expression*) of *The Da Vinci Code* (the *work*).

In most structured retrieval systems, information regarding the bibliographic universe is not recorded exclusively in bibliographic records. Authority records are used to record information about all controlled access points that are currently included in bibliographic records or have the potential to be assigned as access points in bibliographic records. Controlled access points include names of entities identified by FRBR such as members of Group 2 (*persons, corporate bodies*), titles of Group 1 entities (*works, expressions, manifestations* and *items*), and terms for Group 3 entities. In the FRBR model the entities of all three groups are defined, but the main focus is on the first group.

The second IFLA Working Group, the Functional Requirements and Numbering of Authority Records (FRANAR)[2] is charged with the task of continuing the work of FRBR by developing a conceptual model for authority records. In the 2007 draft of the *Functional Requirements for Authority Data* (FRAD) model, the group defines authority records as aggregates of information regarding entities that are assigned as controlled access points in bibliographic records and focuses on Group 1 and Group 2 entities (IFLA, 2007).

A third IFLA Working Group, co-chaired by the authors, was formed in April 2005 and charged with the task of developing a conceptual model for the Functional Requirements for Subject Authority Records (FRSAR)[3]. All controlled access points related to all three entity groups as defined by the FRBR conceptual model have the potential to be the topic of a work. In other words, Group 1, 2 and 3 entities can have an "is-the-subject-of" relationship with *work*. FRSAR's terms of reference are:

- Build a conceptual model of Group 3 entities within the FRBR framework as they relate to the aboutness of works,
- Provide a clearly defined, structured frame of reference for relating the data that are recorded in subject authority records to the needs of the users of those records, and
- Assist in an assessment of the potential for international sharing and use of subject authority data both within the library sector and beyond.

---

[2] IFLA WG on the Functional Requirements and Numbering of Authority Records (FRANAR), http://www.ifla.org/VII/d4/wg-franar.htm

[3] IFLA WG on the Functional Requirements for Subject Authority Records (FRSAR), http://www.ifla.org/VII/s29/wgfrsar.htm

2 Users and Context of Use of Subject Authority Records/Data

During the process of developing an entity-relationship conceptual model of subject authority records, the FRSAR Working Group initially analyzed who the users of subject authority data are, identified contexts of the use of the data, and described some of the use scenarios. Possible subject authority record data user groups include a) information professionals who create metadata, b) reference and public services librarians and other information professionals who are searching for information as intermediaries, c) controlled vocabulary creators, such as catalogers, thesaurus and ontology creators, and d) end-users using information retrieval systems to fulfill their information needs.

The FRSAR Working Group felt strongly that, in order to define user tasks, an actual user study was necessary, and two studies were therefore conducted. The first was a pilot study at the 2006 Semantic Technologies Conference (San Jose, California, USA). Most study participants were either creators of semantic tools, including controlled vocabularies, taxonomies and ontologies, or developers and managers of semantic technology systems. The second study was an international survey sent to information professionals throughout the world during the months of May-September 2007. Participants included authority record creators, vocabulary creators and managers, catalogers, metadata librarians, and reference librarians among others. Participants were asked to describe their work and their use of subject authority data in different contexts, including cataloging/metadata creation, subject authority work, and searching or helping others search bibliographic information. The results of these studies enriched our understanding of subject authority data use and informed and further confirmed the FRSAR user tasks.

Based on the results from our user studies, five subject authority data user tasks, representing uses by all the above user groups, are defined as follows:

Find: To find a subject entity or set of entities corresponding to stated criteria.

Identify: To identify a subject entity based on certain attributes/characteristics.

Select: To select a subject entity.

Obtain: To obtain additional information about the subject entity and/or to obtain bibliographic records or resources about this subject entity.

Explore: To explore relationships between subject entities, correlations to other subject vocabularies and structure of a subject domain.

3 The Conceptual Model

An examination of other models covering subject data and a comparison of the current Group 3 entities served as a starting point. Group 3 entities defined by FRBR include *concept* (an abstract notion or idea), *object* (a material thing), *event* (an action or occurrence), and *place* (a location). This part of the FRBR model has been criticized and several issues regarding Group 3 entities have been raised, particularly the unsymmetrical treatment of space and time and the fact that processes are not modeled.

The working group investigated the approaches of other models, specifically the proposed model by Buizza and Guerrini [2] and <*indecs*> [3] and analyzed possible solutions from conservative (only making minor amendments) to radical (proposing a new model). A small study was performed, in which four students and faculty members at the Kent State University School of Library and Information Science classified existing subject terms used by the NSDL (National Science Digital Library) contributors. These include about 3000 terms assigned based on a variety of subject vocabularies and free keywords. They classified terms into six categories: concrete stuff, abstract stuff, event, time, place, and other. The results show that there is a blurred distinction between concrete and abstract concepts and difficulties in the classification of named instances, which resulted in many terms being put into the 'other' category. This indicates that it would be difficult for any user (end user, librarian or vocabulary developer) to conduct such a task when using subject authority data. These categories also do not seem helpful or necessary to the end users.

As a result, the FRSAR Entity sub-group proposed a more abstract conceptual model:

Fig. 1. Conceptual model

Thema: Anything that can be the subject of a *work*.

Nomen: Any alphanumeric, sound, visual, or any other symbol, sign, or combination of symbols by which a *thema* is known, referred to, or addressed.

This model first confirms what FRBR has already defined: WORK has subject THEMA, and proposes a new part: THEMA has appellation NOMEN. The use of Latin terms is to avoid mapping to an English term (such as subject or concept) that has been understood and translated with a different understanding. The terms for the entities and relationships are subject to change.

Thema therefore includes existing Group 1 and Group 2 entities, and, in addition, all other subjects of works. In a particular application *thema* would normally have implementation-specific types. In the current discussions the most important distinction seems to be the one between named particulars and classes.

In general, *nomen* can be domain-, community- and language-specific. Meanwhile, two important specific types of *nomen* are recognized: *identifier* (name assigned to an entity, which is usually persistent and unique within a domain) and *constructed name* (name constructed in the authority control/vocabulary maintenance process, which usually serves as an access point), for which the term 'controlled access point' is used in FRAD. Attributes of *nomen* serve to carry information about a particular instance and typically include but are not limited to: type, origin/source/system/vocabulary, medium, language, script, transliteration/transcription, time and place of validity, target community, and status.

In addition to the many-to-many relationships between *work* and *thema* and *thema* and *nomen,* as illustrated in Figure 1, there are thema-to-thema and nomen-to-nomen relationships. Particular thema-to-thema relationships are implementation-dependent but the generally applicable ones include Hierarchical (Partitive, Generic, and Instantiation) and Associative (=other) relationships. Some nomen-to-nomen relationships are: Partitive (parts/components of a *nomen* may be *nomen*) and Equivalence (two *nomen* are equivalent, if they have an 'is appellation' relationship with the same *thema*). The equivalence nomen-to-nomen relationship can further be specified. For example, replaces/is replaced by, has variant form/is variant form, has derivation/is derived from, etc.

A good example at the implementation-level is chemical substances. For example, each drug might have its chemical name(s), drug name, trade name, generic name (U.S. Adopted name), system-specific identifiers such as CAS Registry Number, classification code (alpha or numerical), as well as other unique expressions such as a flat structure diagram, structure diagram (include stereo bonds), molecular formulas, etc. The relationships between these names, identifiers, terms, and other expressions are nomen-to-nomen because they are how this same drug is known, referred to, and addressed by specific systems. Meanwhile, in addition to the relationship with other substances, a drug itself contains various compounds and elements, which form thema-to-thema relationships.

The Working Group is currently analyzing attributes and relationships in view of defined user tasks and testing with samples collected from different domains.

4 Interoperability with Other Communities

The final term of reference for the FRSAR Working Group is to assist in an assessment of the potential for international sharing and use of subject authority data both within the library sector and beyond. The challenges in true global sharing and use of subject authority data come from many technological aspects (such as heterogeneous structures), various languages and scripts, diverse construction rules and best practice guides, and dynamically developed and advanced encoding schemas, especially when other communities (museum, archive, science, education) are involved. It is important to separate what we usually call concepts (or topics or subjects) from what they are known by, referred to, or addressed. The potential value of this *work-thema-nomen* model for subject authority data is obvious.

Among the efforts to achieve global sharing and use of subject authority data, some have, in fact, focused on *nomen* (for example, a translated metadata vocabulary, a symmetrical multilingual thesaurus, a multi-access index to a vocabulary, etc.) However, many efforts actually have had to focus on the conceptual level, for example, when mapping between two thesauri. These kinds of efforts usually encounter many more challenges because they deal with the intension and extension of these concepts as well as the relationships among them.

This thema-nomen conceptual model also matches well the encoding schemas such as SKOS (Simple Knowledge Organisation System), OWL (Web Ontology Language), and more general, RDF encoding which uses URIs as the basis of their mechanism for identifying the subjects, predicates, and objects in statements. SKOS defines the classes and properties sufficient to represent the common features found in a standard thesaurus. It is based on a concept-centric view of the vocabulary, where primitive objects are not terms, but abstract concepts represented by terms. Each SKOS concept is defined as an RDF resource. Each concept can have RDF properties attached, including: one or more preferred terms (at most one in each natural language), alternative terms or synonyms, and definitions and notes, with specification of their language [4].

When the DCMI Abstract Model [5] became a DCMI Recommendation in 2007, its one-to-one principle (i.e., each DC metadata description describes one, and only one, resource) was recognized or followed by other metadata standards. Under the one-to-one principle, a record can contain *description sets* that may contain *descriptions* composed by *statements* which use property-value pairs. Consequently, information can be processed, exchanged, referred to, and linked at the *statement* level. This information model is independent of any particular encoding syntax, thus facilitating the development of better mappings and cross-syntax translations (DCMI, 2007). The conceptual model proposed by the FRSAR group corresponds to this abstract model in that it allows any *thema* to be independent of any *nomen*, including any syntax a *nomen* may use. Accordingly, this conceptual model will facilitate the sharing and reuse of subject authority data among subject vocabularies and interoperability of resource metadata.

References

1. Functional Requirements for Bibliographic Records: final report. KG Saur, München (1998)
2. Buizza, P., Guerrini, M.: A Conceptual Model for the New "Soggettario": Subject Indexing in the Light of FRBR. Cataloging & Classification Quarterly 34(4), 31–45 (2002)
3. Rust, G., Bide, M.: The <indecs> metadata framework: Principles, model and data dictionary, Indecs Framework Ltd. (2000), http://www.indecs.org/pdf/framework.pdf
4. SKOS Core Guide.: W3C Working Draft 2 (November 2000), http://www.w3.org/2004/02/skos/
5. DCMI: DCMI Abstract Model (2007), http://dublincore.org/documents/abstract-model/

Mining Police Digital Archives to Link Criminal Styles with Offender Characteristics

Richard Bache[1], Fabio Crestani[1], David Canter[2], and Donna Youngs[2]

[1] Department of Computer and Information Science, University of Strathclyde, Scotland
{r.bache,f.crestani}@cis.strath.ac.uk
[2] Centre for Investigative Psychology, England
dcanter@ukonline.co.uk

Abstract. The partial success in inferring the characteristics of offenders from their criminal behaviour ('offender profiling') has relied on limited data and subjective judgments. We therefore sought to determine if Information Retrieval techniques and in particular Language Modelling could be applied directly to existing police digital records of criminal events to identify significant characteristics of offenders. The categories selected were gender and age group. Results showed that distinct differences in characteristics do exist.

Keywords: Document Classification, Text Data Mining, Language Models, Crime Data, Investigative Psychology, Offender Profiling.

Since the earliest criminological studies it has been clear that, broadly speaking, criminals have characteristics that distinguish them from the general population. There have also been attempts to demonstrate that certain classes of crime are typically committed by people who have similar characteristics. It has also been claimed that what may be called the 'style' of the crime, or the pattern of behaviour, typical of any set of crimes relates directly to subsets of characteristics of offenders. This process of making inferences about significant features of an offender on the basis of the kinds of people who commit crimes in that style has often been called 'offender profiling'. In general such 'profiles' are drawn from the subjective judgement and experience of putative experts with little empirical basis for their claims.

However, the few empirical studies that have been carried out (e.g. [1]) to develop models relating offence style to offender characteristics have relied on intensive content analysis procedures that derive categories from open-ended police and related data sources. Such procedures are both prone to subjectivity and require great human effort making them difficult for police officers to use in the field. However the emerging application of text mining of descriptions available in police digital records [2] provides technologies for performing such analysis automatically. But although it is reasonably straightforward to derive tokens in a systematic, objective fashion from police records, thereby mechanizing the development of the content categories there is still the empirical question of whether the tokens so derived do indeed provide the basis for discriminating between different categories of offender.

D.H.-L. Goh et al. (Eds.): ICADL 2007, LNCS 4822, pp. 493–494, 2007.
© Springer-Verlag Berlin Heidelberg 2007

The present study was therefore set up to establish whether two crucial characteristics of an offender, age and gender, could be reliably indicated using Language Models applied to actual police records, beyond the base rate levels of these characteristics in the offending population. Of course, in principle, this approach can be extended to many other characteristics. The work thus has three possible uses:

1. We can determine whether police records can be examined to reveal behavioural differences between categories of offender.
2. For an unsolved crime with no eye witness, the likely characteristics of the offender can be inferred and thus used to limit the range of possible offenders investigators should consider or to prioritise plausible suspects.
3. Specific features can be linked to offender characteristics. This can inform police investigating crimes as to the likely features to be associated with known categories of offender. This can assist for instance in evaluating witness testimony or hearsay evidence.

It can be argued that the problem addressed here is one of document classification and Probabilistic Language Modelling has been extensively used for this [3, 4, 5]. However, we argue that we are going beyond classification since we are firstly determining if behavioural differences exist between categories at all and secondly we are using the probabilities of terms in the language models to reveal behavioural styles in each group.

The rationale for being able to classify sex or age of offenders from their actions rests on there being significant behavioural differences between these groups and these differences being revealed in the vocabulary used to record the crime within the criminal records. Language Modelling does allow us to firstly establish that there are differences in behaviour between offenders of different sex and age (above and below 18). This is achieved by defining a separate language model per offender category. By exploring the differences probabilities assigned to the terms for, say, the male and female model, we have discovered the terms and thus the behavioural features which are more likely to occur in one group or the other.

References

1. Canter, D., Fritzon, K.: Differentiating arsonists: A model of firesetting actions and characteristics. Legal and Criminal Psychology 3, 73–96 (1998)
2. Bache, R., Crestani, F., Canter, D., Youngs, D.: Application of Language Models to Suspect Prioritisation and Suspect Likelihood in Serial Crimes. In: International Workshop on Computer Forensics (to appear, 2007)
3. Bai, J., Nie, J., Paradis, F.: Text Classification Using Language Models. In: Asia Information Retrieval Symposium, Poster Session, Beijing (2004)
4. Peng, F., Schuurmans, D.: Combining naive Bayes and n-gram language models for text classification. In: Twenty-Fifth European Conference on Information Retrieval Research (2003)
5. Peng, F., Schuurmans, D., Wang, S.: Augmenting Naive Bayes classifiers with statistical language models. Information Retrieval 7(3), 317–345 (2003)

Merging Local and Global Gazetteers

Øyvind Vestavik and Ingeborg T. Sølvberg

Dept. of Computer and Information Science
Norwegian University of Science and Technology
{oyvindve,ingeborg}@idi.ntnu.no

Abstract. Most gazetteers with a global scope contain few local names, and gazetteers with a local scope mostly do not contain foreign names. However, people often use both local and foreign place names in a discourse. We describe some of the challenges in mapping local and global gazetteers that serve different needs and hence may have different structure, granularity and coverage. We pay special attention to the problem of identifying duplicate place descriptions in such registries.

Keywords: Geographic Information Retrieval, Gazetteers, Gazetteer Merging, Standards, Controlled Vocabularies.

People tend to use both local and foreign names in a discourse and in documents. In order to index such documents based on their geographic *aboutness* we need gazetteers with both detailed local coverage and foreign names. We describe a schema level and instance level mapping of gazetteer data from the Alexandria Digital Library Gazetteer (ADL GAZ) and Sentralt Stedsnavnregister (SSR). The findings of this project provide insights into some of the general problems of mapping gazetteers and the resulting gazetteer will be used for indexing Norwegian newspaper articles.

SSR [1] is the official registry of Norwegian place names. It contains information about approx. 600 000 current place names in Norway and is built up around 4 conceptual objects/entities (*Locations*, *Place Names*, *Spelling Variations* and *Occurences* (on maps, road signs etc)). Each entity has a set of attributes and the entities are organized in a tree structure (see figure 1, left side). Each tree describes one place. The ADL gazetteer [2] contains descriptions of approx. 4.4 mill. places and approx. 5.9 mill names used as labels for these places. ADL Standard Reports (See fig. 1, right side) describe a place using a set of key attributes of the place, including *Names*, *Footprints* (location(s)), *Relationships* (to other places) and *Classes* (type of place).

Schema level mapping consists of identifying attributes in the two models that contain the same or comparable information. Schema level mappings are shown in figure 1. Instance level mapping deals with identifying pairs of ADL entries and SSR records that describe the same place (duplicate detection). 4 indicators are used to calculate a score for the probability that a given ADL Standard Report and a given SSR record describe the same place: 1) Co-location or near co-location. 2) Number of shared place names. 3) Shared or similar feature types

D.H.-L. Goh et al. (Eds.): ICADL 2007, LNCS 4822, pp. 495–496, 2007.

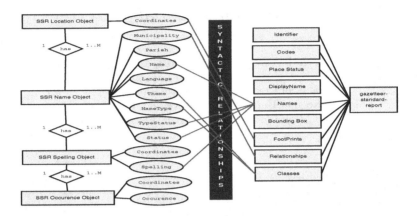

Fig. 1. Schema level mappings between the ADL and SSR data models

(kind of place eg. city, school, mountain etc). The similarity is measured as graph distance in a weighted graph implemented over the two vocabularies describing the feature type of the places described. 4) Part of same larger place. Valid mappings are selected based on best scores and mappings with low scores are discarded.

Initial investigation has revealed some challenges to our approach: 1) Comparing names is difficult for place names containing language specific characters because these names have been transcribed in ADL GAZ. The names *Røa* and *Roa* are for instance both represented as *Roa*. 2) Often, the two gazetteers agrees on only a subset of names for a place. 3) Often, ADL GAZ and SSR do not agree on which names should be considered primary and what names are current. 4) The vocabularies used to describe the feature type of a place are to some extent culturally biased, the set of concepts only partially overlaps, the granularity of similar concepts often differ and there is often only a partial overlap in semantics between similar concepts. 5) ADL GAZ describes a place as being a part of a county whereas SSR describe a place as being part of a municipality. 6) Coordinates describing the location of a place might vary considerably, even for identical places.

References

1. The Norwegian Gazetteer / Sentralt Stedsnavnregister (norwegian only),
   http://www.statkart.no/standard/sosi/html_34/navn/navn.htm
2. Alexandria Digital Library Gazetteer.: Map and Imagery Lab, Davidson Library,
   Santa Barbara CA, University of California, Santa Barbara (1999),
   http://www.alexandria.ucsb.edu/gazetteer

Towards a Hierarchical Framework for Predicting the Best Answer in a Question Answering System

Mohan John Blooma, Alton Yeow-Kuan Chua, Dion Hoe-Lian Goh,
and Zhiquan Ling

Division of Information Studies, Wee Kim Wee School of Communication & Information,
Nanyang Technological University
{bl0002hn,AltonChua,ashlgoh,ling0032}@ntu.edu.sg

Abstract. This research aims to develop a model for identifying predictive variables for the selection of the best quality answer in a question-answering (QA) system. It was found that accuracy, completeness and relevance are strong predictors of the quality of the answer.

Keywords: Question answering systems, Answer quality, Information Retrieval, Multiple Regression, Prediction model.

We developed a theoretical model from past studies on the dimensions of information quality for information resources on the web. The model developed was tested using actual data collected from a user-oriented QA system –Yahoo! Answers. The aim of the paper is to investigate if the hypothesized predictive features on the quality of the answer actually support the prediction of the best answer in the QA system.

Table 1. Predictive features of the quality of answer with their respective significance

Features			Coefficients	Sig.
Level 1	Level 2	Level 3	Beta	Std. Error
Non-textual Features	Question	Category	.028	.178
	Answerer	Reputation	.008	.701
		Authority	.012	.579
	Asker	Reputation	.010	.624
		Authority	-.066	.002
Textual Features	Answer	Accuracy	.230	**.000**
		Completeness	.375	**.000**
		Length	.027	.188
		Language	.039	.063
		Reasonableness	.370	**.000**

Features examined in this study are divided into textual and non-textual. Non-textual features cover three main factors, namely, the category of the question, the asker of the question and the answerer whose answer was selected as the best answer by the asker. Reputation of the asker, authority of the asker, reputation of the answerer, authority of the answerer, category of the question are the five dimensions

D.H.-L. Goh et al. (Eds.): ICADL 2007, LNCS 4822, pp. 497–498, 2007.

of non-textual features used as predictive variables in this study. Textual features represent different dimensions of the answer itself. Accuracy, completeness, language, reasonableness, and length of the answer are five dimensions of textual features used for evaluating the quality of answer.

Reputation and authority were derived from the user ratings while the textual features were evaluated using two human volunteers. A multiple regression analysis was used to model the dependent value quality based on its linear relationship to one or more predictors. Textual features were found to be significant predictors (Table 1.). In particular, the answer's accuracy, completeness, and reasonableness were the most significant predictors for the quality of the answer. Non-textual features were found to be least significant.

It was found that the quality of answer was most influenced by the textual features of the answer itself, rather than the non-textual features. Accuracy, completeness and reasonableness were found to be the most significant textual features highlighting the fact that the user is looking for accurate, understandable, complete and relevant information. Among these variables, completeness was found to be the strongest predictor followed by the reasonableness and accuracy of the answer. The same features were shown to be the indicators of quality of websites and information resources by various studies [1]. However, length and language were not found to be significant predictors. This study thus reveals that irrespective of the length and language of the answer, users prefer the completeness and understandability of the answer. The non-textual features like reputation of the asker, authority of the answerer, reputation of the answerer and the category were also found to be non-significant predictors. Thus these findings confirm that the user values, the understandability of the answer and the extent to which the answer satisfies his/her information needs, more than the reputation and authority of the answerer. However, recent work on evaluation of question answering system by [2] highlights the fact that the current trend points to the centrality of users' perception. This study also reveals that the satisfaction of user needs is more important than the source of the information. By promising to deliver answers, not just documents, question answering systems can more effectively fulfill users' information needs.

In conclusion, our study offers a framework for evaluating answer quality that could be used on related systems. As our study is on a community-based QA system, the results could be extended to other community-based web services and for enhancing knowledge access to digital libraries. Using this framework, researchers will be able to better understand the significant features in developing answer extraction modules for the retrieval of good quality answers in a QA system.

References

1. Fricke, M., Fallis, D.: Indicators of accuracy for answers to ready reference questions on the internet. Journal of the American Society for Information Science and Technology 55(3), 238–245 (2003)
2. Wacholder, N., Kelly, D., Kantor, P., Rittman, R., Sun, Y., Bai, B., Small, S., Yamron, B., Strzalkowski, T.A.: Model for quantitative evaluation of an end – to end question answering system. Journal of the American Society for Information Science and Technology 58(8), 1082–1099 (2007)

2 Directional 2 Dimensional Pairwise FLD for Handwritten Kannada Numeral Recognition

K. Chidananda Gowda, T.N. Vikram, and Shalini R. Urs

International School of Information Management, University of Mysore, Manasagangotri,
Karnataka-570006, India
kcgowda@sancharnet.in, {shalini,vikram}@isim.ac.in

Abstract. In this paper a two dimensional two directional pairwise Fisher's linear discriminant (FLD), ($2D^2$ pairwise FLD) is proposed which is employed for representation and recognition of Kannada numerals. The proposed methodology is robust as it is an extension of 2D pairwise-FLD[3] which is theoretically more efficient than conventional FLD.

1 Introduction

Kannada is the official language of the south Indian state of Karnataka. With a population of 44 million Kannada speaking people in south India, and with large presence of bilingual documents in Karnataka in Hindi/English – Kannada format, an OCR for Kannada numeral recognition becomes imperative. An efficient OCR is important for any DL initiative to succeed. 2D variant of pairwise FLD is proposed by Guru and Vikram[3] for face recognition in the literature and on similar lines we propose the Alt.2D and $2D^2$ pairwise FLD for handwritten Kannada numerals in this paper. Kannada characters are highly irregular in nature and hence any subspace method would out perform structural feature based method.

2 Proposed Model

2D Pairwise FLD [3] is modeled similarly on the lines of 2DFLD[1]. Let A_i^j be an image of size $a \times b$ representing the jth sample in the ith class. Let C_i be the average image of the ith class. The image between class scatter matrix G_b and within class scatter matrix G_w are computed as follows:

$$G_b = \frac{1}{N} \sum_{i=1}^{T-1} \sum_{j=i+1}^{T} k_i k_j (C_i - C_j)^T (C_i - C_j) \tag{1}$$

$$G_w = \frac{1}{N} \sum_{i=1}^{T} \sum_{j=1}^{k_i} (A_i^j - C_i)^T (A_i^j - C_i) \tag{2}$$

D.H.-L. Goh et al. (Eds.): ICADL 2007, LNCS 4822, pp. 499–501, 2007.

The Fisher's criterion thus is as follows.

$$J(E) = \frac{E^T G_b E}{E^T G_w E} \tag{3}$$

Similarly let another image between class scatter matrix H_b and within class scatter matrix H_w be computed as follows:

$$H_b = \frac{1}{N} \sum_{i=1}^{T-1} \sum_{j=i+1}^{T} k_i k_j (C_i - C_j)(C_i - C_j)^T \tag{4}$$

$$H_w = \frac{1}{N} \sum_{i=1}^{T} \sum_{j=1}^{k_i} (A_i^j - C_i)(A_i^j - C_i)^T \tag{5}$$

$$J(F) = \frac{F H_b F^T}{F H_w F^T} \tag{6}$$

Projection of a training image onto these optimal projection axes results with a feature matrix of the respective training image. That is if X_i^j represents the feature matrix of X_i^j, then $X_i^j = F^T A_i^j E$ (7)

Euclidian nearest neighbor classifier is employed for recognition.

3 Experimentation and Conclusion

In order to create a large handwritten numeral dataset we asked one hundred volunteers in the University of Mysore campus, to write all the Kannada numerals from 0-9, 5 times. Images were cropped to 20 X 15 pixels. Features obtained by projecting images on J(F) (eq. 6) is referred to as Alt. 2D Pairwise-FLD and that from eq. 7 is referred as 2D^2 Pairwise FLD. 300 training samples were selected with atleast 10 images per class and tested on the remaining 4700 images. The results along with comparative study is given in Table 1. The proposed methodology outperforms the its FLD variants as observed and hence its theoretical efficiency is corroborated.

Table 1. The best recognition performances of the proposed subspace methodology on the Kannada numeral dataset

Methodology	Dimension of feature vector	Optimal recognition rate (%)
Alt. 2D-FLD[2]	7×15	87.00
2D^2-FLD[2]	4×3	84.34
2DPairwiseFLD [3]	5× 20	91.08
Alternative 2D Pairwise –FLD	**4×15**	**93.17**
2D^2 Pairwise-FLD	**5 × 5**	**94.23**

References

1. Yang, J., Zhang, D., Yang, X., Yang, J.: Two-dimensional discriminant transform for face recognition. Pattern Recognition 38(7), 1125–1129 (2005)
2. Nagabhushan, P., Guru, D.S., Shekar, B.H.: (2D)2 FLD: An efficient approach for appearance based object recognition. Neurocomputing 69, 934–940 (2006)
3. Guru, D.S., Vikram, T.N.: 2D Pairwise FLD: A robust methodology for face recognition. In: IEEE Workshop on Automatic Identification Advanced Technologies (2007)

A Hybrid Approach of Noun Phrase Translation in Cross-Language Information Retrieval

Thanh C. Nguyen, Hieu V. Nguyen, and Tuoi T. Phan

Computer Science Faculty, HCMC University of Technology,
268 Ly Thuong Kiet, HCMC, Vietnam
thanh@cse.hcmut.edu.vn, nguyenvanhieu_danang@yahoo.com,
tuoi@hcmut.edu.vn

Abstract. At present, many researches of noun phrase translation are proposed in Natural Language Processing field, but most of them are in dictionary-based with word-by-word translation and similarity selection. The paper proposes a hybrid approach for noun phrase translation, by combining the set theory and grammar's pattern, and its algorithms, to apply to Vietnamese–English translation. The finding also has good experimental results when applying on Vietnamese noun phrases.

Keywords: Set theory, grammar's pattern, hybrid approach.

1 Introduction

The noun phrase translation from L_s to L_t language often hits difficulties because of its depending on complexity of language's grammar, then it impacts on result quality. There have been published papers addressing approaches of noun phrase translation. The grammar rule-based approach (a), is applied in machine translation in long-term. J. Gao [1] published researches with better result for pair of Chinese-English. With set theory-based approach (b), F.L. Ostenero [2] proposed their solution in 2004, but in which they did not mention their disambiguation enough. The difficulties, which impact on effect of translation process, are many source noun phrases cannot be translated because of 'out-of-dictionary' issue, complexity of language's grammar in grammar rule-based, and number of calculation in set theory-based.

2 Proposed Hybrid Approach of Noun Phrase Translation

The paper proposes a hybrid approach, which applies solutions of (a) and (b), to translate a noun phrase and resolve ambiguity of its result (in below flow chart).

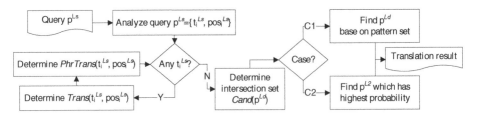

D.H.-L. Goh et al. (Eds.): ICADL 2007, LNCS 4822, pp. 502–503, 2007.
© Springer-Verlag Berlin Heidelberg 2007

In initial iteration, result of our experiment, which includes bilingual dictionary from [7] and data from [3][4][5][6], achieves 72% of precision in Vietnamese-English noun phrase translation. Detailed result is shown as follows:

Group 1 (2 words per phrase): 171 phrases, 86% of precision
Group 2 (3 words per phrase): 126 phrases, 64% of precision
Group 3 (4 words per phrase): 53 phrases, 43% of precision

3   Conclusion and Future Work

In the paper, we propose a hybrid approach of noun phrase translation, which combine both methodologies of set theory-based and grammar rule-based, and its algorithms. Especially, from source noun phrase, these algorithms focus on two cases of searching its existing correspondent translated noun phrases destination noun phrases or analyzing bi-gram probabilities to select best results. In initial iteration, we apply the approach for Vietnamese-English translation and retrieve many propitious results. Our next step is to focus on reducing the calculation time of intersectional set also system resource to improve the speed and processing-time of these algorithms, which, we believe, will make significant contributions to NLP community.

Acknowledgments. We would like to thank all members in the BK-NLP group of HCMC University of Technology (Vietnam) for their enthusiastic collaboration. This work was supported by the key project of HCMC NU -Vietnamese Information Retrieval project (B2005-20-01-TD).

References

1. Gao, J., Nie, J.-Y., Xun, E., Zhang, J., Zhou, M.: Changning Huang: Improving Query Translation for Cross - Language Information Retrieval using Statistical Model. In: ACM (2001)
2. López - Ostenero, F., Gonzalo, J., Verdejo, F.: Noun phrases as building blocks for cross - language Search Assistance. Information Processing and Management: an International Journal archive 41(3), 549–568 (2005)
3. Vietnamese Ministry of Trade, http://www.mot.gov.vn
4. Text REtrieval Conference (TREC), http://trec.nist.gov
5. PC World, http://www.pcworld.com
6. GATE, A General Architecture for Text Engineering, http://gate.ac.uk
7. HO NGOC DUC, Open-source dictionary, http://www.informatik.uni-leipzig.de/duc/Dict/

Deriving Tree-Structured Network Relations in Bibliographic Databases

Alisa Kongthon and Niran Angkawattanawit

Human Language Technology Laboratory
National Electronics and Computer Technology Center (NECTEC)
112 Thailand Science Park, Phahon Yothin Rd.
Klong Luang, Pathumthani, Thailand 12120
{alisa.kongthon,niran.angkawattanawit}@nectec.or.th

Abstract. This paper presents a new algorithm called "tree-structured networks" that can automatically construct parent-child (hierarchical structure) and sibling relationships (non-hierarchical structure) among concepts from a set of documents without use of data reduction or standard clustering techniques. The algorithm is applied to bibliographic databases such as INSPEC and EI Compendex toward the objective of enhancing research and development (R&D) management. Deriving tree-structured networks of research topics is an important goal in R&D management study. Parent-child relationships can help identify emerging areas in an existing field of research. Sibling relationships are interesting as well since they could represent interdisciplinary structures among related topical areas. Based on the initial testing on a set of publication abstracts, the proposed algorithm promises to offer richer structural information on relationships in text sources over the standard clustering techniques.

Keywords: Tree-structured networks, text mining, association rule mining, bibliographic databases, research and development management.

1 Introduction

Clustering text data in high-dimensional space is one of the most interesting topics among many other text mining applications. This paper presents the use of the object-oriented association rule mining (OOARM) technique [1] to automatically cluster related concepts and discern tree-structured networks in bibliographic databases such as INSPEC and EI Compendex. Tree-structured networks capture important aspects of both parent-child hierarchies (trees) and sibling relations (networks). It appears that most standard information retrieval and bibliometric analysis approaches using vector spaces or data reduction (e.g., Principal Components Analysis (PCA) or Latent Semantic Indexing) are able to identify relationships but not hierarchy.

2 The Proposed Algorithm

The proposed tree-structured network algorithm is implemented based on association rule mining. Kongthon et al (2007) introduced the new algorithm called Object-Oriented

D.H.-L. Goh et al. (Eds.): ICADL 2007, LNCS 4822, pp. 504–505, 2007.

Association Rule Mining (OOARM) to effectively discover association rules from text data [1].

The basic tree-structured networks algorithm works as follows:

1. Find all *frequent term-clusters*
2. For each cluster, generate association rules with any other clusters
3. To obtain Parent-Child relations, find association rules that satisfy:
   $confidence(X \Rightarrow Y) = P(Y|X) \geq minParent$, where $minParent \leq 1$
   and
   $support(X) < \varepsilon * support(Y)$, where $0 < \varepsilon < 1$
   Y is then said to be 'parent' of X
4. To obtain sibling relations, find association rules between term clusters with
   ($minSibling \leq confidences < maxSibling$)
   where $0 < minSibling < maxSibling = minParent$

where

- X and Y are set of terms and $X \cap Y = \varnothing$
- An *itemset* is collection of one or more items
- Each *frequent term-cluster* is each k-frequent itemset that is generated by the OOARM algorithm
- Support is frequency of occurrence of an itemset
- Confidence of the rule $X \Rightarrow Y$ is the conditional probability of Y given X.

3   Conclusion

In this paper, a tree-structured networks algorithm was proposed. The algorithm applies the association rule mining technique to discern conceptual relationships (parent-child and sibling) from text data sets. The results from the proposed algorithm were compared with the Principal Component Analysis (PCA) and the Hierarchical Agglomerative Clustering (HAC) approaches. Tree-structured networks promise to offer richer structural information on relationships in text information. Some remarkable features of tree-structured networks that are worth noticing include:

- Tree-structured networks do not require transformation of input data, instead it uses raw occurrences between input data.
- Parent-child and sibling relations can be derived from a set of documents without the use of data reduction or standard clustering techniques.

Reference

1. Kongthon, A., Mueller, R., Porter, A.L.: Object-Oriented Data Structured for Text Association Rule Mining. In: Proceedings of the 2007 Electrical Engineering/Electronics, Computer, Telecommunications and Information Technology (ECTI) International Conference, pp. 1276–1279 (2007)

The Efficacy of Tags in Social Tagging Systems

Khasfariyati Razikin, Dion Hoe-Lian Goh, Elizabeth Kian Cheow Cheong,
and Yi Foong Ow

Wee Kim Wee School of Communication and Information,
Nanyang Technological University
{khasfariyati,ashlgoh,w060022,w060050}@ntu.edu.sg

Abstract. Social tagging systems are a popular means for sharing resources. However, social tagging depends on individual knowledge. We evaluate the effectiveness of tags in describing the resources using support vector machines via classification. We achieved precision and recall at 90.22% and 99.27% respectively, with an average accuracy of 89.84%. Our results show that tags may help users' group resources into broad categories.

Keywords: Social Tagging, Support Vector Machines, Machine Learning.

1   The Study

Social tagging is a process of annotating a resource with keywords by anyone without being limited to a set of vocabulary [1]. Users are able to share their tags and their associated resources with others [2]. The tagging process depends largely on an individual's knowledge and might not follow a standard taxonomy. The effectiveness of tags as document descriptors is thus very much dependent on the individual. Additionally, a social tagging system does not have the same characteristics found in a taxonomic classification system. As such, an evaluation of their efficacy in a social tagging system is worth exploring.

Our goal is to determine the ability of the tags to describe their associated documents in a social tagging system. A total of 20 tags and 1385 English-language web pages associated with each tag were randomly downloaded from del.icio.us, a popular social tagging site. Pre-processing involved stop word removal and stemming. TFIDF was used to weight the remaining terms. Two-thirds of the data were used for training a SVM classifier[1], and the remainder were for testing. Table 1 shows the results from the classifier. Precision rates of more than 80% were obtained indicating that the documents were very relevant to their associated tags. In terms of recall, all tags provided highly pertinent pages. On average, we obtained 90.22% for precision, 99.27% for recall and 89.84% for accuracy.

---

[1] http://svmlight.joachims.org

D.H.-L. Goh et al. (Eds.): ICADL 2007, LNCS 4822, pp. 506–507, 2007.
© Springer-Verlag Berlin Heidelberg 2007

Table 1. Results from experiment

Term	Accuracy (%)	Precision (%)	Recall (%)
Mobile Learning	87.50	87.50	100.00
Text classification	82.09	81.54	100.00
Disneyland	100.00	100.00	100.00
Tennis player	93.22	96.43	96.43
Decision making	96.97	96.97	100.00
Tourism in Egypt	100.00	100.00	100.00
LCD TV	100.00	100.00	100.00
Apple iPod	100.00	100.00	100.00
Dementia	81.82	81.82	100.00
Interferometry	100.00	100.00	100.00
Coffee production	72.73	80.00	88.89
Tin Toy	75.00	72.73	100.00
Car market	80.00	80.00	100.00
Nutritional science	90.00	90.00	100.00
Children abuse	90.00	90.00	100.00
Fashion paris	90.00	90.00	100.00
Journalism	87.50	87.50	100.00
Tsunami	90.00	90.00	100.00
Electronic Game Market	90.00	90.00	100.00
Internet programming	90.00	90.00	100.00
Average	**89.84**	**90.22**	**99.27**

2 Conclusion

Our results suggest that users typically use tags that appropriately characterise the content of the associated documents. Tags may thus help users group their documents into broad categories, and in turn, users may rely on tags to retrieve relevant content.

Similarly, Brooks and Montanez [3] analysed the value of tags for classifying blog entries from Technorati. However, our investigations diverge based on the domain. With blogs, the documents are more sentiment based. This is in contrast with the pages in our dataset, where it is more general.

A limitation is that the sample size for the study is small and generic. On-going work includes analysing with a larger data set and including tags from other domains.

References

1. Macgregor, G., McCulloch, E.: Collaborative tagging as a knowledge organisation and resource discovery tool. Library Review 55(5), 291–300 (2006)
2. Marlow, C., Naaman, M., Boyd, d., Davis, M.: HT06, tagging paper, taxonomy, Flickr, academic article, to read. In: Proceedings of the seventeenth conference on Hypertext and Hypermedia, ACM Press, Odense, Denmark (2006)
3. Brooks, C.H., Montanez, N.: Improved annotation of the blogosphere via autotagging and hierarchical clustering. In: Proceedings of the 15th international conference on World Wide Web, ACM Press, Edinburgh, Scotland (2006)

Synopsis Information Extraction in Documents Through Probabilistic Text Classifiers

Jantima Polpinij[1] and Aditya Ghose[2]

[1] Faculty of Informatics, Mahasarakham University, Mahasarakham 44150 Thailand
`jantima.p@msu.ac.th`
[2] School of Computer Science and Software Engineering, Faculty of Informatics,
University of Wollonong, Wollongong, 2500 NSW, Australia
`aditya@uow.edu.au`

1 Introduction

Digital Libraries currently use several advanced information technologies to organize information and make it easy accessible to users. Current digital library trends to be dynamic digital library [1]. It is possible that business rules also can be approached for improving dynamic digital library. Business rules [2] are statements that define or contain some aspects of IT systems by providing a foundation for understanding how an IT system functions. At present, the need for automated business rules is becoming more essential because of the increasing usage of IT systems. However, it is not easy to extract business rules because they are written in a natural language structure and much of it is ignored. Therefore, one important question in this research area is how to automatically extract a business rule from a document? Based on this, information extraction (IE) [3] typically can be applied. Basically, IE is to transform text into information that is more readily analyzed. We believe that if the content of a document is decreased, the accuracy of rules extraction may be increased logically. With this assumption, if irrelevant information is filtered from the document, it is possible to easily extract business rules from the rest. Therefore, this research proposes a method based on probabilistic text classifier to extract synopsis information. It could be said that this work is the pre-processing of a business rules extraction methodology.

2 Research Methodology and Results

Before learning method, a text document collection must be transformed into a representation which is suitable for computation. The ordinary way of document representation is usually as a structured "bag of words" [4]. Then it will contain each unique word that becomes a feature, including the number of times the word occurs in the document. In addition, we applied the Chi-squared technique (χ^2) [5] to reduce feature size. Afterwards, finding term word weighting is critical in term-based retrieval since the rank of a document is determined by the weights of the terms. We certainly use the popular term weighting for our work. It is called *TF-IDF* [5].

D.H.-L. Goh et al. (Eds.): ICADL 2007, LNCS 4822, pp. 508–509, 2007.

For learning task, the Naïve Bayes [6] classifier is applied. It uses a set of training documents to estimate parameters, and then use the estimated model to filter information in documents. Suppose that we have documents $D = \{d_1,...,d_{|D|}\}$, where Φ is parameter and we use the notation $c_j \in C = \{c_1,...,c_{|c|}\}$ to indicate both j-th component and j-th class. We assume that the documents are generated by a mixture model and there is a one-to-one correspondence between the class labels and the mixture component. Each document d_i is generated by choosing a mixture component with the class prior probabilities $P(c_j; \Phi)$, and having this mixture component generate a document according to its own parameters, with distribution $P(d_i \mid c_j; \Phi)$. So, we can identify the likelihood of a document as a sum of total probability over all generative components:

$$P(d_i \mid \Phi) = \sum_{j=1}^{|C|} P(c_j \mid \Phi)P(d_i \mid c_j; \Phi)$$

To this end, we have to choose $\mathrm{argmax}_j P(c_j \mid d_i; \Phi)$ as the best class of the document. In this research, we also choose the probability that is better of the two probabilities: the "*relevant*" and "*irrelevant*" classes. Finally, the "*relevant*" is information that is used in the process of business rule extraction.

We used only the misc.forsale topic of the 20 newsgroup dataset for our work. We randomly selected 700 documents for training and 300 documents for testing. After the classifier model finished, we evaluated the results of the experiments by using *F-measure* [4]. We used the classifier model to filter irrelevant sentences in each document. If some sentences in a document are in irrelevant classes, they are filtered from the document. The model shows accuracy at 77%. As a result, it demonstrates that this method can provide more effectiveness for filtering irrelevant information from a document.

References

[1] Walker, A.: The Internet Knowledge Manager, Dynamic Digital Libraries, and Agents You Can Understand, D-Lib Magazine, http://www.dlib.org/dlib/march98/walker/03walker. html

[2] Zsifkov, N., Campeanu, R.: Information technology: Business rules domains and business rules modeling. In: Proceedings of the 2004 international symposium on Information and communication technologies (ISICT) (2004)

[3] Turmo, J., Ageno, A., Catalá, N.: Adaptive information extraction, ACM Computing Surveys (CSUR). ACM Press, New York (2006)

[4] Baeza-Yates, R., Ribeiro-Neto, B.: Modern Information Retrieval. The ACM Press, New York (1999)

[5] Yang, Y., Pederson, J.O.: A Comparative Study on Features selection in Text Categorization. In: Proceedings of the 14th international conference on Machine Learning (ICML), Nashville, Tennessee, pp. 412–420 (1997)

[6] Nigam, K., Maccallum, A.K., Thrun, S., Mitchell, T.: Text Classification from Labeled and Unlabeled Document using EM. Machine Learning 39(2/3), 103–134 (2000)

Small World Phenomenon and Author Collaboration: How Small and Connected Is the Digital Library World?

Monica Sharma and Shalini R. Urs

International School of Information Management, University of Mysore, Mysore, India
{monica,shalini}@isim.ac.in

Abstract. We present here the findings of our research to study the "Small World Phenomenon" and the scientific collaboration of authors in the Digital Library domain by using Social Network Analysis metrics. Co-authorship network of prolific authors is created and Social Network Analysis is carried out using UCInet.

1 Introduction

Using Social Network Analysis, it is possible to study the structural features of academic communities by examining their publications [1][3]. Liu et al. [2] studied the co- authorship network in Digital Library which focused on the DL conferences. The present work is an attempt to study the network characteristics of Digital Library as reflected in author collaboration and to study the small world phenomenon as expressed in Social Network Metrics.

2 Data and Analysis

Publication count and coauthor count of authors from World of Science (WoS) are used for identifying the front-runners in the DL domain. A search on 'digital libraries' yielded 1838 records. Eliminating authors with less than 5 papers gave us a set of 52 authors. DBLP record count and coauthor count of these 52 authors were also recorded. The 11 authors who scored high on all the three parameters were identified as the core members of the DL community. Co authorship network of these 11 authors with their coauthors (1053) was constructed using UCInet (Fig 1.).

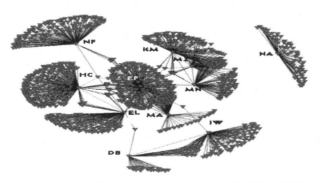

Fig. 1. Co authorship Network of authors in the field of Digital Library

D.H.-L. Goh et al. (Eds.): ICADL 2007, LNCS 4822, pp. 510–511, 2007.
© Springer-Verlag Berlin Heidelberg 2007

Different Parameters such as Betweenness centrality, Degree centrality, Clique, etc. were studied. All the measures strongly suggested that Edward A. Fox and Hsinchun Chen are the most active members in terms of co authorship and record count. The network forms 2 components, Giant (94% authors) and Small component (6%). Table 1 shows various metrics of Social Network Analysis of top 5 authors under study (Values in parenthesis are of study carried out by Liu et al. [2]). It can be interpreted from the results of both the studies that Edward A. Fox (maximum number of cliques) and Hsinchun Chen hold an important position in the network in terms of centrality.

These measures indicate that they play a powerful role in the network and have a great influence over the flow and dissemination of resources in the network. The results indicate that the network is highly connected and that Digital Library community has significantly high international collaboration.

Table 1. Results of Social Network Analysis of co authorship network

Author		Degree	Betweenness	Closeness	Size	Ties	2Step reach
Edward A. Fox	(EF)	261(55)	342150.5(83163.9)	1.540(0.251)	261	194	72.05
Hsinchun Chen	(HC)	209(59)	167245.3(89250.9)	1.532(0.259)	209	56	52.95
Kurt Maly	(KM)	134	116338.6	1.525(0.212)	134	210	42.11
Ee-Peng Lim	(EL)	123	105330.0	1.531	123	50	53.04
Norbert Fuhr	(NF)	121	87432.5	1.514	121	0	11.79
Ian H. Witten	(IW)	97(18)	85283.3	1.507	97	56	11.5

Geodesic or the shortest path is the path between a pair of nodes which involve minimum number of nodes in between which connect the 2 nodes. Elmacioglu and Lee [1] studied the 6 degrees of separation in database community (DB) and found that average distance of coauthors in DB world is 6 whereas the network of Liu et al. [2] work yielded average distance of 3.6, which is approximately equal to our value. In our network the Average Geodesic between the reachable pairs is 3.5. The results reveal that compared to DB community, DL community is a small world and highly connected suggesting that dissemination of information can be fast in the network. This could be attributed to the fact that DL community is a small community.

3   Conclusion

The digital library community is indeed a small world lead by Edward A. Fox and Hsinchun Chen, having a significantly high international collaboration with an average geodesic distance of 3.5 between two pairs.

References

1. Elmacioglu, E., Lee, D.: On Six Degrees of Separation in DBLP-DB and More. ACM SIGMOD Record 34(2), 33–40 (2005)
2. Liu, X., et al.: Co-Authorship Networks in the Digital Library Research Community. Infor. Proces. & Mgt.-An International Journal 41(6), 1462–1480 (2005)
3. Newman, M.E.J.: Co authorship Networks and Patterns of Scientific Collaboration. PNAS 101, 5200–5205 (2004)

Evaluating the Comprehensiveness of Wikipedia: The Case of Biochemistry

Brendan Luyt, Wee Tin Kwek, Ju Wei Sim, and Peng York

Nanyang Technological University, Singapore
Brendan@ntu.edu.sg

Keywords: Wikipedia, Encyclopedia Britannica, evaluation, reference sources.

1 Overview

In recent years, the world of encyclopedia publishing has been challenged as new collaborative models of online information gathering and sharing have developed. Most notable of these is Wikipedia. Although Wikipedia has a core group of devotees, it has also attracted critical comment and concern, most notably in regard to its quality. In this article we compare the scope of Wikipedia and Encyclopedia Britannica in the subject of biochemistry using a popular first year undergraduate textbook as a benchmark for concepts that should appear in both works, if they are to be considered comprehensive in scope.

2 Method

The aim of the research was to compare the scope of coverage between Encyclopedia Britannica (online version) and Wikipedia. The scope of coverage is considered by librarians to be a key measure of the value of an information source and so an important element for the evaluation of the quality of any reference tool. Biochemistry provided the subject for the study. However, instead of just comparing the two works against each other it was decided to benchmark each against a first year undergraduate textbook recommended by instructors in the subject at Nanyang Technological University (Biochemistry, 3rd edition by Christopher K. Mathews). A checklist of concepts/keywords was derived from the textbook and applied to the content of both Wikipedia and Britannica. Given the length of the textbook and the scarcity of time and labour power, one chapter was randomly chosen from each of its five sections. Fortunately, this particular textbook bolded all important concepts in the text, making it relatively easy to pick them out, even for non-experts. Once the list was compiled, a search was conducted. Each concept was first searched for in the Biochemistry page. If it was found, a one was written in the corresponding table entry. Concepts not found on the Biochemistry page, triggered a search within the entire Wikipedia or Britannica site. The process was repeated for all concepts in the four selected chapters. Searching took place during the months of September and October, 2006. For the Enyclopedia Britannica, the research was done using the content available with a premium membership.

D.H.-L. Goh et al. (Eds.): ICADL 2007, LNCS 4822, pp. 512–513, 2007.
© Springer-Verlag Berlin Heidelberg 2007

As data collection progressed it was realized that three more options had to be taken into account when determining whether or not a concept was included. For Wikipedia there was the possibility that the concept was found in a "stub", a short article that is considered by Wikipedia management incomplete and in need of expansion. There was also the possibility that the concept was covered under a synonym, in which case Wikipedia provided a re-direct feature. And finally, for both Wikipedia and Britannica, if the search string does not match any term in the database then alternate articles believed to closely match the term are listed instead along with a percentage estimate of their relevancy.

3 Findings

The number of concepts found within the topic page for both Britannica and Wikipedia are almost the same (23 versus 22 percent). However, a much larger number of concepts were to be found in separate articles in Wikipedia (33 percent) than in Britannica (14 percent). Thus, overall it is not surprising that the number of concepts not found in Wikipedia amounted to only 19 percent of the total as compared to 33 percent for Britannica. Is the difference between the number of concepts found and not found in the two reference works statistically significant or could it be due to random variation? Two chi square tests were conducted to answer this question. The first tested the hypothesis that the level of comprehensiveness (defined as the number of concepts occurring within the topic page as opposed to those found in separate pages) in content coverage of Wikipedia was higher than in Britannica. In this case the hypothesis had to be rejected: $\chi^2 (1, N = 485) = 2.88, p > .05$. In the second test the breath of coverage (defined as the number of concepts found somewhere in the reference tool as opposed to the number not found) was tested with the hypothesis again being that the breath of coverage of Wikipedia was higher than Britannica. Here, the hypothesis could be accepted: $\chi^2 (1, N = 654) = 17.62, p > .001$.

4 Discussion

Our comparison of the main topic page for biochemistry in Wikipedia and Encyclopedia Britannica has shown that both reference works are similar in the scope of their coverage. For the subject of biochemistry at least and in terms of scope of coverage there appears no ground for discrimination against the online collaborative encyclopedia. When we consider, moreover, that Wikipedia actually covers a greater number of concepts than Encyclopedia Britannica, although still falling short of the undergraduate textbook used for the benchmark, this conclusion is strengthened. Of course, our study has nothing to say about the clarity of exposition or even the accuracy of the material presented. These are separate issues that in a comprehensive evaluation of the two information sources would also need to be investigated.

ICADL: The Prolific Contributors

Cheng Hong Lim and Chu Keong Lee

Nanyang Technological University
ascklee@ntu.edu.sg

Abstract. This paper identifies the prolific authors at the International Asian Conference on Digital Libraries (ICADL) from 2002 to 2006. To provide a holistic picture, three methods of counting were used, namely whole counting, fractional counting, and first author counting.

Keywords: Scientometrics, whole counting, fractional counting.

The research productivity of academics has been an active area of research for a long time. This is because the main currency for an academic is his reputation, often measured by the number of research papers he publishes. Although the job of most academics consist of four components, namely, teaching and assessment, research and scholarship, administration, and community service, it is mainly on his research and scholarship that that his reputation is built. Becher (1989) further stressed that it is not the mere *conduct* of research and scholarship that earns an academic his reputation, but it is the *publication* of the results of his research and scholarship that is important in securing his reputation and influence. He ranks the publication of an academic's research findings as being far more important than his ability to teach, stating that "excellence in teaching counts for little towards recognition by established colleagues in the field" (p. 53). The factors that impact scientific productivity are also of keen interest, especially to administrators. Ramesh Babu and Singh (1998) identified persistence, resource adequacy, access to literature, initiative, intelligence, creativity, learning capability, stimulative leadership, concern for advancement, external orientation and professional commitment as being the most important ones.

In this research, the research papers that were presented at the International Asian Conference on Digital Libraries (ICADL) from 2002 to 2006 were analyzed to identify the prolific authors who contributed to this conference over the five-year duration. These five years were selected because of the ready availability of the data from the SpringerLink database. The earlier proceedings also proved difficult to obtain, and was therefore not included in the analysis. It was thought that this would be a timely occasion to take stock of the papers published as this is the tenth year ICADL is being organized. Three measures were used to identify the prolific authors, namely, whole counting, fractional counting, and first author counting. In whole counting, each author of a paper is given a count of 1, regardless of the number of authors. In fractional counting, each author of a paper is given a count of $\frac{1}{n}$, where n is the number of co-authors. In first author counting, all authors except the first, is ignored, and the first author given a count of 1.

D.H.-L. Goh et al. (Eds.): ICADL 2007, LNCS 4822, pp. 514–515, 2007.

Results

The results show that although the exact rank order of the authors differ according to the method of counting used, a group of authors consistently appear among the top authors in each table. These are Edward Fox, Hsinchun Chen, Dion Goh, Yin Leng Theng, Lim Ee Peng, Ian Witten, and Sally Jo Cunningham.

Whole Counting

Rank	Author	Count
1	GOH, DHL	12
2	THENG, YL	9
3	LIM, EP	7
3	FOX, EA	7
5	CHEN, CC	5
5	CHEN, HC	5
5	KIM, SH	5
5	FOO, SSB	5
5	YANG, CC	5
10	NIEDEREE, C	4
11	TANAKA, K	4

Fractional Counting

Rank	Author	Count
1	FOX, EA	6.81
2	CHEN, HC	4.08
3	GOH, DHL	4.05
4	WITTEN, IH	3.91
5	THENG, YL	3.40
6	YANG, CC	3.33
7	LIM, EP	3.31
8	FOO, SSB	2.99
9	ADACHI, J	2.60
10	CHEN, CC	2.50

First Author Counting

Rank	Author	Count
1	FOX, EA	7
2	THENG, YL	5
3	CUNNINGHAM, SJ	4
4	JEONG, CB	3
4	FU, L	3
4	CHEN, HC	3
4	LEE, SS	3
4	LEE, KS	3
9	HSUEH-HUA CHEN	2
9	JUN ADACHI	2
9	CHEN, CC	2
9	CHEN, CC	2

References

1. Becher, T.: Academic tribes and territories: Intellectual inquiry across the disciplines. Open University Press, London (1989)
2. Ramesh Babu, A., Singh, Y.P.: Determinants of research productivity. Scientometrics 43(3), 309–329 (1998)

Author Index

Lecture Notes in Computer Science

Sublibrary 3: Information Systems and Application, incl. Internet/Web and HCI

For information about Vols. 1– 4504
please contact your bookseller or Springer

Vol. 4715: J.M. Haake, S.F. Ochoa, A. Cechich (Eds.), Groupware: Design, Implementation, and Use. XIII, 355 pages. 2007.

Vol. 4714: G. Alonso, P. Dadam, M. Rosemann (Eds.), Business Process Management. XIII, 418 pages. 2007.

Vol. 4704: D. Barbosa, A. Bonifati, Z. Bellahsène, E. Hunt, R. Unland (Eds.), Database and XML Technologies. X, 141 pages. 2007.

Vol. 4690: Y. Ioannidis, B. Novikov, B. Rachev (Eds.), Advances in Databases and Information Systems. XIII, 377 pages. 2007.

Vol. 4675: L. Kovács, N. Fuhr, C. Meghini (Eds.), Research and Advanced Technology for Digital Libraries. XVII, 585 pages. 2007.

Vol. 4674: Y. Luo (Ed.), Cooperative Design, Visualization, and Engineering. XIII, 431 pages. 2007.

Vol. 4663: C. Baranauskas, P. Palanque, J. Abascal, S.D.J. Barbosa (Eds.), Human-Computer Interaction – INTERACT 2007, Part II. XXXIII, 735 pages. 2007.

Vol. 4662: C. Baranauskas, P. Palanque, J. Abascal, S.D.J. Barbosa (Eds.), Human-Computer Interaction – INTERACT 2007, Part I. XXXIII, 637 pages. 2007.

Vol. 4658: T. Enokido, L. Barolli, M. Takizawa (Eds.), Network-Based Information Systems. XIII, 544 pages. 2007.

Vol. 4656: M.A. Wimmer, J. Scholl, Å. Grönlund (Eds.), Electronic Government. XIV, 450 pages. 2007.

Vol. 4655: G. Psaila, R. Wagner (Eds.), E-Commerce and Web Technologies. VII, 229 pages. 2007.

Vol. 4654: I.-Y. Song, J. Eder, T.M. Nguyen (Eds.), Data Warehousing and Knowledge Discovery. XVI, 482 pages. 2007.

Vol. 4653: R. Wagner, N. Revell, G. Pernul (Eds.), Database and Expert Systems Applications. XXII, 907 pages. 2007.

Vol. 4636: G. Antoniou, U. Aßmann, C. Baroglio, S. Decker, N. Henze, P.-L. Patranjan, R. Tolksdorf (Eds.), Reasoning Web. IX, 345 pages. 2007.

Vol. 4611: J. Indulska, J. Ma, L.T. Yang, T. Ungerer, J. Cao (Eds.), Ubiquitous Intelligence and Computing. XXIII, 1257 pages. 2007.

Vol. 4607: L. Baresi, P. Fraternali, G.-J. Houben (Eds.), Web Engineering. XVI, 576 pages. 2007.

Vol. 4606: A. Pras, M. van Sinderen (Eds.), Dependable and Adaptable Networks and Services. XIV, 149 pages. 2007.

Vol. 4605: D. Papadias, D. Zhang, G. Kollios (Eds.), Advances in Spatial and Temporal Databases. X, 479 pages. 2007.

Vol. 4602: S. Barker, G.-J. Ahn (Eds.), Data and Applications Security XXI. X, 291 pages. 2007.

Vol. 4601: S. Spaccapietra, P. Atzeni, F. Fages, M.-S. Hacid, M. Kifer, J. Mylopoulos, B. Pernici, P. Shvaiko, J. Trujillo, I. Zaihrayeu (Eds.), Journal on Data Semantics IX. XV, 197 pages. 2007.

Vol. 4592: Z. Kedad, N. Lammari, E. Métais, F. Meziane, Y. Rezgui (Eds.), Natural Language Processing and Information Systems. XIV, 442 pages. 2007.

Vol. 4587: R. Cooper, J. Kennedy (Eds.), Data Management. XIII, 259 pages. 2007.

Vol. 4577: N. Sebe, Y. Liu, Y.-t. Zhuang, T.S. Huang (Eds.), Multimedia Content Analysis and Mining. XIII, 513 pages. 2007.

Vol. 4568: T. Ishida, S. R. Fussell, P. T. J. M. Vossen (Eds.), Intercultural Collaboration. XIII, 395 pages. 2007.

Vol. 4566: M.J. Dainoff (Ed.), Ergonomics and Health Aspects of Work with Computers. XVIII, 390 pages. 2007.

Vol. 4564: D. Schuler (Ed.), Online Communities and Social Computing. XVII, 520 pages. 2007.

Vol. 4563: R. Shumaker (Ed.), Virtual Reality. XXII, 762 pages. 2007.

Vol. 4561: V.G. Duffy (Ed.), Digital Human Modeling. XXIII, 1068 pages. 2007.

Vol. 4560: N. Aykin (Ed.), Usability and Internationalization, Part II. XVIII, 576 pages. 2007.

Vol. 4559: N. Aykin (Ed.), Usability and Internationalization, Part I. XVIII, 661 pages. 2007.

Vol. 4558: M.J. Smith, G. Salvendy (Eds.), Human Interface and the Management of Information, Part II. XXIII, 1162 pages. 2007.

Vol. 4557: M.J. Smith, G. Salvendy (Eds.), Human Interface and the Management of Information, Part I. XXII, 1030 pages. 2007.

Vol. 4541: T. Okadome, T. Yamazaki, M. Makhtari (Eds.), Pervasive Computing for Quality of Life Enhancement. IX, 248 pages. 2007.

Vol. 4537: K.C.-C. Chang, W. Wang, L. Chen, C.A. Ellis, C.-H. Hsu, A.C. Tsoi, H. Wang (Eds.), Advances in Web and Network Technologies, and Information Management. XXIII, 707 pages. 2007.

Vol. 4531: J. Indulska, K. Raymond (Eds.), Distributed Applications and Interoperable Systems. XI, 337 pages. 2007.

Vol. 4526: M. Malek, M. Reitenspieß, A. van Moorsel (Eds.), Service Availability. X, 155 pages. 2007.

Vol. 4524: M. Marchiori, J.Z. Pan, C.d.S. Marie (Eds.), Web Reasoning and Rule Systems. XI, 382 pages. 2007.

Vol. 4519: E. Franconi, M. Kifer, W. May (Eds.), The Semantic Web: Research and Applications. XVIII, 830 pages. 2007.

Vol. 4518: N. Fuhr, M. Lalmas, A. Trotman (Eds.), Comparative Evaluation of XML Information Retrieval Systems. XII, 554 pages. 2007.

Vol. 4508: M.-Y. Kao, X.-Y. Li (Eds.), Algorithmic Aspects in Information and Management. VIII, 428 pages. 2007.

Vol. 4506: D. Zeng, I. Gotham, K. Komatsu, C. Lynch, M. Thurmond, D. Madigan, B. Lober, J. Kvach, H. Chen (Eds.), Intelligence and Security Informatics: Biosurveillance. XI, 234 pages. 2007.

Vol. 4505: G. Dong, X. Lin, W. Wang, Y. Yang, J.X. Yu (Eds.), Advances in Data and Web Management. XXII, 896 pages. 2007.